BRILL'S STUDIES IN INTELLECTUAL HISTORY

VOLUME 89

THE AGE OF TWO-FACED JANUS

The Comets of 1577 and 1618 and the Decline of the Aristotelian World View in the Netherlands

BY

TABITTA VAN NOUHUYS

BRILL
LEIDEN · BOSTON · KÖLN
1998

This book is printed on acid-free paper.

Library of Congress Cataloging-in-Publication Data

Nouhuys, Tabitta van.
The age of two-faced Janus : the comets of 1577 and 1618 and the decline of the Aristotelian view in the Netherlands / by Tabitta van Nouhys.
p. cm. — (Brill's studies in intellectual history, ISSN 0920-8607 ; v. 89)
Includes bibliographical references and index.
ISBN 9004112049 (cloth : alk. paper)
1. Comets—1577. 2. Comets—1618. 3. Astronomy—Netherlands—History. 4. Janus (Roman deity) 5. Aristotle—Knowledge—Astronomy. I. Title. II. Series.
QB724.N68 1998
509.492'09'032—dc21 98–36297
CIP

Die Deutsche Bibliothek - CIP-Einheitsaufnahme

Nouhuys, Tabitta van:
The age of two-faced Janus : the comets of 1577 and 1618 and the decline of the Aristotelian view in the Netherlands / by Tabitta van Nouhys. – Leiden ; Boston ; Köln : Brill, 1998
(Brill's studies in intellectual history ; Vol. 89)
ISBN 90–04–11204–9

ISSN 0920-8607
ISBN 90 04 11204 9

PRINTED IN THE NETHERLANDS

To Honza, Laura, Schelte, Odile, and Mischa

Ἐπεὶ δὲ περὶ τῶν ἀφανῶν τῇ αἰσθήσει
νομίζομεν ἱκανῶς ἀποδεδεῖχθαι κατὰ τὸν λόγον,
ἐὰν εἰς τὸ δυνατὸν ἀναγάγωμεν,
ἔκ τε τῶν νῦν φαινομένων ὑπολάβοι τις ἂν ὧδε
περὶ τούτων μάλιστα συμβαίνειν.

ΑΡΙΣΤΟΤΕΛΟΥΣ *ΜΕΤΕΩΡΟΛΟΓΙΚΩΝ* Α.

We consider that we have given a sufficiently rational explanation
of things inaccessible to observation by our senses
if we have produced a theory that is possible:
and the following seems,
on the evidence available,
to be the explanation
of the phenomena now under consideration.

Aristotle, Meteorology *I.vii.*

CONTENTS

ACKNOWLEDGEMENTS

I wish to thank all those without whose patient listening and cheerful support the writing of this book would have been infinitely much harder: Marianne Roobol, Mariëtte Roelvink, Veerle Heijndrikx, Anna Boerstra, Carine Damen, Nir Weijl, Heleen Naber, Anusche Mearadji, Marjan Balkestein, Anne Hecox, Peter Ganz, and the two hairy stars, Printje and Janus.

Most heartfelt thanks is due in particular to Marianne Roobol and Mariëtte Roelvink, whose tireless support and assistance have been indispensable.

I would like to express special gratitude to my parents, whose presence and support have constituted a necessary condition for the genesis of this work.

Above all, I thank Nicolette Mout for her continued encouragement.

CHAPTER ONE

INTRODUCTION: *OPINIO LOQUITUR, RATIO RESPONDET?*

Aristotle did comets a great wrong. At least, such was the opinion of the nineteenth-century Dutch astronomer F. Kaiser, who in a short popularizing work on the impending appearance of Encke's comet in 1838 strongly inveighed against the human race for wrongfully turning the hairy stars into celestial scapegoats:

> However peace-loving man may be, one surreptitious hostility nonetheless often reigns in his heart. . . . Everyone seems to desire something inimical among the persons or things that specifically concern him, and even the entire human race seems from a very early stage to have striven to discover general enemies among those works of nature that are the same for everyone and affect everyone to the same extent. Perhaps man thought it too futile to fulminate only against things so closely surrounding him: up there in the realm of the celestial lights, too, he needed something to take issue with, if only verbally. He searched, and thousands of years ago already he found what he was looking for, or rather, he made a random choice, and unfavourable Fate made him single out the comets, who were as innocent as they were defamed.[1]

Most heavily responsible for this unwarranted vilification of the hairy stars had been Aristotle,.

> . . .who, without provocation, overtly took arms against the comets, and who made every possible attempt to discredit these bodies with contemporaries and later generations alike.[2]

---

[1] F. Kaiser, *De komeet van Encke en hare naderende verschijning* (Leiden 1838) 1-2: "Hoe vredelievend de mensch wezen moge, heerscht er toch gewoonlijk eene heimelijke vijandschap in zijn gemoed. . . . Iedereen schijnt iets vijandigs te begeeren onder de personen of zaken, die hem in het bijzonder aangaan, en zelfs het geheele menschelijke geslacht schijnt er al vroeg op uit geweest te zijn, om algemeene vijanden onder die voortbrengselen der natuur te zoeken, die voor allen dezelfde zijn en allen evenzeer betreffen. Het was den mensch misschien te nietig, om alleen tegen dingen te twisten die hem zoo van nabij omringen, hij moest ook daar boven, in het rijk der hemellichten iets vinden, waartegen hij, al was het dan alleen met woorden, kampen kon. Hij zocht en heeft reeds voor duizenden van jaren gevonden of liever eene wispelturige keus gedaan, en het ongunstig lot deed deze keuze vallen op de even onschuldige als gelasterde kometen." (My translation).

[2] Ibidem, 2: ". . .die zich, zonder uitgedaagd te zijn, openlijk tegen de kometen in

He did so by

> begrudging, contesting and denying them the rank among the heavenly lights that was theirs by right, and he ended up loathing and detesting the comets, thinking them hideous, cumbersome, corruptible lumps that dared not venture any higher than the earth's low atmosphere.[3]

And, on top of all, the reputation of comets was lastingly damaged because

> the incisive Stagyrite, all of whose opinions met with general approval, . . . was faithfully parroted also on this issue. . . . Even two thousand years after the scythe of death had cured the first enemy of comets of his slandering, one was forced not only to adopt, but even to solemnly profess adherence to his opinion concerning those objects, if one were to avoid being considered a disgrace to a decent professorial chair.[4]

The above fulmination by an eminent and well-read scientist, is outstandingly representative of the spirit of the age in which it was written. According to a view coined by historiographers of science during the late eighteenth and nineteenth centuries, but still surprisingly resilient today, Aristotle was the great culprit whose overwhelming predominance during the dark centuries of the Middle Ages had brought scientific progress to a virtual standstill. It was suggested that only when his views were abandoned in favour of the 'real science' propagated by such men as Galileo, Descartes and Newton, based on mathematics and empirical observation instead of the idle hair-splitting of the Aristotelians, could mankind firmly and resolutely embark on the path of reason and modernity.

The case of comets was considered to be especially well suited to illustrate the alleged seventeenth-century triumph of progress. First of all, cometology was one of the areas in which Aristotle's theories were most blatantly proved wrong during the period when modern

---

het harnas joeg en die alle pogingen aanwendde, om deze ligchamen bij tijdgenoot en nakomeling in een' kwaden reuk te brengen."

[3] Ibidem: "...haar den rang te benijden, te betwisten en te ontzeggen, die haar van regtswege onder de hemellichten toekwam, en hij eindigde met de kometen te verfoeijen en te verachten, als afschuwelijke, logge, vergankelijke klompen, die zich niet hooger dan onzen lagen dampkring durven wagen."

[4] Ibidem, 2-3: "...[werd] de snedige Stagiriet, die in alles bijval vond, ... ook hierin getrouwelijk nagepraat. ... Ook nog twee duizend jaren, na dat de zeissen des doods den eersten vijand der kometen zijn lasteren had afgeleerd, moest men zijn gevoelen aangaande die voorwerpen niet alleen aannemen, maar zelfs plegtig erkennen, wilde men gerekend worden een fatsoenlijk leergestoelte niet te zullen onteeren."

science emerged. Especially after Newton and Halley had established the periodicity of comets and the fundamental similarity between their paths and those of the planets, demonstrating that the hairy stars, too, described elliptical orbits with the sun in one of the focuses of the ellipse, it became totally implausible to uphold that comets were nothing but burning terrestrial vapours, belched out by the earth and set on fire in the higher reaches of the atmosphere well below the moon, as Aristotle had taught. Thus, one of the Stagyrite's most famous explanations of natural phenomena had been completely invalidated by the advent of modern science.

However, the refutation of Aristotelian cometary theory by early modern scientists was considered important, by Kaiser and his contemporaries, for much more far-reaching reasons. Roughly a century before Halley's and Newton's cometary investigations, in 1577, the Danish astronomer Tycho Brahe had calculated the altitude of the comet of that year, and had demonstrated it to have been situated far beyond the moon. From the point of view of traditional historiography, this rational and scientific establishment of the comet's superlunary position, based on highly accurate observations and mathematical proof, marked one of the most momentous transitions ever to have taken place in the history of science. For not only had Tycho shown that the Stagyrite's theory of comets as filthy, atmospheric phenomena was erroneous; in addition, he had also demonstrated that the Aristotelian world view was untenable. For how could it be maintained that the celestial region was incorruptible and unchangeable, if new bodies such as comets were observed to appear high above the moon? And how could the existence of solid spheres carrying the planets round in their ever-revolving orbits be upheld, if comets pursued their erratic paths right through these allegedly concrete planet-carrying spheres? Surely, after Tycho's work on the comet of 1577, no one in his right mind could continue to endorse the Aristotelian world view, with its rigid dichotomy between celestial and terrestrial regions, divided by the sphere of the moon, and with its solid orbs.

Some historians in addition linked the discovery of cometary superlunarity to the acceptance of the heliocentric system of Copernicus. After all, it was hard for them to believe that, once the early modern scientists' eyes had been opened to the truth on the issue of comets, they should continue to endorse blatant fallacies with regard to the location and mobility of the earth. Hence the exasperation of the late eighteenth-century astronomer Jean-Sylvain Bailly, who thought it incredible that Tycho himself, of all people, should have contrived a geoheliocentric world system instead of being converted to Copernicanism by his cometary observations.[5] Luckily, another blow was dealt the Aristotelian world view just a few decades later, when, in 1609, Galileo pointed his telescope skyward and made momentous discoveries that strongly reinforced the Copernican point of view that the sun was the centre of the universe, as well as once again demonstrating that the Aristotelian dogma of celestial unalterability was untenable.

If, then, historiographers like Bailly realized that the comet of 1577 had not succeeded in instantly turning all early modern scientists into Copernicans, they did believe in a very rapid transition from staunch Aristotelianism before 1577 to rational, well-substantiated heliocentrism after the telescopic discoveries of Galileo. To them, those early modern scholars who did not seriously reconsider their world views during the decades around 1600, were retarded fools whose ideas did not deserve treatment in works of serious historiography.[6]

This nineteenth-century account has had an enormous impact on subsequent historiography, and still underlies the way in which the history of cometary appearances is approached[7]; it also inspired the embryonic stages of the present study. Given the importance traditionally accorded the comet of 1577, and the ensuing telescopic discoveries of Galileo, in bringing down the Aristotelian world view, and in ushering in Copernicanism, it seemed viable to investigate the way in which Netherlandish scholars reacted to 'Tycho's' comet of 1577, and to its 1618 successor, the first notable comet to have appeared since the publication of Galileo's telescopic findings.

The hypothesis appeared warranted that, if anywhere, the reconsideration of world views brought on by the novel discoveries should be especially pronounced in the Netherlands. After all, many of those who had reacted favourably to the Copernican world system as expounded in *De Revolutionibus* had been Dutchmen: Gemma Frisius, physician and mathematician at Louvain, who had expressed a posi-

[5] See below p. 23.

[6] For a more detailed examination of positivistic historiography of cometary thought and its long-lived impact, see chapter II.

[7] See chapter II § 4.

tive interest in Copernicus' work even before it was published[8]; Nicolaus Mulerius, professor of medicine and astronomy at Groningen, who was responsible for the third, improved edition of *De Revolutionibus,* which appeared in Amsterdam in 1617[9]; Philip van Lansbergen, who in 1629 published a book on the earth's movement and later tenaciously defended the Copernican hypothesis against the geocentrism of Libertus Fromondus; and Simon Stevin, who some twenty years earlier expounded the heliocentric world system at length in his *Wisconstighe Ghedachtenissen* ("Thoughts on Mathematics", 1605-1608), a textbook written for Prince Maurice of Orange. Recent research on the reception of heliocentrism in the Dutch Republic has indeed shown that Copernican ideas met with an early and relatively favourable reception among a small cultural elite, and that real controversy on the issue of heliocentrism started only after the condemnation of Galileo in 1633.[10] While these later stages of the reception of Copernicanism in the Netherlands, which in the 1640s and 1650s became linked to the dogged conflict between the orthodox Calvinist Gisbertus Voetius, on the one hand, and the followers of Descartes on the other, have been the subject of several studies[11], the Dutch[12] tracts written on the cometary appearances of 1577 and 1618 have never been taken into consideration. Thus, in view of the way in which these comets allegedly facilitated the adoption of Copernicanism, it seemed possible that an investigation of Dutch cometary tracts from the period 1577-1620 might be able to throw

[8] Lynn Thorndike, *A history of magic and experimental science* V (New York 1941) 410-411; A. de Smet, 'Copernic et les Pays-Bas', *Janus* 60 (1973) 13-23:13.

[9] *Nicolai Copernici Torinensis Astronomia Instaurata, Libris Sex Comprehensa, qui* de Revolutionibus orbium caelestium *inscribuntur. Nunc demum post 75 ab obitu authoris annum integritati suae restituta, Notisque illustrata, opera & studio D. Nicolai Mulerii Medicinae ac Matheseos Professoris Ordinarij in nova Academia quae est Groningae* (Amsterdam 1617).

[10] R. H. Vermij, "Het copernicanisme in de Republiek: een verkenning", *Tijdschrift voor geschiedenis* 106 (1993) 349-367; see also Jonathan Israel, *The Dutch Republic. Its Rise, Greatness, and Fall 1477-1806* (Oxford 1995) 582-583.

[11] Theo Verbeek, *René Descartes et Martin Schoock. La querelle d'Utrecht* (Paris 1988); K. van Berkel, *In het voetspoor van Stevin. Geschiedenis van de natuurwetenschap in Nederland* (Amsterdam 1985); F. Sassen, *Geschiedenis van de wijsbegeerte in Nederland* (Amsterdam 1959); J.A. van Ruler, *The crisis of causality. Voetius and Descartes on God, nature and change* (Leiden/New York/Cologne 1995); for a survey of the debate, see Israel, *Dutch Republic,* 581-587.

[12] The term 'Dutch' throughout the present study refers to both the northern and the southern Netherlands, unless accompanied by an explicit statement to the contrary.

some additional light on the question of the early reception of heliocentrism in the Netherlands.

Therefore, initially, the aim of the present study was to answer the question of what role the cometary appearances of 1577 and 1618 might have played in enhancing the plausibility of the heliocentric world system in the Netherlands, given the fact that, traditionally, they were credited with having facilitated the acceptance of Copernicanism, and that the Copernican hypothesis had made a favourable impression on quite a number of Dutch scholars. However, this formulation of the problem soon proved to be unworkable, for a number of reasons. First of all, it became clear that the association of the comets, and the telescopic discoveries, with the reception of Copernicanism was for the most part unwarranted. Although it has recently been suggested that Michael Mästlin's work on the comet of 1577 may have inspired his, and hence also Johannes Kepler's, Copernicanism (Mästlin was Kepler's teacher)[13], this hypothesis needs to be be seriously qualified[14], and in any case it only refers to one single instance of the role played by the comet in the adoption of Copernicanism. A much more typical reaction to the superlunary location of comets, as well as to the telescopic discoveries, seems to have been to modify, as slightly as possible, the existing Aristotelian world view in order to be able to incorporate the new findings without having to abandon the basic tenets of traditional cosmology.[15] This need not surprise us if we take into consideration the fact that the establishment of a superlunary position for a comet does not in any way logically lead to the conclusion that the sun is the centre of the universe instead of the earth; it much more seriously challenges that other Aristotelian dogma, the unalterability of the heavens. Therefore, an investigation of the way comets influenced early modern world views should focus also on the issue of celestial incorruptibility, the nature of celestial matter, and the strict dichotomy be-

---

[13] By Robert S. Westman, "The Comet and the Cosmos: Kepler, Mästlin and the Copernican Hypothesis" in Jerzy Dobrzycki ed., *The Reception of Copernicus' Heliocentric Theory. Proceedings of a Symposium organized by the Nicolas Copernicus Committee of the International Union of the History and Philosophy of Science, Torun, Poland, 1973* (Dordrecht/Boston 1972) 7-30.

[14] See my discussion of Westman's article in chapter II, p. 33, and chapter IV, p. 131-132.

[15] See for example Roger Ariew, "Theory of Comets at Paris during the Seventeenth Century", *Journal of the History of Ideas* 53 (3) (1992) 355-369.

tween sublunary and superlunary regions, rather than exclusively on the reception of heliocentric cosmology.

Thus, the problem to be investigated became redefined in the following way: To what extent did the observation of the comets of 1577 and 1618 induce Netherlandish authors to abandon certain basic tenets of the Aristotelian world view, including, but by no means confined to, geocentrism? However, on closer scrutiny, this question in its turn raised a host of additional queries. It appeared to find itself, as it were, at the centre of an imaginary onion, around which were layered several strata of related subject matter of increasing generality, that needed to be investigated before a well-founded answer to the question at hand could be formulated. Four such layers of the onion skin presented themselves: the first pertained to the precise nature of the much-employed concept of the 'Aristotelian world view'; the second addressed the influence of Renaissance humanism, with its restoration to dignity of ancient ideas at variance with those of Aristotle; the third was concerned with the astrological and divinatory aspects of the early modern world view; and the fourth involved an assessment of the political situation in the Netherlands during the period 1577-1620, and the influence this exercised on the intellectual life of the authors of the cometary tracts.

Firstly and most obviously, then, any attempt to answer the question of how a particular world view crumbled or was abandoned, logically provokes the concomitant query of what exactly is to be subsumed under that world view. It soon turned out that, in the case of the Aristotelian world view, as in the case of the connection between comets, the telescope, and Copernicanism, nineteenth-century historiography, basing itself exclusively on the writings and opinions of a very small number of seventeenth-century authors such as Francis Bacon and René Descartes, had fabricated a distorted account of historical reality. According to the standard view, which has left a profound mark on the historiography of science even to the present day[16], the Aristotelian world view was a monolithic block of

---

[16] See below, chapter II. See also Charles B. Schmitt, *Aristotle and the Renaissance* (Cambridge Mass./London 1983) 7: "By the time Hobbes's sentiments [strongly denouncing Aristotle and his followers] were written the peripatetic domination of the schools was being seriously questioned, and it was to last only for another generation or so. Descartes, Gassendi, and many others joined the chorus of criticism. Such has been the mainstream of critical opinion regarding the use of Aristotle in the Renaissance schools down to the present day."

incorrect tenets that were all rigidly dependent on one another: the heavy, sluggish mass of earth was located at the centre of the universe, and was surrounded by several layers in which were situated, in increasing order of altitude, the sublunary elements of water, air, and fire. Beyond the sphere of fire came the lunar sphere, which constituted the impassable boundary between the sublunary realm, ever-changing and corruptible, and the superlunary regions, perfect and unalterable, in which the planets, each embedded in a solid orb, pursued their circular and eminently regular paths. This, allegedly, was the cosmology that had been contrived by Aristotle and unthinkingly adopted by all medieval scholars; and this was the system that collapsed within a few decades around 1600.

The above account has led a very long life; only quite recently has it been subjected to critical scrutiny, and been found seriously wanting. Charles B. Schmitt, one of the first scholars to have drawn attention to the unwarranted simplification of Renaissance Aristotelianism, has formulated the problem in the following way:

> Practically all students of the period recognize the continued presence of Aristotle in Renaissance cultural and intellectual life, but such a presence has nearly always been interpreted as the conventional and steadfast backdrop against which the novel and progressive elements are to be evaluated.[17]

He himself therefore intended, among other things, to show that

> the Aristotelianism of the fifteenth, sixteenth, and seventeenth centuries had an internal development of its own, and was itself dependent upon external influences for its own continuing efficacy.[18]

Thus, in Schmitt's view, the Aristotelian world view was not by any means as rigid and conservative as has often been assumed, and was well capable of incorporating new ideas in order to adapt to the many changes with which it was faced. The same idea has been advanced on many occasions by Edward Grant[19], who has minutely investigated the fortunes of Aristotelian cosmology from its earliest reception in the Latin West until its eventual extermination in the final decades of the seventeenth century. Like Schmitt's, Grant's re-

---

[17] Schmitt, *Aristotle and the Renaissance*, 7.

[18] Ibidem.

[19] In a great number of articles, and especially in his magisterial compendium of scholastic cosmology, *Planets, Stars, and Orbs. The Medieval Cosmos 1200-1687* (Cambridge 1994).

searches have shown that no single 'Aristotelian world view' in fact existed, either during the Middle Ages or during the Renaissance, and that many personal idiosyncracies could be, and were, incorporated into the world views of individual scholars.

To Schmitt, Renaissance Aristotelianism "was in effect eclectic, insofar as it was capable of appropriating other philosophical and scientific doctrines for its own purposes."[20] Eclectic, that is, in two different ways: Schmitt mentions "the general impulse of some Aristotelians to draw material from non-Aristotelian sources [such as Platonism, Stoicism, or the Christian tradition] simply because they thought that the insights found among the texts from other traditions could usefully be applied to strengthening their own philosophy", on the one hand, and "the tendency to accept *new* developments, particularly in science and the formalistic disciplines, which clearly offered superior doctrine to that derivable from normal Aristotelian sources" on the other.[21] To Schmitt, then, the Aristotelian world view was capable of incorporating both the innovations brought on by the humanist rediscovery of rival ancient theories, and the new scientific findings made during the Renaissance. The second of these two forms of eclecticism clearly has immediate bearing on our topic: the superlunary position of comets and the telescopic discoveries of Galileo obviously qualify as 'new scientific findings'.

Thus, after this brief overview of the first layer of the onion, the questions to be answered by the present study can be modified: we now want to know, not only whether the comets of 1577 and 1618 induced Dutch authors to abandon certain constituents of their Aristotelian world views, but also whether in fact they succeeded in actually incorporating the consequences of the new findings concerning comets (insofar, of course, as they chose to accept these findings) into their already existing world views.

However, in the light of very recent research, it appears quite probable that the second kind of eclecticism defined by Schmitt (the adoption into Aristotelian world views of material from non-Aristotelian sources), too, played a role in moulding the Netherlandish ideas about the comets of 1577 and 1618. In his introduction to a recent collection of articles tellingly entitled *The Uses of Antiquity*, Stephen Gaukroger has pointed out that his aim in editing the volume was to

---

[20] Schmitt, *Aristotle and the Renaissance*, 89.
[21] Ibidem, 92.

show how, during the period between Copernicus and Newton, "an image of antiquity was constructed and put to use in contemporary debates."[22] Gaukroger goes on to chide Enlightenment historiography for depicting the Renaissance as a straightforward clash between the 'ancients', slavishly relying on authorities from the past, and the 'moderns', who broke free from those same authorities. Much more typical of the spirit of the age, according to Gaukroger, was the attitude of Joseph Glanville, who "[spelled] out his modernist opposition to Aristotle in a way which, far from pitting the moderns against antiquity, uses antiquity to rebut Aristotle"[23], and who severely reproached the Stagyrite for having denounced the wisdom of the ancients, rhetorically asking "How fit then think you is it that the World should now be obliged to so *tender* and *awful* a respect to the *Libeller* of the most Venerable *Sages,* as that it should be a crime next *Heresie* to endeavour, though never so *modestly,* to weaken his *textuary* and *usurp'd authority?* when my *veneration* of the greater *Antiquity* extorted from me those strictures against the *proud Antagonist* of all the *ancient* and more *valuable Wisdom?*"[24]

Combining the ideas of Schmitt and Gaukroger, it seems reasonable to suppose that, whereas not all early modern authors will have gone as far as Glanville in completely abandoning Aristotle in all his manifestations, some of them may well have attempted to incorporate some of the venerable, non-Aristotelian ideas of Antiquity, that were clearly held in high esteem, into their world views whenever they were faced with a phenomenon that could not be explained in terms of the Stagyrite's world system. The recent work of Peter Barker and Bernard Goldstein on the role played by Stoic ideas during the sixteenth and early seventeenth centuries confirms this hypothesis.[25] In his latest article on the issue, Barker has pointed out

---

[22] Stephen Gaukroger, "Introduction: The Idea of Antiquity" in Stephen Gaukroger ed., *The Uses of Antiquity: The Scientific Revolution and the Classical Tradition* (Dordrecht/Boston/London 1991) ix-xvi: ix.

[23] Ibidem, x.

[24] Joseph Glanville, *A Letter to a Friend Concerning Aristotle,* appended to *Scire/i tuum nihil est: or, The Author's Defence of the Vanity of Dogmatizing* ..., (London 1665), quoted by Gaukroger, *Uses,* x.

[25] Peter Barker & Bernard R. Goldstein, "Is seventeenth-century physics indebted to the Stoics?", *Centaurus* 27 (1984) 148-164; Peter Barker, "Stoic contributions to early modern science", in Margaret J. Osler ed., *Atoms, 'pneuma', and tranquillity. Epicurean and Stoic themes in European thought* (Cambridge 1991) 135-154; and Peter Barker, "Jean Pena (1528-58) and Stoic physics in the sixteenth century" in Ronald H. Epp

that "Stoic ideas [were] eclectically combined with other ideas. ... [They] therefore re-entered scientific debates not as a complete system, distinguishable from an Aristotelian or an Epicurean system, but in eclectic combinations with ideas that in antiquity had been regarded as antithetic."[26]

Barker himself has investigated the role played by Stoic ideas in shaping one particular Renaissance scientific debate: that on the substance of the heavens.[27] Although he has also examined early modern cometary observations[28], Barker has nowhere explicitly argued that the observers of the comets of 1577 and 1618 (or, indeed, of earlier comets) may have employed Stoic cometary ideas, incorporating them into their world views in the eclectic way that was typical of the Renaissance. Yet, there are good reasons for maintaining precisely this hypothesis. We already know that the ideas that competed with Aristotle's during Antiquity were used by early modern authors to render more respectable their own views, whatever these might be; we also know that it was common practice to replace certain elements of the Aristotelian world view with other, for some reason more plausible ones. Add to this the fact that, in the field of cometology, the most formidable ancient rivals of the Stagyrite had been the Stoic authors Manilius and, especially, Seneca, and it becomes very tempting to suggest that, possibly, Stoic theories of comets may have assumed heightened significance during the period around 1600.

Having subjected the second layer of the onion, the influence of Renaissance humanism, to a short overview, then, it has become clear that we may well have to ask ourselves, not only whether our Dutch authors incorporated new findings on comets into their world views, but also whether in so doing they drew on Stoic cometary ideas. This hypothesis seems the more plausible if we take into account the fact that many of the Netherlandish authors who committed to paper their thoughts on the comets of 1577 and 1618 were

ed., *Spindel Conference 1984: Recovering the Stoics* (Supplement to *The Southern Journal of Philosophy* 23 (1985)) 93-107.

26 Barker, "Stoic contributions", 154.

27 See his "Stoic contributions to early modern science" and his "Jean Pena and Stoic physics".

28 Peter Barker, "The optical theory of comets from Apian to Kepler", *Physis* 30 (1993) 1-25; Peter Barker and Bernard R. Goldstein, "The role of comets in the Copernican Revolution", *Studies in the history and philosophy of science* 19 (1988) 299-319.

closely connected with, or at least well versed in, Stoic thought. Stoicism made an especially deep impression in the Netherlands: it was propagated by the great humanist Justus Lipsius, who held professorships both at Leiden and at Louvain. Five of the seven scholarly authors whose cometary views are examined in the present study were more or less closely linked with the Neostoicism coined by Lipsius: Erycius Puteanus, who was Lipsius' most promising student, and later his successor at Louvain; Johannes Heurnius and Cornelius Gemma, who, like Lipsius, sympathized with a secret spiritualist sect, the so-called *Family of Love*; Nicolaus Mulerius, who was one of the great humanist's students at Leiden; and Libertus Fromondus, who, after Lipsius' death, undertook to re-edit the eminent scholar's Seneca-edition, to which he added his own comments on the seven books of the *Natural Questions*—the last of which exclusively deals with comets.

We have now pinpointed two important areas of Renaissance intellectual life that, until quite recently, have been neglected by historiographers of science: the adaptability of Renaissance Aristotelianism and the role played by humanism in shaping 'scientific' ideas. Another important constituent of early modern intellectual life which has traditionally been disregarded by historians of science, is the role played by astrology and divinatory beliefs. Although, recently, interest in these subjects has surged, especially since the work of Frances Yates, the astrological components of cometary thought have not yet been very extensively studied, with historians of cometology choosing to focus on the more traditional subject matter of astronomy and cosmology instead.[29] Yet, comets are eminently well-suited to demonstrate the interplay between 'rational' ideas on the structure of the universe and 'irrational' investigations of God's purposes and his use of natural signs as warnings, that was so characteristic of the intellectual climate of the sixteenth and early seventeenth centuries. For this reason, astrological and divinatory beliefs, which as we shall see played an extremely important role in the cometary writings of our Dutch authors, are certainly to be accorded their share of attention, and therefore constitute the third layer of the onion.

And so we reach our outermost stratum of investigation, one which leads us onto the level of historical events, rather than ideas: the present study would not be complete without a certain measure

[29] For a detailed substantiation of this statement, see chapter II, § 4.

of attention accorded to the political situation in the Netherlands during the period 1577-1620. It is my firm conviction that it is impossible to write any kind of intellectual history without taking into consideration the social and political circumstances at the time and place where the ideas to be studied were developed. This is true *a fortiori* for a study of cometary thought in the Netherlands around 1600. First, because the extreme state of turmoil in that region made a profound impression on all authors to be treated, but especially on those writing in 1577-78. Their entire outlook was pervaded by a feeling of insecurity and fear, and the imminent sense of judgement brought on by the gruelling experience of the Dutch Revolt. The appearance of a comet, a phenomenon traditionally endowed with dire connotations of doom and punishment, served to confirm people's darkest forebodings, and inspired some of our authors to lengthy moralistic digressions.

In addition, on a more down-to-earth level, the period under consideration was one of the most important and decisive episodes in the history of the Netherlands. In 1577, the unity of the northern and southern Netherlands had not yet been broken; despite considerable confusion and differences of opinion, serious efforts were made to decide on a common policy for the entire territory of the Low Countries. In 1618, by contrast, North and South had become separated, with the South remaining under Spanish rule while the North had in the end opted for a republican form of government. Because of the novel political situations that presented themselves during the turbulent years around 1600, the period was marked by a good deal of reflection on the ever-changing state of affairs. Since comets invariably provoked political predictions, it seems well worth investigating whether traces of such reflection can be found in the cometary tracts. Also, for the year 1618, it would be interesting to see whether any marked contrast between North and South can be discerned in such possible political allusions in the cometary tracts.

At present, having taken into consideration all four of the abovementioned layers of the onion skin, we are in a position to overlook the entire onion, and to formulate the final version of the questions we would like the present study to answer. Our core question, constituting as it were the heart of the onion, still remains: How did the appearance of the comets of 1577 and 1618 affect the world views of our Dutch authors? Did they place the comets above the moon, and if so, did this lead to the disintegration of their Aristotelian world views, or to their adoption of Copernicanism? However, this core question can now be further refined by adding the concomitant queries that we have been able to distill from the outer layers: How flexible was our authors' Aristotelianism? Was it abandoned all at once, or was it modified to a greater or lesser extent, and if so, how? Were the novel scientific findings incorporated into their previous world views? How did they legitimate such deviations? Did they resort to ancient, non-Aristotelian ideas to replace any Aristotelian tenets that had become implausible? Can any influence of Stoic cometary ideas be detected in their writings? To what extent and in what way did they address the question of the comets' significance, and how did their treatments of this issue relate to their cosmological ideas? To what extent were their opinions affected or even shaped by the turbulent state of affairs in their fatherland? Did the political situation influence their cosmological or astrological beliefs?

These are the queries the present study aims to answer. It is an attempt at redirecting the attention of historians of cometology, which has long been focused almost exclusively on the astronomical side of the cometary question. Although today, the view expounded by Kaiser would never be so categorically upheld, it still underlies the methodology of many contributions to the history of cometary thought, especially of those works concerned with the comets that appeared around 1600.[30] Our aim in the present study is to relate the appearance of the comets of 1577 and 1618 to recent scholarship on Renaissance Aristotelianism, rival ancient philosophies, and astrological beliefs, and in so doing to make a modest contribution towards qualifying the traditional view. But it is also an attempt at sketching, however incompletely or gropingly, the mental world of intellectuals during the turbulent early decades of the Dutch Revolt.

---

[30] See chapter II.

CHAPTER TWO

COMETS AFTER THE CESSATION OF DIALOGUE: A HISTORIOGRAPHICAL INTRODUCTION

The assessment of the role played by Tycho Brahe's observations of the comet of 1577, and by the discoveries made with the newly-invented telescope by Galileo in 1609, in bringing down the traditional 'Aristotelian world view' and ushering in modern science, has remained remarkably constant throughout the centuries. Ever since the latter half of the seventeenth century, the belief that Tycho's and Galileo's achievements were of paramount importance in bringing down the fortress of Aristotelian cosmology has been widespread. Since it is the aim of the present study to investigate whether, in the case of the Netherlands, such a view can be maintained, it is worthwhile, before turning our attention to the Low Countries themselves, to elucidate how this notion came into being. The present chapter will show that the revolutionary status still often accorded to the comet of 1577 and the telescopic discoveries first emerged during the eighteenth century, in the wake of the publication of Newton's *Principia*. Also, it will demonstrate how this 'Newtonian' view continued to reign throughout the nineteenth century, and how even the majority of modern studies have to a certain extent been affected by it.

1. *Dialogue*

In the opening chapter of the first book-length study of the historiography of science, H.F. Cohen's *The Scientific Revolution. A Historiographical Inquiry* (Chicago 1994), the author discusses the "boundary lines" of his work. Because he is interested in "historians' ideas about the Scientific Revolution as a historical problem"[1], the focus of his book is on twentieth-century historiography, when modern science

[1] H. Floris Cohen, *The Scientific Revolution. A Historiographical Inquiry* (Chicago 1994) 10.

emerged as "a historical issue in its own right."[2] The first author Cohen treats at some length is Immanuel Kant, who, at the end of the eighteenth century, was one of the first to pave the way for "a conception of the birth of early modern science that is at once systematic, analytical, broadly interpretive, and concerned with the past in its own right."[3]

Cohen's temporal starting point is hardly surprising, given the definition of his problem: to trace the way in which modern science has been treated as a historical issue in its own right. After all, there was a time when such a conception simply did not exist. The earliest forms of historiography of science were almost indistinguishable from science itself, which consisted mostly of commentaries and analyses of the classical works. Typically, a scientific author would expose and assess the views of other, older writers on the same subject, after which he would sometimes voice his own opinion. This is not to say that these 'pre-modern' authors never criticized their predecessors, because they did: on a variety of issues a variety of opinions was put forward, and sometimes fierce controversies ensued. The point to be made here is not that they never dared contradict each other, but that they spoke to each other at all. A continuing dialogue rose through the centuries. Past authors never died; their views always remained factors to be taken into consideration in the ongoing debate.

In this sense, then, for pre-modern scientists, history did not exist. As Rupert Hall put it, "Before the dynamic movement of the 17th century a knowledge of current science was almost exactly equivalent to a knowledge of the history of science."[4] In each branch of learning, a complex of well-known and frequently cited arguments and points of view came into being through the centuries. Every practitioner of a particular science was thoroughly acquainted with the opinions aired by his predecessors: these were the tools he worked with. An expression of the universally known complex of opinions in all branches of learning, as it had developed by 1500, can be found in the *Margarita philosophica* by Gregor Reisch (1503). In Book IX, chapter 23, this encyclopaedic compendium of learning includes a

[2] Ibidem, 2.
[3] Ibidem, 24.
[4] A.R. Hall, "William Wotton and the History of Science", *Archives internationales de l'histoire des sciences* 28 (1949) 1047-1062: 1049.

discussion of comet lore, featuring a set of assumptions about comets' genesis, substance, and consequences that would have been instantly familiar to any scholar between Aristotle and Newton.

When the *Margarita philosophica* was published, the world of learning found itself on the brink of probably the most turbulent upheaval it had ever known. The period between the publication of Copernicus' *De Revolutionibus* in 1543 and that of Newton's *Principia* in 1687 undeniably saw a profound change of outlook, both in the world of scholarship and in society at large. But age-old attitudes do not disappear overnight. During the sixteenth and seventeenth centuries, history was still regarded as an integral part of scientific knowledge, and classical authorities were still used as opponents in scholarly debates.[5] Thus, William Wotton published his *Reflections upon Ancient and Modern Learning* in 1694 as an attempt to evaluate the scientific advances of his century and to compare its exploits with those of antiquity, in order to defend the Royal Society and its methods against attacks by such champions of antiquity as Sir William Temple. Although Wotton obviously thought much higher of the moderns, he mentioned the exploits of the ancients just as frequently, and took great pains to demonstrate the superiority of the new scientific insights.

The fact that Wotton, along with many other Royal Society apologists, still deemed it necessary, on the threshold of the eighteenth century, to present in his book all the ancient authorities (albeit in the role of antiheroes), in itself proves the extent to which the abandonment of the old doctrines was still in need of justification. The fact that Wotton himself "had more comprehension of the works of the ancients than of the latest discoveries"[6] shows how close his affinity to the old body of ideas was.[7] Collingwood's insight, that "an intense polemic against a certain doctrine is an infallible sign that the

[5] Helge Kragh, *An Introduction to the Historiography of Science* (Cambridge 1987) 2.

[6] Hall, "William Wotton", 1048.

[7] The absence of a sense of radical break with the past during the seventeenth century is also argued by David C. Lindberg, "Conceptions of the Scientific Revolution from Bacon to Butterfield: a preliminary sketch" in David C. Lindberg and Robert S. Westman eds., *Reappraisals of the Scientific Revolution* (Cambridge 1990) 1-26: 5, where he states that "In the battle between the ancients and the moderns, the point at issue was not whether the ancients were deserving of admiration but how much progress moderns had made toward equaling or surpassing the ancient achievement."

doctrine in question figures largely in the writer's environment and even has a strong attraction for himself,"[8] clearly applies here.

Not all the 'Moderns' were as modern as Wotton would have them. The Bolognese astronomer, Giovanni Battista Riccioli, one of the Moderns mentioned by Wotton, in 1651 published a voluminous treatise on astronomy, the *Almagestum novum*, the eighth book of which is devoted to comets and novae. In it, the author displays an attitude basically similar to that of Gregor Reisch, 150 years earlier: he treats the opinions of ancient and recent writers on a substantially equal footing. When, for example, he discusses comets' distance from the earth, he distinguishes three different points of view: the view that comets are sublunary phenomena, the view that they are superlunary phenomena, and the view that some are above and some below the moon. In each category, Riccioli cites the names of ancient and modern authors indiscriminately: Xenophanes figures alongside Scaliger, Pythagoras alongside Descartes.[9] Throughout the book, such, to us rather unlikely, combinations are to be found. Still, at times Riccioli displays a certain impatience with the ancients, especially as far as their lack of workable cometary observations is concerned[10]: because he is chiefly interested in the kind of quantifiable data the ancients cannot provide him with, on the whole he cites the moderns more often.

A similar preoccupation with measurement is to be found in the work of the Danzig astronomer Johannes Hevelius, also mentioned by Wotton. His 1668 *Cometographia* marks a significant step in the historiography of comets, as it contains the first systematic attempt at compiling a survey of all past cometary appearances. To this end, the author lists all allusions to comets he has been able to find, whether in the writings of "historians, philosophers, or astronomers".[11] Although Hevelius' work is remarkably devoid of sympathy for the many non-quantifiable aspects of ancient cometology—his sole aim

[8] R.G. Collingwood, *The Idea of History* (Oxford 1946; paperback edition Oxford 1994) 22.

[9] Ioannes Baptista Riccioli, *Almagestum novum* (Bologna 1651) Book VIII, 28.

[10] Ibidem, 27: "Sed quoniam incuria observatorum veterum, plura de Cometarum via dicere vetat, transeamus ad alias conditiones." The uncertainty of cometary observations leads Riccioli to concede only the probability, but not the certainty, of supralunar comets; ibidem, 119.

[11] Johannes Hevelius, *Cometographia* (Danzig 1668) title page: "Accessit, Omnium Cometarum, à Mundo condito hucusque ab Historicis, Philosophis, & Astronomis annotatorum, Historia. . .".

in compiling his list being to test his hypothesis on cometary motion—, he still quotes a number of Peripatetic views on comets, albeit only to refute them.[12]

Great respect for the accuracy of Hevelius' own observations was displayed by the French *intendant des fortifications*, Pierre Petit, in a book on comets published at the request of Louis XIV in 1665. Petit agreed with Riccioli and Hevelius, that ancient observations were so bad that they led to falsity rather than truth.[13] Unlike Hevelius, Petit at length discussed the astrological aspects of comet lore. Although he found most predictions from cometary appearances ill-founded nonsense[14], the fact that he mentioned them at all shows his affinity with the ancient way of theorizing about natural phenomena, in which it was usual to deal with both the matter and motion of a comet, and its consequences.

The same concern for the significance of comets was displayed by the Polish theologian Stanislaus Lubieniecki in his *Theatrum cometicum*, which appeared in Amsterdam in 1668. The subtitle of the work, *Opus Mathematicum, Physicum, Historicum, Politicum, Theologicum, Ethicum, Oeconomicum, Chronologicum*, shows the author's pretensions to provide his readers with a complete account of cometary phenomena. Although he also treated cometary observations and physical characteristics, the emphasis of his work was on the compilation of a history of comets, not in order to calculate their motions (like Hevelius) but in order to answer the question of whether comets can be said to signify anything. He concluded that each comet was followed by both good and bad events, and his statement that they portended good for good people and evil for bad people comes very close to a denial of the divinatory power of comets.

Although Pierre Bayle's famous *Pensées diverses sur la Comete*, printed in Rotterdam in 1683, had as its main objective to speak out against Roman Catholic superstition, using the appearance of a comet in

[12] See for example ibidem 328-29, 347, 459, 511.

[13] Pierre Petit, *Dissertation sur la nature des cometes au Roy. Avec un Discours sur les Prognostiques des Eclipses & autres Matieres curieuses* (Paris 1665) 76.

[14] This attitude is perhaps partly to be explained by the fact that Petit was writing at the king's behest: he especially stresses the falsity of one of the standard ancient predictions associated with comets, the death of rulers. Apparently there were still enough people giving credence to this presage for Louis to go to the trouble of having it refuted. See Petit, *Dissertation*, 4 and 142, and page 147 where Petit asks the king to have divinatory practices prohibited.

1680 as a pretext, the book also contained a wide range of arguments against cometary divination. Bayle displayed a phenomenal knowledge of the ancient authors, and put their views to a variety of uses. Often, he cited passages from ancient authors like Pliny or Livy to demonstrate the degree of superstition prevalent among the pagans; but sometimes, he showed himself in agreement with ancient authors.[15] Most importantly, his method of refuting the 'science' of astrology, far from being a novel one, owed a great deal to a number of predecessors such as Cicero, Saint Augustine, Oresme and Pico della Mirandola.[16] A similar work, likewise published in the Netherlands and also inspired by the comet of 1680, was written by the Amsterdam minister Balthasar Bekker.[17] He, too, went to great pains to demonstrate the futility of cometary superstition, and asserted that comets were neither signs nor causes of future ills.

By the end of the seventeenth century, then, the attitude of scholars towards the past seemed to be wavering, not quite knowing which direction to take. There were those, like Hevelius, who went reasonably far in disregarding the ancient body of ideas and were only interested in the past insofar as it had produced acceptably accurate observations; there were those, like Lubieniecki and Riccioli, who still stood in the tradition of the old 'dialogue', and who continued treating both the physico-mathematical and the astrological side of the cometary question, using arguments that had accumulated over centuries; and there were those, like Bayle and Bekker, who manifestly felt the need to take a stand against an astrological world-view clearly still very much in existence—as had Cicero, Oresme, and Pico before them.

2. *The Cessation of Dialogue*

The publication, in 1687, of Isaac Newton's *Principia mathematica philosophiae naturalis* had a profound effect on both science in general and cometary science in particular. The harmony of the Newtonian system, in which all heavenly motions, including those of comets, were

[15] As in § 83, where he agreed with Seneca that the heavenly motions did not exist for the benefit of the earth.

[16] On this continuous tradition, see below ch. VI.

[17] Baltasar Bekker, *Ondersoek van de Betekeninge der Kometen* (Amsterdam 1683).

satisfactorily explained, appealed to many, and gave rise to a veritable passion for calculating cometary orbits. Its most famous exponent was undoubtedly Edmond Halley, who not only made the prediction, posthumously proved successful, of the return of the comet now bearing his name, but also compiled a voluminous collection of *Astronomical Tables, with Precepts both in English and Latin for computing the places of the Sun, Moon, Planets, and Comets.*[18] This work opened with a "Synopsis of the History of Comets" whose sole purpose was to explain which of the past observations of comets were suitable for calculating cometary orbits. The rest of the book consisted of tables full of astronomical data on past comets, without reference to any of the theories held by the ancients.

In Halley's footsteps followed a succession of astronomers obsessed with the progressively precise calculation of cometary orbits. The Frenchman Alexis Claude Clairaut occupied himself with the computation of the effects on the course of comets of the gravitational pull of the superior planets[19]; the Dutchman, Nicolaas Struyck, like Halley, tried his hand at the compilation of historical data on cometary orbits[20]; his countryman, J.A. Fas, attempted to predict the return of one of the comets from Halley's list[21]; and the Englishman, Blythe Hancock, wrote one of several books popularizing cometary science, aimed at "render[ing] the process of calculation plain and easy to the not very learned in mathematics".[22] Insofar as these works contained any comments on the history of cometology, these served only to sing the praises of Newton, who had expounded the true system of the universe, and of Halley, who had subjected comets to Newton's laws.[23]

This new approach to the history of comets, which had been foreshadowed by the work of Hevelius in the previous century, showed

[18] Latin version published posthumously, London 1749; English version 1752.

[19] Notably in his treatise *Théorie du mouvement des comètes. Dans laquelle on a égard aux altérations que leurs orbites éprouvent par l'action des Planètes. Avec l'application de cette théorie à la Comète qui a été observée dans les années 1531, 1607, 1682 & 1759* (Paris, s.d.).

[20] Nicolaas Struyck, "Korte beschrijving van alle de Comeeten of Starrtsterren die ik in de geschiedenissen heb kunnen vinden" in: *Inleiding tot de algemene Geographie, benevens eenige sterrekundige en andere Verhandelingen* (Amsterdam 1740).

[21] J.A. Fas, *Verhandeling over twee Comeeten uit de berekeningen van de Heer Halley* (Leiden 1787).

[22] Blythe Hancock, *The Astronomy of Comets. In Two Parts* (London 1786) xv.

[23] See for example Dionis Du Séjour, *Essai sur les Cometes en général, et particulièrement sur celles qui peuvent approcher de l'orbite de la terre* (Paris 1775) ii-iv, and Hancock, 12-13.

respect for pre-Newtonian writings on comets only insofar as these could be of help in the pursuit of contemporary interests (i.e., could provide data to be used in orbit computation). This attitude in writings on comets matches the mood prevalent in eighteenth-century historiography of science in general: history came to be used as a tool for situating the current level of progress in the natural sciences in its historical context, and for surveying the problems yet to be solved by scientists.[24] The Age of Enlightenment equated the history of science to the history of progress.

After Newton, then, a large-scale defection took place from the body of ideas that had prevailed during all of the previous centuries. Never before had scholars refused to communicate with their predecessors; never before had they failed to take them seriously. The 'combined approach' still taken by Riccioli and Lubieniecki, treating astronomical and astrological aspects of comets in a single volume, became more and more obsolete. The only consequence of comets seriously discussed by astronomers in the Newtonian tradition, was the possibility of a comet colliding with the earth and the physical results of such an encounter; the likelihood of which, however, was demonstrated to be extremely small.[25] And, though the purpose for which comets might have been created was still a popular topic for discussion among Newtonians, the ancient view of comets as divine messengers was left to pamphleteers to repeat, and was replaced among astronomers by an exploration of the utility of comets for practical or scientific ends.[26] Scholars turned scientists from now on went their own way, increasingly leaving the history of science to be written in separate volumes, not interfering with the pursuit of science proper.

Almost invariably, the authors of such histories were professional

---

[24] Kragh, *Introduction*, 3. Kragh cites Joseph Priestley's *The History and Present State of Electricity* (1767) and *History and Present State of Discoveries Relating to Vision, Light and Colours* (1772) as examples. Other famous examples are Montucla's *Histoire des mathématiques* (1758) and Delambre's *Histoire de l'astronomie* (Paris 1819-1827).

[25] The most important writings on this subject are Jérôme de Lalande, *Réflexions sur les Comètes qui peuvent approcher de la terre* (Paris 1773) and Pierre-Louis Moreau de Maupertuis, *A letter upon comets* (London 1769).

[26] E.g., to sustain life in the universe by means of circulating vital spirits, or to function as a home for the blessed; see Sara Schechner Genuth, "Devils' hells and astronomers' heavens: Religion, method, and popular culture in speculations about life on comets" in Mary Jo Nye, Joan L. Richards & Roger H. Stuewer eds., *The invention of physical science. Intersections of mathematics, theology and natural philosophy since the seventeenth century* (Dordrecht/Boston/London 1992) 3-26, especially pp. 19-21.

scientists, whose objectives were scientific and methodological: they wanted to derive from them the proper way of practising science and of attaining progress, and showed an interest in past scientific thought only insofar as it could be used as a guide to future discoveries.[27] An early exponent of such progress-related historiography of science was the *Histoire de l'astronomie moderne* by J.S. Bailly, which appeared in Paris in 1782. In it, the author chided Tycho Brahe for retarding the progress of astronomy by introducing his untrue world system at a time when all efforts should have been directed at propagating Copernicus's true one[28], and for adhering to the astrological superstitions of his time: "his eyes were closed to the truth."[29]

Bailly's countryman J.E. Montucla, in the second edition of his *Histoire des mathématiques* (Paris, 1799-1802) showed himself rather more appreciative of Tycho's achievements. Although the Dane did not quite shake off the yoke of superstition in connection with the earth's movement, at least he succeeded as far as cometary theory was concerned: he freed the world of the "despotic reign" of Aristotle's theory, and dealt scholastic philosophy a dangerous blow. Before Tycho, the few correct opinions on comets (notably those of Apollonius and the Pythagoreans), were "seeds of truth, stifled by the weight of superstition, especially the authority of Peripatetic physics."[30] The defenders of traditional cometary theory against Tycho's discoveries, not surprisingly, found no favour in Montucla's eyes: Scipio Claramontius, the author of the *Antitycho* (1621) "seems to have enjoyed a very long life only to retard the progress of the new discoveries as much as he could"[31], and Peripatetics who, after Tycho's eye-opening work, still adhered to the Aristotelian theory,

[27] See for example Ernst Mach, *Die Mechanik in ihrer Entwickelung historisch-kritisch dargestellt* (2nd edition, Leipzig 1889) 6: "Wir wollen nun auf unsern Gegenstand näher eingehen und hierbei, ohne die Geschichte der Mechanik zur Hauptsache zu machen, die historische Entwickelung so weit beachten, als dies zum Verständniss der gegenwärtigen Gestaltung der Mechanik nöthig ist, und als es den Zusammenhang in der Hauptsach nicht stört." Or William Whewell, *A History of the Inductive Sciences, from the Earliest to the Present Time* (3rd ed., London 1857) I, 4: "The examination of the steps by which our ancestors acquired our intellectual estate, may make us acquainted with our expectations as well as our possessions;—may not only remind us of what we have, but may teach us how to improve and increase our store."

[28] J.S. de Bailly, *Histoire de l'astronomie moderne depuis la fondation de l'école d'Alexandrie jusqu'à l'époque de 1782* (Paris 1782) xv and 413-416.

[29] Ibidem, 431-441.

[30] J.E. Montucla, *Histoire des mathématiques* (2nd ed. Paris 1799-1802) II 621.

[31] Ibidem, I, 662-664.

displayed "their ignorance, or their obstinacy to close their eyes to the truth".[32] Montucla's book does include a small section on astrology, because one can learn from past errors, because astronomy owes a lot to astrology, and simply because it is amusing.[33]

Not all eighteenth-century authors had completely abandoned the traditional body of ideas. One scientist who concerned himself extensively with comets was Alexandre Guy Pingré, who wrote a two-volume *Cométographie* consisting of a history of cometary theories, a list of all recorded cometary appearances, and an account of modern views on comets.[34] In his historical exposé, he still deemed it necessary to refute with serious arguments such obsolete cometary theories as those of Aristotle, Anaxagoras, and Democritus; and he was respectful of the Ancients' exploits, recognizing that he and his contemporaries were "dwarfs on giants' shoulders" and that it was unfeasible for the Ancients to achieve anything like the modern results, seeing that they had had much less experience with comets.[35] However, on the whole Pingré's work was permeated with the excitement of progress: finally, comets had been recognized to be stars, and received into the beauty of the solid Newtonian principles, whereas before, they had been relegated to the base class of meteors, and regarded with unfounded superstition.[36]

Throughout the nineteenth century, the tendency to chide the past for not realizing comets to be celestial bodies, and for believing them to carry portentous meaning, made itself felt in historiography. Arthur Berry in 1898 wrote a *Short History of Astronomy* in which he judged that Aristotle's cometary theory was "founded on no solid evidence and of little value"[37]; the comet of 1472 was "the first comet which appears to have been regarded as a subject for scientific study rather than for superstitious terror"[38]; and Kepler's theory of comets' tails was a "curiously correct anticipation of modern theories" on the subject[39]. Rudolf Wolf in his 1877 *Geschichte der Astronomie* praised

---

32 Ibidem, II, 622.

33 Ibidem, II, 369-378.

34 Alexandre Guy Pingré, *Cométographie ou Traité historique et théorique des Comètes* (Vol. I Paris 1783, vol. II Paris 1784).

35 Ibidem, I, 54.

36 Ibidem, iii.

37 Arthur Berry, *A Short History of Astronomy. From Earliest Times Through the Nineteenth Century* (Cambridge 1898) II § 30.

38 Ibidem, III § 68.

39 Ibidem, VII §146.

Tycho Brahe for "saving comets from lowliness"[40]. Amédée Guillemin in a popularizing work on the history of comets devoted his first chapter to superstitions connected with comets, the purpose of this account, according to the author, being to make the reader understand how the transition from "the most extravagant prejudices" to the "serene and reassuring conceptions of contemporary science" took place.[41] And A.D. White wrote "A History of the Doctrine of Comets", in which he maintained that, by the end of the eighteenth century, scientific advance had been able to "[clear] away one more series of those dogmas [cometary superstition] which tend to debase rather than to develop man's whole moral and religious nature", and had brought about the "emancipation from terror and fanaticism".[42]

3. *The Comets: Stock-Taking*

Although the denunciation of pre-Newtonian cometary writings did not abruptly cease with the expiry of the nineteenth century, nonetheless a more nuanced approach came to make itself felt during the earlier decades of the twentieth. Between 1923 and 1958, the American historian of science, Lynn Thorndike, produced an impressive series of eight bulky volumes dealing with the *History of Magic and Experimental Science.* Thorndike vigorously denied the claim that the first thirteen centuries of our era had been an age of retrogression or stagnation.[43] He defended medieval scholarship, which not only preserved Greek science, but also added new observations and discoveries, and tended to rely on rationalism and experimental method, so that "on closer scrutiny and comparison" many of the notions with

---

[40] Rudolf Wolf, *Geschichte der Astronomie* (Munich 1877) 409: "... aus der Erniedrigung errettet".

[41] Amédée Guillemin, *Les comètes* (Paris 1887) xi: "Dans le livre qu'on va lire, on trouvera quelques chapitres consacrés sommairement à leur histoire, ou à ce qu'on peut appeler l'astrologie cométaire. C'est l'introduction obligée de la partie purement astronomique, sans laquelle il serait difficile de comprendre comment on est passé des préjugés les plus extravagants aux sereines et rassurantes conceptions de la science moderne."

[42] A.D. White, "A History of the Doctrine of Comets", *Papers of the American Historical Association* 2 (New York 1887) 147.

[43] Lynn Thorndike, *A History of Magic and Experimental Science* II (New York 1923) 970.

which the sixteenth century is usually credited might be found to date back to classical or medieval authors.[44] This may well have been due to the practice of magic, since "it is possible that magicians were the first to experiment, and that natural science took over experimental method in a crude form from magic."[45] However, in due course science had to free itself from the error of previous "fantastic theories and false data"[46]; in the introductory chapter to volumes V and VI of his *History,* which deal with the sixteenth century, Thorndike declares that ". . .investigations based upon telescope and microscope ushered in a new age of science. . . Like Moses we have brought the reader through the wilderness to within sight of the promised land of modern science."[47] This statement reveals the limits of Thorndike's historicity: despite his sympathetic appreciation of medieval science, he nonetheless believed that the end of the sixteenth century saw a marked transition to the age of modern science, an age far superior to the previous magic-tainted ones.

Yet, despite this and similar value-judgments and attempts at pinpointing scientific ideas among the magical ones, Thorndike's work is tremendously valuable. It contains a very readable, very complete account of the ideas on issues in the occult sciences of both famous and lesser-known scholars from Roman antiquity to the seventeenth century. Obviously, comets, having through the ages been associated with astrological effects, often figure in Thorndike's narrative. He provides the reader with an extensive account of many different cometary tracts that were published over the centuries, tirelessly describing their contents in much detail. The case of comets is especially illustrative of Thorndike's approach, because they have always been connected with both astronomical ('scientific') and astrological ('magical') ideas. Thus the author, in a chapter specially devoted to "The New Stars" of 1572 and 1577, points out the "faultiness of an historical method which endeavors to pick out from the writings of past authors sentences or passages in which they are supposed to have marked the first appearance of a fruitful idea or to have heralded the development of some future discovery, without estimating the meaning and value of this utterance in relation to its context and

---

[44] Ibidem, 971.
[45] Ibidem, 977.
[46] Ibidem, 979.
[47] Ibidem, V (New York 1941) 15.

to the main aims and totality of the author's work."[48] So, although he makes no secret of his preference for modern science over ancient superstition[49], Thorndike is always careful to expound in detail not only the modern, but also the ancient aspects of his protagonists' thought, led by the realization that, in their minds, the distinction did not exist.[50]

While working on his *History*, Thorndike also lent a helping hand to the author of a book concerned exclusively with the history of comets: Clarisse Doris Hellman's *The Comet of 1577: Its Place in the History of Astronomy*. In this book, Hellman proposes "to illustrate the effect of the comet of 1577 on astronomical thought.. . . I want to show that the observations of the comet of 1577 were instrumental in bringing about a change in the opinions regarding comets, and that they mark an increased interest in and a greater number of observations of comets and a decided advance in cometary theory."[51] This formulation of her aim shows her concern with demonstrating the progress of science: in this sense, her outlook is not much different from that of her nineteenth-century colleagues. Sure enough, the spectre of progress pops up again and again in her account both of the ideas propounded by the authors of tracts on the 1577 comet and of cometary theory during antiquity and the Middle Ages. Hellman regularly remarks that a particular author or period "added nothing" to cometary theory[52], or that a particular opinion "must have been a deterrent to progress"[53], or was "valueless"[54]. She shows the same respect for accurate observations and orbit computation as a prerequisite for scientific progress we have seen displayed by authors like

---

[48] Ibidem, VI (New York 1941) 94.

[49] For example ibidem, VII (New York 1958) 406: "Despite his superior views as to comets, tides and the Copernican hypothesis, and his rejection of solid spheres, the outlook of Burgersdyck otherwise still seems sufficiently antiquated, credulous and superstitious."

[50] Ibidem, VI, 81: "The case of Mästlin shows that we cannot divide men into two camps with lines sharply drawn between them, but that we ever find all sorts of intermediate shadings and combinations of opinion. . . . For him, science and astrology go together: in his mind there is not warfare but concord between science and theology."

[51] Clarisse Doris Hellman, *The Comet of 1577: Its Place in the History of Astronomy* (New York 1944) 9.

[52] Ibidem, 34, 41, 65, 216.

[53] Ibidem, 103, 311, 233.

[54] Ibidem, 233, 243, 282.

Guillemin, Wolf and Berry, or even by the authors of the Newtonian era.[55]

However, possibly due to the influence of Thorndike, Hellman's work is itself not as valueless as she believed many of the cometary tracts to have been. Like Thorndike's *History,* it derives much of its worth from its very complete and accurate rendition of the contents of a great number of tracts on comets: not only those from 1577, but also many earlier ones. As far as can be judged without having read all the works discussed, Hellman renders them faithfully: even though the astrological parts have no scientific value, they receive their due of attention. Another important asset of the book is the Appendix, which lists all European treatises devoted specifically to the comet of 1577.

Another book concerned with the reactions to one specific cometary appearance was published some thirty years before Hellman's. As the title indicates, James Howard Robinson's *The Great Comet of 1680. A Study in the History of Rationalism* is one of several works crediting the comet discussed by Newton, Bayle, and Bekker with ushering in the age of Enlightenment.[56] Unlike Hellman, Robinson is not so much interested in scientific theories about comets, as in the gradual abandonment of cometary superstition. In his opinion, the appearance of the comet of 1680 caused a struggle between the theologico-astrological and the philosophical and scientific view.[57] Although Robinson adduces many examples of what he calls 'gross credulity' and 'semi-superstition' in connection with the comet, several enlightened authors, especially in France and the Netherlands, put forward a sceptical view. However, in the end it was the astronomers who ensured the "Victory of Science and Reason". After Newton and Halley, growing rationalism and specific results of scientific observation led to the dissolution of superstition: "It was

[55] See especially her "Conclusion", ibidem 307-317, notably p. 317: "... it becomes evident that in volume and accuracy of observation the year 1577 marks a tremendous leap forward."

[56] The most famous of these is probably Paul Hazard, *La crise de la conscience européenne 1680-1715* (Paris 1935), which contains a separate chapter on "The Negation of Miracle: Comets, Oracles, and Sorcerers" in which the author asserts that, about 1680, a rational system began to be elaborated by people such as Bayle and Bekker that was to lead to the extinction of superstition, a system of which the belief that comets announced no ills whatsoever formed an integral part.

[57] James Howard Robinson, *The Great Comet of 1680. A Study in the History of Rationalism* (Northfield, Minn. 1916) 29.

with the appearance of the comet of 1680 that the rapid dissolution of this old and hallowed superstition had begun."[58]

In addition to such detailed descriptions of the contents of cometary tracts, the early decades of the twentieth century also saw the first editions of source material on the history of cometology. In 1911, two German authors published works on comet broadsides between the fifteenth and eighteenth centuries. Both books contained descriptions and reprints of a large number of these pamphlets.[59] The year 1922 saw the publication of a volume very important to the history of cometary theory: part four of J.L.E. Dreyer's edition of the complete works of Tycho Brahe, containing the Dane's writings on the comets of 1577 and 1585, as well as his polemic with Craig the Scot.[60] Finally, Lynn Thorndike in 1950 published the integral texts of twelve medieval cometary tracts.[61]

By the 1950s, the approach taken by historiographers of cometary theories had changed in comparison with that of the late nineteenth century. Although most authors were still more or less searching the past for vestiges of future scientific progress, and at times adopted a condescending attitude towards the less 'scientific' ideas displayed by many of their protagonists, at least a genuine interest in the whole complex of past thought on comets began to surface. Especially Thorndike's insight, that the clear distinction in the minds of modern authors between 'scientific' and 'non-scientific' ideas had no counterpart in the minds of past writers, was to prove its worth in subsequent historiography.

4. *Recent Historiography of Comets*

Thus, we reach the most recent phase in the historiography of comets. Probably due to the variety of issues these phenomena have been associated with over the centuries, present-day historians employ a

---

[58] Ibidem, 119.

[59] Wilhelm Hess, *Himmels- und Naturerscheinungen in Einblattdrucken des XV.-XVIII. Jahrhunderts* (Leipzig 1911) and F.S. Archenhold, *Alte Kometeneinblattdrucke* (Berlin 1911).

[60] J.L.E. Dreyer ed., *Tychonis Brahe Dani Opera Omnia* IV (Copenhagen 1922).

[61] Lynn Thorndike ed., *Latin Treatises on Comets between 1238 and 1368 AD* (Chicago 1950).

number of different and often mutually exclusive approaches in their attempts to elucidate past cometary thought.

By far the most numerous are the popularizing accounts that flood the market, especially during the months preceding the expected arrival of a famous or spectacular comet. Many indeed were the publications on comets and their history that appeared in the year or so before the return of comet Halley in 1986.[62] Most of these books and articles contain more or less identical, rather superficial information, inherited from nineteenth-century historiography. The emphasis of such popularizing works, often written by professional astronomers, is usually on contemporary insights about comets; in this sense, they are the direct heirs of such nineteenth-century accounts as Guillemin's. But the summaries of cometary history that they usually include by way of an introduction to the modern theories, also regularly resemble the nineteenth-century historiography of progress. They always mention a number of standard notions, making up the following well-worn exposé (reminding one of the monotonous lists of disasters to be found in early modern cometary tracts): some ancient authors correctly assumed the superlunary position of comets; however, Aristotle's authoritative idea that they were sublunary meteors came to reign supreme for two millennia and suppressed scientific investigation; medieval man was terrified of comets and did not investigate them in any meaningful way; then, Regiomontanus and especially Tycho with their superior observation techniques proved comets to be superlunary, so that no reasonable person could thereafter believe Aristotle anymore; Newton and Halley supplied the finishing touch by demonstrating the periodicity of comets; after which superstition was on its way out (although irrational fears of cometary disasters still haunted some science fiction books until well into the twentieth century).[63] This version of events has proved extremely tenacious. As we

[62] Among many others Ruth S. Freitag, *Halley's Comet: A Bibliography* (Washington 1984); Helen S. Hogg & David E. Hogg, "The Tale of Halley's Comet", *Queen's Quarterly* 92 (2) (1985) 283-292; Mark Littman & Donald K. Yeomans, *Comet Halley: Once in a Lifetime* (Washington 1985); Raymonde Barthalot, "Le retour de la comète à tête d'or", *Historama* 21 (1985) 12-19; Roger-A. Blondeau, "Kometengeschiedenis van Aristoteles tot Halley" *Technologia* 9 (2) (1986) 118-132; Richard O. Norton, "Mr Halley, your comet has arrived", *American West* 23 (1) (1986) 46-50; Carl Sagan, "On the Eve of the Comet", *Proceedings of the Royal Institution of Great Britain* 58 (1986) 169-185; André Koeckelenbergh, "À propos de la comète de Halley: grandeurs et terreurs cométaires", *Technologia* 9 (2) (1986) 108-117.

[63] For examples of such accounts, see Peter Lancaster Brown, *Comets, Meteorites and*

have seen, for much of the twentieth century professional historians like Hellman have been affected by it, and even today it is not to be found exclusively in popular works written for the general public. Not that the recent historiographical refinements described in the previous paragraph have passed by historians of cometary thought unnoticed. Undeniably, the obsessive search for past anticipations of modern theories has much declined during the past few decades, to be replaced by a genuine interest in scientifically less fruitful views. However, in the classification of their subject matter, modern historians are still treading in the footsteps of their predecessors. The strict division that originated in the eighteenth century between, on the one hand, comets' astronomical aspects and, on the other, astrological considerations usually associated with them, is still much in evidence in present-day treatments of early modern cometary ideas.

a. *Astronomy: Solving the Copernican Problem*

For a number of authors, research into past cometary thought is still very much tied up with the problem of the early phases of the Scientific Revolution. In 1957, the philosopher of science, Thomas Kuhn, wrote in his influential account of the Copernican Revolution: "After their superlunary character had been deduced from observation and then widely debated, the evidence that comets provided for the mutability of the heavens could not indefinitely be ignored or distorted. Once again the Copernicans were the gainers."[64] He thus accorded comets an important function in connection with the, to him, revolutionary upheaval in sixteenth-century astronomy that was to lead to the birth of modern science. Because comets and novae provided evidence in favour of the Copernican theory, it became more and more difficult to doubt the truth of Copernicus's innovation; and consequently, since the heliocentric system was incompatible with

Men (London 1973) chapters 1 and 10; John C. Brandt and Robert D. Chapman, *Introduction to Comets* (Cambridge 1981) chapters 1 and 10; Patrick Moore, *The Comets. Visitors from Space* (Cambridge 1973) (e.g. p. 17: "In ancient times a comet was believed to be a luminous body floating in the Earth's upper air. This weird notion persisted until 1577, when the great Danish astronomer Tycho Brahe showed that all comets lie well beyond the Moon." Only slightly more historically sensitive, but at least according some attention to medieval views on comets, is Donald K. Yeomans, *Comets. A chronological history of observation, science, myth, and folklore* (New York 1991).

64 Thomas S. Kuhn, *The Copernican Revolution. Planetary Astronomy in the Development of Western Thought* (Cambridge, Mass./London 1957) 208.

Aristotelian physics, scientists like Galileo, Kepler, and Newton strove to replace the latter with a new physics in agreement with the new astronomy. Thus, in Kuhn's view, the appearance of the 1572 nova and the comet of 1577 contributed to the plausibility of the Copernican system, and therefore to the birth of modern science.

As early as 1944, Doris Hellman had accorded comets a key role in breaking down the Aristotelian world picture and contributing to the acceptance of the Copernican doctrine.[65] In a number of later publications, she elaborated this belief by asserting that the increasingly accurate measurements of cometary positions from the fifteenth century onwards, culminating in Tycho's observations of the comet of 1577, led to both the untenability of the Aristotelian cosmology and to Kepler's successful efforts to derive the laws of planetary motion, a feat which, in its turn, was to pave the way for the Newtonian system of the Universe.[66] A similar, though somewhat less pretentious, account, concentrating on fifteenth-century comet observations but including an epilogue on the sixteenth century, was published in 1985 by Jane L. Jervis. Although Jervis does not explicitly claim that fifteenth- and sixteenth-century observations had any direct implications for the reigning world view, her approach is similar to Hellman's in that she judges the merit of past scientific achievements by modern standards: to her, the greatest contribution of fifteenth-century observers was "the integration of comets into the scope of astronomical research".[67] She seems to regard Kepler's theory, that comets moved in straight lines, as a temporary hiccup in an otherwise fertile development initiated by her fifteenth-century astronomers and culminating in the work of Newton and Halley.[68]

The preoccupation with progressive elements of past theories, that

---

[65] Hellman, *Comet of 1577*, 9-10: "The conclusions of these men [the observers of the comet of 1577] affected, not only the theory of comets, ... but also ... the formulation of new systems of the world, encompassing a break-down of the Aristotelian point of view, and imparting an impetus to the Copernican doctrine."

[66] C. Doris Hellman, "The Rôle of Measurement in the Downfall of a System: some examples from 16th century comet and nova observations" in A. Beer ed., *New aspects in the history and philosophy of astronomy* (Oxford 1967) 43-52, especially p. 51-52; and C. Doris Hellman, "The gradual abandonment of the Aristotelian universe: a preliminary note on some sidelights" in *Mélanges Alexandre Koyré* (Paris 1964) 283-293.

[67] Jane L. Jervis, *Cometary Theory in Fifteenth-Century Europe* (Dordrecht/Boston/Lancaster 1985) 128.

[68] Ibidem, 126: "The important result [of Kepler's views on comets] for cometary theory was that comets were once again removed from planetary astronomy where they had been for such a very short time, and again were considered as special kinds

is still much in evidence in the work of historians dealing with the astronomical aspects of cometary thought, has also produced a number of studies relating the problems raised by early modern cometary appearances to the dissatisfaction of certain well-known scientists with the Aristotelian-Ptolemaic cosmology. Thus Robert Westman has examined the role of the comet of 1577 in inducing Mästlin's, and hence Kepler's, Copernicanism.[69] In an attempt to trace the "precise nature of Mästlin's influence"[70] on Kepler's Copernicanism, Westman proposes to "reconstruct an important discovery made by Mästlin that was to affect both his own and Kepler's acceptance of the Copernican hypothesis. That discovery concerned the orbit of the comet of 1577."[71] Westman recognizes that the physical consequences of the superlunary position of comets, as determined by Tycho (such as the rejection of crystalline spheres and the mutability of the heavens), were not sufficient conditions for the overthrow of the Ptolemaic world view. However, he maintains that "in an interesting and unforeseen way... the comet of 1577 did force a reconsideration of the old order of the planets."[72] Mästlin's efforts to compute the comet's orbit (in themselves a "fundamental step in the right direction"[73]) are examined at length by Westman, who reaches the conclusion that Mästlin thought that the appearances of the comet could only be saved by adopting the Copernican order of the planets. Thus, his cometary investigations provided his "young and impressionable" pupil, Kepler, with evidence in favour of the Copernican system.[74]

Another famous astronomer whose efforts at finding an orbit for the 1577 comet have recently been investigated in connection with his rejection of the Ptolemaic system, was Tycho Brahe. In 1979, an

---

of bodies, now celestial to be sure, but bodies which were governed by different laws than were the planets. It remained for Edmond Halley and then for Isaac Newton, late in the seventeenth century, to integrate comets into a dynamic planetary system."

[69] In his article "The Comet and the Cosmos: Kepler, Mästlin and the Copernican Hypothesis" in Jerzy Dobrzycki ed., *The Reception of Copernicus' Heliocentric Theory. Proceedings of a Symposium organized by the Nicolas Copernicus Committee of the International Union of the History and Philosophy of Science, Torun, Poland, 1973* (Dordrecht/Boston 1972) 7-30.

[70] Ibidem, 7.

[71] Ibidem.

[72] Ibidem, 11.

[73] Ibidem.

[74] Ibidem, 26-27.

article by Victor Thoren appeared in which the author proposed to explain the historical links between two "parallel challenges to the cosmological assumptions of Aristotle and Copernicus": the Tychonic system of the world and the comet of 1577, both treated in Tycho's *De mundi aetherei recentioribus phaenomenis* (1588).[75] Thoren reconstructs the process that led Tycho to his geoheliocentric system, of which his deliberations on the comet (whose supralunar position ruled out the existence of solid orbs, and hence made possible the intersection of the orbits of the sun and Mars required by the Tychonic arrangement of the planets), formed an integral part.[76]

Accounts like the above, investigating the role of cometary observations in bringing about revolutionary changes and ushering in the acceptance of new world systems, tend to concentrate on the exploits of a small number of prominent astronomers during the second half of the sixteenth century. However, the past decade has seen several reassessments of this rather limited approach. In 1988, Peter Barker and Bernard R. Goldstein offered their view of "The Role of Comets in the Copernican Revolution".[77] In a highly original approach to the well-known problem, Barker and Goldstein attempt a "Tychonic Reconstruction of the Copernican Revolution": they try to reformulate Kuhn's account of the Copernican revolution by considering Tycho, rather than Copernicus, as its central figure.[78] In this hypothesis, dissatisfaction with the Aristotelian cometary theory is supposed to have induced parallax measurements by such fifteenth-century astronomers as Toscanelli and Regiomontanus, which in turn led to the more precise measurements by Tycho, establishing the untenability of the Aristotelian distinction between the terrestrial and the celestial realm.[79] In order to maintain this Tychonic account, Barker and Goldstein contend, it is necessary to demonstrate that the anomaly posed by the nature of comets was explained by means of a new paradigm (in this case: distance determination by parallax) discontinuous with the previous paradigm.[80] And this is precisely where the

[75] Victor E. Thoren, "The Comet of 1577 and Tycho Brahe's System of the World", *Archives Internationales de l'Histoire des Sciences* 29 (1979) 53-67: 53.

[76] Ibidem, 65.

[77] Peter Barker and Bernard R. Goldstein, "The Role of Comets in the Copernican Revolution", *Studies in the history and philosophy of science* 19 (1988) 299-319.

[78] Ibidem, 300.

[79] Ibidem, 307.

[80] Ibidem, 308.

"Tychonic reconstruction" founders: the parallax measurements of Toscanelli and Regiomontanus, and hence of Tycho, formed part of a continuous tradition of observing comets in the same way, and with the same techniques, as the planets for astrological purposes. Thus, "medieval scientists were clearly capable of arriving at the technique often attributed to Regiomontanus well before the period in which we could plausibly begin to expect innovations associated with a discontinuous transition from medieval to early modern science."[81] Clearly, then, it cannot be maintained that Tycho's parallax measurements were responsible for the fall of the Aristotelian world view. Rather, some time before Tycho's appearance on the scene, the location of comets had, according to Barker and Goldstein, already become an open question, since the view that comets were spherical lenses, obviously incompatible with the Aristotelian theory, was already generally accepted around the middle of the sixteenth century.[82]

The traditional, Kuhnian discontinuity view has also been adjusted by Roger Ariew. Whereas Barker and Goldstein emphasize that Aristotle's cometary theory was already under attack quite some time before the general acceptance of Copernicanism, Ariew concentrates on the fortunes of the Aristotelian theory after Tycho's parallax measurements and Galileo's telescopic discoveries are supposed to have rendered Aristotelian cosmology untenable.[83] In a fashion reminiscent of the work of Schmitt and Grant, Ariew demonstrates that seventeenth-century scholastics could account for many of the new celestial discoveries, including the superlunary position of comets, by modifications of the Aristotelian theory. Thus, the Jesuit Jacques Grandamy of the College of Clermont in Paris admitted that comets, being above the moon, were of the same kind as the stars and planets; however, in order to maintain the incorruptibility of the heavens, he stated that the changes comets underwent were not substantial, but accidental, in keeping with the idea that there is no substantial generation or corruption in the heavens.[84] Another solution was adopted by Pierre du Moulin, who in 1644 asserted that there were

---

[81] Ibidem, 313.
[82] Ibidem, 316.
[83] Roger Ariew, "Theory of Comets at Paris during the Seventeenth Century", *Journal of the history of ideas* 53 (3) (1992) 355-369.
[84] Ibidem, 366-367.

two kinds of comets: sublunar fiery exhalations à la Aristotle, and celestial, miraculously created ones, that could not be accounted for by naturalistic means.[85]

Thus, several recent studies have reconsidered the Kuhnian account of comets as important factors in the Copernican paradigm switch, and tentatively argued for a more balanced account, in which the connection between comets and the new world view is much less clear-cut, and in which it is recognized that, for a long time, there was a considerable overlap between old and new cometary theories. However, one methodological outlook all the discussed works have in common: they concentrate exclusively on astronomical and cosmological issues, disregarding the astrological speculations which, as we have seen, only began to decline after the publication of Newton's *Principia,* and were therefore still much in evidence during the late sixteenth and early seventeenth century. In this sense, the works discussed in this section are still under the influence of the positivism that has left a long-lasting mark on the historiography of science. Although these days, astrology is by no means a neglected field of study, it has for a long time been treated in relative isolation from its more scientifically fecund sister, astronomy.

b. *Astrology: Reinstating the Discarded Image*

As we have seen, eighteenth- and nineteenth-century historians of science scorned the past insofar as it had been preoccupied with what they called 'prejudice' and 'superstition'. During the first half of the twentieth century, this attitude remained widely prevalent, despite a more balanced view put forward by Thorndike. The real breakthrough in connection with the reinstatement of astrology as a subject worthy of study began in the 1960s with the work of Frances Yates. Whereas the debate on the Yates thesis tended to concentrate on the role of astrology in connection with the rise of modern science, during the decades after her most influential publications a good many studies dealing with astrology as such appeared, elucidating such topics as the methods used by medieval and Renaissance astrologers, the position of astrology as an integral part of the early modern world view, and the great number of astrological pamphlets

---

[85] Ibidem, 365.

that circulated.[86] Remarkably enough, most of these works contain no more than a few cursory remarks on the astrological significance attributed to comets.

Rather, the very limited number of studies dealing exclusively with comets as part of the astrological tradition, are to be situated in the context of a more wide-ranging debate about the epistemological function of prodigies, marvels, and miracles in early modern thought. One of the first to analyze the role of 'monsters' in Renaissance thought, and the changes this concept underwent during the sixteenth century, was Jean Céard[87], who signalled a rupture at the end of the sixteenth century: whereas earlier, any rare event, whether natural or counter-natural, was seen as having special significance worthy of being studied, around 1600 an attitude began to prevail in which more attention was paid to the proximate causes of rare phenomena, rather than to their meaning.[88] Lorraine Daston, in a number of articles devoted to similar issues[89], has outlined a rather more gradual development, in which preternatural events like comets, earthquakes, or monstrous births were first gradually 'naturalized' during the late sixteenth and seventeenth centuries, losing their previous religious meaning as 'signs' in the process, after which, under the aegis of Baconian empiricism, they became the first scientific facts, subject to natural causes and "stubbornly insignificant."[90]

In an unpublished doctoral dissertation tellingly entitled *From Mon-*

---

[86] For example, C.S. Lewis, *The Discarded Image* (Cambridge 1964); Don Cameron Allen, *The Star-Crossed Renaissance. The quarrel about astrology and its influence in England* (New York 1966); Keith Thomas, *Religion and the Decline of Magic* (London 1971); Bernard Capp, *English Almanacs, 1500-1800: Astrology and the Popular Press* (London/ Boston 1979); Eugenio Garin, *Astrology in the Renaissance. The zodiac of life* (Bury St Edmunds 1984); Paola Zambelli ed., *Astrologi hallucinati. Stars and the end of the world in Luther's time* (Berlin/New York 1986); S.J. Tester, *A History of Western Astrology* (Bury St Edmunds 1987); Patrick Curry ed., *Astrology, Science and Society. Historical essays* (Bury St Edmunds 1987); Patrick Curry, *Prophecy and power. Astrology in early modern England* (Cambridge 1989).

[87] In his *La nature et les prodiges. L'insolite au XVIe siècle, en France* (Geneva 1977).

[88] Ibidem, 488-89.

[89] Katharine Park and Lorraine J. Daston, "Unnatural Conceptions: the Study of Monsters in Sixteenth- and Seventeenth-Century France and England", *Past and Present* 92 (1981) 20-54; and especially Lorraine Daston, "Marvelous Facts and Miraculous Evidence in Early Modern Europe", *Critical Inquiry* 18 (1) (1991) 93-124, and Lorraine Daston, "Wunder, Naturgesetze und die wissenschaftliche Revolution des 17. Jahrhunderts" in *Jahrbuch der Akademie der Wissenschaften in Göttingen* (1991) 99-122.

[90] Daston, "Marvelous Facts", 112.

strous Signs to Natural Causes, Sara Schechner Genuth examines one particular instance of Daston's 'from-signs-to-causes development': the way in which comets, traditionally deemed to be monstrous signs of calamitous events, slowly but surely came to be seen, by more and more authors, as the natural causes of those same events.[91] Genuth offers the reader a comprehensive account of ancient and medieval cometary theories, but the bulk of her book is devoted to Newton, Halley, and post-Newtonian cometary theories. Her main argument is, that, rather than having died out after the discovery of the periodicity of comets, traditional comet lore was given new life by Newton's and Halley's theories.[92] Whereas comets were now no longer thought capable of effecting revolutions on earth, their traditional functions were henceforth "elevated from a local to a global theater"[93]: they were seen as agents bringing about apocalyptic revolutions in the cosmic order, and thus "weaned people away from views of a fixed, immutable creation and accustomed them to change within the heavens."[94] Thus, although she considerably refines the traditional view that the advent of Newtonian astronomy eradicated ancient comet lore, Genuth still tries to connect past scientific theories to present-day ones. Her interest in the past in its own right is rather meagre: though she occasionally mentions political circumstances as factors advancing or retarding certain scientific beliefs, still the emphasis of her work is on theories and ideas, rather than situations.

Another author who has stressed the continuity of Newton's cometary theories with the age-old prophetic and millenarian tradition, is Charles Webster.[95] However, his account is more sensitive to the political overtones of cometary theories than Genuth's. He discusses Tycho Brahe's eschatological speculations, occasioned by the

---

[91] Sara Schechner Genuth, *From Monstrous Signs to Natural Causes: the assimilation of comet lore into natural philosophy* (Unpubl. diss. Harvard 1988; University Microfilms, 1990).

[92] Another author who has recently nuanced the traditional, positivistic view of post-Newtonian cometary theories is Simon Schaffer (in his articles "Newton's comets and the transformation of astrology" in Patrick Curry ed., *Astrology, science and society,* 219-243; and "Authorized Prophets: Comets and Astronomers after 1759", *Studies in Eighteenth-Century Culture* 17 (1987) 45-74).

[93] Schechner Genuth, *Monstrous Signs,* iv.

[94] Ibidem.

[95] See his *From Paracelsus to Newton. Magic and the Making of Modern Science* (Cambridge 1982), especially chapter 2, "Prophecy".

comet of 1577, and points out that the millenarian concerns evoked by comets were "at each stage... adapted for political purposes, whether by the Puritans to spur on their revolutionary effort, or by the Latitudinarians to promote resistance to James II and to justify the Glorious Revolution".[96] Astrology did not become obsolete with the arrival of the new science. Rather, at the hands of the Baconians it gained credibility. They devised a means whereby it could be made subject to precise experimental inquiry, and thus made the use of astrological imagery, much-needed for effective political and religious polemic, more plausible than before.[97]

Webster's account is representative of a recent rise of interest in the political implications of astrological practices. Studies devoted to this issue tend to concentrate on the early modern profusion of astrological pamphlets containing implicit or explicit political allusions. Keith Thomas has devoted some attention to the subject in his *Religion and the Decline of Magic;* Bernard Capp has published a volume devoted exclusively to early modern astrological almanacs in England[98]; and Nicolette Mout has called attention to the political and moralistic undertones in a number of Dutch prognostications of the last decades of the sixteenth century, a few of which she considers "veiled propagandistic pamphlets" originating from the rebels in the Dutch Revolt.[99]

Although they all mention cometary appearances as having provided fitting occasions for flooding the broadside market with dire predictions, none of these authors have treated prognostications from comets as a distinct genre, apart from regular astrological pamphlets. Only one article seems to have been devoted exclusively to the political connotations of a prognostication occasioned by a comet; to wit, J.R. Christianson's study of Tycho Brahe's German treatise on the comet of 1577.[100] In it, the author attempts to paint as coherent a picture as possible of Tycho's scientific outlook at a crucial time in his intellectual development. He takes into consideration Tycho's

---

[96] Ibidem, 36.

[97] Ibidem, 31-32.

[98] Capp, *Astrology and the Popular Press.*

[99] M.E.H.N. Mout, "Prognostica tijdens de Nederlandse Opstand" in C. Augustijn et al. eds., *Kerkhistorische opstellen aangeboden aan Prof. dr. J. van den Berg* (Kampen 1987) 9-19: 16.

[100] J.R. Christianson, "Tycho Brahe's German Treatise on the Comet of 1577: A Study in Science and Politics", *Isis* 70 (1979) 110-140.

cosmological and methodological views (radical empiricism based on mathematical analysis) his devotion to astrology and its connection to his Philippist theological beliefs (strong emphasis on rationality and free will), and his use of Copernican concepts, as well as such external influences as the "affinity between Tycho Brahe, Philippism, and the foreign policy of peace based on alignment with the *politique* powers".[101] Christianson concludes that "Tycho's work does take on new significance when seen in its social and political milieu as well as its specifically Danish intellectual context"[102].

Leaving aside for the moment the question whether Christianson has been successful in attaining the goals he set himself, what interests us here is his original approach to the problem posed by the multi-faceted early modern comet treatises. Of all the authors discussed in this section, Christianson is the only one to have allowed himself to be led by the multiformity of the tracts themselves, instead of making a selection and treating, for example, only the astronomical aspects in order to solve the Copernican problem, or only the astrological ones to demonstrate how resilient prophetic millenarianism proved to be, or indeed to show how even these tenacious 'old-fashioned' ideas came to contribute to our modern cosmological outlook. Only a combined approach can do justice to the multiplicity of early modern treatments of comets.

The belief in astrology, being an entity that has developed considerably over the centuries, is particularly well-suited to clarify the difference in mentality between past and present. Most of the above-mentioned authors have exploited this suitability with notable success; their studies have taught us a great deal about the early modern frame of mind. However, as far as comets are concerned, they suffer from one flaw. As we have seen, the age-old body of ideas in connection with comets aimed at providing an all-encompassing explanation of these phenomena: 'astronomical' and 'astrological' aspects were parts of an inseparable whole, in keeping with the Aristotelian postulate that, in order to arrive at a full explanation of a phenomenon, all four causes (material, formal, efficient, and final) had to be elucidated. By contrast, modern historians trying to write the history of cometary ideas, have almost invariably concentrated on *either* as-

[101] Ibidem, 131.
[102] Ibidem.

tronomical aspects (roughly corresponding to the material, formal, and efficient causes) *or* astrological aspects (final cause).

Such a division does not do justice to the early modern frame of mind. As the present study will show, comets are especially well-suited to demonstrate the correlation of astronomical, astrological, and political ideas in the minds of Renaissance thinkers. Only by following the lead of our early modern authors themselves, and by treating the same variety of interconnected subjects as they do, can we hope to arrive at an understanding of their ideas on the subject of comets.

CHAPTER THREE

COMETS BEFORE THE CESSATION OF DIALOGUE

1. *Introduction*

The Dutch treatises about the comets of 1577 and 1618 came into being during an age still very much in touch with the variety of opinions advanced with regard to the hairy stars during the previous centuries. Dialogue with past scholars had not yet ceased, so that it was only natural for any author writing a tract, on whatever topic, to hark back to the opinions his predecessors had proposed on the subject. Indeed, the period under consideration in the present study was marked by an unprecedented culmination of 'dialogue'. Around the year 1600, scholars were confronted with a clash of opinions: on the one hand, they were well familiar with the philosophical and cosmological assumptions that had, during the later Middle Ages, come to constitute what is generally known as the "Aristotelian world view". On the other hand, however, the period was characterized by large-scale efforts to uncover the knowledge of Antiquity in its pristine state, stripped of its medieval accretions. This search for pristine wisdom was responsible for the re-emergence to prominence of a great number of non-Aristotelian authors, whose opinions collided with those that the sixteenth-century authors had inherited from their medieval predecessors.

The way in which this collision took place, during the sixteenth and early seventeenth centuries, is examined at greater length in our fourth chapter. However, in order to be able to understand how the authors of the Dutch cometary tracts of 1577 and 1618 coped with the variety of options with which they were faced, and how they solved the conflicts between them, it is necessary first of all to gain a certain measure of acquaintance with these options. The cometology of Antiquity can be divided into two strands: a naturalistic strand, based above all on Aristotle and Ptolemy, and a more divinatory or teratological strand based principally on the writings of Stoic authors.[1] Both these traditions were well known to the authors of the

[1] For a more detailed discussion of the background of these two different outlooks on cometology, see chapter VI.

Dutch cometary tracts. The most important sources of the naturalistic strand, Aristotle's *Meteorology* and Ptolemy's *Tetrabiblos*, both went through numerous sixteenth-century editions.[2] Moreover, both works had been well known and frequently quoted during the later Middle Ages, so that they formed part of the direct heritage of the sixteenth century. The main propagators of the Stoic outlook (Seneca in his *Naturales Quaestiones*, Manilius in his *Astronomica*, and Pliny in his *Naturalis Historia*) were also very popular during the Renaissance, and their works were frequently printed. Thus, significantly, the first printed edition of the *Astronomica* appeared at the very early date of 1472, and was edited by one of the most famous astronomers of the time, Regiomontanus; in 1484 there appeared an edition with commentary by Bonincontrius; and the years 1579, 1600 and 1655 saw publications of a vastly improved text with commentary by Joseph Justus Scaliger.

In addition to the Aristotelian-Ptolemaic and the Stoic outlooks, sixteenth-century authors were also familiar with even more ancient cometologies, mentioned and criticized by later authors such as Aristotle, Seneca, pseudo-Plutarch, Diodorus of Sicily and Johannes Stobaeus. Thus, both Diodorus and Seneca discussed the cometary ideas of the ancient Chaldaeans: according to the former, the Chaldaean notion of comets held them to be atmospheric phenomena, while the latter recorded two mutually exclusive 'Chaldaean' views, one of which agreed with Diodorus' account while the other maintained that, to the Chaldaeans, comets had been a special kind of planets.[3] A similar difference of opinion marked the views of Greek Presocratic thinkers, as they came down to the sixteenth century in the writings of Aristotle, Seneca, and pseudo-Plutarch, who recorded that several of the Presocratics (such as the Pythagoreans and Apollonius of Myndus) regarded comets as real celestial bodies[4], while a number of others chose to account for them as optical effects (Anaxa-

[2] For the *Tetrabiblos*, see F.E. Robbins's Introduction to the Loeb Classical Library edition of the *Tetrabiblos* (Cambridge Mass./London 1940) xii-xvii.

[3] See Diodorus of Sicily, *Library of History* I.18.6, and Lucius Annaeus Seneca, *Naturales Quaestiones*, VII.4.1. Several editions of Diodorus of Sicily's work were available to scholars around the turn of the sixteenth century: Poggio Bracciolini's Latin translation of Books I-V, published at Bologna in 1472 and later reprinted at Paris, Venice and Lyons; Vincentius Opsopoeus' Greek edition of Books I-V, XI-XX, and some fragments of Books XXI-XL, Geneva, 1559; and L. Rhodoman's Greek edition with a Latin translation, printed at Hanau in 1604.

[4] See Aristotle, *Meteorology* I.vi.342b; pseudo-Plutarch, *De placitis philosophorum* III.893b; Seneca, *Naturales Quaestiones* VII.17.1-2.

goras, Democritus, Artemidorus, Hippocrates of Chios, Aeschylus)[5] or as atmospheric phenomena (Xenophanes of Colophon, Metrodorus)[6].

Clearly, the devices put forward to account for cometary phenomena by natural philosophers prior to Aristotle were very diverse. A comet could be either an actual celestial body, an optical effect produced by the light-rays emanating from stars or planets, an illusion created by sunlight permeating the atmosphere, or a real atmospheric fire. The authors of the Dutch cometary tracts of 1577 and 1618 were certainly familiar with all these views, and as we shall see, they employed them on several occasions in their own writings. However, much more important to them were the two conflicting outlooks of Aristotle and Ptolemy on the one hand, and the Stoics on the other; it is to these that we must now devote our attention.

2. *Aristotle's Choice*

In the opening chapter of his *Meteorology,* Aristotle wrote:

> There remains for consideration a part of this inquiry which all our predecessors called meteorology. It is concerned with events that are natural, though their order is less perfect than that of the first of the elements of bodies. They take place in the region nearest to the motion of the stars. Such are the milky way, and comets, and the movement of meteors.[7]

This introduction makes it clear that the Stagyrite considered comets to belong to the same class of phenomena as meteors (the word 'meteor' in his terminology meaning something akin to our modern concept of 'weather conditions', comprising rain, storm, thunder and lightning, hail, snow, wind, earthquakes, etcetera). Since, according to his own cosmological system, no change could take place in the heavenly region beyond the moon, which was formed out of the perfect and everlasting celestial aether, Aristotle had no real choice with regard to the status comets were to be assigned. They were obviously a volatile and unpredictable sort, suddenly appearing, fol-

[5] Aristotle, *Meteorology,* I.vi.342b and I.vi.343a; Seneca, *Naturales Quaestiones,* VII.12.1 and VII.13.1; pseudo-Plutarch, *De placitis,* III.893A.

[6] Pseudo-Plutarch, *De placitis,* III.893B.

[7] Aristotle, *Meteorology,* I.i.338a-b (translated by E.W. Webster, revised by J. Barnes, cited in J.L. Ackrill ed., *A New Aristotle Reader* (Oxford 1987) 158).

lowing their own wayward courses, eluding any attempt at predicting where they would go, and vanishing as capriciously as they had arrived. Thus, it was impossible to maintain, under the vigour of Aristotelian cosmology, any variety of the theory that a comet was a real celestial body.

After refuting both the "celestial body" theory and the "celestial optical illusion" theory put forward by some of his predecessors, the Stagyrite expounded his own explanation of the hairy stars' appearances, which ran along the following lines. The sphere of fire, being composed not of real fire but rather of a hot, dry, and highly inflammable exhalation, is carried round by the revolution of the heavenly spheres on which it borders. This movement frequently causes it to catch fire. Whenever "as a result of the upper motion there impinges upon a suitable condensation a fiery principle which is neither so strong as to cause a rapid and widespread conflagration, nor so feeble as to be quickly extinguished, but which is yet strong enough and widespread enough; and when besides there coincides with it an exhalation from below of suitable consistency; then a comet is produced, its exact form depending on the form taken by the exhalation. . .".[8] Thus, according to Aristotle, there are two requirements for a comet to be generated: a suitable kind of fiery disposition in the sphere of fire, and a suitably constituted earthly exhalation rising to meet it. When the exhalation catches fire, a comet is produced, its shape depending on the form of the exhalation: "if it extends equally in all directions it is called a comet or long-haired star (κομήτης), if it extends lengthwise only it is called a bearded star (πωγωνίας)".[9]

A second way in which comets could be produced, according to Aristotle, was when one of the stars, by its own movement, formed an exhalation in the sphere of fire. When such an exhalation caught fire, it could be seen around the star like a halo, and the star became a comet. Unlike independently formed comets, which followed the movement of the terrestrial sphere, this kind of comet appeared to follow the same course as the star with which it was connected.[10]

Unlike most of his predecessors, Aristotle paid some attention to the kind of occurrences that might accompany the appearance of a comet. In a rather ambiguous passage, he stated that "we may re-

---

[8] Aristotle, *Meteorology*, I.vii.344a (translated by H.D.P. Lee, Loeb Classical Library edition, Cambridge Mass./London 1962, 51).

[9] Ibidem.

[10] Ibidem, I.vii.344b.

gard as a proof that their constitution is fiery the fact that their appearance in any number is a sign of coming wind and drought."[11] This seems to mean that from the fact that they were often followed by wind and drought it could be logically deduced that, therefore, comets themselves were fiery. However, Aristotle went on to say that it was evident that comets owed their origin to the presence of an abundance of hot and dry exhalations, which made the air drier and dissolved any moisture that might have been present in the air, leading to winds and drought. In this passage it is rather unclear which follows from which. In any case, it is evident that, in Aristotle's opinion, both comets and phenomena like wind and drought were consequences of one and the same predicament: an abundance of hot and dry exhalations in the air. As such, comets could be said to be 'signs' of coming storms or drought. However, the natural character of this signifying function needs to be stressed: in Aristotle's account, comets are part of a logically deducible, causal chain of events, all of them meteorological and explicable by natural means. As will become clear below, this is a far cry from later views of comets as 'signs'. To Aristotle, a comet was just another meteor.

3. *The Stoic Outlook*

Aristotle's explanation of cometary appearances did not meet with much approval during Antiquity. Although views resembling the Stagyrite's do appear to have made some headway among later adherents of Stoicism[12], nonetheless, the Stoic outlook, which became especially prominent in imperial Rome, was rather different from the Aristotelian. According to Stoic cosmology, the cosmos consisted only of the four elements, earth, water, air, and fire; nothing comparable to the Aristotelian aether, or to the corresponding fundamental gap between the sublunary and the superlunary regions, was to be found. The Stoics saw the whole cosmos as a living animal, and therefore as an organic whole.[13] This explains why Seneca's account of Stoic cometary theory is so unclear about the exact location of the

---

[11] Ibidem.

[12] See Seneca, *Naturales Quaestiones*, VII.20-21, and pseudo-Plutarch, *De placitis*, 893B.

[13] David E. Hahm, *The Origins of Stoic Cosmology* (Ohio 1977) 91-93; 101-102; S. Sambursky, *Physics of the Stoics* (Princeton 1959) passim.

burning impressions called comets: they belonged to the category of various kinds of fires conceived "on high" ("in sublimi"); this is rather a vague expression, but from the fact that some of them were considered to be quite near to the sun[14], it becomes clear that, apparently, the Stoics did not confine them to the sublunary sphere.

Thus, the impression Aristotle's cometary theory made on his immediate successors was not really very considerable. A number of philosophers continued to endorse views indistinguishable from the earlier Presocratic theories; and the Stoics, who did adhere to an opinion resembling Aristotle's, embedded their explanation in a substantially different context. They held the whole cosmos to be a living being, permeated by a kind of world soul, which could be either fire, air, or pneuma, and which held everything together.[15] Once it had completed one life cycle, it underwent a conflagration (ekpyrosis) changing everything to fire. Then, after a certain period, a new world order would come into being, and would experience exactly the same events as had the previous one. Thus continued a never-ending series of generation, destruction and rebirth.

In Rome, Stoicism was very much associated with divination. Thus, in his dialogue *On Divination,* Cicero all but equated the defence of divination with that of the Stoic doctrine of fatalism and divine providence: "Their [the Stoics'] view is that the world was from its beginning set up in such a way that certain things should be preceded by certain signs, some in entrails, others in birds, others in lightning, others in portents, others in stars, others in dream impressions, others in frenzied utterances. Those who properly perceive these are rarely deceived."[16] The famous Stoic Posidonius, Cicero's teacher, was certainly well-disposed towards astrology, and his views "have often been seen as decisive in gaining favour for the discipline among elite Romans."[17] According to Diogenes Laertius, Posidonius saw the heavens as the commanding faculty of the organism of the world.[18]

However, Tamsyn Barton is certainly correct in pointing out that Stoicism cannot have been the only factor responsible for the Ro-

---

[14] Seneca, *Naturales Quaestiones,* VII.20.4.

[15] Hahm, *Origins,* chapter V.

[16] Marcus Tullius Cicero, *On Divination,* I.118; cited in English by Tamsyn Barton, *Ancient Astrology* (London/New York 1994) 35.

[17] Barton, *Ancient Astrology,* 35.

[18] Diogenes Laertius, *Lives of the Philosophers,* 7.138-9.

mans' favourable attitude towards astrology. Her argument, that the simultaneous rise of astrology and of monarchical Empire in Rome was no coincidence, is convincing.[19] All but one[20] of the Julian emperors of Rome were in some way associated, by their historiographers, with the emergence of a comet. As mentioned above, Julius Caesar's ascension into heaven was thought to have been accompanied by the comet that appeared during his funerary games in 44 BC, as we learn from Suetonius' *Julius Caesar.*[21] And while Caesar was still alive, the war in which he was to fight Pompey was preceded by many portents, including a comet.[22] Cassius Dio reported that the deaths of both Cicero and Agrippa had been announced by a comet, as well as the demise of Augustus in 14 AD[23] Furthermore, a comet was among the omens warning of Claudius' poisoning in 54 AD.[24] Without a doubt the most ominous of first-century comets, however, was the one that was seen during Nero's reign in 64 AD. Its appearance gave rise to a host of rumours among the populace about a possible successor, it being popularly known (and confirmed by examples in the recent past) that such a sight foreboded the death of a ruler. Nero panicked, and, at the instigation of his court astrologer Balbillus, he resolved to try to avert the comet's wrath from his own head onto those of his eminent aristocrats[25]; Tacitus wrily commented that, as usual, the emperor atoned for a cometary appearance by noble blood.[26] Ironically, in the purge that followed, Seneca was one of the victims.

In Imperial Rome, astrological practitioners gained much in importance, thriving in their new role as imperial advisers about all issues involving the future. Especially with regard to questions of succession to the throne, they could exercise considerable influence, as they did in the year of crisis 69 AD, when four emperors were aiming for power, and their respective astrologers urged them to

---

[19] Barton, *Ancient Astrology,* 38-45.

[20] The exception being the fourth one, Gaius Caligula.

[21] Gaius Suetonius Tranquillus, *Julius Caesar,* 88. See also Pliny the Elder, *Natural History,* II.24.93-94.

[22] Lucan, *Pharsalia,* I.526-529: "Ignota obscurae viderunt sidera noctes/ Ardentemque polum flammis caeloque volantes/ Obliquas per inane faces crinemque verendi/ Sideris et terrae mutantem regna cometem."

[23] Cassius Dio, *Roman History,* 45.17.4; 54.29.8; 56.24.3-4; 56.25.5; 56.29.3.

[24] Suetonius, *Claudius,* 46.

[25] Suetonius, *Nero,* 36.

[26] Publius Cornelius Tacitus, *Annals,* XV.47.

action on the basis of their predictions.[27] Thus, both the prevalence of Stoicism and the political atmosphere at the imperial court heightened the popularity of astrology in Rome, and this was the environment in which one of the most frequently quoted texts on comets came into being.

Marcus Manilius was a Stoic poet who, between 9 and 15 AD, wrote a long didactic poem on astrology, the *Astronomica*, probably modelled on Virgil's *Georgics*. In the final hundred-odd verses of Book One, Manilius considered the subject of hairy stars or comets. He began by expounding his views on the way these phenomena came into being. Genuth is mistaken in asserting that Manilius "subscribed to the Aristotelian theory of comets as inflammable terrestrial vapors"[28]; on several points of vital importance, his views display a marked deviation from the Stagyrite's. It is true that Manilius mentioned the possibility that comets were created by vapours breathed forth by the earth, which, having been overpowered by an abundance of hot and dry air, were seized by fire and "a flame takes hold of the matter that suits its nature"[29], thus creating short-lived, blazing fires in the air, which were variously shaped according to the form of the exhalation that fed them. This is quite an accurate rendition of the Aristotelian theory. However, a bit further on in his account, Manilius admitted that this was only one possible explanation of cometary appearances; they could equally well be "dim stars that shine in the heaven with meagre flames", but are attracted by the Sun, absorbed into its fire, and then released.[30] Thus, like his fellow Stoic predecessors, Manilius was not at all specific about the exact location of comets (in the heavens or in the atmosphere), nor does he seem very concerned about it. The Aristotelian dichotomy is strikingly absent from his account.[31]

Moreover, at the very beginning of his section on comets, Manilius stated that "In times of great upheaval rare ages have seen the sudden glow of flame through the clear air and comets blaze into

---

[27] Barton, *Ancient Astrology*, 38-45.

[28] Sara Schechner Genuth, *From Monstrous Signs to Natural Causes. The Assimilation of Comet Lore into Natural Philosophy* (Unpubl. diss. Harvard 1988; University Microfilms, 1990) 29.

[29] Marcus Manilius, *Astronomica*, I.817-822; translated by G.P. Goold, Loeb Classical Library (Cambridge Mass./ London 1977) 71.

[30] Ibidem, I.868-871.

[31] This is confirmed by Manilius' statement that "fire is found mingled with every part of the universe" and that "all nature abounds with fire"; I.852, I.858.

life and perish."[32] Thus it would appear as if he regarded the presence of "times of great upheaval" as a prerequisite for the formation of comets: if no upheaval, then no comets. This seems to be confirmed later on in the section, when Manilius suggests that "It may well be that by means of these moods and conflagrations of the sky Heaven in its pity is sending upon earth tokens of impending doom, for the fires wherewith the heavens blaze have never lacked significance. . .".[33] One could not wish for a more unambiguous expression of the Stoic outlook: divine providence is omnipresent in the organic whole of the universe, and signs in one part of the cosmos always have their repercussions elsewhere. Divine providence in its goodness tries to warn us; if we were not so obtuse and would give heed to the heavenly warnings, the disasters foretold by comets would be averted.[34]

Manilius went on to provide his statement with an empirical underpinning: from the investigation of historical precedents we learn that comets have invariably been followed by agricultural sterility and barren crops, by mortal illnesses and plagues devouring whole nations, by wars and sudden insurrections, by treachery and civil strife. The poet furthermore entertained his audience with several historical examples, such as "the plague which ravaged Erechtheus' folk and bore forth ancient Athens to an unwarlike grave... and none were left to bury, none to weep the dead"; the recent military disaster, "when, its oaths forsworn, barbarous Germany made away with our commander Varus and stained the fields with three legions' blood"; and the battles of Pharsalus, Philippi, and Actium.[35] Wisely, our author concluded the section with words of wellwishing for the reign of the present emperor, Augustus: "Unconquered be the father of our fatherland; may Rome serve none but him. . .".[36]

A similar view of the ominous nature of comets is to be found in the *Naturalis Historia* by that other great Stoic, Pliny the Elder. Pliny, like all Stoics, believed in a universe governed by natural laws, which often remained hidden to human investigation, unworthy of pene-

---

[32] Manilius, *Astronomica*, I.814-816.

[33] Ibidem, I.874-876.

[34] Ibidem, I.904-906: "Wonder not at the grievous disasters which betide man and man's affairs, for the fault oft lies within us: we have not sense to trust heaven's message."

[35] Ibidem, I.884-891, 899-900, 908-918.

[36] Ibidem, I.925.

trating too deeply into nature's grand design. In connection with other meteors such as 'torches', 'beams', and 'chasmata'[37], Pliny explicitly put forward as his own view that they "take place at fixed dates owing to natural forces, like all other events, and not, as most people think, from the variety of causes invented by the cleverness of human intellects; it is true that they were the harbingers of enormous misfortunes, but I hold that those did not happen because the marvellous occurrences took place but that these took place because the misfortunes were going to occur, only the reason for their occurrence is concealed by their rarity, and consequently is not understood as are the risings and the setting of the planets described above and many other phenomena."[38]

There is no justification for Genuth's hesitancy[39] in extending the vigour of this passage to comets as well as other meteors. First of all, as we have seen, it was common practice among historians of Pliny's time to list comets along with various kinds of other portents, including meteoric phenomena, as ominous events taken to signify changes on earth. Furthermore, in his discussion of comets, Pliny, like most Stoics, was not at all specific about their location. At first, he stated that they "suddenly come to birth in heaven itself"[40]; but later on, he summed up, without committing himself to one of them, two rival theories, one of which maintained that comets were eternal bodies having their own orbits, while the other one upheld that they were formed of chance moisture and fiery force, and were dissolved again after a while.[41] Also, in a discussion of historical cometary appearances, he casually remarked about the one that was seen over Egypt under king Typhon, that it was "not really a star so much as what might be called a ball of fire."[42] Thus, he was clearly not much

---

[37] These were the names given to phenomena like shooting stars and the aurora borealis.

[38] Pliny the Elder, *Natural History*, II.26.97 (translated by H. Rackham, Loeb Classical Library edition, Cambridge Mass./London 1938).

[39] Genuth, *Monstrous Signs*, 36-37 tentatively proposes in connection with this passage that, "Since Pliny's argument was fairly general, it is tempting to suggest that he perhaps applied it to cometary portents, which were commonly perceived to be meteors. Pliny, however, remained silent on this point."

[40] Pliny, *Natural History*, II.22.89.

[41] Ibidem, II.23.94: "Some persons think that even comets are everlasting, and travel in a special circuit of their own, but are not visible except when the sun leaves them; there are others, however, who hold that they spring into existence out of chance moisture and fiery force, and consequently are dissolved."

[42] Ibidem, II.23.91-92.

concerned about the exact location of hairy stars; in his view, comets might equally well be eternal stars as fleeting fiery meteors.

Therefore, in all probability, the passage quoted above about the regularity with which meteoric phenomena take place can, in the context of Pliny's work, be taken to apply to comets also. Then it becomes clear that our author, like Manilius, believed them to obey natural laws and to occur for a special reason. They somehow announced misfortunes, but because of their rarity, mankind was unable to discover their exact purpose. Genuth suggests that Pliny might mean that ". . .the physical conditions predisposed to foster the disorders also produced the meteors, or perhaps some divinity prearranged the natural sequence of events."[43] The latter suggestion seems by far the more likely, given Pliny's remark about a comet appearing in the western sky, which is "usually a terrifying star *and not easily expiated*" [my italics]. This seems to presuppose the idea of the comet having been sent by a god of some sort to punish humankind or to put them to the test. Such an interpretation is consistent with the Stoic view of divine providence and cosmic correspondence, another instance of which is provided by Pliny a bit later on, when he praises Hipparchus for "having prove[d] that man is related to the stars and that our souls are a part of heaven".[44]

Thus, the Stoic outlook invested comets with powers hitherto unthought of. Whereas, to most Presocratics, they had been little more than optical illusions, and to Aristotle, just another meteor, on a par with rain and wind, the Stoic vision of a fundamental unity of the universe induced authors like Manilius and Pliny to assign comets no specific location in the universe, but to regard them as symptoms of the divine providence that supervised the cosmos. Moreover, under the imperial government in Rome, it had become common practice to employ astrology as one of the tools with which to play the political game. This may well explain why Manilius was so keen to introduce historical examples from the recent past into his narrative, stressing the treachery of Augustus's enemies and his own valiance. In so doing, he set an example that would prove to be extremely influential, and, as we shall see below, his outlook was to become very popular during the Renaissance.

---

[43] Genuth, *Monstrous Signs*, 36.

[44] Pliny, *Natural History*, II.24.95.

4. *Seneca's Choice*

Arguably the most famous of the ancient Stoics, the eminent sage Lucius Annaeus Seneca has provided us with the most extensive, and during the sixteenth century most influential, remaining account of comet lore to have been created by pagan antiquity. In the very first section of his exposé on comets in the seventh book of his *Naturales Quaestiones*, he condemns the "inane superstition" of people who are terrified whenever something faintly irregular, like an eclipse, occurs in the heavens.[45] Comets, also, invariably give rise to dread and apprehension, Seneca regretfully adds: without fail, at the appearance of a comet, lots of people present themselves "who create terror and predict dire meanings from it", leading the common herd to ask anxiously "whether it is an omen or a star."[46]

Such an exaggerated fascination with rare celestial events is not at all in keeping with Seneca's own perception of the universe. On numerous occasions throughout his account, he alludes to what, to him, is the most extraordinary circumstance of all, worthy of the greatest wonder: the marvellous regularity and harmony with which god has endowed the cosmos.[47] The beauty of the universe and the divine goodness encompassed in it make it worthwhile to investigate, for example, whether comets have the same nature as the planets and the stars. Such an investigation may even contribute to our knowledge of "whether the universe travels around while the earth stands still or whether the earth turns while the universe stands still", Seneca ambitiously announces.[48] Even though this last objective was clearly beyond him (later on, in his refutation of Epigenes' cometary theory, Seneca assumes the whirling motion of the sky to be an es-

---

[45] Seneca, *Naturales Quaestiones*, VII.1.2.

[46] Ibidem, VII.1.5-6. (translated by Thomas H. Corcoran, Loeb Classical Library edition, London/ Cambridge Mass. 1972).

[47] See for example ibidem, VII.1.6: "But, by Hercules, no one could study anything more magnificent or learn anything more useful than the nature of the stars and planets. . ."; or VII.30.1: "If we enter temples with composure, if, when we are about to approach a sacrifice, we lower our eyes, draw in our toga, if we assume every sign of modesty, how much more ought we to do so when we discuss the planets, the stars, the nature of the gods, lest in our ignorance we assert something rashly, impudently, or even lie knowingly!"

[48] Ibidem, VII.2.3.

tablished fact)[49] his inquiry into the nature of comets leads him to an interesting point of view.

After thoroughly refuting the theories of Epigenes, Anaxagoras and Democritus, Artemidorus, Apollonius of Myndus, and his fellow Stoics, Seneca concludes that, according to him, comets are truly celestial bodies: "I do not agree with our Stoics. For I do not think that a comet is just a sudden fire but that it is among the eternal works of nature."[50] Remarkably enough, the arguments he adduces in support of this position are strikingly similar to those that made Aristotle adopt exactly the opposite view. Seneca agrees with the Stagyrite that all things the atmosphere creates are short-lived, because they are produced in an unstable and changeable element. Moreover, he endorses the Aristotelian opinion that "fire either goes where its own nature leads it, that is, upwards, or in the direction that its fuel attracts it, to which it clings and on which it feeds."[51] However, Seneca, unlike Aristotle, concludes from these premises that comets, which "move, preserve their continuity, and are uniform", are *not* to be located in the sublunary atmosphere, and do *not* consist of fire, or at least, not of sublunary fire. In a truly Stoic vein, he states that "no fires have any duration except in their own element. I refer to these divine fires which the universe maintains as eternal fires because they are parts and works of it."[52] Thus, we are faced with the paradoxical situation that similar observations, and more or less identical theoretical presuppositions, drove two great thinkers to diametrically opposed conclusions.

With regard to possible predictions to be made from cometary appearances, the views of our two protagonists were at odds as well. We have seen that, to Aristotle, the apparition of a comet was a sign of coming wind and drought, because both were symptoms of an abundance of dry exhalations in the air. Obviously, it was impossible for Seneca to subscribe to this reasoning, since, to him, comets had nothing to do with any kind of exhalation. However, he instead advanced a view that was no less favourable to prediction from comets. When discussing Aristotle's opinion that comets announce wind and

[49] Ibidem, VII.10.1: "What further is so unbelievable as a whirlwind of long duration, especially when the motion of the whirlwind is overcome by the contrary motion of the sky? For that region has its own whirling, which carries the sky along. . .".

[50] Ibidem, VII.22.1.

[51] Ibidem, VII.23.1.

[52] Ibidem, VII.23.2.

rain, he belligerently declared: "Well, what of it? Do you not judge that that which foretells the future is a celestial body?"[53] Evidently, he assigned predictive power to the stars, an opinion he had already expounded in full in the second book of his *Natural Questions*. This book dealt with diverse branches of natural divination and augury, all of which Seneca endorsed, believing the various signs studied by these practices to have been instituted by divine agency, and even asserting that "Whatever happens is a sign of something that will happen."[54]

Seen in the context of the *Natural Questions* as a whole, therefore, what may appear at first sight to be a chance remark in book 7 assumes a marked significance. Seneca states that, in his opinion, "a comet is not a sign of storm in the same way that there is a sign of coming rain when the oil in the lamp sputters. . . or in the same way that there is a forecast of rough sea if the sea-coots play on dry land. . . but in the way that the equinox is a sign of the year turning to hot or to cold, or as the things the Chaldaeans predict, the sorrow or joy that is established at people's birth by a star."[55] In other words, the predictive power assigned to comets by Seneca was much wider than that accorded them by Aristotle. To Aristotle, indeed, a cometary appearance was comparable to the oil in the lamp: both were consequences of one underlying cause: the condition of the atmosphere. Seneca, however, believing comets to obey permanent laws and to pursue regular courses, compared their predictive power to that of other eternally recurring celestial movements, whether it be accorded them by astronomers (the equinox) or by stargazers (the Chaldaean horoscopes). Defending his theory against possible Aristotelian objections, he assured his readers that "the rising of a comet does not immediately threaten wind or rain, as Aristotle says it does, but makes the entire year suspect."[56] And, affirming once again his Stoic belief in the harmonious unity of the universe, he concluded that comets did not draw from their immediate neighbourhood the

---

[53] Ibidem, VII.28.1.

[54] Ibidem, II.31-50, esp. II.32.3-4; see also Genuth, *Monstrous Signs*, 44, who cites many more instances of Seneca's belief in the predictive power of celestial phenomena, ordinary as well as extraordinary.

[55] Ibidem, VII.28.1. The examples of the oil in the lamp and the sea-coots are quoted by Seneca from Virgil's *Georgics*.

[56] Ibidem, VII.28.2.

signs they gave for the immediate future, but instead had them "stored up and linked to the laws of the universe."[57]

Seneca's laws of the universe formed the foundation on which he based his belief in the possibility, indeed, the commendability, of forecasting the future from natural omens.[58] In fact, therefore, the Senecan point of view opened up a wide new range of possibilities for those endeavouring to determine the consequences of cometary appearances; whereas to an Aristotelian they were little more than weather indicators, to an adherent of Seneca's view they had turned into integral parts of cosmic unity, messengers from the gods, descending to predict the particulars of a universal design.

5. *Ptolemy: A Choice of Choices*

In 29 BC, in one of the best-loved and most frequently cited poems ever written, the great poet Virgil devoted two quite distinct passages to comets. In the first book of his *Georgics* he counted them among the trustworthy signs that "the father himself" had provided man with so that he might know when to keep the cattle indoors or when a tempest would rise. Just like the water-fowl returning to dry land from the sea, just like the sea-coots playing on the shore, and the heron leaving the marshes and flying high above the clouds, Virgil tells us, the sight of stars dragging long tresses of fire shooting through the sky warns farmers of an impending storm.[59] The poet here regards comets in much the same way Aristotle had, namely, as a natural indication of impending windy weather conditions, comparable to our modern barometer. Yet a hundred-odd verses later, he emphasizes the dire and ominous meaning that comets might also carry, as they did in the year of Julius Caesar's assassination. Here, their peers are not waterfowl, sea-coots, or herons, but talking animals, volcanic eruptions, and weeping statues.[60] Never before, says Virgil, making use of his poetic licence, had so many frightful comets blazed in the sky. This passage is much more reminiscent of the Stoic outlook than

[57] Ibidem.

[58] See also Schechner Genuth, *Monstrous Signs*, 42-43.

[59] Publius Vergilius Maro, *Georgics*, I.351-369.

[60] Ibidem, I.464-488.

of the Aristotelian: nature produces unusual circumstances in order to alert people to impending changes.

Not always, then, were the two approaches wholly incompatible. Comets might be both natural weather forecasters and messengers sent by divine providence. A convincing synthesis of both was wrought in the second century AD by Claudius Ptolemy (c. 100-178 AD). Unlike the Roman Stoics, Ptolemy was in no way ambiguous about the location comets were to be accorded. To him, they were no celestial bodies; if they were, he would have treated them in his highly acclaimed astronomical *magnum opus,* the *Almagest,* in which the word 'comet' does not once occur. Instead, he devoted them rather a meagre amount of attention in his equally influential compendium of astrology, the *Tetrabiblos* (or, in Latin, *Quadripartitum*). There, he upheld that comets foretold droughts and wind, a statement coming straight from the Aristotelian tradition. However, Ptolemy went a step further in asserting that they also "naturally produce[d] the effects peculiar to Mars and to Mercury—wars, hot weather, disturbed conditions, and the accompaniments of these."[61] Clearly, in this passage he combined the theoretical framework of the Stoic tradition (i.e. the harmonious correspondence between all parts of the universe, in casu between celestial bodies (Mars and Mercury) and terrestrial conditions (wars, hot weather, etcetera)) with the popular belief, mentioned by Pliny, that individual comets were ruled by specific planets and that these planetary rulers determined the consequences the comet would have. Comets thus came to be assimilated to the nature of celestial bodies, and endowed with similar qualities.

In combining these different elements, Ptolemy added an important new ingredient of his own. To previous authors, whether belonging to the Aristotelian or to the Stoic tradition, the question of whether comets were to be regarded as signs or causes of future events had not been much of an issue. From an Aristotelian point of view, they were signs indicating bad weather; for the Stoics, the question of whether they were signs or causes was rather futile, since the cosmic whole acted according to intertwined laws that it was wellnigh impossible to disentangle. Ptolemy, however, paved the way for a view of celestial bodies (and hence comets, being assimilated to the nature of Mars and Mercury) as causes of terrestrial events. In his

[61] Claudius Ptolemy, *Tetrabiblos,* II.13, translated by F.E. Robbins, Loeb Classical Library (London/ Cambridge Mass. 1980) 193.

introduction to the *Tetrabiblos,* he assumed the stars' movements to actually cause earthly occurrences. There, he stated that ". . .it is so evident that most events of a general nature draw their *causes* from the enveloping heavens." [My italics]. This was not to say that there was no way of influencing them: ". . .certain things, because their effective causes are numerous and powerful, are inevitable, but others for the opposite reason may be averted."[62] Thus, the most blatant effects of celestial causation could be mitigated by taking the appropriate precautions.

Once the validity of such a point of view was acknowledged, the way was paved for a genuinely divinatory treatment of comets. Whereas Stoics like Seneca and Pliny had been doubtful about the viability of trying to interpret the meaning of cometary appearances, associating them with laws of nature that were for the most part hidden from human perception, Ptolemy optimistically included a series of precepts to be used in divination from comets. Important indications for ascertaining in what regions the comet's consequences would take place were the location of the comet's head as it passed through the constellations and the directions to which its tail pointed. Similarly, the shape of the head determined the kind of misfortune and who would fall victim to it, its duration indicated the duration of the calamities, and its position relative to the sun, how soon these would begin to make themselves felt: if a comet appeared in the east, this would be very soon, if in the west, somewhat later.[63]

Similar rules for cometary prediction were laid down in a collection of a hundred astrological aphorisms that under the name of *Karpos* or *Centiloquium* circulated widely during the Middle Ages, often attached to copies of the *Tetrabiblos.* Although probably spurious, it was held to have been written by Ptolemy and consequently carried the weight of his authority. The last five aphorisms dealt with comets and stated that, if a comet was eleven zodiacal signs from the sun, it portended the death of a prince in power; if in a place past the centre, the kingdom's treasury would thrive, but there would be a new governor; if in a place past the meridian, it foretold illness and sudden death; if it moved from west to east, it meant a foreign invasion, but if it stood still, an domestic enemy.

Ptolemy's work constituted the culmination of ancient astronomy

---

[62] Ibidem, I.3. See also below, ch. VI.
[63] Ibidem, II.9.

and astrology. As we have seen, his comet lore managed to combine in a convincing and coherent way the several different attitudes toward the nature and function of comets that had come into being during the long centuries of Greek and Roman antiquity. Notably, he blended Aristotelian meteorology, popular divination techniques, and the Stoic outlook on divine providence and cosmic harmony, to arrive at a synthesis that was to stand the test of many ages, and that it would prove well-nigh impossible to disentangle.

6. *Comets and the Early Christians*

The advent of Christianity made a great impact on ways of thinking about the world, nature, and man, and comet lore was not to remain exempt from these developments. Early Christian scholars were equipped with a cultural outlook made up of the different pagan philosophical traditions thriving during late antiquity, which they somehow needed to christianize. As demonstrated above, comet lore in late antiquity had been especially influenced by two currents of thought: first of all, the Stoic and Platonic outlook of a living universe permeated with cosmic correspondences, and, secondly, the synthesis of Ptolemy, who had combined the Stoic perspective with popular astrological beliefs to create a view of terrestrial events being caused by celestial motions. The early Christians thus saw themselves faced with the task of making these traditions compatible with the tenets of the Christian faith.

The Stoic-Platonic correspondence theory was not hard to christianize. Scholars such as the Church Father Origen readily admitted the power of the stars to foretell the future. Comets, by which Origen understood "one of the stars of the sort that occasionally come into being, meteors in the form of hair, beams, beards, or casks", could also easily be regarded as omens, as long as it was recognized that both comets and stars had been instituted by God, who had created them to signal changes on earth.[64] This view is reminiscent of the Stoic outlook of a God who had framed a universe in which every occurrence announced the imminence of a corresponding change elsewhere. Origen also accorded comets a special role in the Chris-

---

[64] Origen, *Against Celsus,* V.12, quoted by Lynn Thorndike, *A History of Magic and Experimental Science* I (New York 1923) 458.

tian history of salvation, in optimistically informing his readers that they could occasionally signal changes for the good. He cited Chaeremon the Stoic for this opinion[65], and enthusiastically added that, if comets portended important changes on earth and heralded new dynasties, why should not a comet also have arisen at the birth of the most important new Ruler ever?[66]

Moreover, Origen endowed the stars with reasoning powers, and he peopled the skies with angels acting as intermediaries between, on the one hand, the supreme Deity and, on the other, both nature and human society.[67] Thus, his universe was as much alive as that of the Stoics, the main difference being that, where thinkers like Manilius and Seneca had conceived only of an abstract divine power permeating the cosmos, Origen concretized this conception in the form of actual divine messengers.

However, a much more acute problem was posed by the function of comets in the Ptolemaic synthesis. The Christian doctrine of salvation hinged on the crucial assumption that men were able to choose freely to do good, or evil, as they pleased. Otherwise—if, for example, all human actions had already been determined beforehand by divine foreknowledge, or by the motion of the celestial bodies—there would be no reason for God to condemn some people while saving others. This would imply God's injustice—obviously an unthinkable consequence. Clearly, adherence to the Ptolemaic view of celestial bodies (and comets, through their assimilation to the nature of Mars and Mercury) as causes of earthly events would bring Christian authors dangerously close to heresy. Because of the enormous influence that the authoritative views of Ptolemy exercised, Origen felt an urgent need to stress categorically that under no circumstances could any celestial constellation rob human beings of their free will. He expressly stated that the stars were merely signs, *and not causes,* of future events; in no way could they impose necessity.

Origen was by no means the only one to utter this warning against heterodoxy. Almost all Christians writing about astrology or comet lore stressed the merely signifying power of heavenly phenomena, and the human capacity to resist their indications. Origen's fellow

---

[65] Chaeremon's treatise *De cometis,* which Origen professed to have read, has not come down to us.

[66] Origen, *Against Celsus,* I.59.

[67] Thorndike, *History,* I, 452, 457.

Church Father, John of Damascus, in the seventh century expressed his views on the signifying power of the stars in general, and comets in particular, in his voluminous *Exposition of the Orthodox Faith.* Certain Greeks, he said, held that all our actions are determined by the motions of the sun and moon, and the principles of astrology; we, however, believe that they can be nothing more than signs of coming weather conditions, such as cold and heat, drought and rain, and winds. Never, though, can they foretell any human actions; as human beings we have been provided by God with free will and a rational soul capable of reflection, and therefore are the masters of our own deeds. Thus John accorded to the motions of sun and moon a portentous power very reminiscent of that accorded comets by Aristotle: they could indicate weather conditions and nothing more. The role John allotted to comets, however, was a very different one. They were special portents, appearing only to signify the death of kings, and did not belong to the stars created in the beginning, but were newly called into being by God at the appropriate moment, later to be dissolved again. Such had been the star of the Magi, which had demonstrated its irregular nature by moving now north, now south, now east, now west.[68] Whether John believed comets to be created by God naturally or supernaturally, his account does not make clear, but later authors usually took him to have meant the latter.

In the Latin West, too, the Stoic outlook on cometology prevailed during the centuries prior to the recovery of the Aristotelian corpus. Neither Isidore, bishop of Seville (560-636) nor the Venerable Bede (672-735) was very specific about the location of comets, although both authors seem to have regarded them as real celestial bodies rather than fleeting atmospheric meteors.[69] Moreover, they both invested the hairy stars with portentous significance. Isidore, in his *Etymologies,* informed his readers that whenever a star of this kind appeared, it announced pestilence, famine, or war, an opinion he repeated in his cosmological work *De natura rerum.*[70] Similarly, Bede,

---

[68] John of Damascus, *Exposition of the Orthodox Faith,* Burgundionis versio, 21.11.

[69] This can be gathered from the place comets occupied in the authors' treatments of the cosmos: Isidore mentioned them in the middle of a discussion about the names of the constellations (in his *Etymologies,* Book III, chapter 71) and Bede discussed them among celestial phenomena such as solar and lunar eclipses (Bede, *De natura rerum,* chapter XXIII).

[70] Isidore of Seville, *Etymologies,* III.lxxi.

who believed comets to be flaming stars that suddenly came into being, thought cometary consequences to be of two different kinds: on the one hand, they could consist of changes of government, pestilence, or war; on the other, they could be winds and drought.[71]

As to the reasons why he thought comets to signify pestilence, famine, and war, Isidore remained stubbornly silent. He did, however, at the end of the chapter in which he treated them, fulminate against those who believed the observation of the heavenly bodies to lead to knowledge of the future: such an opinion was blatantly contrary to Christian faith, and ought to be ignored by true Christians. Even the pagans Plato and Aristotle had condemned such an attitude. For, if human beings were constrained by necessity to perform specific actions, how could it be possible for the good to deserve praise or for the evil to be exposed to vengeance?[72] Thus, although Isidore was prepared to grant the heavenly bodies a certain influence on earthly events (he notably held that everything in nature grows according to the waxing and waning of the moon)[73] he was firmly opposed to judicial astrology as a doctrine infringing upon the free will given to man.

On the whole, we can conclude that the impact of Christianity on the comet lore of early Christian scholars was not very considerable. The belief in a universal correspondence between heaven and earth, which had been widespread among Stoic and Platonic authors, could easily be reconciled to Christian tenets, as long as human free will was safeguarded by positing that celestial bodies only signified, and never caused, future events. Christian authors, like their Stoic predecessors, were more concerned with the prophetic function of comets than with their physical characteristics, and their accounts were not very outspoken as to the exact location of the hairy stars. More important to them was the significance of one particular comet: the star of the Magi, investing tailed stars with a special place in the Christian history of salvation.

---

[71] Bede, *De natura rerum*, XXIIII.

[72] Isidore, *Etymologies*, III.lxxi.

[73] In his *De natura rerum*; see Thorndike, *History*, I, 633.

7. *Aristotle and Christianity: A Problem-Fraught Love Affair*

However, one very influential strand of ancient comet lore had come dangerously close to eliminating human free will so dear to Christians. Although, as we have seen, Ptolemy himself had admitted the possibility for human beings to avert or mitigate several of the terrestrial effects brought about by celestial causation, yet within the Ptolemaic synthesis lay contained the seeds of determinism. They were grown by the Arabs.

Thus, the famous Arab scholar Albumasar (Abū Ma'shar) (8th-9th century) believed that man's soul, having descended from the sphere of light to the 'hylic', corruptible, sublunary sphere, must strive to return to union with the divine. To attain this aim, however, he needed the assistance of intermediaries in the form of the celestial spheres, whose deities had to be worshipped in order to arrive, in the end, at the One. This worship was dependent on astronomy and astrology, these being the sciences studying the attributes, qualities, and conditions of the intermediaries.[74] In Albumasar's view, the power exercised by the superior spheres on the inferior did not operate by motion alone, as Aristotle had believed; instead, the Arab posited a kind of celestial influence cognate to Ptolemy's idea of a certain power (δύναμις) emanating from the ether and causing changes in the sublunar elements.[75] This force, according to the Ptolemaic view, depended on the zodiacal position of the planet emanating it, on the relative position of the planet and the subject it was supposed to affect, and on the rising and setting of planets in relation to that subject.[76] Albumasar, relying on this Ptolemaic account as well as on Aristotle's remarks on the terrestrial effects wrought by the motions of Sun and Moon, framed a theory according to which the action of the heavenly bodies on the sublunar elements was transmitted through an invisible intermediary, like a magnet attracting iron. The terrestrial objects, in this view, needed to display a natural receptivity to the influences emanating from the

[74] See David Pingree, "Abū Ma'shar al-Balkhī, Ja'far Ibn Muhammad" in C.S. Gillespie ed., *Dictionary of Scientific Biography* I, 33-34.

[75] See John D. North, "Celestial influence-the major premiss of astrology" in Paola Zambelli ed., *Astrologi hallucinati. Stars and the end of the world in Luther's time* (Berlin/ New York 1986) 45-100: 50.

[76] Ibidem, 51.

celestial bodies. The sun and moon, for Albumasar, received their movement from the superior world and by their rays transmitted particular influences to the sublunar world.[77]

Needless to say, such a theory would pose a considerable danger to a Christian world view. Any kind of star-worship, albeit only as a means of attaining a union with God, was obviously inadmissible from a Christian point of view.[78] Moreover, the Ptolemaic theory of astral influences causally determining earthly events, elaborated by Albumasar to include the physically plausible concept of a universe seen as a single vast machinery, in which these influences were transmitted to the sublunary region by means of planetary rays, posed an evident threat to the cherished Christian dogma of human free will.[79]

Nonetheless, once discovered, the Arab amalgam combining Aristotle and Ptolemy with Indian and Persian influences became tremendously popular among the scholars of the Latin West. The thirties of the twelfth century saw no less than two independent translations of Albumasar's most influential work, the *Introductorium Maius in Astronomiam:* one by John of Seville in 1133, and one by Hermann of Carinthia in 1140. Furthermore, John of Seville also rendered into Latin all of Albumasar's other known works.[80] Through the same channels as Albumasar's works, and around the same time, Ptolemy's *Almagest* and *Tetrabiblos,* as well as spurious works wrongly attributed to him, like the *Centiloquium,* were likewise made available to the Latin West as part of the same translating activity. According to Richard Lemay, the translations of Albumasar's (and, to a rather lesser extent, Ptolemy's) works were the main factor responsible for the renewed interest in astronomy and astrology in twelfth-century Western Europe: "Other works of medieval astrology, all derived from the Arabic, which obtained wide currency, still could not rival

[77] Ibidem, 57.

[78] Obviously it was no more acceptable, in theory, to Islam; an explanation of how Albumasar sidestepped this problem is, however, far beyond the boundaries of the present study.

[79] Of course, Albumasar was not the only Arabic philosopher to have influenced the West whose notions were pervaded with determinism. One more clear example of such a figure was Avicenna, very influential to medieval thought, whose theories were similar to Albumasar's in positing the necessity of intermediaries to reach God, and likewise implied astral determinism.

[80] Richard Lemay, "The true place of astrology in medieval science and philosophy: towards a definition" in Patrick Curry ed., *Astrology, science, and society. Historical essays* (Bury St Edmunds 1987) 57-73: 69.

in doctrinal influence and mentality the works of Ptolemy or Abū Maʿshar, of whom Abū Maʿshar ranks the first by far."[81]

The works of Albumasar and Ptolemy were not the only ones to be made available to the Latin-speaking scholarly community in the twelfth and thirteenth centuries. The feverish translation activity from Arabic and Greek into Latin, associated first and foremost with the names of Gerard of Cremona (c. 1114-1187) and William of Moerbeke (c. 1215-c. 1286) included a host of ancient Greek and medieval Arabic authors[82], the most important of whom was undoubtedly Aristotle.

Only a few of the Stagyrite's logical works had been known in western Europe prior to the twelfth century. Due to the unavailability of Aristotle's extensive physical writings, natural philosophy in the schools up to the twelfth century centred on texts of Stoic-Platonic inspiration, such as Plato's *Timaeus,* Seneca's *Natural Questions,* Cicero's *The Nature of the Gods* and Macrobius' *Commentary on the Dream of Scipio.* As we have seen, the Stoic-Platonic framework was quite easily reconcilable with Christian tenets, and the views of Isidore and Bede on comets were remarkably similar to those of Seneca or Pliny. But when Aristotelian works like the *Physics, On Generation and Corruption, On the Heavens,* and the *Meteorology*[83] made their appearance on the western European scene, things changed rather dramatically. The revival of Aristotle met with a very mixed reaction; yet for centuries to come, the Stagyrite's teachings were to cast a firm spell over the educated world.

As soon as a sufficient number of Aristotle's texts had been translated into Latin to realize what the Stagyrite's doctrines were about, attempts were made by Rome to root them out. As early as 1210, the archbishop of Paris forbade the study of the Aristotelian scientific corpus; five years later, at the Fourth Lateran Council, pope Innocent III banned many of Aristotle's works, and at the instigation of the papal legate, the new statutes of Paris university were made to include a like censure. Similar condemnations were pronounced by

---

[81] Ibidem, 69-70.

[82] See the survey of their translations edited by Edward Grant, "The translation of Greek and Arabic Science into Latin" in Edward Grant ed., *A Source Book of Medieval Science* (Cambridge Mass. 1974) 35-41.

[83] All these were translated from the Arabic by Gerard of Cremona; *On the Heavens* and the *Meteorology* were furthermore translated from the Greek by William of Moerbeke.

Gregory IX in 1231, by Innocent IV in 1245, and by Urban IV in 1263[84]; yet the very number of them testifies to their inefficacy. Aristotle would not go without a struggle.

In the end, the archbishop of Paris, Étienne Tempier, and the pope, John XXI, had recourse to their most powerful weapon: in 1277 they condemned 219 propositions under pain of excommunication. The condemnation of 1277 was intended to subvert the philosophical necessitarianism and determinism derived from Greco-Arabic sources, especially Aristotle and Averroes (Ibn Rushd).[85] The main concern of Tempier, and of the theologians, was to safeguard God's omnipotence; to them, it was unacceptable that the regularities of the Aristotelian world system, so vigorously adhered to by the arts masters, should deprive God of his absolute power to do anything he pleased on account of his being bound by these regularities. Thus, one of the condemned propositions read "That in efficient causes when the first cause [God] ceases [to act] the second [or secondary] cause does not cease its operation since it could operate in accordance with nature."[86]

Likewise, the condemnation fulminated against intolerable infringements on God's omnipotence posed by astrological determinism and independent operation of the celestial spheres: "That God could not have made prime matter without the mediation of a celestial body"; "That a motive intelligence of the heavens influences the rational soul just as a celestial body influences the human body"; "That when all celestial bodies have returned to the same point - which will happen in 36,000 years - the same effects now in operation will be repeated"; "That nothing happens by chance, but all things occur from necessity and that all future things that will be will be of necessity, and those that will not be it is impossible for them to be; and that upon considering all causes nothing happens contingently"; "That from the different [zodiacal] signs of the sky diverse conditions are assigned in men, both with respect to spiritual gifts and temporal things"; "That the effects of the stars on free will are

---

[84] For relevant texts and their translation, see Edward Grant, "The Reaction of the Universities and Theological Authorities to Aristotelian Science and Natural Philosophy" in his *Source Book of Medieval Science*, 42-52.

[85] Edward Grant, "The Condemnation of 1277, God's Absolute Power, and Physical Thought in the later Middle Ages", *Viator* 10 (1979) 211-244: 212.

[86] Tempier's Condemnation, article 199, translated by Grant, "Reaction of the Universities", 50.

hidden"; and "That our will is subject to the power of the celestial bodies"[87] were all deemed reprehensible propositions, and accordingly condemned.

Much has been made by historians of the supposed impact of the 1277 Condemnation on the progress of science. Although today no one would go as far as Pierre Duhem, who proclaimed 1277 the date of the birth of modern science, recently scholars such as Edward Grant have called attention to the fact that, in stressing "the contingency of God's operations and his omnipotence to do as he pleased short of a logical contradiction", the condemnation favourably influenced scientific development in stimulating philosophers to contemplate possibilities such as a plurality of worlds or the existence of a vacuum.[88] Such far-reaching questions need not concern us here. For our purposes, the condemnation of 1277 can be taken to testify to the wide currency of ideas about astrological determinism and elimination of contingency, based on the Arabic-Aristotelian tradition, and to the revulsion with which such ideas were regarded by the ecclesiastical authorities.

8. *Comets: The Dominicans*

Like all other areas of natural philosophy, twelfth and thirteenth century cometology was deeply influenced by the recovery of Aristotelian physics and Arabic astrology. To a much greater extent than their early Christian predecessors, high scholastic authors such as Albert the Great and his fellow Dominicans were interested not only in comets as signs of divine providence, but especially in the place they were to be assigned in the chain of natural causes and effects that brought about changes in the sublunary region. The ancient Stoic idea of Seneca and Pliny, that it was not viable for human beings to try to unravel the divine laws of the universe, was aban-

---

[87] Ibidem, articles 38, 74, 6, 21, 143, 161, and 162; Grant, "Reaction of the Universities", 48-50.

[88] Edward Grant, "Condemnation of 1277"; see also his discussion of the Condemnation in his *Planets, Stars, and Orbs. The medieval cosmos 1200-1687* (Cambridge 1994) 53-56 and passim; and Edward Grant, "The effect of the condemnation of 1277" in Norman Kretzmann, Anthony Kenny & Jan Pinborg eds., *The Cambridge History of Later Medieval Philosophy. From the Rediscovery of Aristotle to the Disintegration of Scholasticism 1100-1600* (Cambridge 1982) 537-539.

doned by thirteenth century authors, who operated under the aegis of Aristotelian naturalism.

One of the earliest medieval tracts to deal exclusively with cometology was written in Spain by an anonymous author in the first half of the thirteenth century (about AD 1238, according to Thorndike).[89] It offers an interesting amalgam of Aristotelian physical theory and Arabic-Ptolemaic astrology; next to Aristotle and Ptolemy, the author quoted from a host of Arabic authors such as Albumasar, Haly Abenragel[90], al-Kindi (al-Kindī), Rasis[91], and Mashallah (Māshā'allāh), as well as the Indian sage Buzahen (or Buthahen) whose lore was probably transmitted to Spain by the Arabs. The treatise consisted of sixteen chapters, and displayed a structure that was to become the standard way of treating cometary appearances up to and including the seventeenth century: firstly, the author elucidated what comets were, how they came into being, and how many different kinds of them existed; next, he treated their significance according to characteristics like colour, heavenly location, position with regard to the planets, size, etcetera; then, he informed his readers what the consequences of cometary appearances were, and where these would be felt; and finally, he concluded with a number of examples of past comets and their effects.

The author's view of the physical nature of comets was clearly a fusion of Aristotelian and Arabic-Ptolemaic ideas on the subject. Following Ptolemy's *Centiloquium* and Haly's commentary on it, he asserted that although comets were formed of a substance different from that of the planets, they nonetheless came into being by means of planetary motions. They therefore had powers similar to those of the planets, and likewise influenced sublunary events.[92] He next expounded the Aristotelian theory, that comets were dry earthly exhalations that ignited in the sphere of fire; because the matter of which they were formed was greasy and thick, they remained burning for quite a long time. Since comets were not everlasting, and followed irregular courses, they were not celestial bodies. Similar views were

[89] Lynn Thorndike, *Latin Treatises on Comets Between 1238 and 1368 A.D.* (Chicago 1950) 9.

[90] I.e., Ibn Abi al-Rigial, an eleventh-century Arabic astrologer.

[91] I.e., Abu Bakr Muhammad ibn Zakariya al-Razī, whose *Book of Almansor, Book of Divisions,* and *Short introduction to medicine* were translated into Latin by Gerard of Cremona.

[92] Anonymous, *Liber de significatione cometarum* (Eastern Spain, c. 1238) in Thorndike, *Latin treatises,* 17-18.

set forth by Ptolemy and Rasis, whom our author also quoted, as well as a slightly deviant version adhered to by Buthahen of India, who believed comets to be lights descending from heaven that set on fire a subtle terrestrial vapour.[93]

The next fourteen chapters of the treatise are exclusively astrological. Relying chiefly on Ptolemy's *Centiloquium* and *Tetrabiblos,* on Albumasar's *De magnis coniunctionibus,* and on Buthahen of India, our author paid extensive attention to the nine different kinds of comets[94], to the significance of their colours (linking them to specific planets; thus a red comet is of the nature of Mars) and forms, to their general significance and their specific significance when appearing in particular triplicities or zodiacal signs, and to the rules for determining what terrestrial regions were to be affected, how long the cometary effects were to last, and who were to be their chief victims. In these astrological chapters, the author was influenced more by Ptolemy and the Arabs than by Aristotle; although, like the Stagyrite, the author admitted that comets were of a dry nature, he immediately went on to relate this fact to their correspondence with the natures of Mars and Mercury, and to deduce from this affinity that comets have the same significance as these two planets, to wit, wars, death, robberies, theft, arson, fever, drought, and famine.[95] Thus our author's outlook was that of the Ptolemaic synthesis: comets have the same effects on earth as do planets, and as such they are part of a system of celestial influences affecting terrestrial events. Blatantly astrological was his treatment of the consequences of comets in particular triplicities (based on the dicta of the Indian sages) and in particular zodiacal signs (based on Albumasar): without any concern for possible underlying physical causes, the author listed an enormous profusion of dire repercussions, such as death of noblemen, scandals in the realm of the Christian king, discord among sages and lawyers, and the death of children, pregnant women, and birds.[96]

Likewise influenced both by Aristotle and by Ptolemy and the

---

[93] Ibidem, 18-23.

[94] Ibidem, 24-25. The author professes this classification to be based on Ptolemy and on the Arabic translations of Alkindi; eight of the nine kinds are Veru (Arabic Duamfuc), Tenaculum (Dualaan), Pertica (Duademayn), Miles (Dussuraroch), Azcona (Duarcon), Aurora (Duaduab), Rosa (Huard) and Nigra (Duahad), the ninth one being omitted in the manuscripts used by Thorndike. The author also includes drawings of the different kinds of comet.

[95] Ibidem, 29.

[96] Ibidem, 32-39.

Arabs, Albert the Great wrote a highly original commentary on Aristotle's *Meteorology* in which he displayed a marked interest in the physical nature of comets, and in the way they could be fitted into the pattern of causes and effects linking the heavens and the sublunary sphere. First of all, Albert set forth the several erroneous views on comets quoted by Aristotle and Seneca: some ancient authors thought comets to be conjunctions of several stars, some thought them to be vapours clinging to a planet, and some to be optical illusions, the impressions of stars on humid air.[97] Next, he added some other views he held to be equally erroneous: Seneca's view that comets were real celestial bodies, John of Damascus's view that they were stars newly created by God, and the view of "certain learned moderns" that they were planetary impressions at the meeting line of fire and air.[98]

Not surprisingly, in Albert's opinion the correct explanation of comets was that given by Aristotle: they were "coarse terrestrial vapours, whose parts lay very close together, gradually rising from the lower part of the upper region of the air to its upper part, where it touched the concave surface of the sphere of fire and was there diffused and inflamed and so often seemed long and diffuse."[99] They lasted until their supply of burning material was exhausted. Albert quoted the Arabs Avicenna (Ibn Sīnā), Algazel (al-Ghazzāli), and al-Farghani (al-Farghānī) in support of this opinion: comets were sublunary vapours. However, the Ptolemaic link between comets and celestial bodies was not absent from Albert's account: immediately after asserting that they were "like stars, but not stars", he went on to quote Ptolemy's *Centiloquium,* with the commentaries of Haly, Abraham and Bugaforus, as well as Albumasar's *De coniunctionibus,* in support of the view that comets and other elevated ignited vapours,

---

[97] Albert the Great, *Liber I Meteororum, Tractatus III,* capita 1-3, translated by Thorndike in *Latin treatises,* 62-76: 62-65. Albert erroneously attributed the first view (comets as conjunctions) to Apollonius of Myndus, as quoted by Seneca: "Dixerunt autem quidam physicorum qui fuerunt a natura devii, et sicut dicit Seneca in libro de *Naturalibus quaestionibus,* Apollonius etiam fuit de illa sententia, quod stellae habentes comas sunt stellae multae aggregatae in orbibus inferioribus. . .". As demonstrated above this was not Apollonius's view. Along with a rather corrupt quotation by Albert of Seneca's own view on comets (ibidem, caput 4) this can be taken as evidence that the version of Seneca's *Natural Questions* used by Albert differed on several points from the original.

[98] Ibidem, caput 4; Thorndike, *Latin treatises,* 65-67.

[99] Ibidem, caput 5; Thorndike, *Latin treatises,* 67.

being affected by the planets, were called "second stars" and were to be taken into account in astrological judgments.[100]

Furthermore, Albert summoned the aid of reason in support of Aristotle's view: "it is evident that flame is nothing but kindled fumes; now a comet is a sort of flame, as is apparent to the sight; therefore, it is kindled fumes."[101] Seneca's opinion could be disproved in a similarly commonsensical way: "... what vanishes without setting but within the hemisphere, that is corrupted not completing its course through the hemisphere".[102] Albert's confidence in man's capability of understanding nature's ways came to the fore in his rebuttal of the Damascene's opinion: "The opinion of John of Damascus, that God makes a new star, is not probable, since there is no reason why he should make it in different figures and colors. But a comet is seen in different figures and colors; therefore, it is not a work of God alone. ... Since, therefore, this variety of form is because of some pre-existing matter from which nature works, it is evident that a comet is not then produced by divine work only."[103]

Much the most interesting chapter of Albert's treatment of comets was the last one, in which he investigated the possible reasons why comets could be said to signify the deaths of potentates and wars. With regard to this question, as in his refutation of the Damascene's opinion, Albert excluded God or any supernatural agency from his argument: "...it is evident that a comet has a natural cause not dependent on anything else".[104] How, then, can it be said to have a relation to war or to someone's death? Albert wondered. If such a relation existed, the comet should be either cause, effect, or sign of these calamities. But in itself, a comet is neither cause, nor effect, nor sign. However, following Albumasar, Albert believed comets to be of the character of Mars, their prime mover. Thus, although a comet could not be the immediate cause of the death of kings, "yet its first moving cause denotes the force of Mars, which is a sign of death especially by violence".[105] The prime mover denoted the cause of death, though only inclinatorily, not necessarily; the comet itself should properly be called sign, since it acted "like a counselor whose

[100] Ibidem; Thorndike, *Latin treatises*, 69.
[101] Ibidem.
[102] Ibidem, caput 6; Thorndike, *Latin treatises*, 70.
[103] Ibidem; Thorndike, *Latin treatises*, 70-71.
[104] Ibidem, caput 11; Thorndike, *Latin treatises*, 75.
[105] Ibidem; Thorndike, *Latin treatises*, 76.

advice can be changed by finding better".[106] Comets were associated with Mars, the planet that "signifies the death of inhabitants and disturbances of rulers, as inclining to this by wrath and heat and drought, from which comes animosity or exciting of peoples against one another". Therefore "a comet is said to signify these, although its immediate significance is not connected with them."[107] The reason that comets were said to denote specifically the death of *kings,* not paupers, was that the death of kings was noticed more because of their fame.[108]

Albert's fellow Dominican, Gerard de Silteo, around the year 1264 wrote a *Summa de astris,* including a section on comets that relied very heavily on his great predecessor's treatise, literally quoting large passages from it. In the same manner as Albert, Gerard paraphrased the 'correct', Aristotelian opinion concerning comets: they were fat terrestrial vapours burning in the sphere of fire. And, like Albert, he quoted Avicenna, Algazel and Alfraganus[109], in support of this opinion, as well as the theory of Ptolemy's commentators and Albumasar as to comets as "second stars". Albert's argument disproving John of Damascus was repeated, as were the quotations from Seneca used by Albert to demonstrate the size comets could have, and the different regions of the sky where they could be seen. Interestingly enough, the only point on which Gerard expounded an original view slightly divergent from Albert's, was the question of cometary significance. Although he agreed with his colleague that Mars inclined the elements to heat and drought, and people to unruliness and anger, Gerard laid more stress on the human free will and quoted Saint Augustine, who wrote that astrological investigations were not relevant with regard to salvation, and should therefore not be extensively studied.[110]

A pupil of both Albert the Great and Thomas Aquinas, the Do-

---

[106] Ibidem; Albert has taken this quotation from Aristotle, *On Sleep and Waking.*

[107] Ibidem.

[108] Ibidem.

[109] Or: Alpetragius; the manuscript used by Thorndike, otherwise literally quoting Albert's text, on this point disagrees with the Borgnet edition of Albert the Great's *Opera omnia.*

[110] Gerard de Silteo, *Summa de astris,* Part I, dist. 23, cap. 4; Thorndike, *Latin treatises,* 194-195: "Tamen notandum est quod, cum Mars dominatur super commotionem elementorum, inclinat ad calorem et siccitatem ex quibus pervenit animositas et concertatio populorum adinvicem et hoc accedente libera voluntate. Facta et enim ira contra quosdam animosa concitatione aliquando super quos fertur Mars modo consimili comete ac pacifice minime dicendum est quod isti remunerandi quam illi merite puniendi. Deinde que dicta sunt de asteris in hac prima parte huius operis

minican friar Aegidius from the town of Lessines in Hainault wrote a lengthy and original cometary treatise occasioned by the comet of 1264. It seems to have been rather influential, given the fact that it was translated into French, and that it was still being copied in the fifteenth century.[111] In keeping with his self-professed aim, to collect all that he could find on the subject of comets in previous learned writings, Aegidius quoted a great number of ancient and medieval authors, the most prominent of whom were Aristotle, as translated from the Greek, Seneca, Ptolemy (both the *Tetrabiblos* and the *Centiloquium*), and Albumasar.

With regard to the physical constitution of comets, he followed the Aristotelian theory, refuting several rival hypotheses in the process. Against Seneca's opinion, Aegidius argued that comets were of too brief duration to be heavenly bodies; against that of John of Damascus, he maintained that comets were unlikely to be divine prodigies, having the diversity of appearance of natural phenomena, being shown to faithful and infidels alike, and invariably following planetary conjunctions, thus clearly betraying their natural generation.[112] Like Albert, then, Aegidius was keen to exclude God's agency from a phenomenon that could be explained by natural causation. Agreeing with Aristotle, the "highest of the philosophers"[113], Aegidius held comets to be burning terrestrial vapours in the sphere of fire, usually followed by drought and wind.

However, as to the causes of cometary motions, Aegidius, relying on his observation that the comet of 1264 moved faster than the stars, deviated from the Aristotelian view that the sphere of fire followed the circular motion of the heavens, albeit rather more slowly. Following Albumasar and Ptolemy, our author stated that comets were generated by a heavenly force, especially that of Mars and Mer-

sufficiant ad intelligentiam dicendorum. Nam subtilius aliquid querere talique inquisitione vacare rebus gravioribus et melioribus necessarium semper impendere nec expedit nec congruit, sicut dicit beatus Augustinus *in secundo super Genesi.* Item idem in eodem libro dicit, Multi multum disputant de hiis rebus quas maiore prudentia nostri auctores omiserunt ad beatam vitam non profuturas discentibus et occupantes, quod peius est, multum pretiosa et rebus salubribus impendenda temporum spatia."

111 See Thorndike, *Latin treatises,* 89.

112 Aegidius of Lessines, *De essentia, motu et significatione cometarum* cap. 4; Thorndike, *Latin treatises,* 121.

113 Ibidem, cap. 5; Thorndike, *Latin treatises,* 121: "Summus autem philosophorum in hiis que sub orbe lunari genita sunt cometam dixit esse ignem in fumo terrestri sublimato. . .".

cury. Because these generating forces moved in every direction, so did comets, being 'assimilated' to their natures.[114] However, due to the changeable and corruptible matter of which comets consisted, they were not susceptible to any perfect, celestial form, and thus were unable to assimilate entirely to the nature of their heavenly generators. Because they consisted both of fire, with its natural tendency to move upwards, and earthly matter, tending always to move down, their actual movement was in fact a compound of the two, causing the comet to move sideways.[115]

Like all his contemporaries, Aegidius displayed a vivid interest in the signification of comets. In a lengthy and very complicated chapter on the question of whether, and if so, how comets could be said to signify anything, he discussed the various possible ways in which the hairy stars could be connected to sublunary events. Like Albert, Aegidius endeavoured to determine whether comets could be causes, effects, or signs. They could not be necessary causes, because of the well-worn reasoning that this would deprive human beings of their free will, thus rendering them incapable of being either saved or damned. Nor could they be effects, since effects never preceded their consequences. They could not be signs of superior causes, or else they could only be understood by those with prophetic gifts. Finally, neither could they be signs of inferior causes, for, once again, this would violate the free will. Thus, it seemed impossible to maintain that comets carried any meaning at all. However, Aegidius managed to find his way out of the maze by means of an ingenious piece of reasoning. The human soul, he said, was subject to a threefold division: part of it was divine, part was intellective, and part was animal. The divine part (or rational soul) was immediate from God, and could be affected only by God and free will; thus comets could not

[114] Ibidem, cap. 7, Thorndike, *Latin treatises*, 134-135: "Igitur quantum possibile est rationem dicamus quia huiusmodi diversitas motus non ex loco quin potius ex aliis duobus inesse videtur ipsis scilicet ex generatione et ex materia speciali. Ex generante quidem qui est vis celestis motus generaliter et vi planetarum specialiter, sed vis Martis principaliter et singulariter, ut dicit Albumasar philosophus *in libro octavo de coniunctionibus planetarum, capitulo primo* et *in libro florum suorum, capitulo 27*. Secundum autem Ptholomeum *in Quadripartito* Mercurius in generatione comete communicat Marti. Quia ergo horum generantium non est motus unus sed quandoque ad orientem quandoque ad occidentem quandoque ad aquilonem et quandoque ad austrum, et generans in quantum potest assimilat sibi genitum, inerit et cometis motus diversus secundum virtutem generantis in eo predominantis.

[115] Ibidem.

signify anything to do with this divine faculty. Yet, both other parts of the soul were subject to signification by comets; especially the animal soul, which was easily affected by the stars and whose actions could even be causally determined by them.

After having thus safeguarded human free will, residing in the rational soul, from celestial determinism, Aegidius could proceed to a carefree treatment of the manifold consequences traditionally associated with the appearance of comets. Following Aristotle, he stated that comets, being of a fiery constitution, were signs (*per se*) of desiccation of the air and thickening of terrestrial vapours. These atmospheric conditions in turn (*per consequens*) caused storms, scarcity, infectious air, pestilence, and earthquakes; the heavy winds might furthermore (*per accidens*) cause inundations and thunderstorms. Moreover, Aegidius also subscribed to the reasoning of Albumasar and Ptolemy: comets carried within them the nature of Mars and Mercury, causing heat and choler to abound. An overdose of choler in turn caused noblemen, who were extra delicate and naturally proud, to fly into rage, bringing about changes of government and popular unruliness. Therefore, the signification of comets could be extended to include deaths and wars.[116] Thus, despite his elaborate effort to exempt human will from celestial causation, and to stress that comets should be seen as signs, not causes, Aegidius in fact placed them at the top end of a causal chain carrying down heavenly influences that could seriously affect earthly conditions.

The work of thirteenth-century Dominican authors such as Albert the Great, Silteo, and Lessines, clearly reveals their indebtedness to the recovered Greco-Arabic tradition. As far as the physical nature of comets was concerned, they unanimously agreed with Aristotle's terrestrial exhalation theory. Moreover, they endeavoured to explain rationally why cometary appearances were invariably followed by certain well-defined atmospheric and terrestrial events, making extensive use of the arabicized Ptolemaic theory that comets were assimilated to the nature of the planets that generated them and could therefore be said to signify the effects proper to these planets. Our authors thus naturalized the wars, deaths, and plagues that were traditionally associated with comets, fitting them all into a causal chain of events at the root of which lay the influence with which the celestial bodies affected the earth. Unlike the early Christian scholars,

[116] Ibidem, ch. 8; Thorndike, *Latin treatises*, 160.

who in every comet saw the hand of God at work, the scholastics of the High Middle Ages ventured to exclude divine agency from cometology.

9. *Comets: Measuring Distances*

During the fourteenth century, the study of nature underwent a subtle change of emphasis. Although, like their thirteenth-century predecessors, authors such as William of Ockham and Jean Buridan believed that the physical world was to be approached rationally, they tended to accord more importance to empirical observation and quantification.[117] In the words of Anneliese Maier, "[late scholastic natural philosophy] was the first attempt at an independent analysis of nature . . . The attempt [was] made for the first time to find principles that permit a direct, individual, and empirical perception and understanding of nature, independent of all authority."[118] No longer was it deemed necessary to make a detour by way of Aristotle's authority; rather, the new outlook "led directly from the experience and reason of the individual to nature and its laws."[119]

The new outlook of empiricism and quantification advocated by thinkers like Ockham and Buridan seems to have fallen on fertile ground with several scholars concerned with the appearance and nature of comets. As early as 1299, a marked interest in the course traversed by the comet of that year was displayed by Peter of Limoges, canon of Evreux, who wrote a short tract on the subject. He was the first medieval scientist known to have made use of a measuring instrument, the *torquetrum* or *turquetrum*, to determine the location of a comet. With the aid of that implement, Peter ascertained that the position of the comet at the start of its course was in the 18th degree of Taurus, with a latitude south of the ecliptic of more than thirty degrees. About a month later, "after the feast of Saint Matthew", it was in the 14th degree of Taurus and had a south latitude of five degrees. It remained visible until March 5, gradually diminishing and climbing northward towards the Pleiades and Aldebaran, so that this

[117] Gordon Leff, *Medieval Thought: Saint Augustine to Ockham* (London 1959) 271-285.

[118] Anneliese Maier, "The achievements of late scholastic natural philosophy" in Steven D. Sargent ed., *On the Threshold of Exact Science. Selected writings of Anneliese Maier on late medieval natural philosophy* (Philadelphia 1982) 143-170: 146-7.

[119] Ibidem, 147.

northward movement was always greater than that westward with respect to the fixed stars. Peter also observed that its tail at first pointed at the lesser dogstar but gradually turned northward until it faced Mars.[120] Although these findings were still rather crude, the use of an instrument for measuring cometary positions is in itself a remarkable novelty.

Another tract was occasioned by the comet of 1301, probably also by Peter of Limoges. Once again, he made use of a torquetrum, while devoting almost half of his tract to a description of the comet and its course. As in the case of the 1299 comet, our author only charted a limited part of its course (just one week out of a total of more than a month); also, he noticed once again a gradual movement of the comet's tail, which this time shifted from the north towards the south. Peter thought that, like that of the comet of 1299, the course followed by the 1301 comet and the movement of its tail were influenced by Mars, the planet that generated it.[121]

A similar attitude towards cometary appearances was taken by Geoffrey of Meaux, physician to king Charles IV. He wrote separate tracts on the comets of 1315 and 1337, both of which he personally observed quite scrupulously. The first night on which he saw the comet of 1315, it was near the three stars in the palm of Bootes; on another night, he saw it in between the two stars in the Dragon's tail, and from then on he observed it nightly as it travelled through the back of Ursa Minor and the belly of Draco and on towards the throne of Cassiopeia, on the equinoctial colure, where it had arrived by the twelfth of February. Geoffrey then ceased observing it and applied himself to drawing a circle through all the aforesaid stars, and, on finding that it passed through Virgo, he inferred that the comet had risen in that sign.[122] He followed a similar procedure in the case of the comet of 1337, drawing circles from the pole through a number of well-determined, prominent stars, and finding that the comet was included between the two circles which terminated in Alhayot (Capella, a star on the left shoulder of the Charioteer) and the star on the Charioteer's right shoulder. Next, he waited for six

[120] Peter of Limoges, *Iudicium* on the comet of 1299, in Thorndike, *Latin treatises*, 200.

[121] Anonymous (Peter of Limoges?), *Iudicium de cometis* (1301), in Thorndike, *Latin treatises*, 202-207.

[122] Geoffrey of Meaux, *De stellis comatis* (1315), in Thorndike, *Latin treatises*, 208-214: 212.

nights to be able to determine whether the comet's movement was lateral, only to discover that it was not lateral, but straight towards the pole; from this observation he inferred that it must have risen in Gemini.[123] Although, as far as can be ascertained, he did not use any instrument to obtain his observations, from the above it should be clear that Geoffrey went to a great deal of trouble to determine the comets' courses.

The comets of 1338 and 1402 were both honoured with similar tracts, written by an anonymous French author and by Jacobus Angelus of Ulm, court physician to Duke Leopold of Austria, respectively. Both authors personally observed their comets on several nights, and like Meaux, they recorded the hairy stars' different positions with regard to the fixed constellations. Both also devoted a fair amount of attention to the comets' physical characteristics, describing the form of the tails as well as their colour.[124]

Cometary observations up to and including those of Angelus were still rather crude and unsystematic, and recorded in descriptive terms rather than in accurate instrumental measurements. In the course of the fifteenth century, however, a few scholars (though, it has to be stressed, a very small minority)[125] strove to increase the level of accuracy with which comets were observed. The Florentine Paolo Toscanelli has been hailed by his adulatory nineteenth-century biographer, Celoria, for having "with a firm and wise hand, placed astronomy on the fruitful road of exact observation".[126] His manuscript writings included observations of six comets between 1433 and 1472, which were progressively accurate and systematic. Whereas in 1433, Toscanelli did not record any precise daily observations, but contented himself with indications of the comet's general path, later on he started to draw more and more detailed charts and to compile tables of observations, including dates, times, longitudes, and latitudes. From 1456 onwards, he probably used a torquetrum to deter-

---

[123] Geoffrey of Meaux, tract on the comet of 1337, in Thorndike, *Latin treatises*, 219-225: 223-224.

[124] Anonymous, *Aliud iudicium de cometis* (1338) in Thorndike, *Latin treatises*, 226-233; for a translation and discussion of Angelus' text, see Jane L. Jervis, *Cometary Theory in Fifteenth-Century Europe* (Dordrecht/Boston/Lancaster 1985) 38-40.

[125] See Thorndike, *History*, IV, passim, for a comprehensive discussion of fifteenth-century comet tracts, the overwhelming majority of which were completely traditional and concentrated on comets' physical constitution, their material cause, their shapes and colours, and especially their significance.

[126] Quoted and translated by Jervis, *Cometary theory*, 48.

mine cometary positions.[127] Although they do not include any astrological predictions, Toscanelli's cometary writings do on occasion bear the mark of his interest in astrology; he was one of the foremost Florentine astrologers, and it is probable that the high standard of accuracy he strove to attain was prompted by astrological motives.[128]

None of the scholars discussed so far, not even the meticulous Toscanelli, had ever ventured to determine the distance of a comet from the earth—undoubtedly this was not deemed a very interesting datum to establish, since, from an astrological point of view, it was irrelevant, or at least much less important than a comet's course through the signs of the zodiac. The first man ever to attempt such a measurement was Georg Peurbach, humanist friend of Aeneas Silvius Piccolomini and professor of astronomy in Vienna. His tract on the comet of 1456 was not very noteworthy for its observations of the comet's longitudinal and latitudinal movement, nor for its discussion of comets' physical constitution—they were still hot, dry exhalations ignited in the sphere of fire under the influence of one or more of the planets. Interestingly enough, though, Peurbach went on to examine how the comet's distance from the earth could be determined. The reason for this effort is not obvious; after all, our author already felt he knew where the comet was to be located, to wit, either in the upper region of air or in the fiery region, as Aristotle had taught. Jervis suggests that he may have been trying to ascertain how thick the atmosphere was[129]; in that case, the determination of the comet's position was nothing more than a tool, an intermediary step in a quite different argument. His conclusion included no exact altitude for the comet, but only the statement that it was at least 1,000 miles up[130], after which outcome Peurbach went on to discuss at length the comet's portentous nature.

Whereas Peurbach had made use of nothing more than the crud-

---

[127] Jervis, *Cometary theory*, 52-69, esp. pp. 67-68.

[128] Ibidem, 57, Jervis remarks that Toscanelli's observations of the comet of 1433 display a particular interest in the comet's tail, which in astrology was an important indicator of the geographic region to be affected; the same can be said about a list of cities and their longitudes and latitudes in the manuscript on the comet of 1472 (Ibidem, 67).

[129] Ibidem, 91.

[130] His full argument runs as follows: 1. The distance of the earth's surface to the moon's sphere equals 30,000 German miles; 2. Comets are in the upper region of the air, according to Aristotle; 3. The sphere of fire could not possibly be 27,000 miles thick; therefore, a comet has to be more than 1,000 miles up. See Jervis, *Cometary Theory*, 91.

est estimates in his determination of cometary distances, his pupil and collaborator, Johannes Regiomontanus, tackled the problem in a more systematic fashion. In a highly theoretical treatise entitled *Sixteen Problems on the Magnitude, Longitude, and True Location of Comets* (published posthumously in Nuremberg in 1531) he discussed the determination of the location and size of comets from measurements and quantitative computations of their daily parallax. It was a didactic work, full of diagrams and geometrical proofs, and did not include a single reference to any particular comet; nowhere were the carefully elaborated theories put into practice. Unfortunately for Regiomontanus, his theorems all presupposed a comet without any proper motion, whose tail lies along the radius of the world. Thus, they were not applicable in practice. A treatise on the comet of 1472 usually attributed to Regiomontanus (though Jervis thinks it spurious) did include a determination of the comet's distance from the earth, but this was not based on computations of the sort set forth in the *Sixteen Problems;* instead, the author made use of a crude visual estimate, obtaining a result comparable to that of Peurbach: he found that the parallax could be no more than 6°, and consequently placed the comet no closer than 8,200 German miles from the earth (still well beneath the sphere of the moon, that was estimated at 30,000 German miles). Continuing the Vienna tradition of parallax measurements, Johannes Vögelin in 1532 attributed to the comet of that year the incredible parallax values of 35° 31' 1" and 34° 58' 32" (the correct value being 11.5"). Clearly, Regiomontanus's method was not workable when applied to crude, naked-eye observations.

It has to be stressed that an interest in measurement in no way precluded the astrological considerations usually associated with cometary appearances. Peter of Limoges, Geoffrey of Meaux, the anonymous author of 1338, Jacobus Angelus, and Georg Peurbach all showed themselves convinced adherents of the astrological way of explaining cometary consequences, based on Aristotle's theory of comets as hot, burning exhalations, combined with Ptolemy's and Albumasar's astrological rules. Thus, they all believed their respective comets to have been generated by the influence of one or more of the planets; they all endowed them with the traditional cometary consequences of war, famine, pestilence etc.; and they all applied the rules according to which the effects of a comet could be deduced from such data as the direction in which their tails pointed, or the zodiacal signs in which they appeared.

Two tracts occasioned by the comet of 1368 clearly testify to the importance fourteenth-century authors attached to astrological considerations; especially the question of whether of not comets signified anything, and if so, in what way, was a prominent topic of debate. Arguing in favour of cometary significance was the Bolognese jurist John of Legnano, according to whom the meaning of cometary appearances could be explained in two different ways. One might either explain the dicta of Aristotle by means of an argument based on natural causation, or one might follow the rules expounded by the astrologers. John himself ventured to do both. First of all, he stated that, at first sight, it would seem naturally doubtful that comets should signify winds, inundations, or the death of princes. Since comets and winds are dependent on the same kind of matter for their generation, it seems unlikely that winds should follow cometary appearances, as most of the matter necessary for them to arise would be conglomerated in the comet. Also, comets, being burning exhalations, desiccate the air, and would therefore seem to discourage the occurrence of inundations. And, finally, why should they signify the death of princes any more than that of ordinary people? After all, being a prince did not alter the human complexion.[131]

Yet, despite such evidence to the contrary, John believed that all cometary concomitants (for historical examples bore witness to the fact that, in practice, comets were indeed usually followed by dire consequences) could be explained by natural means. Whenever comets were generated, burning exhalations ascended through the air. Doing so, they left behind both a terrible heat and small particles of their matter, which, being the same matter from which winds were generated, caused winds to arise. Furthermore, sometimes part of the hot, dry exhalations in the interior of the earth, when attracted by the planets to form a comet, were obstructed in their path because the cold of the exterior air blocked the pores of the earth through which they were trying to escape; in that case, their struggle to escape caused an earthquake. Such a displacement of the surface of the earth was capable of displacing the water of the seas, thus causing inundations. Also, the desiccation of the air caused by the presence of burning comets led to crop failure and scarcity. And, finally, the elevated exhalations caused people to become choleric, inciting them

[131] John of Legnano, *Tractatus de cometa* (1368) in Thorndike, *Latin treatises*, 234-259: 247.

to wage war and to topple governments. Princes, leading a more delicate life than others, and playing a leading role in wars, were extra vulnerable to such choleric conditions.[132]

Genuth makes much of Legnano's 'natural' explanation, asserting that he was one of the first to argue that comets were causes, instead of signs, of the events that followed their appearance.[133] Although it is true that Legnano was the first to explicitly separate his natural explanation of cometary consequences from his ensuing astrological interpretation, his outlook was not as novel as Genuth believes. As we have seen, a century earlier, Aegidius of Lessines had already placed comets at the beginning of a completely natural chain of causes and effects, believing them *per se* to signify desiccation of the air, *per consequens* storms, scarcity, infected air, pestilence, and earthquakes, and *per accidens* inundations and thunderstorms. The only difference between the two accounts lies at the beginning of the chain: whereas Aegidius said that "according to Aristotle the cause of [comets'] significance is the fact that they consist of fiery matter, which naturally means the air will desiccate and subtle [humid] vapours will be consumed. . ."[134], Legnano said that "when comets are generated burning exhalations arise, and therefore intense heat and particles of their matter remain behind in the air."[135] Neither account mentions the influence of the planets that generated the comet; in both cases, the ensuing events are explained as the completely natural consequences of the ascending of burning, dry exhalations through the air.

Both Aegidius and Legnano, after their natural explanations, went on to investigate the same subject matter from an astrological point of view. Relying on Ptolemy's *Centiloquium*, with Haly's commentary on it, as well as Albumasar's *Liber de significationibus accidentium et iudiciorum*, Legnano discussed the consequences of comets according to their shapes and colours, and to the zodiacal signs in which they appeared, thus providing his audience with a completely traditional

[132] Ibidem, 248-249.

[133] See her discussion of "Comets as causes in the work of John of Legnano", *Monstrous Signs*, 208-215.

[134] Aegidius of Lessines (1264) in Thorndike, *Latin treatises*, 157: "Idcirco in hac via credimus secundum Aristotelem quod causa huius significationis sit ex eo quod cometes sit ignee consistentie qua significatur naturaliter aerem desiccari et vapores subtiles multos attrahendo consumi. . .".

[135] Legnano, *Tractatus*, 247-48: ". . .quando generantur comete elevantur exhalationes ignite, ergo tunc caliditas intensa et alique partes illius materiei remanebunt in aere ex qua generantur [venti et siccitates] cum ex eadem fiunt materia."

account of cometary significance. I find no evidence for Genuth's assertion, that "John's full-fledged natural treatment of the effects of comets in principle relegated astrological prognostication to the margins."[136]

Much less traditional was Henry of Hesse's *Questio de Cometa*, likewise occasioned by the comet of 1368. In his, quite comprehensive, tract Henry denied almost everything that his predecessors had put forward with regard to the significance of cometary appearances. Not that his scepticism was unprecedented; as we shall see in a later chapter, it was part of a very ancient tradition of astrological scepticism that was almost as old as astrology itself. Yet, Henry was probably the first to specifically apply sceptical arguments to a cometary appearance. In keeping with the fourteenth-century tendency to explain terrestrial occurrences by natural causes only, he endeavoured to prove that any prognostication based on comets was necessarily false. First of all, he did not believe that comets owed their generation to any planetary attraction, or to any special constellation. If this were the case, then a comet generated by Saturn should be a sign of cold and rain, these being effects usually associated with that planet. But this was evidently impossible, since comets were always hot and dry by nature.[137] Furthermore, nothing could possibly be deduced from the comet's colour, since that depended on the density of the matter from which it was formed, not on any affinity with planetary colours.[138] Also, comets did not follow any planet's movements; instead, they stubbornly followed their own courses, regardless of any planetary stations or retrogradations.[139] Henry thought it preposterous to associate a comet with any zodiacal sign: firstly, the names of the zodiacal signs had been assigned them quite arbitrarily by ancient poets, and secondly, the place where a comet was first seen did not coincide with the place where it ignited, so that it could not be determined with which sign it should properly be associated.[140] Systematically, Henry thus invalidated the entire comet lore of the Arabo-Ptolemaic tradition.

Neither did an explanation of cometary effects by means of natu-

[136] Genuth, *Monstrous Signs*, 214.

[137] Henry of Hesse, *Questio de cometa* (1368) in Hubert Pruckner ed., *Studien zu den astrologischen Schriften des Heinrich von Langenstein* (Berlin 1933) 89-138: 103.

[138] Ibidem, 103-4.

[139] Ibidem, 106-7.

[140] Ibidem, 110-111.

ral causation, as advocated among others by Legnano, find favour in Henry's eyes. Although he agreed that a comet consisted of terrestrial exhalations ascending towards the upper layer of air, where they ignited and were burned up, he did not believe that, in the process of ascending and burning, they exercised any action capable of provoking terrestrial effects. The only sure conclusion that could be inferred from the appearance of a comet, was that a warming power was, or had been, operative in the uppermost layers of the air. Such a power did not entail any extra heat in the lower reaches of the atmosphere, let alone on the earth.[141] At most, it could be said that the generation of a comet and atmospheric conditions such as winds, pestilential air, and earthquakes were all symptoms of one underlying cause: an abundance of unhealthy exhalations.[142] As we have seen, this was precisely Aristotle's opinion, which Henry stripped of its Arabo-Ptolemaic accretions.

10. *Sixteenth-Century Heterodoxy*

The fifteenth- and early sixteenth-century preoccupation with measurement and observation achieved another interesting result. Independently of one another, the Paduan astronomer Girolamo Fracastoro and Peter Apian of Ingolstadt, imperial astronomer to Charles V and Ferdinand I, observed that a comet's tail invariably points in a direction opposite to the sun. Apian was a very able and influential astronomer, whose main work *Astronomicum Caesareum* was widely read during the sixteenth century. He thought his discovery of the antisolar direction of comets' tails important enough to adorn the title page of his *Practica* for the year 1532 with a beautiful woodcut representing five different positions of the comet of that year in the sign of Leo, each one with its tail blazing away from the sun.

The discovery made by Apian and Fracastoro was not very spectacular, and its full import was to become clear only with the advent of the Newtonian theory, placing comets in elliptical orbits circling the sun. Yet the antisolar direction of comets' tails inspired several sixteenth-century scholars to abandon certain aspects of the Aristotelian cometary theory. After all, if comets were burning exhalations,

---

141 Ibidem, 98.
142 Thorndike, *History of Magic*, III, 493.

why should their tails have such a close relationship to the sun? Girolamo Cardano (1501-1576), a renowned and prolific writer who published works on medicine, mathematics, physics, music, philosophy, and theology, was probably the person to entertain the most un-Aristotelian views on comets of the earlier part of the sixteenth century. Among his most famous publications were two comprehensive compendia of knowledge, *De subtilitate* (1550) and *De rerum varietate* (1557), both of which included a discussion of comets.

In *De subtilitate,* Cardano, relying on optical arguments, demonstrated that no comet could consist of vapours, since the place in the sky where they were usually seen was much higher than the upper limit to which terrestrial vapours could ascend. Nor could they be burning exhalations situated in the celestial ether, because the superlunary regions did not contain any inflammable material. Furthermore, the objection that comets were vapours attracted by the power of the stars was invalid, since many comets remained visible for several months, so that the whole mass of the earth would not be enough to feed them.[143]

Instead, Cardano believed comets to be located above the moon. Although he mentioned the possibility of determining their distance from the earth by means of parallax measurements, he did not in fact proceed to make the necessary calculations, but instead argued his point in a more qualitative fashion, reminiscent of the scholastic mode of disputation. He corroborated his opinion, that comets were above the moon, by arguing that they moved slower than the moon. This argument presupposed the Aristotelian doctrine, that the further away from the earth an object was located, the slower it moved. Having thus demonstrated to his own satisfaction where comets were to be located, Cardano then addressed the question of what their physical constitution could be, given the fact that they could not possibly be made up of vapours. This is where the discoveries of Fracastoro and Apian come in. It being the case, argued Cardano, that cometary tails are always opposed to the sun, it becomes abundantly clear that comets are nothing other than globes formed in the sky and illuminated by the sun; when the sun's rays shine through the body of the comet, they form a beard or tail.[144]

Convinced by his own theory, Cardano seemed prepared to give

---

[143] Hieronymus Cardanus, *De Subtilitate Libri XXI nunc demum recogniti atque perfecti* (Basle 1554) 117.

[144] Ibidem, 118.

up the Aristotelian incorruptibility of the heavens, stating that "This [a comet] can therefore be generated in the heavens, if there can be any generation there." However, realizing that, to many scholars, this would be an untenable consequence, he added: "If this cannot be admitted, it has to be posited—which is also more likely—that the sky is full of more stars [than are visible to us] which are not very dense, and thus only show themselves to us when the air becomes dry and thin, or as a result of other causes."[145] This hypothesis, reminiscent of Seneca's exclamation, "How many other bodies besides these comets move in secret, never rising before the eyes of men!"[146], conveniently allowed Cardano to maintain the old theory that comets were signs of terrestrial changes. For, if comets become visible to us due to a desiccation and attenuation of the air, clearly the appearance of a comet heralds all the effects associated with dry air. Storms will rise, seas will be tossed by tempests, and nobles and princes, being worn out by cares, sleepless nights, and delicate food, will die. Furthermore, there will ensue drought, crop failure, changes in government and popular uprisings, all caused by the exceptional dryness of the air; the comets which then become visible can only be signs, never causes, of these effects.[147]

Cardano's discussion of comets in his 1557 sequel to *De Subtilitate*, *De Rerum Varietate*, was very similar in content to its predecessor. The author there espoused the Senecan point of view, that a comet could not be located in the perpetually changing sublunary region, because of its long duration and its regular, triple motion. Nor could it be one of the fixed stars, displaying a variety of movements; furthermore, it moved too quickly and was visible too rarely to be a planet. Nor could its motion be accounted for by positing that it was attracted by one of the planets: experience proved that their motions were far too diverse from those of the planets. Therefore, Cardano concluded that they were to be located above the moon, being formed by light collected in one part of the heaven where it increased and appeared

---

[145] Ibidem: "Hic igitur in caelo, si ibi est generatio, fieri potest; si non admittatur, dicere necesse est, quod & verius est, coelum esse sideribus pluribus, sed non admodum densis, plenum; quod cum aer siccescit & attenuatur, vel etiam aliis ex causis, oculis nostris se subijcit." My translation.

[146] Seneca, *Natural Questions*, VII.30.3.

[147] Cardano, *De Subtilitate*, 119: ". . .quae omnia, ut dixi, ab immodica tenuitate ac siccitate aeris fiunt, ut Cometes horum quidem possit esse signum, causa autem minime."

round.[148] The appearance of a tail was produced by the rays of the sun permeating this body of light, "as when a great light permeates a lesser light".[149]

According to Cardano, many more comets were generated than the ones noticed by us; they were only visible when the air was very thin, and were therefore more often seen in summer than in winter. Once again, he expounded his theory that their appearance signified an extreme thinness of the air, with all its concomitants of winds, drought, famine, pestilence, rains, and the death of people (usually princes) with little exercise, great cares, fed on delicate foods, indulging in venereal pleasures, and sleeping little. Furthermore, due to the effects of thin air on the humours of the human body, multiplying choler, people would be much inclined to seditions and wars.[150]

A cometary theory much like Cardano's was held by the royal mathematician in Paris, Jean Pena. From the antisolar direction of comets' tails, he deduced that they were transparent bodies through which the rays of the sun could be refracted. However, he did not state their superlunary position as categorically as his Italian colleague; rather, he held the more limited view that by the principles of optics, *some* comets could be shown to be supralunar.[151]

The 'luminous-body theory' was not the only sixteenth-century deviation from Aristotelian cometology. Due to the Renaissance revival of Neoplatonism, several authors returned to views reminiscent of the Stoico-Platonic outlook of late antiquity. Thus Marsilio Ficino held that, at the death of great men, comets were produced by the ethereal spirits to welcome them home.[152] And the infamous magus, Paracelsus, held the Platonic conviction that the heavens did not consist of a fifth element, but of fire, and were therefore susceptible to generation and corruption. This allowed him to believe that comets were fabricated out of the celestial matter at God's behest by the Superior Penates, who had their abode in the heavens, to be displayed before mankind as signs of future happenings.[153]

---

148 Hieronymus Cardanus, *De Rerum Varietate Libri XVII* (Basle 1557) caput I "Universum, Cometae, ventorum ratio", 2.

149 Ibidem: "...ut cum lux magna minus claram permeat."

150 Ibidem, 3-4

151 C. Doris Hellman, *The Comet of 1577. Its Place in the History of Astronomy* (New York 1944) 96.

152 Thorndike, *History of Magic*, IV, 563.

153 See Christianson's translation of Tycho Brahe's German treatise on the comet

Thus, by the middle of the sixteenth century, cometary theory seems to have come full circle; the whole spectrum of possibilities that had been suggested over the centuries was known and being contemplated by the scholarly community by the time a supernova exploded in 1572. The view of some Greek Presocratics, that comets were, at least partly, optical effects in the sky, had been revived by Apian, Cardano, and Pena; Aristotle's view of comets as purely natural, sublunary indicators of coming weather conditions, had been defended by Henry of Hesse; the Stoico-Platonic outlook of cosmic correspondences that had prevailed in late antiquity and during the long medieval centuries prior to the recovery of Aristotle, had been brought back to life by Ficino and Paracelsus. Yet it has to be stressed that these manifold opinions were held only by a small minority of scholars. The vast majority still clung to the view that had become predominant during the successful offensive of the arabicized Aristotle in the thirteenth century: comets were burning terrestrial vapours in the upper reaches of the atmosphere, attracted by one or more of the planets, whose usually pernicious effects they transmitted to the dwelling-place of man.

---

of 1577, in J.R. Christianson, "Tycho Brahe's German treatise on the comet of 1577: A Study in Science and Politics", *Isis* 70 (1979) 110-140: 133.

CHAPTER FOUR

GREEK MEETS GREEK: THE CULMINATION OF DIALOGUE

1. *Pride and Prejudice*

By the middle of the sixteenth century, the pre-modern 'dialogue' among scholars of all ages was approaching its culmination. While the later Middle Ages had seen an overwhelming predominance of the arabicized Aristotle, the humanist movement of the Renaissance, and the exodus of Byzantine scholars to the West following the fall of Constantinople, brought to the attention of Western scholars an abundance of non-Aristotelian philosophies. For the first time, the West was confronted with the whole plethora of points of view that had come into being during the long centuries of Antiquity. The writings of Plato and the Neoplatonists, the Hermetic Corpus, the sceptical works of Sextus Empiricus, and the Byzantine commentaries on the Aristotelian corpus were all recovered and intensively studied, while the ancient works that had been known for a long time, but had been neglected by late medieval scholastics due to the overzealous concentration on Aristotle (such as the Stoic writings of Seneca and Cicero) were unearthed, subsequently to become extremely popular and influential.

At the end of our previous chapter, we witnessed the re-emergence of non-Aristotelian views on comets during the first half of the sixteenth century. We have seen how several authors (albeit only a small minority) drew upon Stoic, Platonic, and Presocratic ideas to explain the nature of comets, rather than on the prevalent Aristotelian/Ptolemaic account. Scholars like Cardano, Ficino, and Pena proposed novel theories to account for cometary phenomena; yet, the novelty of their explanations owed much to the antiquity of the presuppositions that underlay them. As a true Neoplatonist, Ficino arrived at his theory of comets as produced by the ethereal spirits while searching for ancient wisdom older and more venerable than Aristotle's; and Cardano abandoned the Aristotelian view of comets as meteoric phenomena only to adopt a theory very reminiscent both of the Presocratic view of comets as optical effects, and of the Stoic supposition that the heavens may be full of bodies usually invisible to us. Moreover, he was careful not to topple the Aristotelian dogma of celestial incorruptibility, and he maintained all of the consequences attributed to comets by the Arabo-Aristotelian tradition.

Clearly, by the middle of the sixteenth century, virtually all the ancient and medieval opinions elucidated in the previous chapter had their defenders. Although cometology was for the most part still very Aristotelian in character, Presocratic, Neoplatonic, and Stoic opinions had come back into play as exponents of a venerated, ancient wisdom held in high esteem. At the crossroads of sixteenth-century Europe, several traditions that had remained alive through the centuries met and mingled, and brought the dialogue of past and present authors to an unprecedented culmination. Yet, just one century later, by the middle of the seventeenth century, the picture had changed considerably. As we saw in chapter II, while scholars like Hevelius, Riccioli, and Petit were still very much aware of the ancient body of ideas, and mentioned the opinions of many of their predecessors, they did so mainly in order to criticize them, chiding them for the inaccuracy of their observations, the childishness of their astrological 'prejudices', and the untenability of their cosmological speculations. The question of how, in no more than one hundred years, the outlook of leading scholars could have changed so considerably, has given rise to a seemingly inexhaustible amount of literature discussing the problem of the so-called 'Scientific Revolution'.[1]

Many of the contributors to this debate have fallen victim to a rhetorical snare set by several of the most influential protagonists of the 'Scientific Revolution'. Thus, René Descartes once thanked his teachers "for the fact that everything they taught me was quite doubtful; . . . [otherwise] I might have been content with the smattering of reason which I found in it."[2]; Francis Bacon in his famous *Novum Organum* proposed to transfer the epithet of 'Sophists' to each

[1] For a more detailed discussion of the historiography of the rise of modern science, and of the genesis and fortunes of the concept of a 'Scientific Revolution', and the problems such a concept has posed to historians, see especially H. Floris Cohen, *The Scientific Revolution: A Historiographical Inquiry* (Chicago 1994), and I. Bernard Cohen, *Revolution in Science* (Cambridge Mass. 1985).

[2] In *The Philosophical Writings of Descartes*, trans. J. Cottingham, R. Stoothoff, & D. Murdoch (Cambridge 1985) ii. 411; quoted in Brian P. Copenhaver & Charles B. Schmitt, *Renaissance Philosophy* (*A History of Western Philosophy*, III, Oxford 1992) 60.

and every one of the Greek philosophers, since "...the wisdom of the Greeks was professorial and much given to disputations, a kind of wisdom most adverse to the inquisition of truth... They are prompt to prattle, but cannot generate, for their wisdom abounds in words but is barren of works."[3]; and Galileo, in the guise of his *alter ego* Salviati, utterly ridiculed the most revered tenets of Aristotelian science.[4] These and similar utterances express the pride of a few of the foremost early modern scientists; the fact that they were to be taken at face value for centuries to come reveals the prejudice of their descendants, who were in a position to judge that, with hindsight, these few men were proved to have been right. They have lured generations of historians, from the Enlightenment to the present day, into believing that a radical rupture or 'revolution' occurred, ably conducted by 'great scientists' like Galileo and Descartes, after which the clouds of superstition and ignorance lifted to give way to the brilliance of science.

The success of Descartes', Bacon's, and Galileo's propaganda is well illustrated by the example of Albert Einstein, who, in his foreword to the 1953 Stillman Drake translation of Galileo's *Dialogue Concerning the Two Chief World Systems,* wrote: "A man is here revealed who possesses the passionate will, the intelligence, and the courage to stand up as the representative of rational thinking against the host of those who, relying on the ignorance of the people and the indolence of teachers in priest's and scholar's garb, maintain and defend their positions of authority. His unusual literary gift enables him to address the educated men of his age in such a clear and impressive language as to overcome the anthropocentric and mythical thinking of his contemporaries and to lead them back to an objective and causal attitude towards the cosmos, an attitude which had become lost to humanity with the decline of Greek culture."[5] Einstein here fails to realize the extent to which he himself, along with many others, has fallen victim to Galileo's "unusual literary gift": making brilliant use

---

[3] Francis Bacon, *Aphorisms,* LXXI, in Fulton H. Anderson ed., *Francis Bacon: The New Organon and Related Writings* (Indianapolis 1960) 69-70. Among the Greeks he would classify as Sophists Bacon mentions Plato, Aristotle, Zeno, Epicurus, Theophrastus, Chrysippus, and Carneades.

[4] In his *Dialogue Concerning the Two Chief World Systems* (1632) and his *Dialogues Concerning Two New Sciences* (1638).

[5] Albert Einstein, "Foreword" to *Galileo Galilei: Dialogue Concerning the Two Chief World Systems - Ptolemaic & Copernican* (translated by Stillman Drake, Berkeley/Los Angeles/London 1953; second revised edition 1967) vii.

of the rhetorical devices of humanist culture, the Italian managed to paint a persuasive yet caricatural picture of the scholarly community of his day.[6]

The influence of this seventeenth-century caricature, along with its reception and elaboration by Enlightenment and post-Enlightenment thinkers, has been enormous. As we have seen in chapter II, its stranglehold has extended well into the twentieth century, and the historiography of cometology has been especially affected by it. Authors like Montucla, Pingré, Berry, and Wolf propagated an account of the development of cometary theory that was to become extremely influential, though based on a distortion of the truth. According to this account, all natural philosophers prior to 1577 believed comets to be sublunary vapours. This ridiculous idea, originating with Aristotle and propagated by the scholastics, was overthrown by Tycho Brahe's observations of the comet of 1577, demonstrating that comet to be above the moon. After 1577, therefore, it became unfeasible to adhere to the Aristotelian theory; those who still did, often collectively referred to as 'the scholastics' or 'the Peripatetics', were retarded and superstitious fools. Furthermore, the superlunary position of the comet is supposed to have provided a powerful argument in favour of the Copernican theory: its irregular motion entailed the abandonment of the solid celestial spheres of Aristotelian cosmology, and its location above the moon had as a consequence the abolition of the dichotomy between celestial and elemental regions, thus rendering feasible the Copernican idea of the earth as "just another planet".[7]

Although, in the twentieth century, historians of science have considerably refined this version of events, it still tacitly underlies many of their works.[8] Often, their attention is focused on the few novel theories put forward in the late sixteenth and seventeenth centuries, and on the very limited number of great names associated with these novelties. Influential books such as Thomas Kuhn's *The Copernican Revolution* or Alexandre Koyré's *From the Closed World to the Infinite Universe,* by emphasizing the revolutionary character of the period, have set the tone for most subsequent historians. Although these days, everyone realizes (as, in fact, did Kuhn and Koyré) that the 'revolution' was a long-drawn-out affair, taking at least a century to

---

[6] See Copenhaver & Schmitt, *Renaissance Philosophy,* 60-61.
[7] See for a more detailed account chapter II.
[8] See chapter II.

complete, often the progressive elements are stressed at the expense of the, much more widely diffused, retardative ones.

Thus, the role of the comet of 1577 in toppling the Aristotelian world view, and in rendering viable the Copernican, is still accorded an inordinate amount of attention.[9] Recently, however, both the 'Aristotelian world view' and the nature of sixteenth-century 'Copernicanism' have, at closer scrutiny, proved to have been much less clear-cut entities than has long been assumed. In the light of these findings, it becomes doubtful whether the long-standing status of comets as agents of revolutionary change can survive without a serious reassessment.

2. *A Chameleonic World View*

Conventionally, when historians of science speak of the 'Aristotelian world view', they refer to a well-defined complex of tenets, framed by Aristotle and supposedly adhered to by all medieval scholastics. Its basic constituents were expounded by the Stagyrite in several of his works, most notably the *Physics* and the *De Caelo,* and ran along the following lines. Aristotle's cosmos was a finite and indestructible globe. It knew no beginning and would never come to an end; instead, its divine status entailed that it was to stand firm through all eternity. It was divided into two fundamentally dissimilar regions: the elemental sphere, encompassing everything situated below the moon, and the heavenly region, extending upwards from the moon's sphere up to and including the sphere of the fixed stars.

The whole of the sublunary region consisted of the four elements, earth, water, air, and fire. It was imperfect, and subject to a never-ending cycle of change and decay; by nature of their unstable, terrestrial condition, the four elements continually formed new compounds, while existing ones were dissolved. Plants, animals, and human beings came up and passed away; meteors like storms, rain, clouds, rainbows, and comets roamed about freely, later to be dissolved again. By contrast, beyond the impassable barrier posed by the sphere of the moon, things were dramatically different. There, no change ever occurred; the heavenly region was filled with a celestial fifth element or aether, not susceptible to generation or corruption.

[9] See chapter II.

This perfect substance was the stuff the stars and planets were made of; in fact, the whole of the celestial region was filled with it, leaving no empty space. The seven planets (Moon, Mercury, Venus, Sun, Mars, Jupiter, and Saturn) moved in circular orbits with a perfect regularity, brought about by an ingenious system of solid, concentric spheres carrying them around. Although each planet required several spheres to account for its particular motion, in cursory descriptions of the Aristotelian universe often only one sphere per planet is mentioned, making it possible to speak of, for example, 'the' sphere of Venus. This leaves eight celestial spheres: one for each planet, and one for the fixed stars. The whole system was kept in motion by the Prime Mover, a kind of deity who made the underlying spheres move by being loved by them.

Underlying this cosmological system were the tenets of Aristotelian physics. These entailed that an object could have only one natural motion. The aethereal celestial bodies moved in uniform and circular orbits, circularity being the motion proper to eternity, knowing no beginning or end. Aristotle himself did not address the problem that the planets do not, in reality, describe circles at all, but on occasion stand still and retrograde, thus describing a kind of loops deviating from their circular paths. Yet the astronomical system of Claudius Ptolemy did provide a solution for this intolerable irregularity: he retained the circularity of planetary motion by means of an ingenious system of eccentrics and epicycles.[10] Thus, he provided the Aristotelian world view with the celestial circularity it postulated, while at the same time making possible accurate predictions.

By contrast with the stars and planets, the elemental bodies were naturally endowed with rectilinear motion: if unimpeded, the heavy elements earth and water would move downward, towards the centre of the universe, while the light elements air and fire tended to move upward, towards the periphery. Because earth was slightly heavier than water, the centre of the universe was occupied by a ball of conglomerated earth, with water covering the surface of this earthy globe in the form of seas and oceans. Similarly, fire being lighter

---

[10] In order to make his system work, Ptolemy, though retaining circular motion, was forced to abandon the Aristotelian dogma of uniform motion by introducing the equant point. See for a more detailed discussion for example Michael J. Crowe, *Theories of the World from Antiquity to the Copernican Revolution* (New York 1990) 43-52, or E.J. Dijksterhuis, *De mechanisering van het wereldbeeld* (Amsterdam 1950) 59-67. See below § 3.

than air, it ascended further than the highest of the airy regions, all the way up to the concavity of the lunar sphere, where it formed a sphere of fire. This description of the sublunary region makes it clear why Aristotle's cosmos was essentially geocentric: the inexorable laws of Aristotelian physics entailed that earthy substance would invariably tend towards the centre of the universe, which was the natural place of heavy bodies.

The above survey describes the basic outlines of the Aristotelian world view, or, as Edward Grant has called it, "the overall structural framework of the world as it was popularly conceived in the late Middle Ages".[11] Three basic tenets made up the very core of this Aristotelian cosmology: geocentricity, the fundamental dichotomy between celestial and terrestrial regions, and the uniformity and circularity of celestial motions. Giving up any one of these tenets would pose a serious threat to the system as a whole, and astronomical developments around the year 1600 severely put it to the test. Yet, as we shall see, the skeletal frame of the Aristotelian world view proved more resilient than the rigid interdependence of its constituents would suggest.

In fact, the world system devised by Aristotle never prevailed in medieval Europe in its original form. From the very moment it became known to the thirteenth-century Latin West, it was subjected to revisions and additions aimed at rendering it compatible with a Christian outlook. Thus, Aristotle's Prime Mover was equated with the Christian God, and consequently endowed with the qualities assigned to him by theology, such as immobility, immutability, infinite power, indivisibility, and absence of magnitude.[12] Also, a convenient location for the Christian Heaven, dwelling-place of the Blessed, was devised within the Aristotelian framework: it went under the name of 'Empyrean heaven' and was thought of as an invisible, immobile orb that enclosed the whole world, the final sphere of the Aristotelian cosmos. And often, scholastics assigned the motions of the planetary spheres to the agency of angels or intelligences associated with the celestial orbs.[13]

A more fundamental dilemma was posed to the medieval school-

---

[11] Edward Grant, "Aristotelianism and the Longevity of the Medieval World View", *History of Science* 16 (1978) 93-106: 94.

[12] Edward Grant, *Planets, Stars, and Orbs. The Medieval Cosmos, 1200-1687* (Cambridge 1994) 519.

[13] Ibidem, 523-527.

men by the Aristotelian eternity of the world, a tenet obviously incompatible with the account of the Creation in Genesis. They reacted to this problem in a variety of ways. Some, such as Saint Bonaventure, attempted to show that temporal creation of the world was demonstrable. Some held the heretical view, condemned in 1277, that the matter from which the world was created, was eternal, although, in its present configuration, the world was created. Another solution was put forward by Alexander of Hales and Roger Bacon, who assumed that time began when God created the first motion, and that eternity is equivalent to the whole duration of time. Finally, in an argument that was to prove very influential, Thomas Aquinas asserted that God, in his absolute power, need not precede his effect in duration; therefore, something could be created by God and yet have had eternal existence.[14]

Thus, scholastic authors proved remarkably resourceful in reconciling the Aristotelian world view to Christian dogmas. Many more examples of their inventivity can be found in the work of Edward Grant, who has shown that medieval scholastics were forced by the 1277 condemnation to deviate from Aristotle on a variety of issues which, in their Aristotelian guise, were unacceptable to Christian theology.[15] One striking example of such a departure was the idea held among others by Buridan and Oresme, that God could, if he wished, create other worlds; each world would then have its own physical centre, and each heavy body would fall to the centre of its own world.[16] Hence, from the outset, the reception of Aristotle obliged medieval Christian scholars to contemplate possibilities never conceived of by the Stagyrite, or even flagrantly contrary to Aristotelian principles, and to incorporate these into the 'Aristotelian world

---

[14] See ibidem, 67-77.

[15] See Edward Grant, *Planets, Stars, and Orbs*, passim; Edward Grant, "Ways to interpret the terms 'Aristotelian' and 'Aristotelianism' in medieval and Renaissance natural philosophy", *History of Science* 25 (1987) 335-358; and Edward Grant, "The Condemnation of 1277, God's Absolute Power, and Physical Thought in the late Middle Ages", *Viator* 10 (1979) 211-244. Among the issues incompatible with Aristotelian cosmology that were discussed by scholastics after 1277, Grant ranges the possibility of a plurality of worlds (entailing the idea of a multiplicity of equally privileged physical centres); the existence of extracosmic void space; the possibility of God moving the cosmos with rectilinear motion; the untenability of animated celestial bodies; the system of solid eccentric orbs to replace Aristotle's concentric ones; the existence of interorbicular matter, capable of change; and the spiral, as opposed to perfectly circular, motions of the planets.

[16] Grant, "Ways to interpret", 340.

view'. Although, in the words of Grant, the "overall structural framework of the world" remained the same over the whole period from 1200 to 1687, the "second significant aspect of a world view, namely the details of cosmic operations", were conceived of in a staggering variety of ways.[17]

Apart from the need of adapting Aristotle's cosmos to the postulates of Christian faith, several other factors were conducive to turning Aristotelianism into a multi-faceted body of thought capable of incorporating a variegated range of opinions. In the first place, Aristotle's writings were often unclear and ambiguous; this obscurity induced the scholastics, who were much concerned about details, to supplement the Stagyrite's views with a great number of additions of their own, often altering Aristotle's original intent in the process. Secondly, Latin schoolmen were confronted not only with Aristotle's dicta, but also with the commentaries of their Greek, Byzantine, and Arabic predecessors. These were often at variance with one another, and allowed Latin scholastics either to choose one of the number of solutions already proposed, or else to fashion new solutions of their own.[18] Finally, and perhaps most importantly, the very method of scholasticism favoured the incorporation of many different points of view into the overall framework of Aristotelianism. Scholastic commentaries on authoritative texts usually included extensive lists of thoroughly substantiated arguments on both sides of a particular question. Thus, "by its very nature, the *questio* form encouraged differences of opinion."[19]

The abundance of options scholastics could choose from, along with the fact that each question was usually treated in relative isolation from other, related issues[20], gave rise to a kaleidoscopic range of 'Aristotelianisms'.[21] As Grant has shown, it is very difficult to establish any definite criteria under which all 'Aristotelians' can be sub-

---

[17] Grant, "Aristotelianism and the longevity of the medieval world view", 94-95. On the fashion of hypothetical reasoning among late medieval scholastics, see also Amos Funkenstein, *Theology and the Scientific Imagination from the Middle Ages to the Seventeenth Century* (Princeton 1986) ch. 3.

[18] Grant, "Aristotelianism", 97.

[19] Ibidem, 98.

[20] Ibidem: ". . .what emerged was a series of distinct and often intensively considered problems that remained isolated from, and independent of, other related *questiones*, to which allusions and references were minimal."

[21] The plural form 'Aristotelianisms' to describe the variety of opinions among Latin scholastics was coined by Charles Schmitt in his *Aristotle and the Renaissance* (Cambridge, Mass./ London 1983) 10.

sumed.[22] Each thinker responded in his own unique way to the challenge posed by the tension between Aristotle and Christianity, and between Aristotle and his commentators. Grant therefore proposes to liken Aristotelianism to the biological concept of a zoological species as a population. Thus, "there are as many Aristotelianisms as there were individual Aristotelian natural philosophers. Construed as a species . . . Aristotelianism consists of the whole population of individual Aristotelianisms each of which is the unique product of a single Aristotelian natural philosopher whose ideas, opinions and interpretations are shared with other members of the species, that is, other Aristotelian natural philosophers. An Aristotelian natural philosopher is trained in a tradition and reveals this by repeating some of what he has learned. As a unique member of the species, however, he will also add to, detract from, or alter, however slightly, the sum total of his Aristotelian inheritance... There is no need for a definition of Aristotelianism..., no norm against which to measure whether a departure has occurred. In this sense, *there are no departures or anomalies*."[23]

This Grantian approach is very illuminating. It can contribute towards explaining the longevity of the 'Aristotelian world view': precisely because rival and even incompatible theories could relatively easily be adopted by individual Aristotelians, the overall structure of the Aristotelian cosmos was capable of standing firm. Drawing on centuries of experience in reconciling Aristotelian tenets to external considerations at variance with them, Renaissance natural philosophers strove to incorporate the manifold new discoveries of the age into their individual world views.

Yet, during the sixteenth century, the pliancy of Aristotelianism was stretched to its limit. As we have seen in the previous section, the momentous uncovering operation undertaken by Renaissance humanists once again brought to the centre of attention sources of ancient philosophies other than Aristotelianism, that had either been totally unknown to the West, or else had fallen into oblivion. Preeminent among these rival world views was, interestingly enough, precisely the philosophy that had been the main influence on cometology for centuries: Stoicism.

Peter Barker has shown how interest in Stoicism, and especially in the Stoic ideas found in Cicero's *On the Nature of the Gods,*

---

[22] See his "Ways to interpret the terms 'Aristotelian' and 'Aristotelianism'".

[23] Ibidem, 348; Grant's italics.

surged during the sixteenth century.[24] It has long been recognized that Stoicism exercised a considerable influence on Renaissance politics, ethics, and literature; however, the impact of Stoic ideas on natural philosophy has been largely ignored. This neglect is rather odd, since Stoicism advocated an understanding of nature as basic to an ethically sound life.[25] Or, as J.L. Saunders put it in the introduction to his work on the Stoicism of Justus Lipsius, "Stoicism, as a philosophical system, can be formulated, in a very general fashion, as follows: That morality consists in using every endeavor to obtain an ultimate end which is in accordance with Nature, and that this endeavor, even if it should be unsuccessful, is itself the sole thing desirable and the sole good; that ethics is the chief aim, but not the only concern (for the early Stoics); that physics is a necessary foundation for valid moral theory; and that this central ethical doctrine implies and postulates all physical notions."[26] Whereas, during late Antiquity and the Middle Ages, the ethical elements of Stoic theory were increasingly emphasized at the expense of the physical ones, Renaissance humanism endeavoured to restore Stoicism, like other philosophies, to its pristine state. Although the first to do so systematically was Justus Lipsius, who published his compendia of Stoic philosophy and his edition of the works of Seneca in the early years of the seventeenth century[27], the renewed interest in Stoicism had begun as early as the fourteenth century; by the early sixteenth, Stoic ideas had become widespread.[28]

Like Aristotelianism, Stoicism is very hard to characterize unequivocally due to the manifold differences of emphasis among its adherents. Thus, the very founding fathers of Stoicism, Zeno, Cleanthes, and Chrysippus, disagreed on a number of issues; and thinkers like Seneca and Cicero combined their Stoicism with a range of non-Stoic ideas. Still, the central tenets of Stoic cosmology were usually

---

[24] Peter Barker, "Stoic contributions to early modern science", in Margaret J. Osler ed., *Atoms, 'pneuma', and tranquillity. Epicurean and Stoic themes in European thought* (Cambridge 1991) 135-154.

[25] Ibidem, 137.

[26] Jason Lewis Saunders, *Justus Lipsius. The Philosophy of Renaissance Stoicism* (New York 1955) xiii.

[27] Notably his *Manuductio ad Stoicam philosophiam* and his *Physiologia Stoicorum* (1604), and his edition of the works of Seneca (first volume appeared in 1605).

[28] Barker, "Stoic contributions", 141-142; Barker points out the attention paid by Petrarch to the Stoic ideas contained in Cicero's *Nature of the Gods*, as well as the decidedly Stoic elements in the outlook of Rabelais, who also used the *Nature of the Gods* as a source.

agreed upon; they ran along the following lines. The Stoics' most important deviation from Aristotle's cosmos lay in their rejection of the fundamental dichotomy between heaven and earth. At the heart of Stoic cosmology lay the concept of *pneuma*, sometimes also referred to as *aether* or *fire*. This pneuma was a substance pervading the whole cosmos, and responsible for preserving its different parts. It consisted of a combination of fire and air; vital to its preserving function was the constant tension, within the pneuma, between the warm and expanding qualities of fire, and the cold, contractive qualities of air. Thus, the outward movement stimulated by the fiery constituent of the pneuma was responsible for the external dimensions and qualities of bodies, while the inward movement brought about by the airy constituent explained their cohesion, unity, and internal stability. Due to the all-pervading presence of the pneuma, keeping the cosmic whole together and affecting every part of it, the different parts of the cosmos displayed a mutual 'sympathy'; the Stoic cosmos, far from being divided into two fundamentally different regions, was a single, living entity.[29]

Furthermore, the universe of the Stoics was not eternal in the sense Aristotle's was. The world, taken as the whole of nature, was, indeed, everlasting; but taken to mean 'everything composed of the four elements', it was subject to periodic destruction and change.[30] At the end of each 'Great Year'[31], there would occur a cosmic conflagration, with the divine fire absorbing the whole cosmos and reducing it to a mass of pure fire or pneuma. After a period of time, this fire would change to air, and then to water, from which the familiar four elements of earth, water, air, and fire would be produced. These would then form a geocentric cosmos with earth at the centre, water above earth, air above water, and above that, an increasingly pure pneuma extending all the way to the sphere of the fixed stars. Thus, "the Stoics offer a single, integrated physics of the heavens and the earth, in contrast to the sharp separation of the substance and physics of the heavens from the region below the moon in Aristotle and later Scholastic cosmology."[32]

Moreover, whereas Aristotelian physics postulated that fire, like

---

[29] M.R. Wright, *Cosmology in Antiquity* (London/New York 1995) 120-121.

[30] Saunders, *Justus Lipsius*, 203.

[31] The term 'Great Year' could, among other things, refer to the period it takes the planets to return to exactly the same position they held at an earlier time.

[32] Barker, "Stoic contributions", 138.

the other elements, could only possess rectilinear motion, and that, therefore, the celestial bodies, moving in circular orbits, could not consist of fire, the Stoics maintained that the stars and planets were, in fact, celestial fires. Indeed, the whole of heaven was filled with the fluid pneuma, instead of the solid celestial spheres conceived of in Aristotelian cosmology. The celestial bodies, as well as the cosmos itself, were thought of by the Stoics as living beings, endowed with intellect and rationality; indeed, because they consisted of the most pure aether, they possessed superior intelligence.[33]

The celestial fires drew their nourishment "from the vapours of the earth and the sea and rivers, which rise from the fields and waters when they are heated by the sun."[34] These terrestrial materials were carried out to permeate the heavens during the outward movement of the pneuma, continually circulating from the centre of the cosmos to its periphery and back again; on the inward part of its journey, the pneuma conveyed the influence of the heavens to the inhabitants of the earth, thus providing the physical underpinning for astrology and divination. Once all terrestrial matter had been used up in feeding the celestial fires, the sun absorbed the other heavenly bodies and the whole cosmos returned to undifferentiated pneuma, constituting the cosmic conflagration described above.[35]

These basic tenets of Stoic cosmology were revived by several humanists, for instance Petrarch, from the fourteenth century onward. By the middle of the sixteenth, they had become very widespread in works of literary culture. Moreover, as Barker has shown, Stoic notions, or at least those dealing with the substance of the heavens, also began to gain currency in sixteenth-century 'scientific' debates: such ideas even "...became an important source of alternatives to the Aristotelian position."[36] Due to the influence of the Stoic view that the heavens consist, not of a fifth essence, but of pneuma, that is, a combination of air and fire indistinguishable from the air and fire found in the terrestrial region, a number of natural philosophers, by the middle of the sixteenth century, began to weaken the distinction between celestial and terrestrial matter. Jean Pena, a pupil of Ramus, in 1557 published an essay in which he defended the existence of a

33 Ibidem, 139; Cicero, *On the Nature of the Gods*, II.42.

34 Cicero, *On the Nature of the Gods*, II.117 (translated by Horace C.P. McGregor, Penguin Classics, London 1972) 171.

35 Barker, "Stoic contributions", 139.

36 Ibidem, 143.

single medium, air, extending from the surface of the earth all the way to the fixed stars.[37] Barker believes this innovation to have been suggested to Pena by Cicero's account of Stoicism in *The Nature of the Gods*; both testimonies by Pena's contemporaries and textual evidence support this.[38]

Pena was by no means the only sixteenth-century author to consider the similarities between celestial and elementary matter. Bernardino Telesio argued that the whole universe was filled with an unqualified matter, which was acted upon by the heat of the sun and by the cold of the earth; Telesio saw the universal matter as a vehicle for spreading these qualities to all parts of the world. Thus, he abandoned the dichotomy between heaven and earth, and he did not deem the heavens incorruptible.[39] Girolamo Cardano, who, as we have seen, assigned comets a celestial location as early as 1550, like Pena believed air to be the medium filling the heavens. Also, he denied fire to be one of the four elements; it arose from heat, which was produced by motion, and was not, therefore, a state, but a process.[40] And Francesco Patrizi rejected the traditional four elements in favour of a five-element theory. The new elements, space, light, heat, 'antitypia', and flux, were subject to a constant confluence; Patrizi made no distinction between the celestial and the elementary.

Even a number of otherwise traditional scholastics adapted their Aristotelianism to allow for the material similarity of the celestial and the elemental region. Thus, Andrea Cesalpino in his *Quaestionum peripateticarum libri V* (Florence 1569), claiming to represent the true opinions of Aristotle, stripped of their scholastic accretions, conceived of the heavens as one single continuum, and regarded all matter as of one kind.[41] And the very influential Coimbra Commentators nearly identified celestial with elemental matter.[42]

In so doing, they drew on a respectable scholastic tradition, associated with the names of Aegidius Romanus and William of Ockham, who had already argued that terrestrial and celestial matter were of the same kind. Yet, they modified their predecessors' reasoning,

---

37 Jean Pena, preface to the edition of Euclid's *Optics and Catoptrics* (Paris 1557); see Barker, "Stoic contributions", 143-144.

38 Barker, "Stoic contributions", 144.

39 William H. Donahue, *The Dissolution of the Celestial Spheres 1595-1650* (New York 1981) 50-51.

40 Ibidem, 52-53.

41 Ibidem, 54.

42 Ibidem, 55.

which had rested on purely scholastic arguments invoking such categories as potency, actuality, and privation[43], to arrive at more radical consequences than those dreamt of by Aegidius and Ockham. For whereas the latter had strongly upheld the doctrine of celestial incorruptibility in spite of the identity of celestial and terrestrial matter, some sixteenth-century scholars made the more far-reaching move of contending that not only were the two matters of the same kind, but they also possessed the same properties, including corruptibility.

Between 1570 and 1572, Robert Bellarmine, later to become a Cardinal, delivered a series of lectures at the university of Louvain, one of which included a commentary on Genesis. By means of this highly traditional medium, he aired a number of ideas quite incompatible with the core tenets of Aristotelianism. Relying both upon Stoic writers of antiquity and upon the authority of the Church Fathers, Bellarmine argued that the heavens were corruptible and that they consisted of fire.[44] As a consequence, he likened the stars and planets to living beings, moving of their own volition:

> If we wish to hold that the heaven of the stars is one only, and formed of an igneous or airy substance, an hypothesis which we have declared more than once to be in accord with the Scriptures, we must then of necessity say that the stars are not transported with the movements of the sky, but they move of themselves like the birds of the air or the fish of the waters.[45]

Such ideas clearly betray their Stoic origin. One continuous, fiery heaven, corruptible like terrestrial matter, peopled by living stars moving like fish swimming in the sea: all this is very much in accord with the outline of Stoic cosmology as described above. It is also very alien to what has traditionally been labelled the 'Aristotelian world view'. Evidently, even before Tycho Brahe's revealing observations of the nova of 1572 and the comet of 1577, a number of thinkers

---

43 For the arguments employed by Aegidius and Ockham in support of the identity of celestial and terrestrial matter, see Grant, *Planets, Stars, and Orbs*, 254-259.

44 Among other authors whose work betrays the influence of Stoic ideas on the nature of celestial matter was Cornelius Valerius Veteraquinas (1567); see William H. Donahue, "The Solid Planetary Spheres in Post-Copernican Natural Philosophy" in Robert S. Westman ed., *The Copernican Achievement* (Berkeley 1975) 244-275: 248 n. 23.

45 U. Baldini & G.V. Coyne eds. & trans., *The Louvain Lectures of Bellarmine and the Autograph Copy of his 1616 Declaration to Galileo* (= Studi Galileiani, vol.1, no.2; Vatican City, 1984) 20. Quoted by Barker, "Stoic contributions", 145.

were favourably disposed towards the idea of one single heaven, consisting of corruptible, airy or fiery, matter. By no means all of these authors rejected the whole of their Aristotelian outlooks: although cosmologies like those of Patrizi or Telesio were rather unconventional, scholars such as Cardano, Pena, and Bellarmine continued to adhere to many tenets of the traditional world view, but, on the authority of such highly esteemed authors as the ancient Stoics and the Church Fathers, replaced some of the Aristotelian postulates with allegedly more pristine ideas; in the words of Donahue, often "they felt their modifications to be simple corrections of peripatetic philosophy."[46]

We can conclude that the 'Aristotelian world view', as traditionally conceived by historians of science, never actually existed. Not even during the Middle Ages did there exist one monolithic block of scholars who all adhered to exactly the same, Aristotelian, tenets. Instead, scholastic authors were used from the outset to making their own choices and substituting non-Aristotelian beliefs for Aristotelian ones if the postulates of Christian faith, or the Church, dictated them to do so. It is true to say, however, that the modifications to the Aristotelian ideal type brought on by the Renaissance revival of ancient philosophies, particularly Stoicism, were more far-reaching and potentially lethal to the Aristotelian skeletal frame than the medieval deviations had been. Especially the idea of the corruptibility of heavenly matter came dangerously close to undermining the hard core of scholastic cosmology. If endorsed systematically, it could potentially lead to the overthrow of one of the Aristotelian system's three basic assumptions: the fundamental dichotomy between celestial and terrestrial regions.

3. *"Doubt that the Sun doth Move"*

The dogmas of Christian faith were not the only factors external to cosmology proper to have exercised a strong influence on the way in which the Aristotelian cosmos was conceived. Although its structure was appealing in its simplicity and plausibility, Aristotle's world view from the outset suffered from one important flaw. The Stagyrite's system, composed of regular homocentric spheres all centred on the

[46] Donahue, *Dissolution,* 59.

earth, as such yielded no predictive accuracy. Although, in theory, it conformed remarkably well to Plato's postulate that all the heavenly motions should be accounted for (or 'saved') by means of uniform circular motions, for working astronomers it proved useless. Therefore, a number of Greek astronomers (Apollonius of Perga, Hipparchus, and most notably Claudius Ptolemy), to enhance the predictive value of astronomy without giving up Plato's uniform circular motions, devised ingenious systems to save the planetary phenomena. In doing so they made extensive use of devices such as eccentrics and epicycles; Ptolemy in addition introduced the equant point, thus sacrificing the principle of uniform linear velocity in the circumference of the deferent circle.[47]

Thus, the Ptolemaic system, accounting for the astronomical phenomena, was incompatible with the Aristotelian system prevailing in natural philosophy. In late antiquity, this difficulty was overcome by denying the Ptolemaic eccentrics and epicycles any physical reality. Scholars such as Proclus, Simplicius, and John Philoponus, as well as Ptolemy himself, all held that there was no need for astronomical hypotheses to conform with physical fact: natural philosophy dealt with physical considerations such as the substances of the heavenly bodies and the causes of celestial phenomena, while astronomy was only concerned with predictive, though imaginary, hypotheses.[48] This distinction was reflected in the curriculum of the medieval universities, where astronomy, as one of the four sciences of the *quadrivium,* was taught by mathematicians, while the physical components of the cosmos were treated by professors of natural philosophy, who enjoyed a higher status.

Yet, the separation of the two sciences should not be exaggerated. Nicholas Jardine has pointed out that many medieval scholars, following the lead of Thomas Aquinas, recognized astronomy to be a *scientia media,* "a branch of mathematics having special affinities with

[47] The equant point is a point about which a planet travels at uniform *angular* velocity. Because the equant was not situated in the centre of the circle on which the planet moved, the planet's *linear* velocity was not uniform.

[48] See Ralph M. Blake, "Theories of hypothesis among Renaissance astronomers" in Ralph M. Blake, Curt J. Ducasse and Edward H. Madden eds., *Theories of Scientific Method* (Seattle 1960) 22-49: 23; and Michael J. Crowe, *Theories of the World from Antiquity to the Copernican Revolution* (New York 1990) 69-73. See also Pierre Duhem's famous essay, *To Save the Phenomena: An Essay on the Idea of Physical Theory from Plato to Galileo* (trans. Chicago 1969; first published in French 1908).

natural philosophy."[49] This point of view had ancient precedents; in his *Hypotheses planetarum*, Ptolemy himself had made an attempt at embedding his geometrical system in physical reality, and Aristotle, too, had realized the intermediary status of astronomy. In the later Middle Ages, a compromise was devised to actually, physically, incorporate the Ptolemaic eccentrics and epicycles into the concentric spheres of Aristotelian cosmology. This middle course, referred to by Jardine as the 'Theorica Compromise' because it was advocated in Georg Peurbach's influential handbook of astronomy, the *Theoricae novae planetarum,* held that the Ptolemaic epicycles and eccentrics were real substantial spheres called 'partial orbs', with the partial orbs of each planetary model embedded in an equally substantial 'total orb', a spherical shell centred on the earth. These total orbs were then identified with the Aristotelian concentric orbs.[50]

Thus, within the framework of the Aristotelian world view efforts were made to reconcile devices known to be predictively accurate to physical reality. Yet, in the course of the sixteenth century, dissatisfaction with the Theorica compromise grew. Once again, an important factor inducing some scholars to break with their immediate predecessors was the humanistic spirit of return to the pristine purity of classical precedent. Thus, in the early decades of the sixteenth century, scholars like Girolamo Fracastoro and Giovanni Amici proposed a restoration of Aristotle's original, homocentric world system. They abandoned the Ptolemaic eccentrics and epicycles, but retained their belief in the physical reality of the total orbs. Some of their contemporaries preferred a more radical approach. Gioviano Pontano, minister to the Neapolitan kings, under the influence of the sixteenth-century Stoic and Neoplatonic revival advocated an astrological cosmology, according to which "the spiritual, 'formal' and rational heavens are the domain of the stellar gods, who move of their own free wills in accordance with the dictates of a providence immanent in nature."[51] Like such authors as Patrizi, Cardano, and

---

[49] Nicholas Jardine, "Scepticism in Renaissance Astronomy: A Preliminary Study" in Richard H. Popkin & Charles B. Schmitt eds., *Scepticism from the Renaissance to the Enlightenment* (Wiesbaden 1987) 83-102: 85.

[50] Ibidem; see also Nicholas Jardine, "The Significance of the Copernican Orbs", *Journal for the History of Astronomy* 13 (1982) 168-193: 170; and Edward Grant, "Cosmology" in D.C. Lindberg ed., *Science in the Middle Ages* (Chicago 1978) 265-302: 280-84.

[51] Jardine, "Scepticism", 89.

Pena, Pontano believed the heavens to be a single whole, with the planets moving of their own volition without the need of any physical devices to keep them going.

Clearly, by the middle of the sixteenth century, cosmology and astronomy, and especially the area where they met, the theory of the celestial substance and the nature of the orbs, were in a state of turmoil. Although most natural philosophers probably still adhered to the Theorica compromise, radically different, non-Aristotelian theories of the substance of the heavens were being put forward. Moreover, speculative cosmologers were not the only ones to defect from the Theorica compromise. Working astronomers increasingly adopted the stance, advocated by Ptolemy and some of his immediate successors, but largely forgotten due to the predominance of the Theorica compromise, of denying astronomical hypotheses any physical reality whatsoever. They held that, in order to 'save the phenomena', it was permitted to construct planetary models which were, physically speaking, untrue, since the real description of the heavens was given by means of Aristotle's concentric spheres. This attitude became more and more widespread in the course of the sixteenth century; Jardine calls it the 'pragmatic compromise' because it was "designed to stave off conflict [between Aristotle and Ptolemy] and, in didactic contexts, to forestall the asking of awkward questions."[52]

It was not just the conflict between the Aristotelian concentric spheres and the Ptolemaic eccentrics and epicycles that caused a certain measure of unrest among sixteenth-century astronomers. Perhaps an even more pressing problem was posed by the increasingly inaccurate predictive value yielded by the astronomical tables based on the Ptolemaic system. This flaw had serious repercussions in a variety of practical contexts, but first and foremost it posed a serious obstacle to the long overdue reform of the calendar. In this light it is hardly suprising that, in order to propagate his system, Copernicus, in his preface to the Pope at the beginning of *De Revolutionibus,* wrote: "To them [astronomers] my work too will seem, unless I am mistaken, to make some contribution also to the Church, at the head of which Your Holiness now stands. For not so long ago under Leo X the Lateran Council considered the problem of reforming the ecclesiastical calendar. The issue remained undecided then only because

[52] Ibidem, 93-94.

the length of the year and month and the motions of the sun and moon were regarded as not yet adequately measured."[53]

During the thirty-odd years Nicholas Copernicus spent contriving and refining his heliocentric world system, between roughly 1510 and 1543, he, like all his contemporaries, saw himself faced with a large measure of uncertainty in matters celestial. Still quite widespread was the late medieval Theorica compromise, 'physicalizing' the Ptolemaic eccentrics and epicycles to enable natural philosophers to incorporate them into their Aristotelian world views; this middle course, however, was gradually ceding ground to the fictionalist view, that the eccentrics and epicycles were nothing but calculatory devices, and that, in reality, the heavens were filled exclusively with Aristotle's homocentrics; and, finally, a new direction was taken by a number of thinkers influenced by the Stoico-Platonic revival, who abandoned any form of Aristotelian sphere and, instead, advocated a single, fluid heaven, in which the planets moved of their own volition, and in which cosmic harmonies and providence played a large part.

These, then, were the strands of thought Copernicus saw himself faced with. It is obviously extremely difficult to pinpoint exactly what induced him to give up the dogma of geocentricity cherished by virtually all his contemporaries as constituting one of the core tenets of the Aristotelian, as well as the Stoic, world system; yet, several contributory factors can tentatively be advanced. Copernicus himself in his preface tells us that he was encouraged in giving up the traditional stability of the earth by "the realization that astronomers do not agree among themselves in their investigations of this subject."[54] The data yielded by their calculations are quite uncertain, and, moreover, they employ a variety of methods at odds with each other, with observations, and with the first principles of natural philosophy. Some employ homocentrics, a physically irreproachable system; yet "by this means they were unable to obtain any incontrovertible results in absolute agreement with the phenomena."[55] On the other hand, those employing eccentrics did arrive at observationally more or less correct data; "but meanwhile they introduced a good many

---

[53] Nicholas Copernicus, "Preface to Pope Paul III" in *De revolutionibus orbium coelestium;* translated by Edward Rosen in Jerzy Dobrzycki ed., *Nicholas Copernicus On the Revolutions* (Baltimore 1978); quoted in Crowe, *Theories of the World,* 105.

[54] Copernicus, ibidem, quoted in Crowe, *Theories of the World,* 107.

[55] Ibidem.

ideas which apparently contradict the first principles of uniform motion", and, moreover, they fail to elicit "the structure of the universe and the true symmetry of its parts."[56] Contemplating this "confusion in the astronomical traditions"[57], and taking into consideration the fact that, according to Cicero and Plutarch, several of the ancients had believed the earth to be in motion, Copernicus, too, began to consider the mobility of the earth; after all, he says, "others before me had been granted the freedom to imagine any circles whatever for the purpose of explaining the heavenly phenomena".[58]

Copernicus, then, tried to render his procedure acceptable by affiliating it with several of the traditions flourishing during the sixteenth century. He claimed to be able to demonstrate the Stoico-Platonic structure and harmony of the universe, while at the same time improving the accuracy of astronomical predictions; furthermore, he appealed to the authority of the ancients, and to the fictionalist view, current in contemporary astronomy, that any mathematical model was acceptable to explain the heavenly motions. Of all the arguments in favour of his theory, Copernicus probably regarded the 'structure and harmony' argument the most important, this being the only ground on which he could base his realist interpretation of the heliocentric hypothesis.

Copernicus ventured to bring to an end the confusion reigning in astronomy by demonstrating that the order and size of all the planets and spheres, and the correlation of the different parts of the universe, could be explained by linking the motions of the other planets to the orbiting of the earth.[59] In his opinion, the heliocentric hypothesis yielded several important assets. First of all, it dispensed with the Ptolemaic equant. From Copernicus's point of view, the introduction of equants into astronomical models was an inadmissible intervention, violating the cherished Aristotelian dogma of uniform circular motion, and derogating from the universal harmony. Ptolemy had introduced the equant point for the sole purpose of rendering more accurate the predictions yielded by his system; in so doing, he had sacrificed the uniformity of the planets' linear velocity. By situating the sun, rather than the earth, in the centre of the universe, Coper-

[56] Ibidem.
[57] Ibidem.
[58] Ibidem, quoted in Crowe, *Theories of the World*, 108.
[59] Ibidem.

nicus was able to save the uniformity and circularity of celestial motions, a postulate that obviously meant more to him than that other core tenet of Aristotelianism, geocentricity. Furthermore, the Copernican system was capable of not only accounting for, but also explaining the phenomenon of bounded elongation, the observed fact that the zodiacal positions of Mercury and Venus are always within 28° and 46° respectively of the sun. Also, by placing the sun in the centre of the universe, the apparent retrogression of the planets could be both saved *and* explained.[60]

Despite all these advantages, Copernicus for a long time felt hesitant about publishing his work. As he himself was later to admit in his preface to the pope, he feared the scorn of the philosophers on account of the novelty and unconventionality of his opinion.[61] This is not to say that the Copernican system was novel because it contemplated the possibility of situating the sun in the centre of the universe. As Amos Funkenstein has shown, both the motion of the earth and the abandonment of one absolute centre of the universe were actively discussed during the Middle Ages.[62] Both Buridan and Oresme reduced the arguments for and against the diurnal rotation of the earth to the level of probability alone, not absolute certainty, so that "the impossibility of an axial rotation for the earth became a mere improbability."[63] And the possibility of a plurality of worlds, leaving the universe without an absolute centre, was very popular with medieval scholastics, and induced Ockham, for one, to argue that, even within an orthodoxly Aristotelian universe, so-called 'natural places' were relative, because "a body moving 'naturally' upwards towards a point within its natural environment (a periphery) can be said to move away from the opposite points therein."[64]

Yet, if, to medieval Schoolmen, the motion of the earth and the relativity of place were not impossibilities, they certainly were improbabilities; no-one thought it likely that the earth really was in motion, or that our world really was only one of many.[65] What Co-

---

[60] See Crowe, *Theories of the World*, 92-94.

[61] Copernicus, "Preface", quoted in Crowe, *Theories of the World*, 106.

[62] Amos Funkenstein, "The Dialectical Preparation for Scientific Revolutions. On the Role of Hypothetical Reasoning in the Emergence of Copernican Astronomy and Galilean Mechanics" in Robert S. Westman ed., *The Copernican Achievement* (Berkeley 1975) 165-203.

[63] Ibidem, 193.

[64] Ibidem, 195.

[65] Ibidem, 198.

pernicus did, and what made him fear the scorn of his contemporaries, was to assert that what had always been regarded as highly improbable was actually true. Like the idea of the similarity of celestial and terrestrial matter, that had been discussed and accepted during the Middle Ages without taking the obvious step of granting the corruptibility of the heavens, the medieval arguments in favour of the earth's rotation and the relativity of motion were taken to their logical conclusion. Where, traditionally, questions about celestial corruptibility, the movement of the earth, and the absence of one absolute centre of the universe had been responded to with a resolute "no", at least some sixteenth-century thinkers now endeavoured to answer them in the affirmative.

However, the latter approach was still very much a minority view. Copernicus himself, and his adulatory pupil and propagator, Georg Joachim Rheticus, were for a long time to remain the only ones to regard the heliocentric system as physically true. In view of the intellectual context in which the *De Revolutionibus* was published, it would have been strange indeed if everyone had rashly accepted the Copernican hypothesis as an accurate description of the true state of affairs in the heavens. For centuries, scholars had been putting forward arguments in favour of ideas that they did not regard as true; moreover, in astronomy, an influential tradition endorsed the framing of hypotheses as phenomena-saving devices, which astronomers could make use of without in any way having to judge them on truth content. For these reasons, it is hardly possible to speak in an absolute sense of scholars 'accepting' or 'rejecting' Copernicanism. Like any theory, Copernicanism could be conceived of in a variety of ways. In the words of Robert Westman: "Acceptance may connote provisional use of certain hypotheses (without commitment to truth content), or acceptance of certain parts of the theory as true while rejecting other propositions as false, or the overzealous belief that all propositions of the theory are true even in the absence of sufficient confirmatory evidence, or acceptance of the theory as true without regarding it as a program for further research."[66]

During the decades immediately following the publication of *De Revolutionibus*, most scholars opted for the "provisional use" of the heliocentric system, "without commitment to truth content". Often,

[66] Robert S. Westman, "The Melanchthon Circle, Rheticus, and the Wittenberg Interpretation of the Copernican Theory", *Isis* 66 (1975) 165-193: 185.

historians have blamed the Lutheran theologian who wrote the foreword 'Ad Lectorem' to Copernicus's work, Andreas Osiander, for having retarded 'the acceptance of Copernicanism' by presenting the heliocentric hypothesis as just that: a hypothesis (which word in the sixteenth century referred to "suppositions not known to be true, but from which empirically verifiable consequences could be deduced"[67]). It should, however, be borne in mind that Osiander's assertions reflected two views widely diffused among sixteenth-century scholars: the idea of the 'pragmatic compromise', that astronomical hypotheses need not be submitted to the tests of physical consistency, and the conviction that certainty, or truth, could only be attained by divine revelation.[68]

Robert Westman has shown how, immediately after the publication of *De Revolutionibus* in 1543, a group of young Wittenberg astronomers around the influential figure of the *Praeceptor Germaniae*, Philip Melanchthon, framed a "common methodological outlook or style, a consensus on how to read the newly published *De Revolutionibus*."[69] The interpretation of the Copernican theory adhered to by this informal circle of scholars, whose membership, apart from Melanchthon himself, included Erasmus Reinhold, Caspar Peucer, and Johannes Praetorius, was to prove extremely influential. They stood at the cradle of Johann Kepler's 1609 dictum, that there had come to exist two Copernicuses: "the one, author of a truly sun-centered cosmology which was supported by a few physical arguments and a new, simpler argument for the ordering of the planets; the other, an inventor of planetary models, calculated on the assumption that the planets revolve about a sun which is eccentric to the true centre of the universe."[70]

In fact, the 'Wittenberg Interpretation' was a child of its time as much as the Copernican theory itself. Melanchthon received the new world system with mixed feelings: on the one hand, he embraced the medieval point of view, that the motion of the earth, although it *could* be interpreted as a real one, was, in fact, an absurdity, violating

[67] See John L. Russell, "Catholic Astronomers and the Copernican System after the Condemnation of Galileo", *Annals of science* 46 (1989) 365-386: 385.

[68] See Bruce Wrightsman, "Andreas Osiander's Contribution to the Copernican Achievement" in Robert S. Westman ed., *The Copernican Achievement* (Berkeley 1975) 213-243: 235-237.

[69] Westman, "Melanchthon Circle", 166.

[70] Ibidem, 168; Westman quotes Johannes Kepler, *Astronomia nova* (1609), *Gesammelte Werke*, ed. Max Caspar, III (München 1937) 237.

Aristotelian physics and Scripture; on the other hand, he admired the harmony and beauty of the Copernican lunar theory; and on the one hand, he insisted that students be taught the Ptolemaic point of view, while, on the other, he himself used several Copernican data.[71] Despite his admiration for a number of features of Copernicanism, however, Melanchthon in the end settled for the 'pragmatic compromise': he would "permit and even encourage some articulation of the conceptual and empirical components of the Copernican theory: the improvement of observations, systematization of tables, discussion of the problem of precession, and use of certain Copernican planetary models *in a geostatic framework.*"[72]

Following Melanchthon's cautious use of the heliocentric hypothesis, his disciples also pragmatically regarded the Copernican theory as a convenient calculatory device. They gladly made use of the possibilities opened up by the new system, such as the opportunity of dispensing with the cumbersome Ptolemaic equants and the framing of new astronomical tables; however, they maintained strict neutrality on the subject of the truth of the heliocentric world system. Reinhold, who drew up the *Prutenic Tables*, the first set of astronomical tables to be based on the Copernican system, showed a "notable lack of commitment with respect to the heliocentric astronomy"; in fact, his tables were "carefully framed so that they [were] essentially independent of the mobility of the earth."[73] And in Wittenberg textbooks of astronomy, Copernicus's name was mentioned on several occasions, for example in the treatment of the size of the earth, and especially for his numerical values; however, when it came to a discussion of the question of whether the earth moved, all the traditional arguments were put forward, without any hint that the same Copernicus who had provided such good calculations, had proposed the mobility of the earth.[74] Owen Gingerich concludes that, although Copernicus's name often appeared in print, his heliocentric system was never discussed in depth.[75]

The Wittenberg Interpretation proved very influential during the first decades after 1543. At universities all over Europe, beginning

[71] Westman, "Melanchthon Circle", 173-174.

[72] Ibidem, 174; my italics.

[73] Owen Gingerich, "From Copernicus to Kepler: Heliocentrism as Model and as Reality", *Proceedings of the American Philosophical Society* (1973) 513-522: 515.

[74] Ibidem, 516.

[75] Ibidem, 520.

students were exposed at most to the Copernican numerical values and the Prutenic Tables[76]; more advanced astronomers did recognize the fact that Copernicus had assigned a triple motion to the earth, but they emphasized the possibility of 'transferring' Copernicus's models to a geostatic reference frame[77]; and, in general, they were not very interested in the issue of cosmological choice, regarding Copernicus as the *reformer* of Ptolemaic astronomy in an essentially conservative sense, rather than as the advocate of a realistically interpreted heliocentric world system.[78]

Yet, according to Robert Westman, by the 1570s a new generation of astronomers stepped forward, who showed "a new appreciation of the work of Copernicus".[79] They were more sensitive to cosmological issues, and began to realize that it had been Copernicus' intent to put forward a realist system of astronomy, a harmonic whole in which it was impossible to alter any of the planetary distances without affecting all others.[80] Westman sees Johannes Praetorius, Tycho Brahe, and Michael Mästlin as the most prominent members of this "generation of the 1570s". And among the factors reinforcing the re-direction of their attention towards the cosmological implications of the Copernican theory, he counts the supernova of 1572 and the comet of 1577. "Appearing at a time when the network of astronomers was just commencing to grasp the broader meaning of the Copernican hypothesis, the observations and interpretations of these two great occurrences were swept into a new framework, a framework in which the *choice* between the Ptolemaic and the Copernican theories took on competitive significance for the first time in the eyes of contemporaries."[81]

Obviously, Westman believes the nova and the comet to have been instrumental in the overthrow of the old, geocentric world view

---

[76] See for a general survey Lynn Thorndike, "Post-Copernican Astronomy" in his *History of Magic and Experimental Science* VI (New York 1941) 3-66.

[77] Westman, "Melanchthon Circle", 180; see also Thorndike, *History of Magic* VI, 56-57 for a discussion of Giovanni Antonio Magini's attempt, in his 1589 *New Theories of the Celestial Orbs Agreeing with the Observations of Copernicus*, to devise a theory of the planets which would conform to Copernicus's observations without making use of his absurd heliocentric hypothesis.

[78] Westman, "Melanchthon Circle", 181.

[79] Robert S. Westman, "Three Responses to the Copernican Theory: Johannes Praetorius, Tycho Brahe, and Michael Mästlin" in Robert S. Westman ed., *The Copernican Achievement* (Berkeley 1975) 285-345: 288.

[80] Ibidem, 287-289.

[81] Ibidem, 289.

in favour of a realistically interpreted heliocentric one. At the end of this chapter, we shall be able to discuss his arguments at greater length, and to assess their plausibility. In order to do so, however, we need first of all to make a detour by way of the optical theory of comets.

4. *The Comet of 1577: Its Proper Perspective*

a. *The Optical Theory of Comets*

During the earlier decades of the sixteenth century, Aristotle's cometology, like his world system in general, came under attack from a number of sides. As we have seen, the Viennese school of astronomy, led by Regiomontanus, despite developing a method of parallax determination to calculate comets' distances from the earth, did not arrive at any values inconsistent with Aristotelian theory. But the renewed interest in non-Aristotelian philosophies about the constitution of the cosmos inspired a number of cometary theories more immediately threatening to Aristotelian cometology than the parallax measurements.

As we have seen in the previous chapter, the discovery by Peter Apian and Girolamo Fracastoro of the antisolarity of comets' tails inspired some scholars to depart from Aristotle's theory, that comets were burning terrestrial vapours in the upper reaches of the air. Thus, Cardano believed comets to be luminous bodies located above the moon, with their tails formed by sunrays shining through the comet's body.[82] This theory closely resembles some of the Presocratic views on comets as heavenly bodies usually invisible to us, with tails produced by the optical effect of reflection; it also betrays some Stoic influences. Yet, it should not be regarded as a mere repetition of ancient, non-Aristotelian ideas by Renaissance authors anxious to return to the pristine purity of classical predecent. Although Cardano's, and similar, 'optical' cometary theories were clearly, from a sixteenth-century perspective, inspired in part by the humanist tradition of recovery of the ancients, a number of recent studies have invested them with a more far-reaching significance. With the benefit of hindsight, Peter Barker and Bernard Goldstein have found that

[82] See chapter III, § 10.

the "optical theory of comets" played an important part in shaping the conceptions of Tycho Brahe and Kepler; hence, they have re-evaluated the traditional emphasis on the role of Tycho's 1577 parallax observations in bringing down the Aristotelian world picture, and have found it lacking on several grounds.[83]

In a 1993 article[84], Barker provides us with a detailed account of the genesis and subsequent elaboration of the optical theory of comets. In 1531, 1532 and 1533, three conspicuous comets appeared; among the scholars observing them were Peter Apian of Ingolstadt and Girolamo Fracastoro of Padua. Independently of one another, they both found that all three cometary tails pointed directly away from the sun. Fracastoro made no attempt at explaining this antisolarity; Apian, however, suggested that, in the case of the comet of 1531, the appearance of the tail was caused by the light of the sun. Still, Apian did not attempt to generalize this explanation; he realized that some comets were surrounded by rays on all sides, and thought others had tails pointing towards the ground. Thus, not all cometary tails could be caused by sunlight; therefore, Apian recognized that at least some of these comets were, in fact, Aristotle's "vapors or exhalations, hot and dry, fat and gluey, raised by the strength of the stars to the highest layer of the atmosphere".[85]

Several scholars in subsequent years accepted and elaborated Apian's discovery of antisolarity. The Louvain professor, Gemma Frisius, supported Apian's findings concerning the comet of 1533 with his own observations. He also explicitly made the connection between the sixteenth-century discovery of antisolarity and the theory of the Presocratics Hippocrates of Chios and Aeschylus, that the tail of a comet is nothing but reflected rays of the sun.[86] Unlike the ancients, however, Gemma thought the cometary tails to be due, not to reflection, but to refraction.[87]

Much the most elaborated version of the optical theory was con-

---

[83] Peter Barker & Bernard R. Goldstein, "The Role of Comets in the Copernican Revolution", *Studies in the history and philosophy of science* 19 (1988) 299-319; Peter Barker, "The Optical Theory of Comets from Apian to Kepler", *Physis* 30 (1993) 1-25; Peter Barker & Bernard R. Goldstein, "Is Seventeenth-Century Physics Indebted to the Stoics?", *Centaurus* 27 (1984) 148-164.

[84] Peter Barker, "The Optical Theory of Comets from Apian to Kepler", *Physis* 30 (1993) 1-25.

[85] Ibidem, 9.

[86] See chapter III, p. 43-44.

[87] Barker, "Optical Theory", 10.

trived by Jean Pena in 1557, partly on the basis of a faulty observation of Gemma's. In his *De Radio Astronomico et Geometrico*, the book that also contained his cometary theory, Gemma claimed to have observed that the diameter of the Moon was no greater when measured at the horizon than higher in the sky, and that the angular separation between two fixed stars did not vary with their position above the horizon. He therefore concluded that there could be no such thing as atmospheric refraction. Pena inferred from this passage in Gemma, that there was no change in the medium between the observer on earth and the—celestial—object observed; hence, the "sublunary" element air had to extend all the way to the fixed stars.[88] As we have seen, Pena was able to corroborate this observational result by an appeal to the authority of Stoic doctrine.

He employed an equally optical argument in his explanation of the antisolarity of comets' tails: in his view, they were caused by refraction of the solar rays, that formed a pyramid of light after passing through the body of the comet. For this to be possible, it had to be inferred that the comet's head was "a transparent body, denser than air, but clear like the purest glass", since such was the only kind of body capable of creating pyramids of refraction.[89] Furthermore, Pena rejected not only Aristotle's theory of comets' constitution, but also his account of the location of at least some of them. In this case, however, his argument was purely Aristotelian, although it led to a non-Aristotelian conclusion: since it had been observed that some comets moved more slowly than the Moon against the background of the fixed stars, it had to be inferred that at least some comets were situated above the Moon, in keeping with the Aristotelian principle that, the further an object was removed from the earth, the slower it moved.

Girolamo Cardano's views on comets were discussed at length in the previous chapter.[90] Here, we need only repeat those elements of interest in connection with Barker's elucidation of the optical theory. In his *De Subtilitate,* Cardano explicitly cited Fracastoro, just before stating that comets' tails were always opposed to the sun, and that the likeness of a beard or tail was caused by sunrays shining through the body of the comet, which was a globe. However, his arguments

[88] Ibidem, 11.
[89] Ibidem, 12.
[90] See chapter III, § 10.

in favour of a superlunary position for all comets were taken, not from Fracastoro's discovery of antisolarity, but from Seneca: comets cannot be fires, because they remain visible for such a long time that there could never rise up enough terrestrial vapours to feed them, and, furthermore, their motions are unlike those of the terrestrial element fire. Cardano concluded that comets were intermediate in nature between the moon and the stars: they were neither capable of allowing the sunlight to permeate it unhindered, like the stars, nor dense enough to reflect it, like the moon.[91]

After describing the cometary theories of Apian, Gemma Frisius, Pena, and Cardano, Barker proceeds to analyze the influence they may have exercised on the views of Tycho Brahe. He states that Tycho "accepted a version of the optical theory"[92], because in his *De mundi aetherei recentioribus phaenomenis* (published in 1588, the book in which he expounded his views on the comet of 1577), he endorsed the view that the tail of the comet of 1577, which he observed pointing constantly away from Venus, was created by light from that planet. More explicitly, in a letter to Rothmann of 1589, Tycho stated that the tail of the 1577 comet could not have been simply an extension of the head: if it were, there would be no need for it to hold any constant orientation, and, moreover, stars could be seen through the tail, but not through the head. Therefore, Tycho was inclined to believe that the tail was caused by light from the Sun or Venus shining through the head, in Barker's view "a clear endorsement of the optical theory".[93] Barker then goes on to discuss the reception of the optical theory by Kepler, who criticized it on several grounds, subsequently to arrive at his own theory, that cometary tails were formed by matter ejected from the head.

It is very difficult to assess the amount of influence the optical theory of comets may have exercised during the final decades of the sixteenth century. Arguing against the commonly accepted view, that Tycho's observations of the comet of 1577 marked the decisive break with Aristotle, Barker and Goldstein contend that "Aristotle's theory had already been replaced by the theory that comets were spherical lenses when Tycho Brahe came on the scene, and he simply adopted

---

[91] Hieronymus Cardanus, *De Rerum Varietate Libri XVII* (Basle 1557) I.1, 2.
[92] Barker, "Optical Theory", 15.
[93] Ibidem, 17.

this established theory."[94] It is, however, doubtful whether such a categorical view can be maintained. In the first place, not nearly everyone had, by the time of the arrival of the comet in November 1577, embraced the optical theory; the vast majority of observers, both before and after 1577, continued to adhere to the Aristotelian exhalation theory.[95] Therefore, Barker and Goldstein are not justified in speaking of the "general acceptance of the view that comets were spherical lenses".[96]

Nor does it seem to be the case that Tycho "simply" adopted the ready-to-use optical theory thought up by his predecessors. It is true that, both in his German treatise on the comet of 1577, written shortly after the comet's disappearance, and in his *De mundi aetherei recentioribus phaenomenis,* published in 1588, Tycho maintained that the tail of a comet was nothing but the rays of the sun (or of a planet) passing through the comet's body, and that consequently this body was of an intermediate nature, neither capable of transmitting rays invisibly nor of reflecting them.[97] This statement, very similar to that proposed by Cardano, can, indeed, be interpreted as a clear endorsement of the optical theory. However, in his tract on the new star of 1572, published in 1573, Tycho still maintained that, because he had conclusively demonstrated the star to have been located in the eighth heavenly sphere, it followed that it could not have been any kind of

---

94 Barker & Goldstein, "Role of Comets", 315.

95 See C. Doris Hellman, *The Comet of 1577. Its Place in the History of Astronomy* (New York 1944) chapters IV-VII; in his later article, Barker himself qualifies his earlier statement by admitting that the optical theory, significant though it was, was "by no means universally accepted." (Barker, "Optical Theory", 25).

96 Barker & Goldstein, "Role of Comets", 316.

97 See Christianson's translation of Tycho's German treatise on the comet of 1577, in J.R. Christianson, "Tycho Brahe's German treatise on the Comet of 1577: A Study in Science and Politics", *Isis* 70 (1979) 110-140: 135: "From this [the tail's antisolarity] it appears that the tail of a comet is nothing but the rays of the sun which have passed through the body of the comet, for this body, not being diaphanous like other stars, cannot transmit the rays invisibly, and not being opaque like the moon, cannot reflect the rays, but since the body of the comet is some medium between rare and dense, it holds a part of the radiance from the sun within itself, and from this comes the light of the head by reason of the resistance of celestial matter of which the head is fabricated, but because it is also somewhat rare and porous, it lets those solar rays pass through which are seen by us as a long tail hanging to the head of the comet." Compare this passage to Cardano, *De Rerum Varietate,* I.1, where the author states that comets' light is not clear enough to let through the rays of the sun (like the stars), nor dense enough to reflect them (like the moon), and concludes that, therefore, a comet is of an intermediate nature, between the stars and the moon.

comet or any other fiery meteor, since all these were generated below the moon. The only one who did not subscribe to this theory, according to Tycho, was al-Battani (al-Battānī), who had claimed that comets originated in the heavens; Tycho was not yet in a position to judge whether the Arab was right, but promised that, if he were to see a comet during his lifetime, he would check this.[98]

Apparently, then, Tycho did not yet subscribe to the optical theory in 1573. Yet, he most certainly knew about its existence; he was very well versed in contemporary astronomy, and had met Philip Apian, Peter's son, at Ingolstadt in 1570. Why, then, did he not accept the theory until after the appearance of the comet of 1577? Possibly, the reason why, in his tract on the nova, he still paid lip service to Aristotle's exhalation theory for comets, was a rhetorical one. After all, he was trying to convince an audience, the overwhelming majority of whom he knew to endorse the Aristotelian cometary theory, of the fact that the nova was situated far above the moon. One of the most obvious arguments he could adduce in favour of this fact, given his audience, was to resolutely deny that the nova was a comet: admittedly, Tycho says, comets are sublunary vapours, but this new phenomenon was definitely nothing of the sort. There was no need for him to uproot his readers' convictions about the nature of comets; indeed, that would probably not have done his cause any good. Much the more tactically sound strategy was to stress the marked contrast between sublunary comets on the one hand and the superlunary nova on the other.

Tycho, then, was cautious about endorsing the optical theory of comets prior to 1577. Given his character and methodology, this is hardly surprising: it would have been unlike Tycho indeed to adopt a theory about a phenomenon he had never actually observed himself. Furthermore, the limited number of scholars who did subscribe to the optical theory were themselves rather wavering about the consequences its adoption entailed for cometary theory in general. Barker and Goldstein rightly point out that the acceptance of the optical theory does not impose any necessity to place comets beyond the moon[99]; instead, comets in their 'optical' guise as spherical lenses

[98] Victor E. Thoren, *The Lord of Uraniborg. A Biography of Tycho Brahe* (Cambridge 1990) 67-68. In a later edition of this work Tycho replaced the name of Albattani with that of the correct propagator of comets above the moon, Albumasar.

[99] Barker & Goldstein, "Role of Comets", 316.

could be located anywhere, either above or below the moon, or at any location in the continuous heaven of Stoic cosmology. It is, therefore, hardly surprising that the early adherents of the optical theory were rather noncommittal about comets' location: Frisius remained silent on the issue; Fracastoro placed them directly below the moon in a special 'sphere of comets'; Apian believed that at least some comets were sublunary vapours, but remained silent about the location of the other, optical variety; and, reversely, Pena and Cardano concluded that at least some comets were superlunary, not on the basis of the optical theory, but on the basis of the Aristotelian 'the slower it moves, the more distant it is located' argument.

Thus, the optical theory was not in itself capable of causing the overthrow of a world system. Instead, it should be regarded as one of the many piecemeal modifications the Aristotelian world view was capable of accommodating.[100] Similarly, the parallax measurements that had been carried out by several astrologers during the late Middle Ages, had never yielded any values inconsistent with Aristotelian cometary theory.[101] Clearly, the technique of parallax computation, very well known to medieval observers of the sky, was as incapable of single-handedly bringing down the Aristotelian dichotomy between heaven and earth as the optical theory. Yet, both the parallax measurements and the optical theory had a special role to play in weakening the Aristotelian world view. In the hands of Tycho, both were to become tools that could be employed to render viable the superlunary position of comets.

Thus, one key question presents itself. We have seen that a number of sixteenth-century authors abandoned Aristotle's exhalation theory long before 1577, and assigned at least some comets a place above the moon; that the Stoic unity of the cosmos and the abandonment of the Aristotelian dichotomy was becoming quite popular with sixteenth-century scholars; and that the technique of parallax computation was well-known from the later Middle Ages onwards. Given

---

[100] See ibidem, 317, where Barker and Goldstein, arguing against the Kuhnian conception of the basic incommensurability of pre- and post-revolutionary paradigms, point out that, unlike the practice of modern science, during the sixteenth century modifications in the Ptolemaic theory were made piecemeal: "When one part was changed no modification of any other part was required precisely because the whole was not yet regarded as subject to the requirement of internal logical consistency."

[101] See chapter III; see also Barker & Goldstein, "Role of Comets", 312-313.

these facts, what was the importance of the comet of 1577? Why was Tycho the first to arrive at parallax values which were actually correct? What was the nature of the change that allegedly occured in cometary theory after 1577, and what was its scope?

b. *The Nova's Novelty*

At first sight, the question of what it was that distinguished the comet of 1577 from its predecessors seems to be easily answerable: it was the parallax measurements conducted by Tycho Brahe. Although, unlike their eighteenth- and nineteenth-century predecessors, modern historians of science realize that Tycho was not the first to apply the technique of parallax computation to comets, the fact remains that he was the first to do so correctly, arriving at values that actually placed the comet above the moon. But what was it that enabled him to measure the comet's distance so much more accurately than his predecessors from the, not even very distant, past?

It might be suggested that Tycho's measurements were extraordinarily good, because he was much better equipped than his predecessors. Insisting on the greatest possible degree of precision, he had surrounded himself in his own observatory at the island of Hven with the very best instruments available at the time, many of them endowed with improvements he had himself invented to meet his high standards of accuracy. However, although obviously Tycho's equipment, along with his tenacity, enabled him to attain higher degrees of accuracy than anyone else, such precision was not, in fact, required to arrive at more or less correct parallax values. Several other observers of the comet of 1577 independently from Tycho obtained reasonably accurate results, prompting them to admit that the comet was located beyond the moon. None of these men had at their disposal instruments that could stand comparison with the ones at Hven.

Their methods of parallax determination were quite crude: Mästlin, who said he did not trust his instruments, preferred to determine the comet's position by means of naked eye observations of its motion relative to four fixed stars of known latitude and longitude, aided only by a piece of thread[102]; Gemma employed a similar method;

[102] C. Doris Hellman, "The Rôle of Measurement in the Downfall of a System: Some Examples from Sixteenth Century Comet and Nova Observations" in A. Beer ed., *New aspects in the history and philosophy of astronomy* (Oxford 1967) 43-52: 51.

and Roeslin placed the comet above the moon on the basis of such qualitative arguments as the purity of its light and the perfection of its motion.[103] This has led Doris Hellman to point out that "Very crude measurements, some made without any instrument or with a thread held before the eyes, sufficed to indicate the supra-lunar position of the star and the comet."[104] If such is the case, inevitably the question presents itself of why the observers of earlier comets had never been able to arrive at any correct parallax values. Clearly, this cannot be attributed solely to Tycho's arrival on the scene, with his high standards of accuracy, since Mästlin and Gemma deduced more or less correct values from much simpler observation techniques.

William Donahue has stressed the importance of Tycho's 'inheritance' to explain the novelty of his experimental results, pointing out that he stood in the confluence of two trends. First of all, he knew about and agreed with the optical theory of comets as expounded by Cardano, and, furthermore, he was favourably disposed towards the Stoic idea of fire filling the entirety of the heavens. Therefore, "unlike his predecessors, he approached the tasks of observation and calculation with the hope of proving this comet to be beyond the moon, which hope caused him to insist on unprecedented accuracy."[105] Thus, Donahue reverses the argument: Tycho did not arrive at correct parallax values because of the accuracy of his observations; rather, his insistance on uncommon precision was prompted by the cosmological preconceptions he had inherited from previous sixteenth-century authors, which led him to expect the comet to be located beyond the moon.

There is much to be said in favour of such an explanation. Although Donahue disregards the fact that several of Tycho's contemporaries arrived at a superlunary position for the comet on the basis of much cruder observations, a somewhat modified version of his theory could be conducive towards clarifying the events of 1577. It seems very probable that both the optical theory of comets and Stoic cosmological ideas that were gaining currency in the learned world of the sixteenth century called the attention of a number of scholars to the possibility that some comets might be supralunar. Acceptance of the fact that at least some comets were spherical lenses with their

[103] Hellman, *Comet of 1577*, 172-173.
[104] Hellman, "Rôle of Measurement", 47.
[105] Donahue, *Dissolution of the Celestial Spheres*, 63-64.

tails formed by sunlight, and of the uninterrupted continuity and material similarity between the heavens and the terrestrial atmosphere, may well have rendered the scholarly climate more receptive of discoveries deviating from Aristotelian postulates. It is a recognized fact that what is actually observed is partly dependent on the observer's personal preconceptions.

Still, this cannot be the whole story. If it were, that is, if all that was needed to make observers recognize the comet's supralunar position on the basis of parallax observations was an inherited background of Stoic and optical ideas, why was not one of the earlier sixteenth-century comets (say, the comet of 1556) the first to be placed decisively above the moon? After all, in 1556, the optical theory had been around for more than two decades, and Stoicism had already gained wide currency. In themselves, these two factors were apparently insufficient to bring about more than incidental conjectures about the supralunar location of comets. Some other agent must have come into play during the years between the two cometary appearances of 1556 and 1577, rendering scholars receptive to the idea of superlunary change in a way no revived ancient theory had been able to accomplish. This agent, widely thought of as a herald from Heaven, arrived in the redoubtable guise of a new star.

In November 1572, a supernova exploded in the constellation of Cassiopeia. A star, that had always been invisible to the naked eye, in a matter of days enormously increased its brightness, thrusting the major part of its matter into space at astronomical speeds. After a short period, the phenomenon started to decrease in intensity, although its fading contours remained visible through the whole of 1573; early in 1574, it disappeared completely. This 'nova', as it was referred to by its observers, caused a tremendous amount of excitement and speculation on both the scholarly and popular levels, and was perceived as a much more shocking phenomenon than any kind of comet, however bright. It gave rise to a vast literature: more tracts were written about it than had been written about any previous comet.[106] Almost every one of Europe's leading astronomers hastened to publish their views on the unusual appearance. John Dee and Thomas Digges in England, Munosius in Spain, Maurolyco in Sicily, Hagecius in Bohemia, Graminaeus, Mästlin, Peucer, and the Landgrave of Hesse-Cassel in the Holy Roman Empire, Clavius in

[106] Hellman, *Comet of 1577*, 111.

Rome, Cornelius Gemma in the Netherlands, and Tycho Brahe in Denmark: scholars from all over Europe contributed to one of the liveliest debates of the sixteenth century.

The bewildered commotion that captivated the scholarly community after the nova's appearance was caused first and foremost by the novelty of the phenomenon. The only new star mentioned in recorded history was the one Pliny reported to have been seen by Hipparchus around 125 B.C.; since then, no phenomenon even slightly similar had shone forth. Indeed, Aristotle had stated that the experience of ages taught that changes never occurred in the heavens.[107] Thus, the nova was from the outset placed in a class of its own, as a mysterious phenomenon worthy of the highest degree of heedfulness. Comets, by contrast, had been known and observed for centuries. Since the reception of Aristotle's writings, in the thirteenth century, the overwhelming majority of scholars had assigned them to the sublunary atmosphere as transitory, fleeting vapours. The adherents of the optical theory had in no way been able or even willing to conclusively demonstrate their supralunar position, while the early parallax measurements, yielding values in accordance with a lowly location for the hairy stars, had done nothing but reinforce the Stagyrite's opinion.

The nova, then, was different. It was not burdened with the heavy Aristotelian tradition that weighed down on its hairy colleagues, and was thus able to serve as a kind of catalyst forcing the scholarly community to reconsider its fundamental tenets. We have already seen how Tycho Brahe, in his tract on the nova, explicitly contrasted it, as a superlunary phenomenon, with the sublunary comets; Thomas Digges, too, placed the nova beyond the moon, while retaining his belief in the elementary, terrestrial nature of comets.[108]

Many of the leading astronomers agreed that the nova was definitely a heavenly phenomenon, located far above the moon, since it showed hardly any observable parallax. Thus, the nova brought about what dozens of comets had failed to accomplish: it enabled its

[107] Aristotle, *On the Heavens*, 270b 12-16: "The truth of it [the unalterability of the heavens] is also clear from the evidence of the senses, enough at least to warrant the assent of human faith; for throughout all past time, according to the records handed down from generation to generation, we find no trace of change either in the whole of the outermost heaven or in any one of its proper parts." (Translated by W.K.C. Guthrie, Loeb Classical Library edition, Cambridge Mass./London 1986) 25.

[108] D. K. Yeomans, *Comets: A Chronological History of Observation, Science, Myth, and Folklore* (New York 1990) 33.

observers to correctly measure its parallax. Now we might well wonder, along with Sagredo in Galileo's *Dialogue Concerning the Two Chief World Systems,* how, "if the matter [of parallax determination] is surrounded with so much confusion, uncertainty, and error, . . . [did] it happen that so many astronomers have so confidently declared the new star to have been very remote?"[109] The expert Salviati's response to this question was, that "Either of two sorts of observations, both very simple, easy, and correct, would be enough to assure them of the star being located in the firmament, or at least a long way beyond the moon. One of these is the equality—or very slight disparity—of its distances from the pole when at its lowest point on the meridian and at its highest. The other is that it remained always at the same distance from certain surrounding fixed stars; especially κ Cassiopeiae, from which it was less than one and one-half degrees distant. From these two things it may unquestionably be deduced that parallax was either entirely lacking, or was so small that the most cursory calculation proves the star to have been a great distance from the earth."[110]

Apparently, then, measuring the nova's parallax was child's play, that could easily be performed without any instruments or complicated calculations. In fact, it was considerably easier than determining the correct parallax value for a comet. Several complicating factors that can serve to render parallax computation very tricky in the case of certain comets, are absent with supernovae. Most importantly, they do not change their positions.[111] Comets always display a special movement of their own, rushing east, west, north or south in accordance with their unique orbits around the sun. As we have seen in our previous chapter, the reason why Regiomontanus's method for parallax determination had foundered miserably had been its inventor's failure to take into account the proper motion of every comet.[112] Moreover, a supernova is much further removed from the

---

[109] Galileo Galilei, "The Third Day" in *Dialogue Concerning the Two Chief World Systems* (Florence 1632, translated by Stillman Drake, Berkeley/Los Angeles/ London 1967) 311.

[110] Ibidem.

[111] See Thoren, *Tycho Brahe,* 124: "Even more problematical was that comets were known to be moving bodies. However anomalous the visitation of the new star had been, its resemblance to the "fixed" stars conveyed an a priori suggestion that any movement it displayed would be a manifestation of parallax. For the comet, any identification of a parallactic motion depended on a preassessment of its intrinsic motion."

[112] See p. 80.

earth than a comet, making it a great deal easier to accord it a place outside the earth's atmosphere on the basis of rough approximations of its correct parallax value. Thus, the comet of 1577 was not very far distant from the earth during its period of visibility; at perihelion, it even found itself inside the orbit of Mercury[113], so that the slightest of mistakes might lead its observers astray and cause them to erroneously place it within the earth's atmosphere.

But perhaps the most compelling reason for according the nova a place among the fixed stars, was simply its star-like appearance itself. It clearly looked like a star, albeit one of exceptional brightness; it stood still and displayed not even the shadow of a tail. On account of these characteristics, along with its lack of parallax, leading astronomers like Tycho, Mästlin, Gemma, the Landgrave of Hesse-Cassel, Peucer, and Digges felt quite convinced that it was, indeed, a heavenly phenomenon, a newly created being beyond the moon. They were much impressed: Gemma, whose soul "was filled with gratitude to God that this splendid phenomenon had come during his lifetime", was inspired by the novel appearance to construct his 'cosmocritical art', and Tycho, who had up till 1572 been chiefly preoccupied with alchemical experiments, felt impelled to direct his attention to the sky rather than the earth, the nova thus marking "the formal beginning of Tycho's career as an astronomer."[114]

Because of the novelty of the phenomenon, and the absence of authoritative opinions on the subject of new stars, its observers were puzzled about its nature and significance. A rather obvious solution adopted by many of them, including Tycho and Mästlin, was to pronounce the new inhabitant of heaven a "miraculous divine new creation."[115] It was therefore thought of as a sign from God beyond the natural order, whose true nature was never to be discovered by human cognition, and which, in accordance with the widespread eschatological expectations of the age, was closely associated with the impending end of the world. Tycho, for one, related the nova to the astrological sign of Aries, a move that invested it with great significance, since Aries was the sign where, in 1582, a great conjunction was to take place at the ending of the watery trigon and the begin-

[113] Yeomans, *Comets,* 33.
[114] Thoren, *Tycho Brahe,* 55; Thorndike, *History of Magic,* VI, 68.
[115] Thorndike, *History of Magic,* VI, 69.

ning of the fiery trigon, thus inaugurating a new astrological age.[116] And Cornelius Gemma even suggested, much to Tycho's horror, that the new star might be an angel, or even God himself, appearing to humankind veiled in a cloak of light.[117]

Thus, the scholars who affirmed the nova's supralunar position were hesitant about offering any natural explanation for the phenomenon. By according it the status of a miracle, they were still able to circumvent questions about the immutability of the heavens. Hence, the nova's appearance had no direct cosmological consequences: its main significance lay in the analogy that was subsequently to be established between it and the many comets that shone forth during the following decades, the first of which, of course, being the one that appeared in 1577. Many of the authors who wrote cometary tracts in the decades after 1572 themselves acknowledged the important role the nova had played in redirecting astronomers' preconceptions. Thus, Tycho, in his German treatise on the comet of 1577, wrote:

> The greater part of succeeding philosophers have accepted this Aristotelian opinion, and have considered it quite impossible that comets or anything new could be born in the heavens, though some have recently doubted this, on the grounds that it was possible to observe and discover that the new star or comet, which appeared four years ago in the constellation Cassiopeia, had no parallax whatsoever and always remained in one place like the fixed stars, and therefore not only could not be under the moon in the element of fire or of air, but that it stood in the uppermost sphere of the fixed stars, among the other [stars], as I have proved and demonstrated sufficiently in my little book on that same star. Various distinguished mathematicians, both in Germany and other nations, have concurred in their observations that the same star stood in the heavens with the other fixed stars. This miracle has made it necessary for us to abandon the opinion of Aristotle and take up another: that something new can also be born in heaven. *Since it could happen in the case of the aforementioned star, it is not impossible that other comets are born in the heavens and are not comprised of dryness and fattiness extracted from the earth.*[118]

[116] Thoren, *Tycho Brahe,* 70.

[117] Germana Ernst, "From the Watery Trigon to the Fiery Trigon: Celestial Signs, Prophecies and History" in Paola Zambelli ed., *Astrologi Hallucinati. Stars and the End of the World in Luther's Time* (Berlin/ New York 1986) 265-280: 271.

[118] Tycho Brahe, *Vonn der Cometten Uhrsprung was die Alten vnnd neuen Philosophi inn denselben vermaint vnnd dauon zuhalten sei,* (written in 1578, but never published during Tycho's lifetime) translated by Christianson, "Tycho Brahe's German treatise", 133. My italics.

Similar utterances were to be repeated time and again during the following decades. The English mathematician, Thomas Hood, in 1590 wrote in his *The Use of the Celestial Globe in Plano:* "That starre was either newly created by Almightie God in the firmament, or els certainly it is possible for a comet to be engendred there"[119]; the Louvain professor, Libertus Fromondus, believed that "After Aristotle, almost all, even the leading, philosophers, among them Ptolemy himself, deviated from invention and truth as if they were already assured of them; until, during this past century, the great occasion of the Cassiopeian star induced most of them to return to the ancient opinion"[120]; and his Leiden colleague, Willebrord Snellius, happily referred to the nova as an ardently wished for opportunity for discovering whether or not the heavens were subject to generation and corruption, because it stood perfectly still, and remained for sixteen months, allowing Tycho to demonstrate its supralunar location.[121]

While the example of the nova served to make the leading astronomers more receptive to the idea of supralunar comets, conversely, the appearance of the comet of 1577 made at least Tycho reconsider his earlier estimate of the nova's nature. Whereas, before, he had believed it to be a miraculous being, about whose true nature nothing certain could be discovered by the human intellect, he was now more inclined to think that, like the comet, it had consisted of heavenly matter, though of a lesser degree of perfection than that of which the ordinary stars were formed. He believed it "resembled the fixed stars as an alchemical artefact does the natural metal."[122]

Donahue's hypothesis about Tycho's 'inheritance', then, can be broadened to include the role played by the new star of 1572. It can

---

[119] Quoted by Paul H. Kocher, *Science and religion in Elizabethan England* (San Marino, California 1953) 177.

[120] Libertus Fromondus, *Dissertatio Cometae anni 1618* (Antwerp 1619) 94: "Post Aristotelem universi fere, etiam Astronomiae Praesides, in quibus ipse Ptolemaeus, quasi inventionis iam & veritatis securi abierunt; donec tandem saeculo nuper lapso occasio maxime Cassiopaeanae stellae plerosque ad respectum veteris opinionis permovit." By "the ancient opinion", Fromondus means pre-Aristotelian cometology.

[121] Willebrord Snellius, *Descriptio Cometae qui anno 1618 mense Novembri primum effulsit* (Leiden 1619) 2-3: "Tandem anno 1572 novum illud & longe splendidissimum sidus in Cassiopejae asterismo positum oportunitatem optatissimam praebuit deprehendendi an in caelo sidera etiam orirentur & interirent: Nam neque loco, in quo primum constiterat, unquam movebatur, & polo propinquum nobis Septentrionalioribus numquam occidebat, & longissimo temporis intervallo per sedecim continuos menses commode spectandum se exhibebat."

[122] Thorndike, *History of magic,* VI, 74.

be said that Tycho and his more advanced contemporaries were influenced by several sixteenth-century developments. They knew about the optical theory of comets and the possibility it entailed, that certain comets might be beyond the moon; they knew about the Stoic theories of the unity of heavenly and sublunary matter, revived by the humanist movement; and they knew about the technique of parallax observation invented by their medieval predecessors. The nova provided the finishing touch by allowing them to correctly apply the old parallax techniques; it also helped them to overcome the psychological barrier posed by the powerful Aristotelian cometary tradition. Come the year 1577, Tycho, for one, was ready for the appearance of a comet. This time, "although he had said that the nova of 1572 could not be a comet because it was in the region of fixed stars, whereas comets were generated below the moon, nevertheless, he started to look out for an astronomical body, not an atmospheric phenomenon"[123], and he duly found the comet to be located far beyond the moon. The same result was reached independently by Tycho's colleague, Michael Mästlin, who, confident of his success, went on to attempt to compute it an orbit.

c. *The Comet and Copernicanism*

The comet of 1577 was the first of its kind for which orbit computations were attempted. Although several ancient authors, most notably Seneca, had suggested that the hairy stars might describe their own proper courses, like the planets, and although Fracastoro, in his homocentric world system, had devised a special sublunary sphere to account for the motions of comets in general, no-one before 1577 had ever tried to construct an orbit for one particular cometary appearance. Encouraged by the outcome of his observations, though, Mästlin grew convinced that, if the comet of 1577 was a heavenly phenomenon, it should be possible to discover the characteristics of its celestial trajectory.

The calculations he carried out in accordance with this belief have been invested with special significance. Robert Westman acknowledges that specific issues directly following from the supralunary location of the comet (such as rejection of the crystalline spheres and

[123] Hellman, *Comet of 1577,* 121.

the imperfection of the heavens) were "not *sufficient* conditions for the overthrow of the ancient [world view]", since "such adverse testimony to the physical foundations of the old cosmos said nothing about its mathematical structure."[124] Yet, Westman argues that "In an interesting and unforeseen way. . . the comet of 1577 did force a reconsideration of the old order of the planets."[125] Thus, Mästlin, after a great deal of laborious calculation and inventive trial and error, arrived at the conclusion that only the Copernican hypothesis was capable of saving the appearances of the comet.

Even before 1577, Mästlin had been favourably disposed towards the Copernican theory. Thus, in his work on the nova of 1572, he had endorsed Copernicus's belief in the "immense Height of the Stellar Orb", and had admired the Pole's demonstration of "the certain distances of all the Planetary Orbits from the Center of the World."[126] Yet, when the fact that all his observations of the comet of 1577 were in the same plane convinced him that the object displayed a certain regularity of motion, he turned first of all to the Ptolemaic system as an aid to computing its orbit. Only after having tried all the planetary spheres of Ptolemaic astronomy, and failed, did he turn to the heliocentric hypothesis, adding that "compelled by necessity, I came to it with reluctance."[127] Despite this initial lack of enthusiasm, however, Mästlin went on to discover that only one heavenly sphere was capable of saving the comet's appearances: the Copernican sphere of Venus. Yet, although the success of a Copernican sphere where all the Ptolemaic spheres had failed, must have confirmed Mästlin's positive inclinations toward the heliocentric theory, he still employed it as nothing more than a calculational tool. His diagrams of the comet showed no indication concerning the physical reality of the model, and his construction included some

---

124 Robert S. Westman, "The Comet and the Cosmos: Kepler, Mästlin, and the Copernican Hypothesis" in Jerzy Dobrzycki ed., *The Reception of Copernicus' Heliocentric Theory. Proceedings of a Symposium organized by the Nicolas Copernicus Committee of the International Union of the History and Philosophy of Science, Torun, Poland, 1973* (Dordrecht/Boston 1972) 7-30: 11.

125 Ibidem.

126 Ibidem, 9.

127 Michael Mästlin, *Observatio et demonstratio cometae aetherei qui anno 1577 et 1578 constitutus in sphaera Veneris apparuit cum admirandis eius passionibus varietate scilicet motus, loco, orbe, distantia a terrae centro etc. adhibitis demonstrationibus geometricis et calculo arithmetico, eiusmodi de alio quoquam cometa nunquam visa est* (Tübingen 1578) 54, quoted and translated by Westman, "Comet and Cosmos", 17.

obvious physical impossibilities[128]; as Westman says, "It was left for Tycho Brahe to take the next logical step: to abolish all crystalline spheres."[129]

Like Mästlin, Tycho had held Copernicus' achievement in high esteem for a number of years before the appearance of the comet of 1577. In September 1574, he delivered a series of astronomical lectures at the university of Copenhagen, in which he made no secret of his admiration for his Polish predecessor, whom he believed to have "restored the science of the celestial motions and considered the course of the heavenly bodies more accurately than anyone else before him."[130] Tycho went on to say that, "although [Copernicus] holds certain [theses] contrary to physical principles, for example, that the Sun rests at the center of the Universe, that the Earth, the elements associated with it, and the Moon, move around the Sun with a threefold motion and that the eighth sphere remains unmoved, he does not, for all that, admit anything absurd as far as mathematical axioms are concerned."[131] In accordance with this conception of the status Copernicus' theory was to be assigned, the Dane in his lectures proceeded to analyze the planetary motions "according to Copernicus' opinion and numbers but referring everything to a stable Earth. . .".[132]

Clearly, then, in 1574 Tycho's reading of the heliocentric hypothesis neatly conformed to the generally accepted Wittenberg Interpretation. Tycho was never quite persuaded by Copernicus' physical arguments in favour of the earth's motion; yet, the symmetry displayed by the planets' distances in their heliocentric arrangement powerfully appealed to the young Dane. During the years following his 1574 lectures, his main cosmological concern was to 'invert' the Copernican model, both for the sun and moon and for the other planets, so that they might be reconciled to a motionless earth at the centre of the universe. And shortly after the disappearance of the comet of 1577, in February 1578, he brought together the various inverted Copernican models into a single scheme. This arrangement

[128] Such as the fact that the comet and Venus, in Mästlin's model, were moving in the same sphere, but in opposite directions.

[129] Westman, "Comet and Cosmos", 25.

[130] See J.L.E. Dreyer ed., *Tychonis Brahe Opera Omnia* I, 149. Translated by Kristian Peder Moesgaard; quoted by Westman, "Three Responses to the Copernican Theory", 307.

[131] Ibidem.

[132] Ibidem; Westman, "Three Responses", 307-308.

combined the advantages both of the Copernican harmony and the physical plausibility of geocentrism. It placed the earth at the centre of the universe, with the sun and the three superior planets, Mars, Jupiter, and Saturn, orbiting it; the two inferior planets, Mercury and Venus, though, revolved round the sun.[133]

Even more satisfying would have been a completely geoheliocentric system, with all planets, except the earth, circling the sun. As yet, however, such an arrangement seemed to Tycho an utter impossibility; after all, when observing the parallax of Mars, he had found that the radius of that planet's orbit had to be smaller than the diameter of the sun's orbit, to account for the smaller distances of Mars at opposition.[134] This meant that the orbits of the sun and Mars had to intersect, an obviously untenable consequence for one who believed in solid spheres. Years later, Tycho expressed his exasperation in a letter to Peucer: "I was still steeped in the opinion, approved and long accepted by almost all, that the heavens were composed of certain solid orbs which carried round the planets, and. . . I could not bring myself to allow this ridiculous penetration of the orbs; thus it happened that for some time this, my own discovery, was suspect to me."[135]

In 1578, then, Tycho could not yet bring himself to abandon the concept of planet-carrying, solid spheres. Within the next few years, however, the comet came to his aid. In his German treatise of 1578, he had already suggested that it probably moved within the sphere of Venus, with that planet revolving around the sun.[136] But it was only when, years later, he sat down to compose the final chapter to his definitive work on the comet, *De mundi aetherei recentioribus phaenomenis*, that he realized that the daily distances his colleague Mästlin had measured for it demonstrated that its path would have cut right through the Ptolemaic spheres of Venus and Mercury, if such had existed.[137] This unexpected discovery quite fortuitously provided him with the evidence needed for the successful propagation of his geo-

[133] Westman, "Three Responses", 326.

[134] Ibidem, 329.

[135] Tycho Brahe to Caspar Peucer, 13 Sept. 1588. Translated by Christine Jones; quoted in Westman, "Three Responses", 329.

[136] J.L.E. Dreyer ed., *Tychonis Brahe Opera Omnia* IV (Copenhagen 1922) 388. See Victor E. Thoren, "The Comet of 1577 and Tycho Brahe's System of the World", *Archives Internationales de l'Histoire des Sciences* 29 (1979) 53-67: 54.

[137] Thoren, "Comet of 1577", 61.

heliocentric world system: the previously "ridiculous" penetration of the orbs of the Sun and Mars had been eliminated, not because the paths Tycho had plotted for those planets were incorrect, but because no such solid orbs in fact existed.

We are now in a position to judge whether, as Westman believes, the comet of 1577 really took on heightened significance because it appeared at a time when the *cosmological choice* between the Copernican and the Ptolemaic systems, as opposed to the pragmatic use made of the heliocentric hypothesis by the Wittenberg Interpretation, came to the foreground. It is, of course, true that the so-called 'Generation of the 1570s' concerned themselves with new and more far-reaching problems than the earlier, 'Wittenberg' generation. They tried their hands at new arrangements of the cosmos, arrangements capable of incorporating the beautiful symmetry of the Copernican ordering of the planets. No longer were they interested merely in the abandonment of the Ptolemaic equant and the borrowing of new parameters.[138]

Yet, Mästlin's attitude towards the comet displayed the standard hallmarks of traditional Wittenbergism. He tackled the problem of computing its orbit in a purely conventional way, reluctantly employing the Copernican planetary arrangement as a mere phenomena-saving device, without in any way declaring himself in favour of its physical connotations. In fact, as early as 1570-72, he had accepted the reality of the Copernican system, not on the basis of any physical considerations, but because of the aesthetic appeal of its mathematical simplicity and the excellent correspondence of its hypotheses with observations.[139] The comet, if anything, may have contributed somewhat towards further confirming him in this view; yet, he continued to treat astronomical problems, those raised by the comet as well as all others, as though they were completely distinct from the reality his aesthetically-minded other self at the same time accorded the Copernican system. Westman concludes, that "Much as a typical Ptolemaic astronomer of the period, who first recites the catechisms of Aristotelian physics and then proceeds to the "real business" of saving the phenomena, Mästlin showed considerable competence and innovative talents as a mathematical astronomer but little happi-

[138] Westman, "Three Responses", 338.
[139] Ibidem, 330-333.

ness for the strange "physical" investigations of his student [Kepler]."[140]

The comet, then, can hardly be said to have influenced Mästlin's cosmological ideas, or to have contributed towards his adoption of Copernicanism. The case of Tycho Brahe was rather different, in that the comet did, in an unforeseen way, confirm him in his belief in the reality of a non-Ptolemaic cosmological system. Significantly, though, that system was not the Copernican: it was not heliocentric and did not upset any of the tenets of Aristotelian physics. It should perhaps rather be regarded as the final consummation of the Wittenberg Interpretation, born of an essentially conservative desire to cling to the postulates of traditional, earth-centred physics while at the same time incorporating the pristine, cosmic harmonies restored by Copernicus. Most of Tycho's efforts had, for years, both before and after 1577, been directed at achieving such a synthesis; the comet can at most be said to have provided him with the final piece of a jigsaw he might well have solved on other grounds.

All in all, then, the part played by the comet in rendering more credible the Copernican system was rather modest. It induced neither Tycho nor Mästlin to give up the postulates of Aristotelian physics; nor did it make them adopt any new world system. Rather, these two men, belonging to the scientific and cultural avantgarde of the late sixteenth century, arrived at their, in any case quite cautious, modifications of their world views as a result of other, mostly aesthetic, incentives. But by far the most conclusive derogation from the comet's alleged revolutionary status lies in the fact that, in discussing its role in connection with late sixteenth-century innovations of the world view, we have thus far only been able to mention the two names of Tycho and Mästlin.

d. *Lesser Gods*

The strong emphasis usually laid on the importance of the comet of 1577 in shaping the world views of leading astronomers such as Tycho or Mästlin, or in stimulating the acceptance of Copernicanism, has done more than just cast a somewhat distorted light on the genesis of their cosmological views. More importantly, it has been

140 Ibidem, 335.

responsible for an over-enthusiastic concentration on the alleged progressiveness of those figures who placed the comet above the moon. While, undeniably, Tycho's cometary observations were to exercise a greater amount of influence on subsequent cometary and cosmological theories than, for example, those of Andreas Nolthius, yet, dealing exclusively with the Dane's observations at the expense of the, very numerous, other tracts published about the comet inevitably results in a wrongfully biased representation of the historical reality of the year 1577.

In order to try to understand with what preconceptions and apprehensions the observers of the comet turned their eyes skyward in November 1577, we need to eliminate hindsight and treat Tycho, Mästlin, and the few others who favoured a superlunary location for the comet as representatives of only one (and probably not even the most plausible one) of many possible positions, a minority view that violated the still most influential theory of the generation of comets.

A large number of authors, among them almost all of the pamphleteers who seized the occasion of a bright cometary appearance to flood the populace with astrological forecasts and dire predictions, continued to adhere to the time-honoured Aristotelian theory.[141] None of the astrologers attempted any determination of the comet's parallax, which, for their purposes, would have been quite unnecessary, since the cometary characteristics they had to take into account in making their predictions included the colour, shape, direction of the tail, and zodiacal position of the comet, but not its distance from the earth. These authors continued to work within the bounds of the Arabo-Aristotelian tradition that had, since it appeared on the northern European scene in the thirteenth century, become firmly entrenched in, both learned and popular, culture, as an integral part of the Aristotelian world view. Whereas, due to the very limited literacy and the absence of printing, most of this rich astrological tradition of the Middle Ages has become invisible to the historian, the enormous number of astrological pamphlets that were printed during the sixteenth century should be regarded as a mere continuation of a tradition that had been alive throughout the later Middle Ages. The astrological tradition was very much bound up with the Aristotelian world view, and those who earned their living by means of issuing prognostications, even if they had known about the possibility of con-

[141] See Hellman, *Comet of 1577,* chapter VIII.

ceiving an entirely new cometary theory with potentially lethal consequences for their astrological world views, would definitely not have been eager to adopt it.

However, the astrologers were certainly not the only ones who were reluctant about allowing comets to move from the filthy regions under the moon to the immaculate heavens above it. Almost all the academics who published their views on the comet of 1577 proved equally averse to the idea. Indeed, their unwillingness may stem in large part from the same considerations as that of the astrologers. It was not uncommon for sixteenth-century scholars, especially doctors, but also university teachers, to write almanacs and prognostications themselves. They were as firmly rooted in the medieval astrological tradition as were their more popular counterparts, and therefore probably equally committed to maintaining the cosmological status quo. Thus Thomas Twyne, a friend of John Dee who had studied medicine at Oxford, was the author of many almanacs and prognostications, among them a tract on the comet of 1577. In it, he propounded a perfectly Aristotelian account of the generation and effects of comets, according especial importance to the comet's position in the zodiac and the direction in which its tail pointed.[142]

Even men who knew Tycho well, corresponded with him, belonged to the same scholarly circle, and occupied important university positions, could not bring themselves to give up Aristotle's cometary theory. Thus, Bartholomaeus Scultetus, who had taught the young Tycho how to use the cross-staff, and whose mathematics lectures Tycho had attended while at Leipzig in 1562, determined a parallax value of 5° 22' for the comet of 1577. Acknowledging that Scultetus was a proficient mathematician, and that therefore his trigonometrical calculations were probably correct, Hellman is inclined to attribute the faulty outcome of his parallax computation to observational errors, and to the fact that he made but a single determination.[143]

Another correspondent of Tycho's, Thaddaeus Hagecius, one of the foremost astronomers of central Europe and imperial physician to Maximilian II and Rudolf II, had placed the nova of 1572 in the heavens. In fact, the results of his measurements of the new star's position were closer to Tycho's than those of any other observer.

---

[142] Ibidem, 286-289.

[143] Ibidem, 216, n. 105.

However, his observations of the comet of 1577 were much less successful. Although Hagecius employed state of the art, well-constructed instruments for his cometary observations, his results were nonetheless inferior to those of Mästlin, who, as we have seen, used only a piece of thread. Applying the theory of Regiomontanus, the Czech arrived at a parallax value of five to six degrees for the comet, placing it decisively below the moon. He criticized Mästlin on several grounds, thinking his data "uncertain and not pertinent to the investigation of parallax" and his hypothesis about the comet's motion "unfounded", adding that he knew no-one who supported Mästlin.[144] However, Hagecius later recognized that he had been wrong: in his work on the comet of 1580, he rectified his errors.[145]

Andreas Nolthius from Einbeck showed no such inclination to admit his own error. He had believed the nova to be a comet, composed of elementary matter, since, trying to determine its parallax trigonometrically, he had found it to be in the region of air, below the moon. Similarly, in his tract on the comet of 1577, he first of all stated the Aristotelian view that comets were sublunary vapours, ignited by Mars, Saturn, and Mercury. Presupposing this definition, he then went on to determine the present comet's parallax, believing that "because the comet was in the air far beneath the moon, the parallax must be determined before anything definite could be said about [its] position."[146] From two observations made at one hour's interval, he concluded, on the basis of a series of complicated but faulty computations (thus, he had no fixed basis for measuring parallax, and he thought that it could be determined by a single observation)[147] that the comet's corresponding parallax values had been 4° 59' and 5° 32' respectively.

Equally error-studded were the calculations of Georg Buschius from Erfurt, son of a Nuremberg physician. For the nova, he had arrived at the incredible parallax value of 22° 40'. As for the 1577 comet, he reported it to have displayed a parallax of 2° 20', thus placing it in the elementary part of the world, between the earth and the moon. However, he gave no indication of the way he had arrived at this large parallax value, saying that it would be too difficult for

[144] Ibidem, 204.
[145] Ibidem, 193.
[146] Ibidem, 222.
[147] Ibidem, 225.

the ordinary man to understand.[148] However, he did include a clear elucidation of what comets were: they were meteors, formed out of hot, sulphurous, saltpeterish and terrestrial materials drawn up by the forces of the sun, moon, and stars.[149] Interestingly enough, he explained the fact that comets' tails are always averted from the sun, not by means of the optical theory, but simply by saying that "the sun rules all the comets."[150]

Tracts like those of Nolthius, Buschius, and Scultetus might just as easily have been written in the fifteenth century. Starting from the idea that comets were sublunary vapours, they were not in a position to detach themselves sufficiently from their preconceptions to subject their observations to correct computation. Significantly, Mästlin chided many of his fellow observers for *wanting* the comet to be elementary, and consequently making every effort to follow Aristotle.[151] Very probably, at least in the case of Nolthius and Buschius, a large part was also played by their limited proficiency at mathematics, and by the fact that they based their calculations on the faulty method of Regiomontanus.

Yet, even those who were capable of executing more or less correct parallax measurements tried their best to make their findings fit their preconceived notions. Thus, Paul Hainzel and Caspar Peucer, both among the very ablest astronomers of their generation, independently of one another reached the conclusion that the nova of 1572 showed no parallax. But both of them went on to say that, nonetheless, it must have been a sublunary phenomenon.[152] Another, equally conservative solution was adopted by Valesius of Covarruvias, physician to Philip II of Spain, who suggested that, although the nova was a heavenly phenomenon, it was not new. Rather, it was an old and faint star that had become brighter because of a change of air in between it and the earth, or because of a condensation in one of the spheres through which its light passed.[153] As we have seen, a similar solution, reminiscent of the Stoic idea that there were more things in heaven than those seen by human eyesight, had been adopted by Cardano to render his version of the optical theory ac-

[148] Ibidem, 229.
[149] Ibidem, 230.
[150] Ibidem, 231-232.
[151] Ibidem, 149.
[152] Ibidem, 115.
[153] Ibidem, 117.

ceptable to those believing in the Aristotelian incorruptibility of the heavens.[154]

Stoic conceptions also proved their mettle in connection with the problems posed by the comet of 1577. A number of astronomers, in the light of recent findings about comets' distances from the earth, felt inclined to adopt the idea put forward by Pena, that the region of air extended from the earth to the fixed stars. This allowed them to save the Aristotelian exhalation theory, while admitting that some comets were situated much higher than the moon.[155] In fact, such a presupposition eliminated the need for making any choice in favour of either the Aristotelian theory or an alternative cometary theory, such as the view that comets were spherical lenses: once it was admitted that the dichotomy between heaven and earth was non-existent, the question of what material comets were made of lost its relevance. Thus, Johannes Praetorius stated that comets were terrestrial vapours, that, however, became visible not because they were on fire, as Aristotle would have it, but because they were illuminated by the rays of the sun.[156] Such a compromise between the optical theory and the exhalation theory was one possible solution to save Aristotle while at the same time allowing for superlunary comets.

Another, quite similar solution, adopted by most of those who recognized that the comet of 1577 was located in the heavens, was to assert that, while some comets were beyond the moon, others were in fact sublunary. Helisaeus Roeslin, who admitted that recent parallax measurements had demonstrated that the comet of 1577 belonged in the aethereal regions, concluded that this meant that Aristotle's doctrine did not apply to all comets: at least some of them were above the moon, while others were terrestrial vapours.[157] The same idea

---

[154] See p. 85-86.

[155] Among those who adopted this solution were Johannes Praetorius, Henry Brucaeus, and Christopher Rothmann, see Thorndike, *History of Magic*, VI, 83-84.

[156] Tycho Brahe, *De mundi aetherei recentioribus phaenomenis* (Uraniborg 1588) in J.L.E. Dreyer, *Tychonis Brahe Opera Omnia* IV, 356: "In hoc tamen Cometa non adeo impense, intermediantibus certis Observationibus, Mathematicorum invictas Demonstrationes consuluit [Praetorius], verum se a communiter approbata Peripateticorum, de Cometarum causis & Generationibus, doctrina, seduci passus est, adeo ut Cometam hunc, molem e vaporibus Terrestribus concretam appellare non dubitarit. Licet vero disputet, an halitus illi accensi, vel a Sole illuminati fuerint, malitque cum Apiano (qui idem in quibusdam Cometis a se animadversis asserebat) sentire, exhalationem illam subtilem, e qua Cometa hic constabat, non fuisse inflammatam, sed potius illustratam a lumine Solari...".

[157] Hellman, *Comet of 1577*, 172.

was propounded by Mästlin, who held that the Aristotelian theory of comet generation could only be applied to comets in the elementary region, implying that such comets in fact existed.[158]

Thus, the immediate reactions to the comet of 1577 were not much different from those elicited by previous hairy stars. The majority of those writing for a popular audience continued to make use of the Aristotelian theory, with all its astrological connotations, and failed to be interested in the question of parallax. Many of the academics who did try to determine the comet's distance from the earth, did so on the basis of Aristotelian preconceptions and used the faulty method of Regiomontanus, arriving at exactly the kind of erroneous parallax values as the members of the Vienna school of astronomy, one century earlier. And finally, those who did measure the comet's parallax correctly, employed similar tactics as the early sixteenth-century adherents of the optical theory: they either admitted that some comets were above the moon while at the same time retaining the Aristotelian, terrestrial variety; or they made use of the well-known Stoic theory of the unity of heavenly and earthly matter to render viable the idea that the sphere of the moon did not pose any impassable barrier, and that comets might well rise up all the way to the heavens.

5. *Conclusion: The Aristotelian World View Revisited*

The world view Aristotle contrived never reigned supreme in the Latin West. From the outset, it had to contend with various usurpers, with whom uneasy alliances needed to be struck. Its earliest rival had been Christianity, several of whose dogmas were at variance with a number of fundamental Aristotelian postulates, such as the eternity of the world or the impossibility of a plurality of worlds. After a number of initial clashes, the most important of which resulted in the famous Condemnation of 1277, the pagan Aristotle and the almighty Christian God made a precarious pact: although Christian scholars were obliged to admit that, in a universe governed by an omnipotent deity, any special or counternatural effect was theoretically possible, they were free to conclude that, in any but the most unusual circumstances, nature would conform to the laws framed for her by Aristo-

[158] Ibidem, 152.

tle's physics. This agreement resulted in the method of scholasticism: arguments on both sides of a particular question were always adduced, and both pros and cons received their due amount of attention; however, in the end the conclusion was invariably reached, that, despite the fact that God was capable of anything short of a logical contradiction, to state that he actually *had* created a plurality of worlds, or that he actually *had* set the earth in motion while stabilizing the heavens, was an absurdity. The laws truly effectual in the realm of nature, were those that were in conformity with Aristotelian postulates.

An equally teetering balance was struck between Aristotle's physics and Ptolemy's astronomy. Coherent and harmonious though it was, the Aristotelian arrangement of the heavenly bodies was in no way capable of accurately accounting for their motions. Several solutions to this thorny problem were proposed: either the eccentrics and epicycles needed to save the appearances were all embedded in the concentric orbs of Aristotelian cosmology, or, conversely, they were denied any reality at all, and regarded as mere calculatory devices whose sole aim was to enable astronomers to make accurate predictions.

During the sixteenth century, the weaknesses of these two compromises, the one between Aristotle and God and the one between Aristotle and Ptolemy, were slowly but surely beginning to be exposed by a number of novel developments. Due to the humanist effort at unearthing the ancients, and the Renaissance predilection for recovering the founts of pristine wisdom that had become muddied and diluted by the intermediary centuries of Aristotelian predominance, a number of scholars enthusiastically turned to ancient, non-Aristotelian philosophies, such as Platonism and Stoicism. Thus, the Stoic idea of continuous, fluid heavens consisting of the same matter as the elementary region, underwent a revival with several sixteenth-century thinkers. They came to deny the idea of any fundamental difference between celestial and terrestrial regions, and, harking back to the authority of the Church Fathers, who had of course been influenced by Platonism and Stoicism rather than Aristotelianism, some of them, like Robert Cardinal Bellarmine, took the logical step of endorsing the idea of celestial corruptibility.

An equally conservative attitude lay at the root of Nicholas Copernicus' solution to the strain in the relations between Aristotle and Ptolemy. Whereas, before, arguments in favour of the earth's motion and the stability of the starry vault had been disposed of as utter absurdities, Copernicus now ventured to affirm them as true. His motives, like those of the advocates of corruptible heavens, were traditionalist: he meant to solve the insufferable tension between the highly complicated devices of Ptolemaic astronomy and the postulates of Aristotelian physics by reverting to the ancient idea of Aristarchus of Samos, that it actually was the earth that moved. The main advantage of such a solution lay in the aesthetically appealing harmony between the distances of the planets, which Copernicus regarded as the restoration of the Stoico-Platonic symmetrical structure of the universe, and the mutual sympathy between its parts.

If the Aristotelian world view had been completely at odds with the new ideas propounded by such men as Bellarmine or Copernicus, Aristotelianism might have been swept away relatively easily once the theories of celestial corruptibility and heliocentrism had proved their mettle. But, in fact, it was not. Despite the fact that both Copernicanism and Stoicism took on quite well during the final decades of the sixteenth century, and appealed to many thinkers, they did so within essentially Aristotelian contexts. Impressed with its accuracy and its harmonious dimensions, astronomers embraced the new heliocentric theory propounded by Copernicus, while robbing it of its heliocentrism: their main efforts were aimed at rendering it compatible with geocentric, Aristotelian cosmology, in the way they habitually treated astronomical hypotheses. Similarly, the adoption of the Stoic notion of the unity of celestial and terrestrial matter in no way obliged them to break with Aristotelian physics in any decisive way.

The comet of 1577 has often been credited with facilitating the break with Aristotle and the adoption of Copernicus. However, the purpose of this chapter has been to show that this is a very oversimplified representation of historical reality. First of all, long before 1577, several authors had adopted the notion that at least some comets were situated beyond the moon, under the influence of the discovery of the antisolarity of cometary tails. Yet, apart from Cardano, none of them asserted that *all* comets were superlunary; although they abandoned Aristotle's account of comets' constitution, this did not necessarily entail the renunciation of more fundamental Aristotelian tenets, such as geocentricity or the dichotomy between heaven and earth. Superlunary comets could easily be accounted for by an appeal to Seneca's idea, that some heavenly bodies might be hidden in space, only to become visible when they approached the earth, or when the atmosphere was sufficiently clear, or by pronouncing them to be divine miracles outside the reach of the ordinary laws of nature.

Nor did the comet of 1577 incite anyone to change their world views, or to embrace Copernicanism. Mästlin, who had already adopted Copernican views on account of their aesthetic appeal, in his treatment of the comet remained completely traditional: he employed the heliocentric planetary spheres as convenient calculatory devices. The comet did provide Tycho Brahe with a means of overcoming the final obstacle to the plausibility of his own, Tychonic world system; however, the role it played was a very minor one, and, moreover, Tycho's geoheliocentric world system was essentially conservative and did not upset any of the tenets of Aristotelian physics. Most importantly, the vast majority of observers of the comet continued the medieval tradition: either they were not interested in the comet's location at all, or, if they were, their calculations were based on an erroneous method that provided them with values consistent with the Aristotelian idea of sublunary comets.

Thus, the Aristotelian world view was capable of incorporating a great variety of positions. Yet, during the sixteenth century, it came under attack from several sides. Around the year 1577, it seemed as if the challenges posed by Stoicism and heliocentrism could be reconciled relatively easily to Peripatetic tenets, as had Christianity and Ptolemaic astronomy before them; but the seeds of change had been sown, and the flexibility of Aristotelianism was being stretched to its limits. As far as comets were concerned, there were several ways of accounting for them. Roughly, three different attitudes toward the hairy stars can be distinguished: there were those who indiscriminately adopted the Aristotelian cometary theory; those who knew about recent findings concerning the nature of the hairy stars, and in various ways tried to reconcile these with their traditional physical conceptions; and those who were taking the first faltering steps on the road that was to lead to the breakdown of those conceptions.

CHAPTER FIVE

SENECA SAVES ARISTOTLE: COSMOLOGY IN THE DUTCH TRACTS

1. *Learned Tracts and Pamphlets*

When, in the early evening of 13 November 1577, Tycho Brahe, while out catching fish for his dinner, was startled by the sight of a great comet, the Low Countries were engulfed in war. For the whole of the sixteenth century, the Habsburg Netherlands had been plagued by strains between the centralizing Brussels government and local authorities anxious on safeguarding their traditional jurisdiction and acquired privileges. Although, during the first half of the century, these tensions had seemed manageable, with the renewed outbreak of war against France from 1540 onwards dissatisfaction had mounted.[1] By the 1570s, it had turned into revolt, and while Tycho was carefully scrutinizing the movements of the comet in the peace of his Hven observatory, civil strife and looting troops were ravaging the Netherlands and disturbing the calm of its scholars.

The sixteenth century had seen a great deal of turmoil, in the Netherlands as elsewhere in Europe. As early as the 1520s, Luther's ideas had become widely known throughout the Low Countries, and, according to Erasmus, his writings were circulating everywhere.[2] Lutheranism, however, was by no means the only unorthodox doctrine to make headway among Dutch believers. Other, doctrinally and organizationally much less coherent religious sects also exercised a considerable appeal, and gave the early Dutch Reformation its distinctive, fragmented character.[3] Many of these dissident sentiments were inspired by the medieval eschatological tradition.[4] The chiliastic

[1] See Jonathan I. Israel, *The Dutch Republic. Its Rise, Greatness, and Fall 1477-1806* (Oxford 1995) 129-137.

[2] Ibidem, 79.

[3] Ibidem, 84-85.

[4] J.J. Woltjer & M.E.H.N. Mout, "Settlements: the Netherlands" in Thomas A. Brady, Heiko A. Oberman & James D. Tracy eds., *Handbook of European History 1400-1600: Late Middle Ages, Renaissance and Reformation. Volume II: Visions, Programs and Outcomes* (Leiden/ New York/ Cologne 1995) 385-415: 390.

yearnings of the Anabaptists led to such excesses as the disastrous attempts at establishing new Jerusalems at Münster in 1534 and at the abbey of Oldeclooster, near Bolsward in Frisia, in 1535. After these experiments had ended in tragedy, Dutch Anabaptists reacted in two fundamentally different ways: Munsterites and Batenburgers continued their aggressive, sometimes even terrorist, stance, preaching the impending end of the world and pillaging the countryside, while Melchiorites, Davidites, and Mennonites advocated a pacifistic attitude.[5] These diverse strands of Anabaptist heterodoxy provided most of the victims of government persecution: although Lutherans and Calvinists were also afflicted, the Anabaptists' defiance of government authority and the established Church rendered them especially vulnerable.[6]

By no means all those entertaining dissident ideas were as easily identifiable as the Anabaptists, though. A far more numerous and potentially threatening group, from the Habsburg government's point of view, was constituted by those who held reformist views in secret, without openly advancing them. Many of these so-called 'Nicodemists' belonged to the country's élite: they were regents, nobles, scholars, or clergymen, and because of their positions they preferred to harbour their dissident opinions covertly.[7] Diverse strands of spiritualism, advocating a dichotomy between superficial outer reality and inner, spiritual truth, made considerable headway among these groups. Especially the emphasis placed by all spiritualists, including David Joris and Hendrik Niclaes, founder of a secret sect called 'Family of Love', on the direct entry of the spirit of God into the hearts of the believers, without any need of mediation by Church officials, proved appealing: it allowed one to seek direct communication with God, while at the same time outwardly conforming to Church rituals.

Religious and intellectual life in the sixteenth-century Netherlands, then, was in turmoil. In the highly urbanized Low Countries, both north and south of the great rivers, literacy was higher than elsewhere in Europe[8], and many of the new religious ideas that were causing more and more unrest among all levels of the population

[5] Israel, *Dutch Republic*, 88-89; Woltjer & Mout, "Settlements", 391.
[6] Israel, *Dutch Republic*, 93.
[7] Ibidem, 94.
[8] Israel, *Dutch Republic*, 686.

were spread in print. The writings of all the great reformers, as well as German spiritualists such as Sebastian Franck and Caspar Schwenckfeld, circulated widely; Menno Simons published a series of works in Dutch urging the general public and the authorities to follow no government but Christ's; and Dirk Volkertsz Coornhert produced many publications advocating individual piety.[9]

Yet, despite the intellectual ferment pervading Dutch society, and the widespread expectations of impending chiliastic cataclysm, interest in potentially ominous cosmological events seems to have been relatively modest in the Low Countries prior to the outbreak of the Revolt. The great conjunction of the superior planets in the year 1524, believed by many to be the harbinger of a second deluge, caused great panic as well as the publication of an enormous number of pamphlets in the German Empire[10]; the neighbouring Low Countries, though, seem to have managed to retain their composure relatively well.[11] In comparison with the proliferation of prognostications and pamphlets describing natural occurrences such as eclipses, conjunctions, or earthquakes, that flooded France and the German Empire, the Netherlands were rather ill-supplied, at least during the first half of the sixteenth century. During the reign of Charles V, a number of pro-Habsburg prophecies appeared, foretelling great deeds the emperor was to accomplish; the printing of such works was often stimulated by the authorities.[12] Frequently, foreign prognostications

[9] Woltjer & Mout, "Settlements", 393; Israel, *Dutch Republic*, 92, 99.

[10] See Paola Zambelli, "Many ends for the world. Luca Gaurico instigator of the debate in Italy and Germany" in Paola Zambelli ed., *Astrologi Hallucinati: Stars and the End of the World in Luther's Time* (Berlin/New York 1986) 239-263; Paola Zambelli, "Fine del mondo o inizio della propaganda?" in Leo S. Olschki ed., *Scienze, credenze occulte, livelli di cultura* (Florence 1982) 291-368; and, for the enormous boom in pamphlet production around the year 1524, Hans-Joachim Köhler, "The *Flugschriften* and their importance in religious debate: a quantitative approach" in Zambelli ed., *Astrologi Hallucinati*, 153-175. The great panic seems to have been caused mainly by the woodcuts on the pamphlets' title pages, since most people tended to look only at the title page. See Helga Robinson Hammerstein, "The Battle of the Booklets: Prognostic Tradition and Proclamation of the Word in early sixteenth-century Germany" in Zambelli ed., *Astrologi Hallucinati*, 129-151: 141.

[11] The only pamphlet on the subject of the conjunction to be found in the great Dutch pamphlet collections was a quodlibetical question by a Louvain master of arts, printed at Antwerp in 1522 (*Quodlibet Magistri Thomae Montis Medici & astrologi de significationibus coniunctionum superiorum planetarum quae erunt anno 1524 in Februario, Per eundem ex diversis auctoritatibus collectum ac responsum in alma universitate Lovaniensi in scolis artium*, Antwerp, 1522; Petit I, 9 (1522)).

[12] M.E.H.N. Mout, "Prognostica tijdens de Nederlandse Opstand" in C. Augustijn et al. eds., *Kerkhistorische opstellen aangeboden aan Prof. dr. J. van den Berg*

were simply translated to supply the Dutch with insights into the future—among them, a translation of a German work on the comet of 1527.[13] Yet, during the 1530s and 1540s, several indigenous, mostly Flemish, astrologers, such as Peter van Moerbeeck, Jan van Musseem, Cornelis Schutte, and one 'Solomon the Jew', also disclosed to the public their calculations concerning the events of the following years.[14] Sometimes, altercations would ensue; thus, in the years 1530 and 1531, Gaspar Laet, physician of the city of Antwerp, and Jan Thibault, physician and astronomer to the emperor, were at loggerheads over each other's competence in matters prognostic.[15]

Another, probably rather more learned controversy, in which not the competence of a particular pair of astrologers, but the legitimacy of astrology as a whole was at stake, took place in the 1550s. The debate over the worth and practicability of astrological prediction was an old and venerable one, that went through an especially vivid phase during the sixteenth century.[16] In the Netherlands, the Bruges physician Francis Rapard represented the sceptical stance: in his *Great and Perpetual Almanach,* published at Antwerp in 1551, he ridiculed many of the revered tenets of the astrological art. Thus, he maintained that predictions of wars or other political events, or even of the weather, were impracticable nonsense, and he denied that medical treatment ought to be determined by astrological means. His Brussels colleague, Peter Haschaert, one year later indignantly replied with an "Astrological Shield Against the Scourge of Francis Rapard", defending the usefulness of astrology, especially in medical matters. In order to convince his audience of the veracity of his

(Kampen 1987) 9-19: 10-11. See for example Petit I, 34 (1531): *Pronosticatie in forme van Prophecie, beginnende vanden Jare ons Heeren 1531 te loouen en ter eeren den... Keyser Carolus V...* (Antwerp 1531). See also G.D.J. Schotel, *Vaderlandsche volksboeken en volkssprookjes van de vroegste tijden tot het einde der 18e eeuw* I (Haarlem 1873) 99-102.

[13] Petit I, 24 (1528): *Vander wonderliker openbaringhe deser teghewoordighe teyken ende cometen des hemels verschenen, in welcker manieren ende ghedaenten die van vele gheloof-achtegher personen gesien sijn... Gheprent Tantwerpen bi mi Willem Vorsterman, voor Jan Haselberch van Constans. Int jaer o.h. 1528.*

[14] Petit I, 41, 43, 49, 87; Knuttel 86, 91-94.

[15] See Petit I, 27 (*Correctorie ofte bewijsinghe des erroren oft fauten beuonden in diuersche Pronosticatien ende Almanacken vanden Jare 1530 door die diligentie vanden ionghen Meester Gaspar Laet Medecijn der Stadt van Antwerpen*); I,32 (*Pronosticatie van Meester Jan Thibault Medicijn ende Astronomijn vander Keyserl. Mai. vanden Jare ons heeren 1531 ghecalculeert op den orizon van ... Antwerpen*) and I,33 (*Defensie Responsijf Astrologick by Gaspar Laet de Jonghe, op die sotte ende dwalende Apologie, tseghen hem doer Jan Thibault printer ende lettersetter, int jaer van 1530 wtghegheven*).

[16] See below, chapter VI.

assertions, Haschaert enumerated both recent and ancient examples of the effects produced by comets and eclipses. Moreover, his interest in celestial occurrences was not short-lived: he went on to issue a series of prognostications, and he was the only Netherlander we know of to have written about the comet of 1556.[17]

Still, despite the activities of a number of physicians issuing annual prognostications, and despite a small contribution to the current international debate on the validity of astrology, interest in celestial occurrences was relatively low among the general public of the Netherlands prior to the appearance of the nova in 1572, judging by the very limited number of pamphlets that appeared on the subject. Nor does the learned environment of Louvain, at the time still the only university of the Netherlands, seem to have been very responsive to events such as the appearance of comets. Virtually the only Dutch scholar to have devoted some attention to the hairy stars during the earlier half of the sixteenth century was the mathematician and physician Gemma Frisius, whom we have already mentioned in connection with the optical theory of comets.[18]

Gemma was born at Dokkum, in northern Frisia, on 8 December 1508; in 1525, at the age of seventeen, he was sent to Louvain university, where he was to spend the rest of his life.[19] He was one of the foremost scholars of his age, friendly with such men as Andreas Vesalius (whom he used to help steal corpses from the Louvain gallows) and Gerard Mercator, who was his pupil, and received at the court of Charles V. According to a historical work by Gulielmus Zenocanus, printed in 1559[20], the appearance of the comet of 1533 gave rise to a controversy between Copernicus, Apian, Hieronymus Scala, Cardano, and Gemma Frisius. Sadly, though, Copernicus's work on this comet was lost, and the dispute in question cannot be

---

[17] See Lynn Thorndike, *A History of Magic and Experimental Science* V (New York 1941) 328-329; also Mout, "Prognostica", 10. Cornelius Gemma also observed the comet of 1556, as well as that of 1558; however, he never published his findings until 1575, in his much more wide-ranging work, *De naturae divinis characterismis* (Antwerp 1575), Book II, chapter 1, 26-35. Johannes Portantius, in his pamphlet on the comet of 1577, (see below) also claimed to have observed the 1556 comet, but, if he ever published his findings at all, they appear to have been lost.

[18] See supra, p. 116.

[19] See for details on Gemma's life and work Fernand Van Ortroy, *Bio-Bibliographie de Gemma Frisius, fondateur de l'école belge de géographie, de son fils Corneille, et de ses neveux les Arsénius* (Brussels 1920; reprint Amsterdam 1964).

[20] *De republica vita moribus gestis. . .imperatoris. . .quinti Caroli. . .Libri septem.*

adequately traced.[21] Luckily Gemma later recorded his own observations of the comet in his *De Radio Astronomico et Geometrico Liber* (Antwerp 1545), where he endorsed Apian's discovery of the antisolarity of cometary tails and expounded the theory that the comet's tail was an effect of refraction.[22] Although Gemma, like almost every physician of his age, was well versed in astrology, his treatment of the comet of 1533, as put down in his *De Radio Astronomico,* was in no way astrological.[23]

One might expect Dutch interest in celestial phenomena to have waxed somewhat with the appearance of the nova in 1572. This highly unusual and awe-inspiring rarity, which as we have seen caused feverish excitement throughout Europe, appeared at an especially portentous moment in the history of the Low Countries. Six years after the iconoclastic fury, five years after the arrival of the Duke of Alva, and only a few months after the capture of the small town of Brill by the rebel Sea Beggars had made the Revolt turn into a civil war[24], one would expect the advent of a new citizen of heaven, with all the biblical connotations it entailed, to provoke the Dutch to put their pens to paper, if nothing else could.

Still, the new star did not elicit many more publications than the comets that had appeared earlier in the century. The unusual occurrence inspired only a few Netherlanders to attempt to explain its nature and meaning, the most important, and most verbose, of these being Cornelius Gemma, the son of Gemma Frisius, who had succeeded his father as professor of medicine at Louvain. It might be no exaggeration to say that the appearance of the nova gave Gemma's life a new turn. He regarded it as a wonderful new creation of God, a divine sign, sent by the Almighty by way of a message for people to decipher, and as a harbinger of the dawn of a new era of wholesale renewal and improvement. Its meaning could not be adequately grasped by means of the traditional sciences of Ptolemy, Plato, and

---

[21] Thorndike, *History of Magic,* V, 410.

[22] See supra, p. 116.

[23] Thorndike, *History of Magic,* VI, 184, suggests that Marcellus Squarcialupus, in his *De cometa in universum atque de illo qui anno 1577 visus est opinio* (1578) may have criticized Gemma Frisius for prolonging the astrological effects of comets for forty years. However, it is much more probable that Squarcialupus's criticism was directed at Gemma's son, Cornelius Gemma; he referred to Cornelius's opinions on numerous other occasions in the same work, and, moreover, Cornelius did indeed prolong the effects of the comet of 1577 for forty years.

[24] See below, § VI.3.a.

Aristotle, Gemma contended; rather, an entirely new science was needed, a lore of "Divine Signs in Nature", which he himself set out to compile, publishing the results of his efforts in his 1575 *De naturae divinis characterismis.* Immediately after the nova had appeared, in 1573, Gemma also composed a work devoted entirely to the new star, which saw several editions, both in Latin and in Dutch, during the 1570s.[25]

In his discussion of the nova towards the end of the second book of his *De naturae divinis characterismis,* Gemma chided his Spanish colleague, Hieronymus Muñoz, for believing the new star to be a celestial comet, located in the sphere of the fixed stars. He argued that, although one might assign to the heavens a certain measure of mutability, it could not be maintained that the law of nature would tolerate so great and sudden a change as was the appearance of the nova. Therefore, Gemma concluded, the generation of the new star was not effectuated by natural means; rather, it was called into being supernaturally by God, without the mediation of any secondary causes.[26]

Gemma had already propounded this idea in the, quite lengthy,

---

[25] *Stellae peregrinae iam primum exortae et caelo constanter haerentis* Φαινομένον *vel observatum, divinae providentiae vim & gloriae maiestatem abunde concelebrans per C.G.* ὁ σοφός μάντης ἄριστος (Louvain 1573); *Stellae peregrinae iam primum exortae et caelo constanter haerentis* Φαινομένον *vel observatum, divinae providentiae vim & gloriae maiestatem abunde concelebrans per D. Cornelium Gemmam. Lovanij Medicinae professorem Regium.* ὁ σοφός μάντης ἄριστος (Louvain 1573); *De nova stella iudicia duorum praestantium mathematicorum. D. Cypriani Leovitii a Leonica. Et D. Cornelii Gemmae, Professoris regij Louaniensis...* . (s.l. 1573); *De peregrina stella quae superiore anno primum apparere coepit, Clariss. Virorum Corn. Gemmae Louaniensis Germani. & Guliel. Postelli Barentani Galli, Ex Philosophiae naturalis, mysticaeq; Theologiae penetralibus deprompta Iudicia* (S.l. s.d.); *Dialexis de novae et prius incognitae stellae invsitatae magnitudinis & splendidissimi luminis apparitione, & de eiusdem stellae vero loco constituendo. . .per Thaddaeum Hagecium ab Hayck. . .Accesserunt aliorum quoque doctissimorum virorum de eadem stella scripta. . .*(Frankfurt 1574). For details see Van Ortroy, *Bio-bibliographie,* 378-386, who, however, fails to mention the Dutch translation of Gemma's *De stella peregrina* which appeared anonymously with the Louvain university printer, Jan Bogaerts, who had also published the Latin edition. It appeared under the title of *Des vremde sterrens nu eerst hem verthoonende, ende vast gheduerighens observatie: Waer door die wonderlijcke crachte der voorsichticheyt Gods, ende die Maiesteyt sijnder glorie, den menschen ooghen haer claerlyck openbaert.* Mout, "Prognostica", 12 ascribes this pamphlet to the Louvain theologian Johannes Molanus, who, however, only signed the imprimatur.

[26] Cornelius Gemma, *De naturae divinis characterismis* (Antwerp 1575), Book II, 270-271: "Primo quod huius gratia caelum mutabile doceat [Muñoz] iisdem argumentis quibus aliquando nos usi sumus: praeclarum sane & elegans paradoxon, sed tamen incertum, an quamvis inordinatae videantur illae centrorum mutationes iudicio mentis humanae, natura tamen certam in omnibus constituat legem, quae longo temporis cursu ad aequalitatem aptissimam redigatur. Ac postremo licet hanc passionis legem

Dutch-language pamphlet he devoted to the nova a few months after its appearance.[27] The new star, he said, was a "prodigy of the Lord" that had shone forth to demonstrate to mankind the divine providence and glory. Gemma was strongly convinced that the nova was not an ordinary star: it shone much brighter than any one of the fixed stars, and was even more brilliant than most of the planets, displaying wide and bright flaming rays, so that those who took it to be one of the regular celestial bodies were gravely mistaken.[28] Moreover, it could not possibly be a star, for the obvious reason that nothing new could be generated in the heavens: no generation or corruption could take place there, and the stars already in existence, fixed in number, were nothing but parts of the heavens formed of the same celestial matter, only somewhat denser.[29]

Nor could the new phenomenon be accounted for by calling it a comet or exhalation: it displayed none of the forms traditionally assigned to comets, "either hair or beard, tail or sword, lance or shield, horn or trumpet, or the figure of a pyramid or column"[30], nor did it manifest any irregular course. In fact, it displayed no movement at all, a feat that filled Gemma with the greatest possible wonder. Conversely, comets, being formed of coarse and greasy terrestrial vapours

---

caelis inesse ponamus; nemo profecto hanc tantam sit facile concessurus, ut spatio annorum duum sensibilis ulla mutatio fiat, qualis contingit in sublunaribus locis.. . .Quamobrem relinquitur ut haec tam praeceps mutatio & tamen in caelo fixorum syderum facta, non naturali lege quam ipsi somniant fieri possit, sed tota sit supra naturam, nec praeter voluntatem Opificis ullam causam obtineat propiorem."

[27] *Des vremde sterrens nu eerst hem verthoonende, ende vast gheduerighens observatie: Waer door die wonderlijcke crachte des voorsichticheyt Gods, ende die Maiesteyt sijnder glorie, den menschen ooghen haer claerlyck openbaert* (Louvain 1573). This pamphlet, written on 1 February 1573, was the second, slightly revised version of an earlier Latin edition that received the imprimatur on 17 December 1572.

[28] Ibidem: "Die maniere ende ghedaente des lichts boven alle andere sterren is seer te verwonderen, want sy alle die vaste sterren verre te boven gaet, *Syrium en procyona* ende oock by nae alle die planeten door haer groote en claer vlammende, straelen, als dat sy wel verre verdoelt moeten syn, die de selve voor een ghewoonlycke sterre aen sien."

[29] Ibidem: ". . .want die ghewoonlycke sterren hem al tsamen vertoonen, ende die van deerste, tweede, derde groette syn oock licht om te tellen, ende gheen niewe sterren en connen daer natuerlyck worden ghegenereert: want die nature des Hemels ingenerabel is, en incorruptibel, ende die sterre niet en sijn anders dan deelen des Hemels van eender substantie, dan dat sy dichter syn van haren schipper in een ghevrocht."

[30] Ibidem: "Vele min canmen daer een Comete oft exhalatie af maken, want waer is oyt Comete ghesien, sij en hadde die ghedaente van eenen haer oft baert, oft steert, oft sweert, oft pijke oft schilt, oft horen, oft trompette, oft die figure van eender pyramis, oft columne."

attracted by the power of the stars and planets, always pursued unpredictable paths, and with their sad and menacing complexions they infected the air, causing pestilence and death.[31]

Rather, Gemma contended, than calling it a star or a comet, the new inhabitant of the heavenly region—for it was definitely to be located beyond the moon, as our author was unable to detect any parallax for the phenomenon[32]—should be regarded as a divine prodigy, and mere mortals should not attempt to discover its meaning, in the knowledge that human reason was incapable of grasping such mysteries.[33] Yet, despite this sceptical stance, Gemma did not feel inhibited about expressing his hopes for the future: the prodigious new star, he suggested, might bring in its wake a great change, as had the star of Bethlehem before it. Possibly, it had come to herald the Lord's second coming, or, at the very least, a transformation on a more human scale, the disappearance of lust, pride, avarice, luxury, and enmity and the accession of peace, grace, and forgiveness, through the mediation of Christ.[34]

Such an impending great alteration in earthly affairs, Gemma believed, was not only a possible way of explaining the nova's meaning, but even a quite likely one. For several reasons, our author invested the star with a special, divine significance. First of all, the nova's place within the constellations was such that it constantly formed the

[31] Ibidem: "Want die comete ten laetsten altijt haer plaetse veranderen, altijt droeve staen ende vervaerlyck dreyghende, nemmermeer glensterende, ende meer oorsaeck syn van groot quaet ende miserie, dan voor sekere teken gheacht, want sy opghetoghen worden wt die grove vetticheyt der aerden, door die crachte van die sonne, stimulerende mitsgaders andere planeten, en sterren des firmaments, ende schickt hem soo in dopperste des lochts...".

[32] Ibidem: "Sekere parallaxim oft diversiteyt van aspeckt, en hebbe ick niet connen bemercken, ten waer omtrent die 4 min. op die morghenstonde, dwelck licht gheschieden mach door schijmelinghe der ooghen over soo cleyne ghescille...".

[33] Ibidem: "Maer Godt gheve van die apparitie wat sij wesen mach (tsij eenich natuerlijck licht, tsij bode oft voorlichter van den tweeden advent des heeren, met een ghedaente van sterre bekleet, om moghelyck die cirkels, ende die herren des werelts van den fondamente te commoveren) ick en wil daer gheen sentencie af besluyten, kennende die cleynicheyt mijns verstants, groot is altijt dat wonder dwelck ons door haer wort bediet, boven alle nature ende begrijp der menschelijcke philosophie."

[34] Ibidem: "Die selve Heer wil ons gunnen dat wij met by nae desghelycke prodigie, desghelycke veranderinghe in ons alle moghen verwerven, dat is (te weten) wtghestorven synde alle begheerlyckheyt, hooverdie, ghiericheyt, luxurie, onghenaede teghen malcanderen etc. Dat wy eens moghen met den niewe Jaer in Christo herboren worden, ende Christus in onslieden die welcke mach oock af vaghende die tranen sijn volcks, peys en ghenade van als, ende verghiffenis gheven."

figure of a lozenge with the three neighbouring stars of Cassiopeia, so that the two diagonal lines that could be drawn within the lozenge appeared in the shape of a cross, with the nova shining at the top of it. To Gemma, this state of affairs was charged with the highest significance, since such a cross-shaped constellation was to be found nowhere else in the heavens. It was reminiscent of the sign sent to Moses in the wilderness, or of the cross that had appeared to Constantine the Great. In the same way that the human soul was capable of comprehending certain prophetic visions, and of transmitting them to the body by way of the spirits, Gemma believed, likewise the power of God, poured forth into the world like a soul, often revealed to mankind the figure of coming events by way of corporeal substances.[35] Furthermore, the nova had appeared at the principal point of the heavens, the intersection of the Milky Way and the equinoctial colure. Because the Milky Way divided the heavens into a left-hand and a right-hand side, and the equinoctial colure into a top and a bottom half, the spot where the nova shone was especially portentous: it marked the division of right and wrong, truth and falsehood, and therefore indicated a great and rare alteration.[36]

A pamphlet of rather a different tenor was published at Antwerp on 15 December 1572 by an anonymous author.[37] Unlike Gemma, who as we have seen expressed a chiliastically inspired hope for renewal and improvement heralded by the nova, his colleague showed himself rather less optimistic. He accorded the new star a much less

[35] Ibidem: "...om dat die generael crachte Godts (als een ziele die werelt inghestort) door compassie van ons gheslachte, dickmael in een lichaemelycke substantie schildert die ghedaente van toecomende dinghen: ghelyckerwys in sommighe particuliere menschen, dickmael die ziele door Godts permissie opghetoghen (tsij bij ruste, oft buyten slape) begrypende is sekere visioenen, die sy eerstemael door den gheest den lichaem, als haerder huysvrauwe communiceert, ...".

[36] Ibidem: "Maer het meeste wonder van allen, hoortmen niet slapende voor bij te gaen, dat is van haerder plaetse de welcke bevonden wort op tprincipale poinct des hemels, in die vergaderinghe van Galaxias, ende die colurus equinoctij verni, dwelck is die principaele duere ende ghewelfsel des Hemels, treckende door die polos, ende beyde die poincten van Aries, en Libra, daer die nachten, ende daeghen haer verghelycken, ende trecht van slincks, thoege vant neder scheyden als duer die weerlycke, ende gheestelycke rechte die waerheyt van dolinghe onder die versaemlinghe der menschen. Ende daerom ghelooflyck schynt dat die selve apparitie meest trecken wilt op een groette, ende selsaem veranderinghe des werelts, ...".

[37] According to the title, this pamphlet, like Gemma's, was also the second, translated edition of a previous Latin version: *Van die wonderlijcke nieuwe Sterre openbarende in den hemel/ die noyt voor desen tijt ghesien is gheweest/ Wt den Latijne in Duytsch ouergheseedt.* (Antwerp 1572).

noble origin than Gemma: it could in no way be said to be a supernatural or divine apparition. Instead, like comets, it was a fiery meteor, consisting of fat, hot and dry exhalations. Moreover, evidently, its matter was even denser and more sluggish than that of which the hairy stars were formed: whereas comets did move, albeit slowly, the nova had not budged from its place of appearance for over a month, and would probably remain stationary for much longer.[38]

Having established to his own satisfaction the similarity in nature between the nova and comets, our author went on to subject it to a perfectly traditional astrological treatment, based on the idea that its constancy and brightness indicated its association with Saturn and Venus. It had therefore worked Saturnine and Venereal effects, such as cold weather with severe frost. Moreover, the nova displayed a certain affinity with the Saturnine comet that had appeared in 1556, standing at the same spot where that comet had vanished. The author therefore believed that the new star indicated evil consequences: "Our forefathers testify that strange heavenly fires have never appeared without any evil meaning, but always portend heat, sterility, pestilence, popular unrest, and the deaths of princes and great lords."[39] However, he ended his pamphlet on a more optimistic note, reminiscent of Gemma's hopeful expectations: God had surely not created this new star without a purpose, and, although it might portend great dangers and wretchedness, it might, on the other hand, herald an end to misery, war, and revolt, and a great,

[38] Ibidem: "Sy wort ghehouden van der selver materien ende substantien ende seer ghelijck te zijne als alle vierige teeckenen des hemels te weten van vette werme ende drooghe exhalatie. . . . Ende van alsulcken materien schijnen te wesen die Cometen, van welcken sy ghemaect ende ghevoedt worden die een wijle tijts huer plachten te openbaren ende eenen traghen voortganck te hebben. Maer dese sterre welck ons nu in de locht schijnt ende openbaert en canmen gheensins ghemercken datse huer verroert oft voortgaet, al ist nochtans dat sy der Cometen natuere seer ghelijck is ende een teecken is wt eenen wermen ende drooghen exhalatien ontsteken: daerom moet nochtans huere materie grover wesen dan van den Cometen ende dat ooc veele, het welck aen hueren voortganck goet is te bemercken, ende is die selve ooc een oorsake huers gheduericheyts, Want sij nu langher als een maent ghestaen heeft, ende mocht noch wel veel langher staen."

[39] Ibidem: "Ende dat dese wonderlijcke sterre ons niet veel beter en sal betuyghen oft voorsegghen. Onse voorvaders betuyghen ons, dat die vreemde hemelsche vieren noyt sonder eenige quade beteeckenisse en hebben gheopenbaert, maer de selve altijt hitte, onvruchtbaerheyt, pestilentien, oproer onder tvolck, Princen ende grooter Heeren doot ons alle tijt presageren."

world-wide reformation and triumph of the Catholic Christian Church.[40]

The two pamphlets described above are the only remaining works published by Dutch authors to deal specifically with the nova of 1572.[41] Apparently, the increasing unrest in the Low Countries following the dramatic events of the previous years, such as the Iconoclasm, the institution of Alva's Council of Troubles, and the exodus of Protestants trying to escape from the harsh Spanish repression, had not rendered the Dutch any more receptive to possible heavenly messengers than had the widespread religious unrest of the previous decades (unless, of course, our vision is being obfuscated by the possible fact that many specimens of prognostic literature have gone missing or been destroyed). Yet, the Dutch writings on the nova that do remain reveal to us several characteristics that also mark the, much more profuse, literature on the many comets that were to appear within the next few decades.

First and foremost, they give an indication of the several different vehicles that were available for communicating to the public one's opinions on heavenly phenomena. The behaviour of the celestial bodies, and especially the appearance of new arrivals in the superlunary region such as comets or novae, was (and is) an outstanding example of a topic that can be of great interest to scholars and the wider public alike. This circumstance is reflected in the means chosen by those describing them. As we have seen, Cornelius Gemma, professor of medicine at Louvain, aired his opinions on the nova in several different ways. He divulged his first reaction to the phenomenon in a short pamphlet in Latin, of which, however, he soon prepared a revised, Dutch translation, like its Latin counterpart printed in pamphlet form by the university printer. This booklet, written in the vernacular in an easily understandable, not very complicated

---

[40] Ibidem: ". . .so en is warachtich niet te vergheefs oock dese sterre van Godt almachtich inden hemel gheschape: maer is ons een vermaninghe ende een seker waerteecken van wonderlijcke toecomende dinghen. Want veele ende groote perijckelen ons te verwachten staen ende gemeyne allende ende groote benautheyt onder die ghemeynte. Oft anders . . . soe sullen wy aller onser allenden, verbaestheyt, armoeden, oorloghen ende oproeren nu een eynde hebben, ende teenemael ontslaghen worden. Ende een nieuwe reformatie over die gantsche werelt, ende der heyligher gemeynder, Catholijcke, Christelijcker kercken Triumphe ende victorie ghenieten."

[41] Nicolaus Bazelius, in his pamphlet on the comet of 1577, stated that he had twice written about the nova of 1572; however, these works have been lost.

style, was obviously intended for the general public. Finally, he also recorded his thoughts about the new star in a learned and lengthy work that dealt with all kinds of divine signs in nature, written in impeccable Latin and clearly aimed at a scholarly audience.

Such a variety of means of expression wielded by one and the same author was not uncommon during the sixteenth century. Especially members of the medical profession tended to be capable of switching from one style of writing to the other with remarkable ease. Early modern medical science depended to a large degree on the observation and interpretation of heavenly phenomena. The idea basic to iatromathematics was, that the different parts of the body were governed by particular celestial signs. The movements of benevolent and maleficent celestial bodies through these signs, and their positions relative to one another, were therefore deemed to exercise a decisive influence on the bodily conditions of earth-dwellers. For this reason, physicians had of old been among the most ardent watchers of the sky, and concomitantly, they were usually the ones who responded most naturally to the general public's demand for astrological predictions. But at the same time, of course, they were members of the world of learning, communicating with their peers by means of scholarly tracts. With good reason, Nicolette Mout has pointed out that at the end of the sixteenth century, there existed a certain measure of interaction between the world of learned tracts and the world of prognostications, especially in connection with comets and novae, and with astrological medicine.[42]

The tendency to associate medicine and astrology (two sciences which had throughout the Middle Ages been closely linked) with the prognostication of future happenings in human history was enhanced during the sixteenth century. To a certain extent, this may well have been due to the influence of Paracelsus, who accorded great importance to prophecy, by which, among other things, he understood the observation of the heavens, in which the plan of history was drawn out for the benefit of humankind. To Paracelsus, astrology served a political purpose, and he produced a significant number of pamphlets in which he interpreted celestial events "to reinforce the political message".[43] Thus, medicine came to be

[42] Mout, "Prognostica", 10.

[43] Charles Webster, *From Paracelsus to Newton. Magic and the Making of Modern Science* (Cambridge 1982) 17.

linked, not only to astrology, but to prophecy and politics as well.[44]

In point of fact, the majority of Dutch authors who published their views on the many comets that appeared between 1577 and 1618 (9 out of a total of 16 non-anonymous publications) were members of the medical profession. Some of them were university professors of medicine, others occupied the more modest position of town physician; but with regard to their approach to the interpretation of the nature and effects of comets, this did not really make much of a difference. Thus, two of the professors, Heurnius and Mulerius, chose to express their views on the new arrivals in the heavens in pamphlets in the vernacular, rather than in scholarly Latin treatises, and, although the contents of these writings were rather more learned than those of similar booklets devoted to the comets by ordinary town physicians, the difference was a gradual rather than a fundamental one, as we shall see below. Perhaps it is justified to say that the vernacular pamphlets in which university professors such as Gemma, Heurnius, and Mulerius dealt with celestial phenomena, constituted a link between the learned world and the outlook of the common man. The existence of such works suggests that, possibly, the dividing line between élite and popular culture was less rigid than might be expected (although it is virtually impossible to assess the extent to which the pamphlets were actually read, and although the role played by financial profit as an incentive to such publications should probably not be underestimated).

Dutch interest in celestial phenomena appears to have risen quite considerably in the years following the nova's expiry. Once again, it is very difficult to determine the factors that might have contributed to this increase. Had the international debate provoked by the new star alerted people to look out for any possible successors to the strange phenomenon? Had the publication of the unusual findings of Tycho and, later, of Galileo rendered their Dutch colleagues eager to discover for themselves what truth was in their assertions? Had the increasingly awkward political situation, and the insecurity and distress brought about by a conflict that came to bear at least partly the character of a civil war, made the Dutch more receptive to heavenly messengers indicating doom (or victory)? Or is the 'increase' due

[44] See Webster, *Paracelsus to Newton,* Chapter 1 ("Prophecy"); see also Wolf-Dieter Müller-Jahncke, "Der Paracelsische Weg zu Astrologie und Magie" in Udo Benzenhöfer ed., *Paracelsus* (Darmstadt 1993) 98-136.

simply to the possible fact that fewer tracts from the time span between 1577 and 1620 have been lost than from the previous decades?

Whatever the reason, Dutch cometary writings of the period 1577-1620 form a profuse, rich, and varied collection. Especially the comets that appeared in 1577 and 1618 were honoured with a substantial amount of attention in the Low Countries. For the comet of 1577, five publications by Dutch authors remain: four pamphlets in the vernacular (three of which, by Johannes Portantius, David van Mauden, and Nicolaus Bazelius, were printed in Antwerp, and one, by Johannes Heurnius, in Cologne) and one lengthy learned tract in Latin, written by Cornelius Gemma and printed at Antwerp by Christopher Plantin. The comet of 1596 was the subject of a Dutch-language pamphlet by one Pieter Heymensz, doctor of medicine from Delft; and the appearance of what was later to become Halley's comet in 1607 was discussed by Nicolaus Mulerius, future professor of medicine and mathematics at Groningen, in one of his annual 'practicas'. However, the great comet of November 1618, the third and by far the most brilliant to have appeared in that year, that was to provoke a prolonged and dogged debate between Galileo and the Jesuits[45], inspired the greatest number of Dutch cometary tracts. Ten publications dealing specifically with this comet were published in the Low Countries: six Dutch-language pamphlets and four Latin tracts. Remarkably enough, all of the pamphlets appeared in the northern Netherlands (at Leiden, The Hague, Groningen, Middelburg, Franeker, and Amsterdam), while all but one of the learned tracts were published in the southern Netherlands, by three professors at Louvain university, Thomas Fienus, Libertus Fromondus, and Erycius Puteanus. The one exception was the treatise written by the professor of mathematics at Leiden university, Willebrord Snellius; it was printed in Leiden by Elsevier.

A second feature the Dutch publications on the nova had in common with the later works dealing with comets, next to the fact that they to a certain extent bridged the gap between learned and popular writings, was the basic methodology that underlay them. Gemma and his anonymous colleague devoted roughly equal amounts of at-

---

[45] See Stillman Drake & C.D. O'Malley eds., *The controversy on the comets of 1618. Galileo Galilei, Horatio Grassi, Mario Guiducci, Johann Kepler* (Philadelphia 1960).

tention to the nova's material constitution, and the course it followed through the constellations, as to its significance and its possible consequences. Such a procedure was of course a perfectly normal one at the end of the sixteenth century; as we have seen, even until the age of Newton, scholars were expected to address the teleological function of natural phenomena, and to elucidate all four Aristotelian causes.[46] Therefore, splitting up the cometary tracts into their constituent parts ('astronomical', 'astrological', and 'political') can be said to be rather an anachronistic procedure. Nonetheless, for the sake of clarity, such a distinction is indispensable; without it, the modern reader would soon lose his way among a profusion of arguments, and connections made between different observations, that, to him, are not at all self-evident. Accordingly, in the following, the treatises will be separated into two basic parts: the present chapter will address the astronomical problems posed by the cometary appearances, such as their constitution, form, location, and motion, while chapter VI will deal with astrological questions, such as the comets' efficient cause, their possible effects (if any), and whether they can be said to be signs or causes of the consequences they entail (if any), as well as with the influence of the political situation on the authors' ideas, and with any possible propaganda value that might be detected in their writings. Wherever possible, the reader will be reminded of the existing interrelationships between the different questions; and, in the Conclusion, an attempt will be made to integrate the various currents of the authors' arguments.

Presently, then, we must concern ourselves with the astronomical problems posed by the appearance of comets. We shall deal first of all with the Dutch reactions to the comet of 1577, as articulated in the pamphlets by Heurnius, Bazelius, van Mauden and Portantius, and the treatise by Gemma. Obviously, the main question we need to ask will be: To what extent had the erosion of the Aristotelian world view progressed in the Netherlands of 1577? Had the Dutch authors broken with any of the basic Peripatetic tenets of geocentricity, dichotomy between the sublunary and the superlunary realm, and uniformity and circularity of celestial motions? Did they express their sympathy, or perhaps even their adherence, to such alternative views as Copernicanism, Stoic cosmology, or the optical theory of comets? Did they try to integrate these novel ideas into some basi-

[46] See supra, chapter II.

cally Aristotelian framework? Did they place the comet above the moon or below it, or were they uncommittal about that issue? In other words, to which of the categories indicated at the end of chapter IV did they belong?[47]

2. *Dutch Reactions to the Comet of 1577*

a. *Order Out Of Chaos*

Before embarking on an investigation of these issues, however, a short outline of the lives and careers of the authors (insofar as these are known) is required. Without a certain degree of knowledge about the, at times quite precarious, circumstances with which they were faced, and about the religious, institutional, and social context in which they lived and worked, we run the risk of misunderstanding or distorting their arguments. Cornelius Gemma, our most important author on the comet of 1577, has fallen victim to such garbled accounts several times already.

As so often happens with historical figures, appreciation for Gemma's intellectual achievements has fluctuated considerably over the centuries. His only biographer to date, Fernand Van Ortroy, writing in 1920, has on several occasions criticized him on completely anachronistic grounds, displaying a marked insensibility to the scholarly attitudes of the period. Thus, he has reproached Gemma for his "passion for astrology and admiration for prodigies", which supposedly led him to "an excess of credulity, incomprehensible in one so learned".[48] Furthermore, Van Ortroy objects to the accepted sixteenth-century practice of combining medical and astrological lore in combating disease, calling it "a truly monstrous union, making a great number of victims", and chiding Gemma for having built his reputation as an astrologer on the "credulity of good souls."[49] Quetelet agrees, that Gemma's writings bear witness to the fact that "he did not always couple his erudition with very solid ratiocination"[50]; and, more recently, Doris Hellman has slightingly remarked

---

[47] See supra, p. 144.

[48] Van Ortroy, *Bio-bibliographie*, 140.

[49] Ibidem, 131.

[50] Ad. Quetelet, *Histoire des sciences mathématiques et physiques chez les Belges* (Brussels 1864) 84.

that "He was much influenced by the errors of astrology, but nevertheless ranks as one of the foremost sixteenth-century astronomers"[51], while Lynn Thorndike has passed an even more severe judgment, writing that "The interest and activity in the field of astrological prediction of Cornelius Gemma were even more intense and prolonged than those of his father, Gemma Frisius, while his intellectual ability and scientific aptitude were distinctly inferior."[52]

All these historians, however, fail to realize (or at least to point out) that, in blaming Gemma, they might as well blame the entire sixteenth century. At the time, Gemma was respected and honoured as one of the foremost scholars of his day. Lavish praise was bestowed upon him by many of his outstanding contemporaries, most notably by Tycho Brahe, who called him "that highly famed Louvain Philosopher, and excellent physician, the blessed heir to his father's erudition, especially as far as the mathematical arts are concerned. . .".[53] Even if half of this and similar panegyrics[54] are to be ascribed to humanist rhetoric, at least they testify to the fact that, undoubtedly, Gemma was a respected member of the international cultural élite of his day. Nor did his engrossment in astrology, and in the science of divine signs in nature, set him apart from his contemporaries in any way. The eagerness to unveil the secrets of nature preoccupied most, if not all, of Gemma's colleagues, including Tycho himself, who was a keen alchemist, and, of course, Johannes Kepler, whose *Mysterium Cosmographicum*, and even his *Astronomia Nova*, were born out of a desire comparable to Gemma's incentive in composing his *De naturae divinis characterismis*. Although Kepler's efforts at uncovering the divine plan in the universe were of an overwhelmingly mathematical nature, and were later to prove partly successful, while Gemma's were based on a magico-astrological methodology that was to be discarded as useless by subsequent generations, the inspiration that led them was common to them both, as was the ancient body of ideas that underlay it. Dialogue between the two strands was still effortlessly possible, not to say inevitable (if, indeed, it is justified to speak of

---

[51] C. Doris Hellman, *The Comet of 1577. Its Place in the History of Astronomy* (New York 1944) 178.

[52] Lynn Thorndike, *A History of Magic and Experimental Science* VI (New York 1941) 30.

[53] Tycho Brahe, *De mundi aetherei recentioribus phaenomenis* (Uraniborg 1588) in J.L.E. Dreyer ed., *Tychonis Brahe Opera Omnia* IV (Copenhagen 1922) 238.

[54] See Van Ortroy, *Bio-bibliographie*, 127-128 for other eulogies of Gemma by his contemporaries.

various 'strands' at all, for a period in which the ideal of the unity of knowledge was still being actively pursued). What distinguished them was their methodology, not their objectives. If Kepler and Gemma represented two different branches, they were definitely branches that grew and were nurtured on one and the same tree.

Cornelius Gemma was born in Louvain on 28 February 1535, at 4.40 a.m., as the son of the eminent mathematician and physician Gemma Frisius. After having received his primary education at the school of a certain Bernhardus at Malines, in 1549 he enrolled in the Faculty of Arts at Louvain university. Three years later he commenced the study of medicine, and in 1569, a year before receiving his doctorate, the Duke of Alva appointed him to the chair of medicine *in Artem parvam Galeni* that had been recently created, with the support of King Philip and the Estates of Brabant, in 1558. The new professor was charged with teaching therapeutic method from Hippocratic *loci*. On 9 November 1574, Gemma in addition obtained one of the two ordinary professorships of medicine.[55]

Gemma himself, in his *De naturae divinis characterismis,* tells us that the old Gemma Frisius did not put much faith in his son's mathematical prowess; at least, he does not seem to have believed that Cornelius would ever match his own skills in the field. However, he did think his son capable of following in his footsteps in the discipline of medicine, and even of surpassing him.[56] Nonetheless, some of Cornelius Gemma's work bears an at least partly mathematical nature; most notably, he completed his father's unfinished work on the astrolabe, *De Astrolabo Catholico Liber,* providing it with a preface, a dedication to Philip II, and the eighteen final chapters. In addition, he of course produced astronomical works on the nova of 1572 and the comet of 1577, as well as a set of meteorological ephemerides over the years 1560-1564, for the compilation of which he employed the Prutenic Tables, in admiration of the "divinum ingenium" of Copernicus.[57]

As mentioned above, however, Gemma's concerns with astronomical subjects need to be considered in connection with his efforts at

---

[55] Ibidem, 124-126.

[56] Ibidem, 127; Cornelius Gemma, *De naturae divinis characterismis,* II, 22: "Quorsum, ait [pater], tantopere nunc illuc aspires? cum ad eam perfectionem quam nos attigimus tute in eadem disciplina minime sis perventurus; in medicina non solum assequi possis, sed si operam des, fortassis etiam superare. . .".

[57] See Van Ortroy, *Bio-bibliographie,* 367-374, and 136 n. 2.

discovering the God's plan with the universe. His two most extensive and original works, *De arte Cyclognomica* (1569) and *De naturae divinis characterismis* (1575) are both on the interface of medicine and astrology, attempting to pinpoint the nature of the connections and sympathies between the celestial bodies, human society, and the microcosm of the human body. Thus, Gemma's work does not display anything like the strict dichotomy that Van Ortroy seems to suggest when he remarks that "It [Gemma's oeuvre] comprises an astronomical and a philosophico-medical part".[58]

Apart from the scanty information that can be gained from a few scattered remarks Gemma himself included in some of his works, hardly anything is known about our protagonist's private life. The few clues that do remain as to his circle of friends all point in the same direction, namely, that of the group of learned men that, during the final decades of the sixteenth century, used to meet at Plantin's printing press at Antwerp, one of the busiest and liveliest meeting places of humanists in Europe.[59] The Plantin press was much frequented by travelling scholars; moreover, since the plan for the printing of a Polyglot Bible had been conceived in 1565, Plantin had surrounded himself with such expert collaborators as the Hebrew scholar Andreas Masius, and the orientalists Guy Lefèvre de la Boderie and Francis Raphelengius, both pupils of the famous Guillaume Postel.[60]

It has been established beyond doubt that Plantin was a member of the secret spiritualist sect known as the Family of Love.[61] The founder of this group had been the Emden merchant Hendrik Niclaes, who regarded himself as a prophet to whom divine knowledge had been revealed, and who published a number of works

[58] Ibidem, 130.

[59] Thus, of the few items of his correspondence that are still available to us, several were directed at or received from members of the Plantin circle: a letter dated 2 April 1572 to Abraham Ortelius, a letter dated 1 May 1570 to Benito Arias Montano, king Philip's supervisor of the Polyglot Bible project; and two letters from Plantin to Gemma, one of 16 July 1567 and one undated. See Van Ortroy, *Bio-bibliographie*, 395-396.

[60] B. Rekers, *Benito Arias Montano (1527-1598)* (London/Leiden 1972) 45.

[61] See M. Rooses, *Christophe Plantin, imprimeur anversois* (Brussels 1883); L. Voet, *The Golden Compasses: A History and Evaluation of the Printing and Publishing Activities of the Officina Plantiniana at Antwerp* (Amsterdam 1969); H. de la Fontaine Verwey, "Het Huis der Liefde en zijn publikaties" in his *Uit de wereld van het boek I: Humanisten, dwepers en rebellen in de zestiende eeuw* (Amsterdam 1975) 85-111; A. Hamilton, *The Family of Love* (Cambridge 1981); Rekers, *Arias Montano*, 70-104.

advocating direct communion with God. The teachings of Niclaes were akin to those of the Anabaptists, especially with regard to their strong chiliastic overtones and the expectation of the imminent end of the world; however, unlike his Anabaptist counterparts, who found their audience for the most part among the lower classes of society, Niclaes won his adherents chiefly in intellectual and commercial circles. One Familist tenet which probably appealed strongly to these groups was Nicodemism, which enabled them to practise an outwardly impeccable conformism to the predominant religion of their place of residence, and to take an active part in social and political life, while at the same time inwardly entertaining unorthodox, spiritualist ideas.

In 1554, Plantin's printing press at Antwerp had been set up with the financial support of some wealthy disciples of Niclaes, and was intended to serve the diffusion of the prophet's writings[62]; however, around the year 1573, a rift occurred, in which Plantin and his circle broke away from Niclaes and chose as their new leader Hendrik Jansen Barrefelt, alias Hiël or "Life of God". The new, 'second' Family of Love was centred around Plantin's press at Antwerp, and included among its members, next to Plantin himself, the famous geographer Abraham Ortelius. Other intellectuals who frequented Plantin's household included the botanists Carolus Clusius, Matthias Lobelius, and Rembertus Dodonaeus, the geographer Gerard Mercator, the classicists Justus Lipsius, Hadrianus Junius and Johannes Goropius Becanus, and also Cornelius Gemma.[63] However, due to lack of evidence, it is impossible to establish whether these men were also members of the Family of Love.

Our second 1577 author, Johannes Heurnius was born on 4 February 1543 in Utrecht as the scion of a prominent family. At the age of eighteen, in 1561, he left his home town for the Southern Netherlands to study the arts, and later medicine, at Louvain university. During the two years he spent there, he stayed at the house of Cornelius Gemma; consequently, the two must have known each other

[62] Rekers, *Arias Montano*, 70.

[63] Ibidem, 74. The fact that Rekers refers to Cornelius Gemma as "Gemma Frisius" could lead to some confusion, since, usually, his father is designated by that name. However, Rekers must mean Cornelius, because he writes that, when Montano was introduced into Plantin's circle, he used to discuss mathematics with "Gemma Frisius"; since the elder Gemma had died in 1555, whereas Montano did not arrive in Antwerp until 1568, he can only have met Cornelius.

quite well. After his stay at Louvain, Heurnius travelled to Paris, where he spent three years studying with the famous detester of Aristotelian learning, Peter Ramus, and with Duretus. Continuing his academic peregrination, he crossed the Alps, and went on to receive his doctorate of medicine at Padua in 1571, after which he entered the service of François Perrenot, Cardinal Granvelle's cousin, as his personal physician. Finally, in 1573, he returned to the Low Countries, to establish himself as a doctor in his home town of Utrecht; he soon acquired a flourishing medical practice.

According to the *Nieuw Nederlandsch biografisch woordenboek*, Heurnius' return to his native country in 1573 is to be ascribed to his alleged conversion to Protestantism.[64] However, more recent publications suggest that Heurnius' life can be seen in quite a different, less clearcut perspective. Time and again, his name appears in the literature dealing with the Niclaes sect, or with late sixteenth-century spiritualism in general. Thus, Alastair Hamilton, in his 1981 study of the Family of Love, has suggested that Heurnius might have been one of Niclaes' twenty-four so-called "Elders". These Elders were part of a priestly hierarchy that Niclaes, spurred on by a vision he had in 1567, had created in order to administer the sect; the prophet himself functioned as the highest bishop, presiding over seven subsidiary orders of priests, among them the Elders.[65] After the rift that occurred in 1573, in which Niclaes lost much of his following to Hiël, the only ones to have remained entirely loyal to Niclaes were his Elders. It is, however, extremely difficult, if not impossible, to ascertain who these might have been, since in Niclaes' writings they only figure hidden behind mysterious pseudonyms. Nonetheless, Hamilton ventures to suggest that Heurnius might have been among them.[66]

One of the reasons he adduces for doing so is the fact that Heurnius and Niclaes used the same printer to publish their works. Obviously, after his break with Hiël, Niclaes could no longer rely on Plantin, one of the foremost Hiëlists, to print his writings. So far, it has proved impossible to establish who took over from Plantin as Niclaes' printer after 1573; however, several of the books and engravings by Niclaes that were published between 1576 and 1580

[64] P.C. Molhuysen & P.J. Blok eds., *Nieuw Nederlandsch biografisch woordenboek* (Leiden 1911-37) IV 745-746.

[65] Hamilton, *Family of Love*, 56-57.

[66] Ibidem, 62.

included the following address: "Printed at Cologne, by Niclas Bohmbargen".[67] No printer of that name can be traced in Cologne for the period indicated; but, remarkably enough, precisely the same address appears on two works by Johannes Heurnius, published likewise between 1576 and 1580: an edition with Heurnius' commentary of pseudo-Aquinas' *Secreta alchimiae magnalia*, printed in 1579, and the work of interest to us in the present study, Heurnius' Dutch-language pamphlet on the comet of 1577.[68]

The fact that Heurnius used the same, otherwise unknown, printer as did Niclaes has been regarded as an indication as to his possible ties with the Family of Love. Furthermore, during the last few decades of the sixteenth century, the newly founded university of Leiden was to become a hotbed of spiritualist, Familist ideas. In 1581, Heurnius obtained the chair of medicine at Leiden, a position he was to fill until his death in 1601. Although the substantial number of works he published in that quality were perfectly traditional, medical treatises, based on Hippocrates and Galen, at Leiden he found himself in an environment greatly receptive to the ideas of direct contact with God, inner rebirth and deification of man with the aid of the Holy Spirit, indifference to outward ceremonies, and longing for peace and order that were propagated by Barrefelt. Many of the scholars who had originally belonged to the Plantin circle at Antwerp moved to Leiden after the foundation of the university in 1575: Lipsius obtained the chair of history and law in the arts faculty, while Dodonaeus became professor of botany and Raphelengius of Hebrew; Clusius initiated the famous Leiden *hortus botanicus*, and in 1583, Plantin set up a press at the new university town. It is, therefore, not unreasonable to suppose that Heurnius, too, was favourably disposed towards spiritualist ideas.[69]

Much less is known about the authors of the other three pam-

[67] De la Fontaine Verwey, "Huis der Liefde", 99.

[68] Ibidem.

[69] See C.C. de Bruin, "Radicaal spiritualisme te Leiden", *Rondom het woord* 17 (3) (1975) 66-81: 73. In 1596, Heurnius published a *De Humana Felicitate* in which he advocated the avoidance of any kind of fixed dogma, as did Lipsius in his famous *De Constantia*; see Hamilton, *Family of Love*, 98 & n. 56. For an edition of Heurnius' *De humana felicitate*, with Dutch translation, see C.L. Heesakkers (with a contribution by W.G. Heesakkers-Kamerbeek), "Petrus Forestus in gedichten en brieven. Een overzicht van het materiaal en een editie van zijn correspondentie met Johannes Heurnius" in H.A. Bosman-Jelgersma ed., *Petrus Forestus medicus* (s.l. s.a. [1996]) 119-243.

phlets on the comet of 1577. Johannes Portantius, a geographer and astrologer from Ghent, spent most of his life at Antwerp. Lodovico Guicciardini, in his description of the Netherlands, called him a learned man, a great mathematician well versed in several other sciences as well. He was one of the collaborators of Abraham Ortelius, contributing a map of Livonia to the famous *Theatrum Mundi*. The unfortunate Portantius was, towards the end of his life, arrested by the Antwerp magistrate, charged with either sorcery or the composition of subversive, anti-government verses. He was tortured more than once, but the sources differ as to the outcome of this adventure: some say he succumbed, while others believe him to have survived the ordeal.[70] His pamphlet on the comet of 1577 was printed at Antwerp by Plantin; in the same year, a German translation appeared with Heussler in Nuremberg.

The last two authors from 1577 were both physicians. David van Mauden was born in Antwerp to an ancient and prominent family. He probably studied medicine at Louvain, after which he became professor of surgery at the Antwerp city hospital. He published a work in Flemish on anatomy, based on the engravings of Vesalius, adding two additional ones to illustrate some discoveries of his own. This book, first printed by Plantin in 1568, appears to have been quite successful: it was reprinted at Amsterdam as late as 1646. Finally, Nicholas Bazelius, who discussed the comet of 1577 in his prognostication over the year 1578, was born in Nieuwkerke, in western Flanders, and became town physician and surgeon in Bergues Saint Winoc.[71] The pamphlet in which he dealt with the comet appeared both in Flemish and in French and Latin translations.[72]

The years 1577 and 1578, when our cometary treatises were written, were decisive ones for the Netherlands. In retrospect, they would prove to have been the final years of unity between northern and southern provinces[73]; at the time, they seemed, perhaps even more

[70] Bergmans, Paul, "Portant (Jean) ou Portantius" in *Biographie Nationale de Belgique* 18 (Brussels 1905) 68-69.

[71] Today the town of Bergues Saint Winoc is in northern France; in Bazelius' day, however, the region in which it is situated was part of Flanders. Even to the present day, Flemish is spoken by many of the inhabitants of this northern French region, known as "French Flanders".

[72] Hellman, *Comet of 1577*, 328-331.

[73] See below, chapter VI.

ominously, to mark the final stages of history itself, the dreaded chaotic reign of Antichrist that necessarily preceded Christ's second coming and the separation of the sheep from the goats. Among intellectuals such as the members of the Plantin circle the idea became prevalent that the chaos of the times could be overcome by charity and mutual support, and by seeking direct contact with the creator with whom rested all the secrets of the universe. A spiritualist communion with the divine architect, they believed, would lead them from darkness to light, from the chaos of the times back to the order that had been lost with the Fall of Man. The appearance of a comet, a body of old associated with divine messages, came at an especially portentous moment; it was investigated with great care.

b. *Cornelius Gemma: "Materies Minime Sublunaris"*

The tract[74] Cornelius Gemma devoted to the apparition of the comet he first saw on 14 November 1577 has earned him a derivative and highly selective fame among historians of cometology. Tycho Brahe, extensively dealing with the phenomenon in his *De mundi aetherei recentioribus phaenomenis,* in the final chapter discussed the observations made by his most distinguished foreign colleagues, with Gemma's account figuring prominently among those who had, like Tycho, accorded the comet a superlunary position.[75] The Dane criticized his Flemish colleague quite severely, time and again reminding his readers that, whereas Gemma's parallax estimates happened to be approximately correct, he had not once applied the mathematical method for accurately determining a comet's distance from the earth.[76] But despite this reproof, the fact that the Fleming was one of only four observers who had, like Tycho, accorded the phenomenon

[74] Cornelius Gemma, *De prodigiosa specie, naturaque Cometae, qui nobis effulsit altior lunae sedibus, insolita prorsus figura, ac magnitudine, anno 1577. plus septimanis 10. Apodeixis tum Physica tum Mathematica. Adiuncta his explicatio duorum chasmaton anni 1575. nec non ex cometarum plurium phaenomenis epilogistica quaedam assertio de communi illorum natura, generationum causis atque decretis supra quam hactenus a Peripateticis annotatum est* (Louvain 1578).

[75] Tycho Brahe, *De mundi aetherei,* chapter 10, in Dreyer ed., *Opera Omnia,* 238-251.

[76] Ibidem, 240-241: "Etsi autem haec de Parallaxeos Quantitate pronunciatio, non secundum Leges Mathematicas, e certorum Datorum per Triangulos in numeros deriuatione, procedit, sed duntaxat ita nude ab ipso Gemma, sine ulla Demonstratione, asseritur... tamen non dubium est, illum, si non satis praecise scopum ipsum attigit, saltem in hoc recte collimasse, quod Parallaxin longe minorem Cometae, quam Luna etiam a Terris remotissima obtineat, assignarit. . ."; 242: "Est quidem haec ratio Parallaxin dijudicandi non satis exacte & Mathematice a Gemma adducta,

a superlunary position, has made some of the fame subsequent generations were to bestow on his Danish colleague reflect on Gemma.[77]

However, Tycho's discussion of Gemma's tract was highly selective. As the Dane himself admitted, there was much more to the work than the limited portion he had deemed relevant for his own investigations. Gemma's first chapter, which dealt with macrocosmic portents in general, and his third, which was concerned with the comet's physical characteristics and its significance, were both honoured with a certain measure of praise by Tycho, who, however, failed to discuss them because, he explained, his own present book dealt only with mathematical and astronomical considerations.[78] Tycho's selectivity has been adopted by all subsequent historians, causing Doris Hellman, for example, to assert that Gemma's second chapter (the one discussed by Tycho, on the comet's physical appearance and motion) was "the most important chapter of the work".[79]

Judging from Gemma's own utterances, however, we can be certain that he would not have agreed with such a categorical statement. In fact, his work, when read in its entirety, conveys the impression of a moral and ethical exhortation, rather than of an astronomical or even an astrological treatise. At the end of the first chapter, just before embarking on his mathematical investigation of the comet's course through the constellations and its distance from the earth, Gemma articulated the purpose his tract was meant to serve, and his reasons for subjecting the comet to a mathematical inquiry at all. Having, in the first chapter, at length discussed the wretched situation throughout Europe, and expounded his theory that the "divine

---

Demonstrataque. . ."; 243: "Et licet haec a Veritate aliena non sint: tamen, ut semel atque iterum dixi, Demonstranda fuisset potius e selectis & separatis Dedomenis, non ita nude referenda, ut in his Mathematica certitudo apertius elucesceret."

[77] See for example Hellman, *Comet of 1577,* 177-183.

[78] Tycho Brahe, *De mundi aetherei,* 238-239: "In quorum [Capitum] PRIMO, non proprie quae ad Cometam solum spectant, proponit, sed ab initio, de Mundi inferioris iam aegrotantis Causis, Qualitatibus, Curationibus, & Signis ac Praesagijs Criticis, suo quodam modo, scite admodum, nec inconvenienter Philosophatur. Postea vero Chasmata bina. . . commemorat; & de eorum Natura ac indicationibus probabiliter ratiocinatur. . ."; ibidem, 250-251: "Quae autem TERTIO & vltimo CAPITE, De Natura, Causis, Principiis, & Viribus, atque Decretis huius Cometae, copiose & erudite in medium adducit, hoc loco in considerationem vlteriorem deducere, non est animus; siquidem haec partim Physica sunt, partim ad Astrologicam dijudicationem proprie pertinent. De quibus hoc Libro nihil tractare, sed saltem Mathematicam & Astronomicam huius Cometae contemplationem absolvere, proposuimus. . .".

[79] Hellman, *Comet of 1577,* 182.

signs" displayed in the natural world were harbingers of an impending great alteration, he now proposed to explain what special significance could be attached to the present comet, a "divine sign" *par excellence.*

Yet, in order to arrive at a correct description of the comet's nature, force, and meaning, it was essential first of all to inquire into the causes of the phenomenon, its generation, its general constitution, and the analogy it displayed with particular stars and celestial signs. This information could only be obtained by means of diligent observation, and an accurate description of the comet's appearance and its course through the heavens: "The only way open to human understanding leads through sense perception".[80] In order to discover the powers exercised by a particular phenomenon, one had to discover its nature; and in order to discover its nature, a complete aetiology was required, an examination both of proximate causes and principles, and of observed facts and accidents perceived by the senses. Before considering a phenomenon's effects and ends, its material and formal causes had to be elucidated.[81] Therefore, Gemma observed the comet on every clear night, determining its position by means of astronomical instruments. However, these observations, described in his second chapter, were made only by way of a preliminary: the real issue to which Gemma's tract was devoted, was the comet's significance and its relevance to conditions on earth.[82]

The fact that Gemma's astronomical considerations served an auxiliary purpose, however, does not render them any less interest-

---

[80] Gemma, *De prodigiosa specie,* 35: "Certe haec una per sensum patet illuc [in caeli tecta] humanis ingeniis via."

[81] Ibidem, 36-37: "Vires cuiusque rei vel ἐνέργειαν minime citra naturae cognitionem percipi posse: naturam vero per causas proximas atque principia definiri, & hanc postremo ab observatis & accidentium quae sensu percipiuntur syndrome Physica vestigari, omnibus qui vel tenuiter doctrinam methodicam sunt amplexi, promptum est intelligere. In causarum genere prima est materiae consideratio, quae nobis & formam ostendit, internam scilicet rei essentiam, adiunctis passim effectibus illis, quae denique & finem rei, effectusque, & universae significationis vim optime repraesentant."

[82] Ibidem, 18-19:"Quapropter si quis naturam, viresque, & significata praesentis phaenomeni volet pernoscere... iuxta communem methodum prius de generationis causis atque principiis, de universa illius constitutione, ut loco, tempore, magnitudine, motu, figura, caeterisque qualitatibus, & ad vicinas stellas, caelique partes analogia diligenter inquirat necessarium est. Huius denique gratia sensu imprimis opera atque industria, quam & nos pro virili nostra impensam ad singulos dies, quibus videre licuit, ut ante iudicium vel apodeixim mathematicam omnes peritiores diligenter explorent, operae pretium esse putavimus."

ing. Our author put a great deal of effort into examining the comet as accurately as possible. With the aid of an astronomical ring, a cross-staff, and a celestial globe, he made observations several times each cloudless night, at four-hour intervals, accurately recording his findings.[83] He first saw the comet on 14 November, when its head displayed a reddish-silvery colour, and was about the size of the planet Venus or Jupiter, so that some people initially thought one of the planets had acquired a head of hair.[84] The tail was at first a blood-red colour because of interjacent vapours; but as soon as the sky cleared, it grew brighter. It extended in an antisolar direction, and displayed a very rare form, the solar rays being refracted not in a straight line, but in a curvature; its maximum length was 15°, with its splendour reaching out even further, easily attaining 20°; its maximum width was hardly 5°.[85]

The comet's head pointed at Spain, while the curvature of its tail was directed at Italy, the tail being exceptionally bright and denser on its northern side than towards its southern extremity, where it was rarer and darker. With regard to its figure the comet belonged to the 'horn-shaped' variety or κερατοειδής, the kind called 'Miles' by Alkindi: its splendour resembled a great horn, a peacock's tail, or a plume on a soldier's helmet, and from the head also issued forth a kind of smaller, second tail that was directed at the south-west.[86] A fortnight after the comet's first appearance, on 28 November, the second branch of the tail became very well visible after the sky had been cleared by a violent thunderstorm; it remained thus for several days, until it was burned up, having grown gradually weaker or become more dispersed.[87] The most curious sight the comet displayed, however, "rare and admirable above all else"[88], took place on 3 De-

---

[83] Ibidem, 21: "Sequitur observatio nobis praecipue per annulum, globum ac radium facta"; and 34: "Interim circa hoc terrificum iubar ea usus sum diligentia, ut eiusdem diei diverso tempore (per interstitium scilicet horarum 4 aut plus eo) . . . non visa sit mihi diversitas maior 40 minutis. . .".

[84] Ibidem, 20.

[85] Ibidem, 20-21.

[86] Ibidem, 21: "Figurae specie κερατοειδής, cuiusmodi fere apud Alkindum militis nomen invenit, quanquam & ἱππιτεύς Graecorum rectius appelletur, quod illi perpetuus quidam processus velut altera cauda deorsum oblique seu versus Euroafricum promineret, reliquum iubar cornu latissimum, vel caudam pavonis, vel plumae militaris effigiem optime prae se ferret."

[87] Ibidem, 25.

[88] Ibidem, 27: "Circa octavam horam accessit huic prodigiosae faci rarum quippiam atque prae caeteris admirandum."

cember: suddenly, the comet's head gaped open wide, and from it issued forth three huge rays like fiery spears, quivering, and pointing in the direction of the Mediterranean; one of them darted towards Italy, the second towards the Herculean coast, and the third towards the westerly regions.[89] A similar spectacle reportedly took place on 6 January 1578, with bright rays darting across the sky.

Gemma's description of the comet's motion and the course it followed through the constellations was even more meticulous than his account of its external appearance. He observed the phenomenon as often as possible, preferably twice every night, in order to be able to determine its parallax. During the time span between the comet's first appearance on 14 November and its disappearance on 18 January, Gemma was able to conduct measurements for it on 35 out of a total of 66 nights; the remaining nights had been too cloudy, or the comet too tiny, for it to be observed. For each clear night, Gemma recorded the comet's distance to two, or sometimes three or four, of its neighbouring stars, and occasionally also its distance from one or two of the planets, by means of his cross-staff and a celestial globe. For most nights, he also calculated the comet's longitude measured against the zodiac, its latitude from the ecliptic, and the direction in which its tail pointed, occasionally adding the tail's length.

One of the most important data Gemma derived from these observations was the comet's own course through the heavens. It was first seen in the beginning of Capricorn, and moved in a straight line, from which it never deflected even by a hair's breadth, through the signs of Capricorn and Aquarius, vanishing around the middle of the sign of Pisces. Despite his inability to corroborate this by means of his own observations, Gemma believed that, in fact, the comet had moved from one tropical circle to the other; although he had first sighted it in the beginning of Capricorn, and had been unable to observe it beyond the middle of Pisces, where it vanished in the chest of Pegasus, he was convinced that its actual generation had taken place in the arrow of Sagittarius, near the tropic of Capricorn, while its real extinction was to be situated in Andromeda, near the tropic

[89] Ibidem: "Visa est etenim multis qui tunc attentius observarunt, repente dehiscens ingentes radios tres velut igneas hastas vibrare sensim ad Mediterraneum fretum, idque constanter asserunt fide digni prius excusso telo recta Italiam petiisse, secundo littus Herculeum, postremo occiduam plagam. . .".

of Cancer.[90] From these data, Gemma was then able to deduce that the comet's maximum declination, like the sun's, had been 23 1/2 degrees, once toward the north and once toward the south.

During its existence, the comet displayed a marked regularity, with regard to its luminosity as well as with regard to its speed, both gradually diminishing with perfect proportionality.[91] When it first appeared, its velocity was at its highest: between Gemma's first observation, on 14 November, and his second, twenty-four hours later, the comet had moved almost 5° in its own proper course. Thereafter, its speed began to diminish fast: 4 1/2° between 15 and 16 November, 2 1/2° between 19 and 20 November, 1 1/2° between 29 and 30 November, and only 30' between 22 and 23 December. The intensity of its brightness underwent a similar decline: at its most luminous when it first appeared, the comet grew smaller and smaller, just as it moved slower and slower, so that many believed that it had disappeared around the middle of January. Gemma himself was, however, able to observe it until the 18th of that month, although with a great deal of difficulty; on that day, only a small trace of it could be seen in the chest of Pegasus, too tenuous to be measured with a cross-staff.[92]

The sections of his tract for which Gemma was to be remembered best were, of course, his parallax measurements. As we have already been able to gather from Tycho's criticism, however, these were not very spectacular. His first parallax measurement was made on 20 November, when Gemma found the distance of the comet's body to the star in the hand of Antinous to be 2° 4', a value that had not changed a few hours later, when the comet was about to set; from

---

[90] Ibidem, 32-33: "Primum apparuisse circa initia Capricorni, Saturno proximum: hinc vero ubi paululum conscendisset, secutum lineam rectam venti Libonoti ad extremum durationis punctum ab ea vix latum (ut sic loquar) culmum deflexisse... Ductus totius linea rursum a Sagittario telo (neque in fine numine certo) in ipsius caput ac verticem summum per latera ac manus Antinoi... per aequinoctialis contactum, & Meridiani circuli polum transeuntis inter Delphina & equum minorem usque ad pegasi sidus, ac si cursum versus miserandae Andromedae caput & vincula direxisset. Quin quamvis extremam illius apparitionem oculorum acies minime sit assecuta, puto tamen, certisque rationibus adducor ut credam, ab uno circulo tropicorum ad alterum pervenisse."

[91] Ibidem, 31: "...ut quantum motus celeritati tantum quotidie & lucis splendori proportione decesserit" and 32: "... proportione servata, ut quam celerrime & uno die circiter 5 graduum longitudinem peragraret, verum ut progressu sensim cum lucis vigore & motus celeritas minueretur...".

[92] Ibidem, 32.

this information, he then went on to deduce the parallax, which he found to be no more than 40'. However, he did not explain to his readers how he had arrived at the value of 40', failing to elucidate what method he had followed and not including any calculations. He did, however, boldly assert that, from 20 November until the beginning of the next year, the parallax had decreased daily; again without offering his readers any evidence.[93] Further on in his tract, he did mention five more parallax measurements he conducted on the nights of 2 December, 13 December, 19 December, 22 December, and 27 December; however, these were as vague as the earlier one. On 2 and 13 December, Gemma merely declared that the comet's parallax on those days had been "very small", while on 19, 22 and 27 December he did state a specific parallax value (27', 20', and 15' at most), but again failed to come up with any relevant calculations.[94]

Despite these omissions, however, Gemma does appear to have accorded a great deal of importance to what he himself termed "mathematical proof". Mathematical evidence, he said, was to be used to determine a comet's distance from the earth; in this particular case, it led to an outcome that was "clearly divine and admirable above all else": the comet was to be situated above the moon.[95] As we have seen, throughout Gemma's treatise mathematical considerations were subservient to the higher ends of gaining an insight into the divine plan; yet, because without an accurate description of the comet's mathematical characteristics, such as its course and distance, no such illumination was to be attained, Gemma applied himself to measurement with self-professed diligence. He tried to determine the comet's parallax to the best of his abilities, by measuring its distance

---

[93] Ibidem, 24: "Stella in manu Antinoi caudae supremas strias exactissime contingebat: . . . distantia quidem a cometae corpore 2 gr. 4 min. minimum differens cum ad occasus cardinem pervenerit: ex quo praecipue parallaxis sive aspectus diversitas colligi debet, quam nos ad summum invenimus min. 40 imo minorem indies usque ad principium alterius anni."

[94] Ibidem, 27: "2 Decemb. Cometae corpus adhuc paulo Occidentalius atque Australius superiore in rictu equi minoris, a quo vix abfuit decem minutis, tum circa Meridianum hora media sextae, tum versus Occasum magis hora 9. Unde constat fuisse minimam parallaxim. . ."; 29: "13 Festo Luciae. . . Stabat. . . sub horam decimam versus occasum non multum aspectus varietate differens aut parallaxi, ab ea proportione qua fuit in medio caeli. . ."; 30: "19 Decemb. . . . Inventa est parallaxis minor 27 minutis. . . 22 Decemb. . . Diversitatem aspectus reperi non maiorem 20. min. . . . 27 Decemb. . . . parallaxis circiter 15 min. . . .".

[95] Ibidem, 34: "Altitudo a terris sive distantia, cuius inventum plane divinum atque prae caeteris admirandum per ἀπόδειξιν Mathematicam, maior quam ipsius Lunae ex praecedentibus definita est."

to the nearest of the fixed stars several times each night; nonetheless, he said, he was still very anxious to learn the results of his colleagues' observations from abroad, impatiently awaiting the findings of all those who had also concerned themselves with the distance of the nova (Hagecius in Bohemia, Muñoz in Spain, Digges and Dee in England, and Tycho in Denmark) in order to be able to verify his own.[96] Still, Gemma was quite confident of the veracity of his conclusion: the comet, for which he had not once found a parallax higher than 40', was to be situated beyond the moon's sphere.

But where, exactly, is a body displaying a parallax of 40' to be located? As we have established in connection with the nova, Gemma said, a body without any parallax at all was much higher up than any of the planets. By contrast, all objects located below Mars displayed a perceptible diversity of aspect: a very small one in the case of the Sun, some three to four minutes more in the case of Venus and Mercury, and the largest in the case of the moon, as was confirmed during lunar eclipses. The moon's largest parallax, according to Gemma, was 1° 6', when it reached perigee; when in apogee its diversity of aspect could still be determined at 50', more than half a degree.[97] The comet, at most displaying a parallax of 40', but often much less, must therefore necessarily be located much further away from us than the moon. Because both its species, its brightness, and the vicissitudes of its motion closely (if not exactly) matched Mercury's, Gemma concluded that it could safely be said to be embedded in the sphere of that planet.[98]

Yet, the fact that this particular comet was clearly superlunary did not induce Gemma to reconsider the findings of his predecessors in the Aristotelian tradition. When compared to comets seen at earlier times, he said, the present one was very different: its figure, its magnitude, and many of its accidental characteristics all set it very much apart from its less elevated congeners. A precedent in the category of superlunary comets had been described by Albatechnus (Al-Battānī), who had once seen a comet in the sphere of Venus[99]; however, a

---

96 Ibidem.

97 Ibidem, 35-36.

98 Ibidem, 36: "Quis itaque non videt huius cometae locum longe fuisse Luna superiorem? id est in orbe Mercurii, cuius & speciem splendoremque & inaequalitatem motus si non exacte, saltem vestigiis proprioribus observavit."

99 Gemma probably means the comet observed by Albumasar. In his *De naturae divinis characterismis* (Book I, p. 213) he does, correctly, attribute the observation of a comet in Venus's sphere to Albumasar, not Al-Battānī.

number of astronomers of whom Gemma thought very highly had reported rather different findings. Thus Regiomontanus had observed a parallax of almost 6° for the comet of 1475, placing it in the highest of the three Aristotelian layers of air, close to, but below, the concave of the moon's sphere. Similarly, the comet Gemma himself had observed in 1556, he stated, was to be located at about the same distance. And, in 1532, Johannes Vögelin had even determined a parallax value of more than thirty degrees, placing that year's comet close to the middle layer of the air.[100]

Gemma did not dispute the accuracy of his predecessors' measurements; after all, he himself had obtained a similar result for the comet of 1556. Therefore, he was quite convinced that some, indeed most, comets were atmospheric phenomena, to be located below the moon in accordance with the Aristotelian doctrine. Yet, the present comet added a new, though perhaps not entirely unexpected, element to cometary theory. Two years earlier, in his *De naturae divinis characterismis,* Gemma had already hinted at the possibility of superlunary comets.

In the eighth chapter of Book One of that work, he paid a substantial amount of attention to comets, including them in the category of "Prodigies that occur in the ethereal region". Most comets, he there stated, were much lower than the sphere of the moon, as could be demonstrated from parallax observations.[101] They were generated by rare conjunctions of the superior planets, especially when these occurred in fiery signs or close to the Milky Way, and they usually emulated the most dominant of these planets with regard to their motion.[102] These were of course perfectly Aristotelian tenets, easily reconcilable with the reigning world view. Yet, Gemma went on to add some modifications that might have raised more than one Peripatetic eyebrow. He announced that the purpose of his explanation of the doctrine of comets was to make sure that nobody would confuse them with other prodigies, either of an inferior or of a superior dignity.[103] To Gemma, then, comets were of an undis-

[100] Gemma, *De prodigiosa specie,* 39.

[101] Gemma, *De naturae,* I, 198: "... quanquam certius ex parallaxi cometarum, hi multo inferiores concavo Lunae quam saepissime deprehendantur."

[102] Ibidem, I, 199, 206.

[103] Ibidem, I, 206-207: "Sed placuit etiam particularis methodi schema in crinitibus facibus explicare: quo doctrina illorum, & ratio exactius constet: neque deinceps quispiam cum cometis, multa naturae prodigia tum inferiora dignitate, tum sublimiora confundat...".

putedly intermediary nature, connecting the divine beings of the celestial realm with the variable elementary creatures. This intermediary nature was reflected in their material constitution and in their location, our author believed.

After having expounded the cometary theories of some of his predecessors (such as the idea of John of Damascus and the Pythagoreans, that they were stars created in the beginning, that remained hidden in space for most of the time, and the majority opinion, that they were corruptible, fiery meteors consisting of burning, greasy and oily matter) and having reminded his readers that, at present, great ambiguity reigned over the issue of comets' distances, he went on to put forward his own, "new and paradoxical opinion on the generation of comets".[104] To one who diligently examines the nature of the universe, he said, there is nothing absurd about the assertion that, in addition to the two traditional universal opposites of mortal and immortal, lucid and opaque, concrete and dissolute substances, there exists also a third substance, partaking of both opposites, situated on the border between both realms, and constituting a middle state between the two extremes.[105] Such a substance, Gemma believed, was to be assigned to its own proper region, the "region of aether". This region was not in any way to be identified with the traditional, Aristotelian sphere of fire, which (if it existed at all) was located in between the ethereal region and the layers of air.[106] To Gemma, then, the aether was not the substance of which the heavens above the moon were made, but rather a stuff not conceived of by Aristotle, filling an area formerly occupied by the Aristotelian sphere of fire, just below the moon's orb. Its nature was neither entirely elementary nor completely celestial; rather, it was intermediary and temperate because in it, the traditional opposites were allowed to meet. Without such a middle substance, Gemma stated, no contact between the heavens and the earth was possible, just as no contact is possible

---

[104] Ibidem, I, 113-115.

[105] Ibidem, I, 115: "Si quis vero diligentius rem examinet, quanquam in universo Deus naturam primis duobus generibus discretam esse voluerit, conditione scilicet mortalis & immortalis substantiae, item lucidae & opacae, concretae materiae firmiter ac dissolutae, proportionis aequalis & inaequalis . . . in utriusque tamen confinio, tertium quiddam amborum particeps, atque inter extrema medium statuisse nihil repugnat, neque a naturae symphonia praestantissima est alienum."

[106] Ibidem: "Id sane . . . vero vocabulo regionem aetheris nuncupabo: non quidem Vulcani illius officinam aut ignis regionem, quae (si uspiam sit) inter hanc potius & connexum aeris statuatur."

between the human soul and the body, without the mediation of the spirit.[107]

Closely corresponding to the intermediate qualities of the ethereal region were the characteristics of comets. They, too, displayed a kind of middle nature, both with regard to their physical constitution and with regard to their motion. Through many observations, it had been established that comets rose up much higher than other meteors; and although it was generally believed that their matter was elementary, the splendour of their light was of a divine kind. Furthermore, their motions were also partly stable and regular, like those of the celestial bodies, and partly volatile and irregular, as befitted sublunary creatures.[108] Also, comets displayed a circular motion, which was not at all in keeping with the fleeting nature of fire. Ptolemy was right when he called comets "corruptible stars", Gemma declared; they were attracted by the planets that dominated them, and thus the circularity of their motions could be accounted for.[109]

Moreover, if comets were burning fires, the antisolar direction of their tails, as observed by Gemma's father and by Apian, could not be satisfactorily explained. It should rather be assumed that, like the Milky Way (which Gemma also located in the ethereal region below the moon), comets were parts of the ethereal orb, that were compressed into the form of a globe, illuminated by the rays of the sun.[110] Gemma believed the sun to be endowed with a vital, generating force, begetting and tempering all things, capable of assembling and dissolving, strengthening and weakening everything, and thus to be the principal cause of cometary generation.[111] He was therefore favourably disposed towards the theory of Hippocrates of Chios, that

---

[107] Ibidem: "Dicam mehercule huic parti non penitus naturam inesse elementarem, neque omnino caelestem, sed quiddam cui . . . naturam facile tribuat ex contrariorum genere temperatam. Primum etenim non recte tulisset universitas rerum disparia membra, semper agentia ac patientia mutuo, tanta naturarum antipathia praedita inter se, absque medio comparati: uti nec anima in corpus agit, nisi per spiritum: atque in caeteris simili modo."

[108] Ibidem: "Deinde cum omnes observatores hactenus in unum confero, colligitur certo, crinita sydera multo sublimius ferri quam caetera meteora: & quamvis materiem ex elementis inferis duci velint, lucem tamen divinitatisque speciem admirantur, & motum ex parte stabilem, caelestibus consonantem, ex parte lubricum, quae materiae fluxae, inordinataeque respondeat."

[109] Ibidem, 116.

[110] Ibidem, 117.

[111] Ibidem, 118.

a comet is nothing but a certain refraction of solar rays, and chided Aristotle for refuting it.[112]

Thus, when he wrote the *De naturae divinis characterismis* in 1575, Gemma was already on his way towards abandoning the Aristotelian account of cometary generation. He had come up with the idea of a separate, ethereal sphere, situated just below the concave of the lunar orb, which was filled with matter of an intermediary nature capable of accounting for the, partly regular, partly irregular characteristics displayed by comets. In imitation of his father, he endorsed the optical theory of comets, believing them to be condensed globes of ethereal matter refracting the solar rays. Like the elder Gemma, he attributed this theory to the Presocratics, Hippocrates of Chios and Aeschylus. Yet, he still firmly believed that comets belonged to the genus of fiery meteors, and were located below the moon, albeit higher up than any of the other meteors. Also, he still adhered to the Ptolemaic idea, that had become closely allied to the Aristotelian theory during the Middle Ages, according to which comets came into being after planetary conjunctions in fiery signs, and in their motions followed the planet that dominated them.

Nonetheless, within Gemma's highly systematized categorization of divine signs in nature, the possibility of superlunary comets could also be accommodated. After his examination of "Prodigies that occur in the ethereal region", he went on to discuss those prodigies that took place "in the celestial [regions], either actually or apparently".[113] The consideration of such occurrences was a very elevated, and, from Gemma's point of view, even a "quite divine" activity. Three kinds of celestial prodigies could be distinguished: the first was in itself to be ascribed to natural causes, but could be called prodigious because of its rarity[114]; the second was of an intermediate nature, between divinity and nature[115]; the third was thoroughly divine and supernatural. This last category included eclipses that occurred at unnatural times, such as the one that took place at the death of Christ; the sun standing still, as it did in the day of Joshua; or the appearance of a new star, such as the star of Bethlehem or the nova

[112] Ibidem, 119.

[113] Ibidem, I, 211: "Decima commemoratio prodig. de his quae caelestibus accidunt vel vere vel secundum apparentiam".

[114] Such as conjunctions of the three superior planets, solar and lunar eclipses, etc.

[115] Such as three suns, stars falling down from the sky, etc.

of 1572. Among the prodigies to be included in this last, supernatural category, Gemma also mentioned the comet seen by Albumasar above the planet Venus.[116]

Clearly, then, in 1575 Gemma was convinced that, in the vast majority of cases, comets were sublunary meteors, located in the ethereal region just below the sphere of the moon. The appearance of a superlunary comet was not ruled out; but if such a body was to shine forth, according to the principles of the doctrine of divine signs in nature, it was to be accounted for by proclaiming it a metaphysical, supernatural phenomenon. Thus, if Gemma were to remain consistent with his own theory, he ought to declare any superlunary comet that was to appear during his lifetime a metaphysical, divine prodigy.

Let us presently return to Gemma's tract on the comet of 1577. As we have seen, his parallax measurements had quite convinced him of the phenomenon's superlunary position. Yet, he was well aware of the fact that the overwhelming majority of past comets had been assigned by their various observers to the elementary region; nor was he prepared to call into question the veracity of these findings. He therefore stood by his previous assertions, that most comets were sublunary; the present phenomenon did not alter his opinion on the subject. It did, however, induce him to refine his earlier statements. Whereas, before, he had indiscriminately assigned all comets to the category of fiery meteors, and had mentioned the superlunary comet of Albumasar only parenthetically, in his 1577 tract he defined three different kinds of comet. These kinds differed with regard to their distance from the earth, the excellence of their splendour and motion, the extent of their significance, and their degree of divinity.[117] The first variety was situated very near the earth's surface; its predictive power was confined to just one state, or to a very restricted area of the world. The second was somewhat further removed; it could be either in the Aristotelian sphere of air, such as the one observed by Vögelin, or near the concave of the moon's sphere, such as the one seen by Regiomontanus. This kind of comet was endowed with a somewhat wider-ranging significance; thus, the comet of 1556, that

[116] Gemma, *De naturae*, I, 213: "Testatum reliquit Albumasar, se cometam vidisse supra Veneris corpus."

[117] Gemma, *De prodigiosa specie*, 40-41: "Illud postremo adiungimus probabili coniectura, iuxta varietatem loci, triplicem esse speciem Cometarum: iuxta remotionis intervallum, lucis quoque & motus praestantia, & significationis amplitudine, ac totius divinitatis ordine differentem."

had also belonged to this species, had been the harbinger of the "French calamities" that had closely followed it. The third kind was most rare; it exercised by far the strongest influence, and was more universally visible because it was located in the celestial region. The further it was removed from the corruptible sublunary regions, the nobler its disposition, especially if in addition it displayed a remarkable magnitude, duration, dignity of location, and brightness.[118]

The comet of 1577 had, beyond the shadow of a doubt, belonged to this last and noblest category.[119] Still, it has to be stressed that such "divine" comets were, in Gemma's eyes, very rare occurrences. He remained as convinced as ever that the vast majority of comets were sublunary. Thus, he again expounded his "new and paradoxical" opinion, first set forth in the *De naturae divinis characterismis,* that the area just below the sphere of the moon was much wider than the sphere of fire of the Peripatetics, and filled, not with any "fiery principles", but with aether. By contrast, the Aristotelian sphere of air should be considered much narrower than was usually assumed, while, this time, Gemma left no room at all for any sphere of fire. He called the ethereal sphere a "luminous region", of which the Milky Way was a more compressed section, endowed with an intermediary nature between "those things that were *entirely* corruptible, and those that *seemed* eternal".[120]

Indeed, with regard to the corruptibility of the heavens, Gemma was prepared to go one step further in 1577 than he had been in 1575. In the section of the *De naturae divinis characterismis* in which he refuted Muñoz' opinion, that the nova of 1572 had been a celestial comet, he had stressed that, even if the heavens suffered a small degree of change, it was not to be assumed that these minute alterations could be detected with the senses, except perhaps after a number of years. But it was not to be believed that such sudden changes as the appearance, from one day to the next, of a new star or a

---

[118] Ibidem, 41: "Tertium genus ut maxime rarum, ita & fortissimum, reliquisque latius patens in aliqua celesti regione conspicuum: imo nobilius tanto quanto a caducis sedibus longius removetur, praesertim si & magnitudo & stabilitas sive duratio, & loci dignitas, & lucis splendor accesserint...".

[119] Ibidem: "...quae cuncta sane concurrunt in nuper descripto phosphoro novo."

[120] Ibidem, 40: "Latissimum contra quod cedat unius aetheris elemento: idemque non ignem esse, ut nostri existimant: sed lucentem aliquam sedem, magisque concretam vi principis formae; in qua consistat velut orbis sui pars densior Lacteus circulus natura videlicet media, inter corruptibilia prorsus, & quae videntur aeterna".

comet could naturally take place in the heavens.[121] Thus, in 1575, Gemma still adhered to the substantial inalterability of the celestial region. However, two years later, in the tract on the comet of 1577, his assertions were much less cautious.

Obviously, he said, we must admit the possibility that God, if he wanted, could by force of his will alone create heavenly phenomena to issue warnings to humankind. Yet, in practice, he often made use of natural causes, intertwining these with the metaphysical ones. Thus, Gemma thought it well possible that, if God chose to confine himself to the "dungeons of nature", he might call into operation certain efficient and instrumental causes capable of compressing and illuminating celestial matter, and thus creating new stars, or offspring of stars facilitated by some celestial conjunction.[122] Such a novel celestial body might then remain hidden for a certain period of time, attached to the body of the star or planet that generated it, like a foetus in its mother's womb, until, at a propitious moment, it was severed from its generator's body and assumed its own course.[123]

In these utterances we can clearly detect the influence of the late sixteenth-century intermingling of Aristotelian and Stoico-Platonic tenets, that was described at length in the previous chapter.[124] Gemma acknowledged the medieval *potentia Dei absoluta,* the power of God to do anything he pleased short of a logical contradiction; he also granted the fact that, often, God chose nonetheless to make use of natural causes instead. Yet, unlike the Peripatetics, who had deduced from these two tenets the doctrine that, if God employed natural causation, he invariably did so in accordance with the postulates of the Aristotelian universe, Gemma reversed the argument: God, being omnipotent, could perfectly well employ natural causes to

---

[121] Gemma, *De naturae,* II, 271.

[122] Gemma, *De prodigiosa specie,* 46: "Quamvis vero caelorum partes in unum cogere, iterumque dispergere, vel solo voluntatis suae verbo ac spiritu sit facillimum, utitur tamen interdum & naturalibus causis, easque metaphysicis velut innectens humanis mentibus universalem formam alicuius praenotionis, sive praesagii subministrat. Ideoque si placet inter hos carceres naturae visibilis permanere, iungantur ex illius sinu συναιτίαι quaedam, vel effectrices & instrumentariae causae, quibus densandi pariter illustrandique vis insita ad radiorum concursus varios efflorescat. Hinc & stellas, stellarum soboles, id est, congressu superiorum genitas provenire dicamus. . .".

[123] Ibidem, 46-47.

[124] See supra, § IV.2.

bring about *non*-Aristotelian effects, such as the generation of new celestial bodies.

Of course, the ideas of the birth of new heavenly bodies above the moon, and of these creatures remaining hidden in space for some time before finally becoming visible to human eyes, were not original with Gemma. This had been one of the most common ways of explaining cometary phenomena among authors such as the Chaldaeans, the Pythagoreans, Artemidorus, Apollonius of Myndus, and Hippocrates of Chios. The latter had, moreover, combined the idea of comets as celestial bodies usually hidden from view with the theory that their tails were formed by optical effects caused by the sun. Such a combination offered a number of advantages eagerly made use of by Gemma: first of all, it enabled him to explain, like his father, the antisolar nature of cometary tails, and, secondly, he could employ it to lend credibility to the Stoic idea of heavens subject to change.

Despite on several occasions stressing the weakness of the human understanding and its incapability of discovering the hidden laws of nature[125], Gemma did venture to suggest that, possibly, each celestial orb was capable of creating new beings in accordance with its kind and dignity; or that somewhere in the heavens, stars might lie hidden, that shone forth to human eyes only at certain preordained times, after which they disappeared once again.[126] Or, perhaps, God sent his messengers as heralds of good or bad events, shaping them in the form of stars by the force of his will.[127] All this is very reminiscent of the Stoic outlook: a universe governed by the benevolent, paternal God, capable of subjecting nature to his will without having to act in accordance with any laws conceived by the human intellect, and a universe endowed with the power to issue warnings to mankind without being obstructed by the rigid rules of the Aristotelian world sys-

---

[125] Gemma, *De prodigiosa specie,* 38, 45.

[126] Ibidem, 45-46: "At quomodo rursus coagmentatio fiat in sublunaribus, explicare non ita difficile est. In caelo rarum quiddam & insolens, ut tamen si vel supra humani ingenii vires nullam naturae partem aut robur effetum aestimaverimus, ad res gignendas quotidie novas: ipsa profecto etiam de novo cogitandi te observationis industria stimulabit. Quid enim absurdum adeo, si & pro caeli cuiusque modo vel dignitate certam subinde generationis speciem in singulis posse fieri concedamus? Quid si latere alicubi stellas, quae prodeunte tempore suo nobis indefinito, sub obtutum veniant tandem, ac rursus ab oculis auferantur?"

[127] Ibidem, 46: "Quid si denique sint certi bonorum malorumque pronuncii ad summi Patris arbitrium sese in speciem sideris exerentes?"

tem. Once again, we are reminded of Seneca's exclamation, "How many other bodies besides these comets move in secret, never rising before the eyes of men!"[128]

Furthermore, Gemma believed the Stoic outlook to be in accordance with Holy Writ. By means of a number of quotations from the Psalms, he attempted to render the idea of alteration in the heavens, and of God using new heavenly bodies as messengers, more acceptable to his readers. Psalm texts like "He was covered in light as in a garment, and spread out the heavens like a robe", or "He inclined the heavens and descended, and darkness was under his feet", to Gemma lent credence to the latter idea, while in favour of alterable or corruptible heavens could be quoted Psalm 101, where it said that "all will vanish, though you remain, all wear out like a garment, like clothes that need changing you will change them."[129] This same text had, a few years earlier, been cited, among many others from both the Old and New Testaments, by Robert Bellarmine in his 1572 Louvain Lectures, in which he had demonstrated that both the Bible and the teachings of the Church Fathers endorsed the idea of corruptible heavens consisting of fire.[130]

Recent findings about celestial alterability, novae, and superlunary comets, then, could be legitimated by demonstrating them to be in accordance with Sacred Scripture and with non-Aristotelian texts from antiquity. As we saw in the previous chapter, this strategy was adopted by an increasing number of scholars in the course of the sixteenth century; in connection with the comet of 1577, Gemma, too, made eager use of it. The unshakeable evidence of the parallax measurements had demonstrated its superlunary location; therefore, it could not be doubted that this particular comet, at least, "rendered untenable the opinion of the Peripatetics on the subject of the gen-

[128] Seneca, *Natural Questions*, VII.30.3.

[129] Gemma, *Prodigiosa specie*, 46; ibidem, 38: "... cum nobis ne quidem hoc demonstratione sit exploratum, an caeli perpetuo servent eandem legem; an etiam corruptioni, generationique obnoxii, ac senescentes sensim ut vestimentum veterascant: cui quidem opinioni nostrae Sacra Scriptura fidem facit, & multiplex huius aetatis observatio suffragatur."

[130] See Ugo Baldini & George V. Coyne eds., *The Louvain Lectures (Lectiones Lovanienses) of Bellarmine and the Autograph Copy of his 1616 Declaration to Galileo* (Vatican Observatory Publications, Studi Galileiani, vol. 1 no. 2, Vatican City 1984), especially pp. 8-11, for relevant passages of both Bellarmine's Latin text and the editors' English translation.

eration of this kind of meteor."[131] Because the location of a phenomenon lent it its form, Gemma said, and because all things were conserved by the regions out of which they were generated, it was indisputable that the matter out of which the present comet was constituted was not at all sublunary. The comet clearly displayed a transparent and firmly compressed disposition, as opposed to the opacity, volatility and corruptibility of the elementary region. Furthermore, its motion was well proportioned, and its head very bright, showing how little it had in common with a burning exhalation, whose nature it was to chaotically follow its fuel. Comets, on the other hand, kept their tails constantly averted from the sun, so that they should be regarded as refractions instead of fires. The present comet's matter, then, was of a much more divine kind than that of ordinary burning vapours; probably, it was a compressed portion of the celestial region in which it moved.[132]

We can conclude that Gemma's explanation of the comet of 1577 can be read as an attempt, possibly unconscious, at reconciling elements from two traditions. Despite the fact that his views on the generation of comets in general, and of the present comet in particular, were not compatible with the cometary theory espoused by the Peripatetics, his tract, if studied carefully, displays a predominantly Aristotelian world view. Gemma remained attached to the most basic tenets of Aristotelian cosmology: the fundamental dichotomy between celestial and elementary regions, the circularity and regularity of celestial motions, geocentricity, the divine, incorruptible quintessence pervading the entire superlunary region, the corruptible and transient nature of the area below the moon, all these were key constituents of Gemma's outlook. Still, he was prepared to follow the lead of his sense perceptions, and to make an effort to fit the obtained results into his basically traditional world view.

The phenomena displayed by the vast majority of comets could be accounted for by means of Gemma's "ethereal region", which should be regarded as a slight modification of, rather than a major deviation from, Aristotelian physics. After all, being located entirely below the

---

[131] Gemma, *De prodigiosa specie*, 19-20: "Rara in illo, atque prae caeteris admiranda quam plurima cuncti intelligent, qui diligentius idem cum cometis aliarum aetatum hactenus tum figurae, tum magnitudinis, & caeterorum accidentium nomine compararint. . . . est tamen huic cometae natura vel maxime singularis, quae Peripateticorum sententiam de generatione istiusmodi μετέωρων prorsus evertit."

[132] Ibidem, 37, 45.

moon's orb, it did not violate the dichotomy between sub- and superlunary realms, with their fundamentally different physical operations. All it did was enable the characteristics displayed by comets, which partook of both celestial regularity (regular, circular motions, constant brightness) and elementary irregularity (sudden appearance and disappearance), and which, according to Gemma, in the vast majority of cases moved below the moon, to be accomodated within an essentially Aristotelian framework. In fact, the function of the ethereal orb in Gemma's outlook was not much different from that of the sphere of fire in Aristotle's: it provided a breeding-ground for comets, meteors that, however close to the sphere of the moon, were still sublunary phenomena.

Yet, in allowing the ethereal region to partake of the celestial nature as well as of the elementary, Gemma indisputably moved some way towards the world view of late antiquity, in which the lunar orb constituted no boundary, and in which the cosmos was one great continuum. He also showed himself favourably disposed towards ideas like those of Hippocrates, the Pythagoreans, or Seneca, that the universe might be full of hidden bodies which only displayed themselves at times preordained by God, and added the suggestion that, at God's instigation, novel celestial bodies might be generated within the celestial orbs by means of natural causes such as planetary conjunctions. Such ideas, apart from the support lent them by prominent ancient authors, received additional weight through the dicta of the Church Fathers, who had taught the corruptibility of the heavens, and through a number of Bible texts that could be interpreted in its favour.

Still, Gemma was not prepared to abandon the celestial incorruptibility taught by Aristotle. Although, in order to account for novae or superlunary comets, he allowed the celestial orbs a certain generative power to create new beings whenever God wished them to, he remained adamant that the stuff of which the heavens were made was of a divine, perfect nature, of incomparably higher dignity than ordinary elemental matter. Superlunary comets, consequently, were endowed with a far nobler nature than those generated in the ethereal region. It needs to be stressed that Gemma believed comets located above the moon to be extremely rare phenomena, which only appeared at very portentous moments. Thus, he was not at all inclined to allow the heavens the same measure of alterability as the earthly regions; but *if* a new celestial body appeared, it could be accounted for by means of ancient, non-Aristotelian theories of high respectability, and be legitimated by means of Bible texts and doctrines of the Church Fathers. Such a strategy carried the obvious advantage of authoritatively incorporating the appearance of new celestial bodies in an otherwise perfectly Aristotelian physical framework.

The appearance of the superlunary comet of 1577 was thus satisfactorily accounted for, without having to abandon the dichotomy between celestial and elementary realms. It posed even less of a problem with regard to the other fundamental tenets of the Aristotelian world view, circularity of celestial motions and geocentricity. Gemma saw no reason to abandon either one of these tenets. He did not once mention the name or theory of Copernicus, and his tract clearly reflected a geocentric world view. Thus, in his discussion of the comet's parallax, he evidently regarded the sun as one of the planets when he said that all bodies below Mars displayed a certain perceptible parallax, these bodies being "the Sun, Venus, and Mercury", an indisputably Ptolemaic arrangement of the planets.[133] Moreover, when expounding his opinion as to the duration of the present comet's effects, Gemma stated that he believed these were to last until "the fortieth orbit of the sun."[134]

In Gemma's universe, the planets were still firmly embedded in their own solid orbs. This becomes apparent from his statement, that the comet of 1577 was situated in the orb of Mercury.[135] His reasons for believing so were, however, rather flimsy. Unlike Mästlin, who as we have seen spent a great deal of effort trying to fit the comet into one of the celestial orbs, Gemma made not the slightest attempt at mathematically corroborating his statement. He was, of course, well aware that the characteristics of the comet's movement did not correspond exactly to the ones that were to be expected of a body embedded in Mercury's sphere, admitting that "its species, splendour, and the inequality of its motion did not exactly correspond to those of Mercury's orb, but at least it did [correspond to it] while retaining its own proper footsteps".[136] Gemma's ascription of the comet to

[133] Ibidem, 36.
[134] Ibidem, 53: "...etiam usque ad quadragesimum Solis anfractum facile revirescet."
[135] Ibidem, 36: "Quis itaque non videt huius cometae locum longe fuisse Luna superiorem? id est in orbe Mercurii...".
[136] Ibidem: "... id est in orbe Mercurii, cuius & speciem splendoremque & inaequalitatem motus si non exacte, saltem vestigiis proprioribus observavit."

Mercury's orb was, in fact, mere guesswork, his main reason being that its intensity of light and its movement roughly corresponded to those of the lowest planet. Much less did he attempt to compute a proper orbit for the comet, as Tycho was to do; Gemma was interested in a different kind of information.

And so we return to our point of departure. To Gemma, the investigation of the comet's mathematical and physical characteristics was not an end in itself, but a means of gaining an insight into the arrangement of the cosmos and the divinely inspired sympathies between its parts. Therefore, much rather than waste his energy on laborious calculations attempting to fit the comet into one of the celestial orbs, he spent his time pondering the meaning of the fact that the comet had moved much faster through the Saturnine signs of Capricorn and Aquarius than through the sign of Pisces, or the significance of the circumstance that its course had been directed exactly from one tropic circle to the other. His main concern was to correctly explain to his countrymen the import of the appearance of a sign of such sublime divinity at such a propitious moment in history. In our next chapter, we shall concern ourselves with his solution to this vital puzzle.

c. *Johannes Heurnius: "In the Astral Element"*

Like Gemma, the Utrecht town physician Johannes Heurnius (to be appointed to the chair of medicine at Leiden only in 1581) was much concerned about the harrowing political conditions with which his country was faced. His pamphlet about "The History, Nature, and Significance of the Horrible Comet, which appeared in the Year of our Lord 1577"[137], was written in the manner of an admonition to his fellow countrymen, counselling and encouraging them, rather than as a treatise on the comet's astronomical and physical characteristics. As is to be expected in a short pamphlet written in the vernacular, it included more astrological predictions, and less elevated, ethical digressions, than Gemma's Latin treatise; however, both works tried to convey the same message: a great alteration was imminent, and humankind needed to prepare for it before it would be too late.

Still, Heurnius, again like his Flemish colleague, thought it neces-

---

[137] Johannes Heurnius, *De Historie, Natuere, ende Beduidenisse der erschrickelicke Comeet, die geopenbaert is int Jaer ons Heren 1577.* (Cologne, s.d.).

sary to include a description, albeit rather a crude one, of the comet's course and its physical constitution; such a description served to explain to his readership why exactly it was that he believed the phenomenon to be endowed with predictive power at all. No astrological predictions were possible without first of all establishing a phenomenon's astronomical features. As in the case of Gemma, then, we have to acknowledge that, whatever Heurnius told his readers about the comet's mathematical and physical characteristics, was told with the underlying purpose of rendering more convincing the astrological and ethical elaborations that formed the bulk of his pamphlet.

Heurnius' description of the comet's course through the heavens was much more crude than Gemma's. He first observed it on 11 November 1577, after four-thirty in the afternoon. At that time, its position was in the fifteenth degree of Capricorn, at a distance of 15° from the brightest star in Aquila. Its declination from the ecliptic amounted to 21° 43', so that it stood not far from the equinoctial colure.[138] Without mentioning any measurements he might have conducted to determine the comet's course, Heurnius then went on to say that its daily movement amounted to one degree towards the northeast; this is obviously an extremely rough estimate, if we bear in mind Gemma's conclusion, that during the first days of its visibility the comet moved 5° daily, but that soon, its speed diminished considerably until eventually it covered no more than 15' a day.

Although he did not mention any systematic observations of the comet, Heurnius does appear to have watched it from time to time, reporting that it moved "past the constellation of the Eagle, through the constellation of the Dolphin, with its tail pointing at the mouth of the small horse, the fowl of Pegasus, and afterwards towards the shoulder of Pegasus; so that on the first of December I found it in between Equiculus and Delphinus."[139] Its colour was initially much more fiery and red than the constellation of Aquila (which, indeed, is of a Martial and Jovial complexion, and therefore quite reddish,

---

[138] Ibidem: "De erschrickelicke Comete die ons openbaerde dit Jaer 1577 den elfften Novembris t'savens na half vyf uren, vinden wy in den 15 graet des Steenbox, ende stont 15 graden van de principael Sterre des Arens, ende wyckt van den Zodiackse Cirkel nae ons thoe 21 graden 43 minuten niet veer van den Equinoctiael...".

[139] Ibidem: "...ende trock syn selven verby het gesterte des Arens, door den gesterte der Vische Delfyn, met syn sterte wisende op de mont des kleine Peert, dat Föllen van Pegasus, ende voorts daer na op den armum Pegasi: so dat ick op den eersten dach Decembris hem vonde tusschen den Equiculum ende den Delphinum."

Heurnius added), but with a sad, murky lustre about it. The tail stretched out from West to East, with its concave side turned southwards and its convex side northwards, in the semblance of a Turkish sabre. After a while, it became wider, and lost its coppery-red colour, turning dark and leaden grey instead.

These rather scanty communications constituted all that Heurnius had to say on the subject of the comet's course through the heavens. As to its distance from the earth, that subject was not even touched upon by the Utrecht physician. His pamphlet did not include a single acknowledgment of the current scholarly debate on the issue; much less did he attempt to determine for himself the diversity of aspect shown by the comet, not even in the approximative manner employed by Gemma. Entirely conventionally, Heurnius informed his readers that the comet had been generated by Mars, the planet that had dominated the conjunction which had preceded the hairy star's birth, and that was affiliated with Aquila, the sign in which the comet had stood for a long time.[140] Next to Mars, Mercury and Saturn were also "Lords" of the comet: Mercury because it reigned over its ascendant and medium caeli, and Saturn because it had stood firmly in its own "houses", the eighth and ninth, at the time the comet was born.

Such a conventional way of explaining the comet's generation, expounded at the very beginning of the pamphlet, might lead us to expect any utterances Heurnius might add concerning his world view to bear an overwhelmingly Aristotelian character as well. After all, the attribution of the birth of a comet to the conjunctions of certain planets in particular signs, or in the locations where previous eclipses had occurred, was one of the most common tenets of the medieval cometary tradition that combined Aristotle's theory of comets as burning terrestrial vapours with the Ptolemaic idea that they were generated by the planets and assimilated to their natures, thus bringing in their wakes the effects associated with the particular planets that had called them into being.[141]

Sure enough, after his brief description of the comet's course and the causes of its generation, Heurnius went on to the "Explication of

---

[140] Ibidem: "...*quia Mars dominus Coniunctionis praecedentis excitavit hunc Cometam, cum pervenit ad locum Eclypsis celebratae in signo Librae, in Aprili, & consortii ratione Martis cum Aquila, circa quam diu effulsit Cometa*...".

[141] See supra, chapter III, p. 57.

this rare *meteoric* impression" [my italics], as an introduction to which he included a brief description of the universe, that bore the standard hallmarks of traditional, christianized Aristotelianism: "In the beginning God created the Heavens and the Earth, the Waters around the Earth, saturating it like a sponge (so that, if God had not provided them with sure boundaries, they would have inundated the Earth); about and through the waters blows the Air; around and through the Air shines the element of Fire (which also permeates it); but around and through the element of Fire glows a heavenly, luminous, transparent substance, in which the Planets and Stars run their courses."[142] Thus far, Heurnius' account was completely traditional: earth, water, air and fire surrounded each other in concentric circles, while beyond the fiery sphere was located a heavenly region filled with a distinct kind of matter. Furthermore, he called the comet a "meteoric impression", thereby suggesting that he adhered to the Aristotelian theory of cometary generation.

Yet, instead of going on to provide his readers with a Peripatetic account of the present comet's engendering, Heurnius chose to cast rather a different light on the question. The comet, he stated, was generated "in the lowest part of the fifth element" (i.e., the above-mentioned luminous and transparent substance that filled the heavens beyond the sphere of fire), which was called the "astral element". Here, then, we find an obvious deviation from Aristotelian cosmology. Where, to the Peripatetics, the superlunary region formed a single whole displaying the same characteristics everywhere, eternal and incorruptible in its entirety, Heurnius' quintessence was apparently divided into two distinct sections, an upper and a lower part, the lower part even being endowed with its own appellation of 'astral element'. Furthermore, this astral element was clearly not as unalterable as would be expected of a quintessential substance in Aristotelian cosmology: in it, comets could be generated.

Indeed, the cause of the present comet's generation had been a condensation or coarsening of the astral element, Heurnius main-

---

[142] Heurnius, *Historie*: "In den eersten schyep Godt Hemel ende Erde, het Water om de erde, hem doordringende als door een spongie, ja dattet de Erde overlopen solde, so veer Godt geen wterste Palen geset hadde: dann om dit Water, ende door dit Water blaest de Lucht: om de Lucht ende door de Lucht schynt het Element des Viers (dat hem oock doordrinckt) maer rontom ende door het Element des Viers, blinckt een hemelsche, lichte, doorschynelicke substantie, daer in dat de Planeten ende Sterren haren loop volbrengen."

tained. For, whenever the heavenly harmony was 'poisoned', the fifth element was wont to degenerate ("*vervalsen*"), coagulating and curdling ("*schift*") to purge itself of the impurities inflicted on it.[143] In this particular case, the poisoning of the heavenly harmony had been due to two eclipses that had occurred during the months preceding the comet's engendering: one in April, in Libra, and one in Aries. The comet had appeared exactly half-way between these two signs, in Capricorn, at the time when Mars arrived in Libra at the spot where the first eclipse had taken place. Heurnius' account remained rather unclear as to the exact reason why such a succession of occurrences constituted a threat to celestial harmony. Possibly, it had something to do with the astrological rule, mentioned by Manilius, that at an eclipse of the moon, the sign which it then occupied, and the sign diametrically opposed to it, both suffered an eclipse of their influences which could last for an indefinite period of time.[144] In the case of the comet of 1577, it could be said that the signs of Aries and Libra had become weakened especially severely, since not only had they both suffered an eclipse themselves, but, being diametrically opposed to one-another, their power was even more reduced because they shared in each other's decrepitude. The subsequent entry of Mars, being an especially malignant planet, into the weakened sign of Libra, then, might have aggravated this imbalanced situation, leading to the "poisoning" mentioned by Heurnius, and causing the astral element to get rid of its impurities.

Whatever the exact reason, Heurnius clearly believed in the capability of the astral element to change its disposition so as to form new heavenly bodies, despite its quintessential nature. Furthermore, he expressly made a connection between the alterability with which his astral element was endowed, and the theory attributed to the Pytha-

[143] Ibidem: "In het leechste deel van dit vyfste Element, geheten het astralisch Element, is dese Comeet gegenereert, wt oorsake dattet vergrooft is, . . . door sulcken vergift der Hemelscher harmonie, vervalst dese quintum Elementum, ende groff wordende schift ofte rundt, om sich selven te purgeren van syn onrein. . .".

[144] Marcus Manilius, *Astronomica* IV. 841-856: ". . .whatever signs have seen the Moon eclipsed, reft of her brother and plunged into the darkness of night, when the Earth comes between to cut off Phoebus' rays and Delia draws not the light wherewith she is wont to shine, those signs too become as stricken as the planet occupying them, bowed the while in grief and deprived of their customary powers, and they mourn Phoebe as though she had been carried out for burial. . . . However, signs suffer together in pairs, not of those situated side by side but of those which shine at each other from opposing quarters. . ." (Translated by G.P. Goold, Loeb Classical Library edition, Cambridge Mass./London 1977) 291.

goreans, according to which, Heurnius believed, "the Comet is nothing else but a substance generated with the Sun, hiding close to the Sun's body, but at certain times separating itself from it to convey a message to us."[145] This was, however, a rather corrupt rendition of the Pythagorean theory, since the authors of Antiquity attributed to the Pythagoreans solely the idea that a comet was one of the planets, appearing only at long intervals and not rising far above the horizon, just like Mercury.[146] This theory implied the existence of only one comet, describing an orbit like that of the planet Mercury; it did not, however, mention the idea, added by Heurnius, that this comet was "a substance generated with the sun". Heurnius himself mentioned Cornelius Gemma's *De naturae divinis characterismis* as his source for this addition; however, all Gemma had said on the subject was, that John of Damascus had thought comets to be stars created in the beginning, but hidden (perhaps under the rays of the sun) before suddenly appearing in a certain region, and that, personally, Gemma believed him to have arrived at this idea from the Pythagorean analogy between the comet and the planets Venus and Mercury (which were often so close to the sun that they could not be seen).[147]

Heurnius, then, allowed himself a certain measure of licence in his interpretation of the Pythagorean theory, in order to make it compatible with his own idea of the astral element purging itself of any poisonous imbalances. His reasoning probably ran along the following lines: because God does not allow anything in nature to happen in vain or without purpose[148], the appearance of a comet necessarily carries a certain significance. Moreover, the sympathetic correspondence between occurrences in the superlunary heavens and the

---

[145] Heurnius, *Historie:* "...daerom hebben hem sommige Pytagorixe Philosophen (so den geleerden Doctor Gemma in syn Cosmocriticis verhaelt) hem niet sonder reeden laten duncken, dat de Comete anders niet en is, dan een gegenereerde substantie met de Son, onder hem schuilende, dan op sekere tyden daer van scheyt, ons yet beduidende."

[146] Aristotle, *Meteorology,* I.vi.342b; pseudo-Plutarch, *De placitis philosophorum,* III.893b.

[147] Cornelius Gemma, *De naturae,* I, 113: "Iohannes Damascenus vult natos esse cometas cum caeteris stellis, nimirum in ipsis primordiis mundi, sed latere (fortassis sub radiis solis) deinde impelli subito ad certam quampiam regionem. Hoc credo a Pythagoricis sumptum, exemplo stellarum Veneris & Mercurij."

[148] Heurnius, *Historie:* "...want een yder dinck volbrengt syn werck daert van Godt thoe geordonneert is... want daer is so kleine geschep niet, het betuicht de Glory des Almachtigen, int volbrengen haerder einde, daerom dat se van Godt geschapen syn."

events on earth the comet is meant to foretell, leads to the conclusion that the comet must be situated in the region permeated by the celestial quintessence, or, in the words of Heurnius, "it lives in the flowerbed where the seeds of natural future occurrences have been sown."[149] Whenever, for astrological reasons, the lower part of this quintessence suffered a disruption of its normal harmonious state, it endeavoured to purge itself by coagulating; due to the correspondence between macro- and microcosm, moreover, a similar disorder was soon to be expected on earth.

Yet, although he did not say so explicitly, the idea that the comet must be superlunary left Heurnius with a physical impossibility: according to the accepted Aristotelian physics, no generation was supposed to take place in the heavens. He therefore came up with the idea of old ascribed to the Pythagoreans, according to which there was only one comet, a planet-like body describing an orbit much like Mercury's. Finally, the idea, implied by this theory and explicitly stated by Gemma, that if it so closely resembled Mercury the comet spent most of its existence hidden in the rays of the sun, provided Heurnius with the final touch: the comet was not really a new celestial body, but rather a substance generated at the same time as the sun (i.e., at the beginning of the world), and so closely allied to it that it remained invisible to human eyes for most of its existence. The purging of the astral element, necessitated by the poisoning of the heavenly harmony, could then be said to consist of nothing more than the separation of the comet's body from the mass of the sun. This way of accounting for celestial novelties is very reminiscent of Gemma's suggestion, that the stars might, prompted by certain planetary conjunctions, be capable of producing offspring that remained hidden in their "wombs" until assuming their proper courses at moments of great significance.[150]

Heurnius' account remained rather unclear as to the precise area covered by his "astral element". His statement, that it consisted of the "lower part of the fifth element", does not offer us much of a clue: nowhere did he speak of any upper or lower limit for it. Notably absent from his description of the structure of the world was the sphere of the moon as constituting a fixed boundary between sub-

---

149 Ibidem: "...daerse woont int bedde daer het saet gestroyt is, van natuerlicke thoekomende dingen...".

150 See supra, p. 183.

lunary and superlunary realms; and he remained equally silent about the exact location of the border between the astral element and the higher section of the celestial quintessence. Yet, it can be maintained with a reasonable degree of probability that Heurnius thought of the astral element as located between the sphere of the moon and the firmament of the fixed stars.

First of all, his above-mentioned account of the structure of the universe along Aristotelian lines was fairly straightforward about the distinction between the four sublunary elements and the fifth, celestial one. Although he did not explicitly mention the lunar orb as constituting the boundary between the two realms, his phrasing does seem to imply the traditional Peripatetic view of distinct regions above and below the moon. Much more uncertainty, though, surrounds the upper limit of the astral region. Especially ambiguous is Heurnius' remark, that "...the astral element ... was, by nature, created half mortal or corruptible, and half immortal, since half of it touches on the unalterable Heaven, and half on the perishable element of Fire."[151] It is not easy to assess what exactly Heurnius meant by this statement. Clearly, he thought of the astral element as immediately bordering on the sphere of fire, the last of the sublunary spheres; moreover, we have already seen that he also referred to it as being "the lowest part of the fifth element". Possibly, Heurnius meant to make a distinction between, on the one hand, the area in which the planets run their courses, and, on the other, the firmament in which the stars are fixed. A hint in this direction can be gathered from his belief, that comets were generated in the astral element, and also very close to the sun; this would seem to imply that the sun, too, was located in the astral element. Moreover, Heurnius believed the astral element to have been coarsened by the effects of two eclipses, which again seems to indicate that both the sun and the moon found themselves in the region thus designated. The "unalterable Heaven" of which our author speaks can, in that case, be identified with the starry vault of the firmament.

Because the comet consisted of a coarsened part of the astral element, whose nature was partly corruptible and partly incorruptible,

---

[151] Heurnius, *Historie:* "In desen acker staet dese Comeet, ick segge in dat astralisch Element, dat half van naturen sterflick ofte verganckelick, ende half onsterflick geschapen is: want het half an den onveranderlicken Hemel raeckt, ende half an het verganckelicke Element des Viers."

it, too, was endowed with intermediary qualities. The celestial quintessence that shone beyond the sphere of fire, Heurnius stated, conferred life on all things, including human beings: "The supreme Beings and Powers are also in all things and in us; for they live thanks to them."[152] There existed a correspondence between the 'quintessence' terrestrial beings carried inside them, and the celestial quintessence, Heurnius believed, with the terrestrial quintessence following the celestial one like a magnet pursuing the pole.[153] It was a "naturally rational element", capable of foretelling future misery because of its close alliance with the substance of the heavens; without it, no thing beneath the firmament could remain alive. Even at untimely deaths, the "astral intellect" remained behind in the human corpse after the other elements had left it; it would then knock and sigh as a token of its sadness at departing, a phenomenon the common people usually referred to as a "Ghost". And, on the approach of a murderer to the corpse of his victim, the human quintessence would recognize him and, in anger, cause the body to start bleeding afresh.[154]

The comet, likewise consisting of quintessential, astral matter, played a similar role in the chain of correspondences between heaven and earth. Being situated in the astral element, it possessed the qualities necessary for predicting the future. Not that comets were in themselves capable of causing future terrestrial occurrences, Heurnius added; rather, they were natural signs, displayed by nature to man-

---

[152] Ibidem: "Hier wt merckt dat de opperste Wesens ende Crachten oick in alle dinck ende in ons zyn, want sy daer door leven."

[153] Ibidem: "Want alle dinck (naest Godt) hangt syn leven an dese quintum Esse, die in hem is, ende onse quinta essentia volcht dem quintam Essentiam des Hemels, als een Magneet den Pool."

[154] Ibidem: "Ende dit is een natuerlick verstandich Element, het welcke (alse na by den hemelschen Lichame wesende) voortast de thoekomende Elende, ende is van so hogen cracht, dat geen dinck onder het Firmament sonder hem leven mach. Wanneer een Mensche ontydelicken door gebreck der ander Elementen sterven sal, ende dese astralische Venuft nochniet verteert is, so geeft dit van bedroeftheit syns scheiden een teicken, klopt, ende sucht (het gemeen Volck heetet een Geest) Ja wanneer der een Moorder kompt, by het Lyck dat hy omgebracht heeft, so sal hem dat noch kennen, ende hem met een natuerlicke bewegenisse vergrammende, den Dooden doen bloeden." The problem of the body bleeding in the presence of its murderer was widely discussed from the twelfth to the early eighteenth century. Thinkers like Mersenne and Descartes accepted the veracity of the phenomenon; for a list of authors who discussed the question, see Bert Hansen, "Science and Magic" in David C. Lindberg ed., *Science in the Middle Ages* (Chicago/ London 1978) 483-506: 494-5.

kind at God's command to warn people of impending changes.[155] Being formed out of a substance that was half immortal and divine, and half mortal and corruptible, they were ideally suited to serve as messengers maintaining contact between heaven and earth.

Heurnius' concepts of correspondence between micro- and macrocosm, and of an intermediate substance, the "astral element", necessary to realize the links between heaven and earth, are reminiscent of Gemma's ethereal orb, in which celestial and elementary qualities were allowed to mingle in order to sustain contact between the divine and the transitory. They also had a distinctly Platonic ring to them. Neoplatonic authors of late Antiquity had personalized the connections between heaven and earth; thus, for example, Apuleius in his *De deo Socratis* conceived of aerial spirits or demons, whose task it was to keep God in contact with his sublunary creatures. These spirits were inconstant and subject to passions, yet immortal, and their abode was the atmosphere, located halfway between heaven and earth.

In Plato's own cosmology, as expounded in the *Timaeus*, the necessity of a mediator between two extremes is also stressed on a number of occasions. Thus, the Demiurge created the World Soul out of a mixture of Being and Becoming, and of the Same and the Different: "From the indivisible, eternally unchanging Existence and the divisible, changing Existence of the physical world he mixed a third kind of Existence intermediate between them; again with the Same and the Different he made, in the same way, compounds intermediate between their indivisible element and their physical and divisible element: and taking these three components he mixed them into a single unity. . .".[156] Out of this substance, he created the universe and the gods, who were themselves not essentially immortal, since they had received their being from the Demiurge.[157] He then went on to charge the gods with the creation of mortal beings other than gods,

---

[155] Heurnius, *Historie*: "Niet dat ick wil segghen dat Cometen oft dergelicke Teickenen, de thoekomende Veranderingen voortbrengen ofte maken: neen geensins, dan het syn natuerlicke Teickenen, dat de Natuer door Gods gebott haer oorsake der veranderinge tot ons neder seint, want die nu ryp syn."

[156] Plato, *Timaeus*, 35, translated by Desmond Lee (Penguin Classics edition, *Plato: Timaeus and Critias* (Harmondsworth 1979) 47).

[157] Ibidem, 41: "Therefore, since you [the gods] have been created, you are not entirely immortal and indissoluble; but you will never be dissolved nor taste death, as you will find my [the Demiurge's] will a stronger and more sovereign bond than those with which you were bound at your birth."

whom he could not call into being himself, since everything he created was divine. The gods therefore framed the human body out of earth, water, air, and fire; yet, the soul they implanted inside this body they had received from the Demiurge himself, and, accordingly, it was immortal and divine.[158]

In Plato's outlook, then, the gods, having been created out of the World Soul with its intermediate nature between Being and Becoming, occupied a middle position between the divine Demiurge and mortal creatures on earth. Their function was to act as a mean between the Demiurge and the sublunary realm, and, having been themselves created by the divine architect, they were of a lesser degree of godliness than he. Moreover, their abode was in the heavens, as Plato identified them with the celestial bodies; the fixed stars were located in the "Circle of the Same" and the planets in the "Circle of the Different". Thus, the characteristics of these Platonic gods were very similar to those Heurnius accorded his astral element: their nature was intermediate between mortal and immortal, they lived in the heavens, and they acted as liaisons between God and men. Moreover, his idea of a human quintessence, naturally rational and superior to the four terrestrial elements out of which human bodies were formed, and closely allied to the quintessence of heaven, is reminiscent of Plato's immortal and divine soul, capable of uplifting man from his earthly state towards heaven, his true home.[159]

Despite such Platonic overtones, however, on the whole Heurnius' tract can be said to be as predominantly Aristotelian as Gemma's. His world view displayed the standard elements of orthodox Peripateticism: the earth was at rest at the centre of the universe, surrounded by the spheres of water, air, and fire; beyond the fiery orb, the heavens began, filled with a fifth substance essentially different from the sublunary elements. Most of Heurnius' cometology, too, was in accordance with the widely embraced Aristotelian-Ptolemaic

158 Ibidem: "And in so far as there ought to be something in them that can be named immortal, something called divine, to guide those of them who are ready to follow you and the right, I [the Demiurge] will begin by sowing the seed of it and then hand it on to you [the gods]; it remains for you to weave mortal and immortal together and create living creatures."

159 Ibidem, 90: "We should think of the most authoritative part of our soul as a guardian spirit given by god, living in the summit of the body, which can properly be said to lift us from the earth towards our home in heaven; for we are creatures not of earth but of heaven, where the soul was first born, and our divine part attaches us by the head to heaven, like a plant by its roots, and keeps our body upright."

tradition of the later Middle Ages: comets were generated by special celestial occurrences, such as eclipses or conjunctions; they were dominated by certain planets; and, as we shall see in our next chapter, they were followed by all the events traditionally described as their concomitants. Still, like Gemma's, Heurnius' account deviated from the conventional pattern on the important issue of comets' location and their material constitution. Both authors held that comets, possessing an intermediate nature between the celestial and the elementary, were formed out of a special kind of matter partaking of both qualities. However, Heurnius went one step further than his Flemish colleague: whereas Gemma had placed this "middle matter" immediately *below* the lunar orb, in the area formerly occupied by the Aristotelian sphere of fire, Heurnius located it immediately *above* the moon's sphere, in the lowest part of the fifth element. His reasons for doing so were not in any way instigated by mathematical findings or measurements; rather, they seem to have arisen from his Platonically-inspired belief in cosmic sympathies operating by means of intermediary entities connecting the celestial and terrestrial extremes.

d. *Nicolaus Bazelius: "Many Fatty Exhalations Mixed with Slimy Vapours"*

The prognostication for the year 1578 'calculated' by the Bergues Saint Winoc town physician Nicolaus Bazelius[160], which included his discussion of the comet of 1577, was so full of doom and gloom that the modern reader has a hard time trying to take him seriously. True, his home country was suffering the consequences of a horrible conflict that divided families and alienated friends from one another; true, the spirit of the age was pervaded with chiliastic expectations; and, true, the consequences of old associated with comets were invariably dire and dreadful. But the extent to which Bazelius imbued his pamphlet with hunger, sword and pestilence, with robbers and highwaymen, traitors and hypocrites, arson and mutiny, tempests and death, was exceptional even if measured by contemporary standards, and betrays the true pessimist. No wonder, then, that, when, on top of all the ominous constellations and dismal eclipses of the earlier

[160] Niclaes Bazelius, *Een nieuwe prognosticatie vanden wonderlijcken ende ellendighen Jare ons Heeren 1578. Midsgaders de beschryvinghe vande Comete des voorleden Jaers. Ghecalculeert by M. Niclaes Bazelius, Medecijn ende Chirurgijn ordinaris der stede ende Casselrye van Sinte VVinox-Bergen in VVest-Vlaenderen* (Antwerp 1578).

months of 1577, in November of that year a comet appeared, Bazelius rose to the occasion with an almost sinister enthusiasm.

The tone of the whole pamphlet was set in its very first sentence, where our author announced that "In order to begin to make known the great miseries, anxieties, horrors, and calamities threatening all people, whatever their states, conditions, or persuasions, throughout this troubled and seditious year, we must first of all explain the nature, condition, and characteristics of the horrible and very long-lasting occultation or eclipse of the moon", that had occurred in the night between 26 and 27 September 1577.[161] This eclipse, Bazelius stated, had lasted exceptionally long, and would consequently bring in its wake very maleficent, extensive, and prolonged effects. Moreover, to make matters worse, its operations would be reinforced by especially wicked constellations that had taken place both before and after the eclipse, so that for the present year we should expect nothing but all kinds of sadness and misery at all levels of society. Our author then went on to a detailed discussion of the occultation's astrological characteristics, which, not surprisingly, turned out to have been extremely bad: bloodthirsty Mars had been Lord of the eclipse, and Mercury, burned by the rays of the sun and poisoned by the venom of the Dragon's tail, had assisted him, while Saturn, the "age-old deadly enemy of the human race", had with his inimical and cruel complexion obfuscated the rays of the, normally benign, Sun and Moon. All these ominous circumstances provided more than enough material for Bazelius to fill several pages expounding their horrible consequences, which we shall discuss in our later chapter.

Yet, Bazelius well realized that, however portentous had been the constellations of the previous year, and however admonishing had been the several eclipses that had occurred, they had been given due attention only by learned astrologers. The common man, not noticing them because of their regular occurrence, had not heeded the warnings contained in the constitution of the heavens, our author maintained, echoing Seneca's idea, that things that occur regularly

---

[161] Ibidem: "Omme te beghinnen te vercondigen de groote ellenden, ancxten, benautheden, ende rampsalicheden die alle menschen van wat state, conditie, ende ghesintheyt hy sy alle desen beroerlicken ende seditieusen Jare nakende zijn: so sullen wy al vooren verclaeren de natuere, conditie, ende ghesteltenisse vanden verschrickelijcken ende seere langhe gheduerende verduysternisse, ofte Eclips vander Mane, de welcke hier op onsen Meridiaen gheschiede int voorleden Jaer tusschen den 26. ende den 27. Septembris."

fail to attract the attention of the populace, however wonderful they may be.[162] Therefore, God, in his boundless mercy and ineffable love for mankind, had sent a more conspicuous messenger of his inescapable punishment: the present "hideous and gruesome comet or tailed star", that first appeared on 14 November 1577 in the highest regions of the air under the firmament.[163]

Bazelius himself first saw the comet on 14 November, shortly after sunset; yet, trustworthy persons had informed him that in the countryside it had been spotted as early as the eleventh of that month. Our author therefore surmised that the phenomenon had probably first shone forth on 10 November, the night of the new moon.[164] According to his own observations, at the time of its first appearance the comet had stood in the sixth degree of Capricorn in the southerly border of the Milky Way, not far from the tropic of Capricorn, more than 3 1/2° to the east of Saturn, and had been moving in the direction of Fidicula, one of the fixed stars. Its declination from the ecliptic had amounted to about three degrees, and from the equinoctial circle a bit less than twenty minutes.[165] Our author, however, seems to have realized the inaccuracy of his own observations, adding that he did not intend them to derogate from the more diligent observations of his more learned colleagues.[166] Perhaps this self-criticism ex-

---

[162] See supra, chapter III, p. 53.

[163] Bazelius, *Prognosticatie*: "Bouen alle voorgaende waerschouwingen die God den Heere almachtich door de veruaerlijcke Eclipsen van Sonne ende Mane, ende ander constellatien ofte versamingen vande Planeten onder malcanderen, ende met de vaststaende sterren int firmament, de welcke aldermeest vande geleerde ende principalijcke Astrologyns, waer genomen syn, ende van de gemeente veracht, om dat de sulcke dicwils gebeurende, niet en ouerdincken wat datter naer volcht. Soo heeft Godt door syn ongrondeloose bermhertichz ende onsprekelijcke liefde die hij tot dat versteent menschelijc geslachte dragende is, eenen anderen bode ofte voorloper willen wtseynden int firmament, als eenen Herault van zynder straffe ende ontvliedelijcke gramschap, het welcke is dese afgryselijcke ende verschrickelijcke Comete oft sterren metten steerte, die haer int hoochste vander locht onder het firmament vertoocht heeft sedert den xiiii. Nouembris...".

[164] Ibidem: "Maer ons dinct dat sy eerst haer mach geopenbaert hebben ofte beginnen schynen metter nieuwe Mane voorghaende te weten den x. Novembris."

[165] These data were hardly accurate; Tycho Brahe corrected them in his *De mundi aetherei recentioribus phaenomenis*, adding that Bazelius was either not at all well versed in the practice of astronomical observation, or else he had wanted the proper equipment to conduct it: "Verum apparet ipsum vel in Astronomicis Observationibus minus exercitatum, vel Instrumenta et media, quibus has perficeret, ad manus non habuisse; ideoque candide nec citra rationem, fatetur, se salva aliorum diligentiori Observatione, haec sua protulisse." (Brahe, *De mundi aetherei*, 362-363).

[166] Bazelius, *Prognosticatie:* "Salua tamen Doctiorum diligentiore obseruatione."

plains why Bazelius determined the comet's location only once, on the night of its first appearance.

On that same night, the comet's tail had been very awesome, stretched out to a great length and width; its colour had been pale, sad, and terrifying, and from time to time it had emitted fiery sparks, like glowing iron being forged by the blacksmith on his anvil. In addition to its regular tail, from the comet's head had also emanated many shorter rays in a southerly direction, in the semblance of a beard, visible especially well in clear weather; so that the phenomenon might deservedly be called Πωγωνίας or "bearded star".[167] This appellation had been coined by Aristotle, who as we have seen distinguished only two kinds of comet: the long-haired κομήτης and the bearded πωγωνίας.[168] However, Bazelius must have either misunderstood Aristotle's distinction, or else have copied it from an intermediary source which included a corrupt rendition of his text: in the *Meteorology,* I.vii.344a, the Stagyrite maintained that a πωγωνίας extended lengthwise only, while a κομήτης emitted rays in all directions; thus, Bazelius' observation, that his comet emitted many short rays around its southern edge, in addition to its regular tail, is more consistent with Aristotle's κομήτης than with his πωγωνίας. This makes it likely that Bazelius did not use the *Meteorology* itself, but one of the medieval astrological tracts quoting Aristotle inaccurately.

Although Bazelius recorded the comet's exact position only on the first night of its visibility, he did offer his readers a rough overview of its further motions through the constellations. Every day, the comet was seen setting in the west at a later time, so that it became clear that the phenomenon displayed an—irregular—eastward movement of its own, which by the beginning of December amounted to two degrees a day measured against the length of the zodiac. As to its latitude, it rose steadily northward every day, in the direction of the zenith; its tail was constantly directed at the East. Thus slowly but

[167] Ibidem: "Nu als dese Comete haer aldereerst vertoogt heeft, heeft sy gehadt haeren steert seer vervaerlijcken lanc ende breet wtgestreckt. . . hebbende een bleecverwich droef ende seer vreeselijc gesichte, by tyden vierige sparcken wtspuwende, gelijck alst gebeurt als den smidt dat gloeyende yser op den aembeelt smedende ende formerende is, wtbreedende oock vele corte raeyen als eenen breeden baert naer het Zuyden, principalijc als het weder seer claer was. . . soo dat daeromme dese Comete met rechte (volgende de opinie van Aristoteles ende ander geleerde Astrologijns) genaemt mach wesen Πωγωνίας, Barbatus, ofte een gebaerde sterre."

[168] See supra, chapter III, p. 45.

surely moving east and north, the comet passed through the constellations of the Eagle, the Dolphin, the Swan, and Pegasus, where it seemed to stand still for eight days near the tropic of Cancer. Finally, not far from the bright star Scheat in the right shoulder of Pegasus, it vanished in about the twentieth degree of Pisces.

However, the most portentous characteristic of the comet's course had been, that it had from the very first directed its motion at the constellation of Cassiopeia, and at the self-same spot where the nova of 1572 had shone. Bazelius therefore believed the present comet to have been closely affiliated with that "most illustrious and alien star ever seen", especially since the hairy star had moved so slowly, passing through less than three zodiacal signs in a period of two entire months, unlike most comets, which (according to Bazelius) always moved very fast.[169] As in the case of Gemma, then, we can see how Bazelius, too, ventured to enhance the comet's miraculous character, and thus its trustworthiness in matters prophetic, by emphasizing its analogy with celestial bodies of high dignity.

Yet, unlike either Gemma or Heurnius, the western Flemish town physician did not feel inclined to in any way 'upgrade' the hairy star's physical characteristics by according it an intermediary position between heaven and earth; his account remained overwhelmingly embedded in the Aristotelian-Ptolemaic cometary tradition. Although he was not quite consistent in his formulation of the comet's distance from the earth, at one point placing it "*in* the firmament", while later on in the same sentence assigning it a location "in the upper air *under* the firmament"[170], on the whole his pamphlet gives us no reason to doubt that his understanding of the comet's generation and constitution was thoroughly Aristotelian. Despite his acknowledgment, at the very beginning of his treatment of the comet's mate-

[169] Bazelius, *Prognosticatie:* "...met die voorseyde sterre dunct ons dese Comete groote gemeynscap te hebben door dat sy soo tragelijc in haren eygen loop geroert is, als binnen twee geheele Maenden geen drie hemelsche teeckenen vande twelfue doorwandelt te hebben ende ooc zeer tragelijc naer het Noorde opgecomen is, want dergelijcke ooc niet vele van te vooren gesien ofte bemerct geweest hebben, want sy altyts seer snellijc (zoo wel inde lengde als inde breede) voorts gegaen, ende geroert hebben."

[170] Ibidem: "...eenen anderen bode ofte voorloper willen wtseynden *int* firmament, als eenen Herault van zynder straffe ende ontvliedelijcke gramschap, het welcke is dese afgryselijcke ende verschrickelijcke Comete oft sterre metten steerte, die haer int hoochste vander locht *onder* het firmament vertoocht heeft sedert den xiiii. Nouembris...". [My italics].

rial cause, of the fact that on the subject of cometary causes great dissension reigned among both ancient and modern astrologers, in which the final word had not yet been spoken, in his own discussion of the issue Bazelius blissfully ignored every opinion even slightly deviant from the majority view.[171]

As we have seen, according to the amalgam of Aristotelian and Arabo-Ptolemaic cometology that had gained wide currency from the high Middle Ages onwards, two circumstances were required for the generation of a comet: a special celestial constellation (such as a planetary conjunction or an eclipse), and an abundance of hot and oily terrestrial exhalations assembling in the atmosphere. Both these conditions had been satisfactorily fulfilled in the case of the comet of 1577, as Bazelius diligently demonstrated. First of all, the comet's efficient cause had been provided by the conjunction of all the planets, with the exception of Venus and Saturn, that had occurred in September in the sign of Virgo, not far from the vernal equinoctial point. The workings of this conjunction had been assisted by the great lunar eclipse already mentioned, which had started in Aries, a fiery sign, in the house of Mars, a fiery planet, and ended in Leo, a sign of an equally fiery nature. "From these data may be gathered how many causes collaborated in order to generate this comet", Bazelius confidently stated.[172]

As to the phenomenon's material cause, this had undoubtedly been supplied by the eight days of hot and dry, summery weather that had followed the eclipse. Such conditions of course indicated the gathering of suitable material for the birth of a comet: if the weather is hot and dry, this means that the atmosphere is full of "very many fatty exhalations mixed with slimy vapours", the ideal constituents for comet-generation.[173] These exhalations had been attracted by the

---

[171] Ibidem: "Vande cause materiale van deser Comete ende ooc van alle andere, is groote disputatie onder alle oude soo wel als ionge Astrologijns, gehouden: vande welcke het proces noch niet gesloten en is."

[172] Ibidem: "Hier wt machmen mercken hoe vele causen ouer een gecommen syn totter generatie van deser Comete dienstelijc."

[173] Ibidem: "Bouen dien dient genoteert, dat meer dan acht daghen lanck, naer dat den Eclips gepasseert was, het weder soo heeft schoone ende drooge, heet geweest als oft Somer geweest hadde, den windt wt den Zuydt-Oosten blasende. Alle het welcke ten sy dat onseker sy al dat vanden Philosophen soo wel als vanden Astrologins vande materiale cause vande Cometen geschreuen, totter generatie van deser Comete ende tot geduerscheyt ouervloedighe materie toe gheschict heeft, seer vele vette exhalatien met slymige vapeuren ondermingelt. . .".

celestial powers of stars and planets, Bazelius stated, in complete accordance with the Ptolemaic tradition. When they had arrived in the upper air, they had caught fire and formed a comet. To Bazelius, then, comets were fiery meteors, terrestrial exhalations burning in the atmosphere under the firmament.

This perfectly Aristotelian idea was reinforced by the stories of a number of people who had told Bazelius that, on 6 January, they had seen a great number of smouldering coals falling from the sky, some small, some large, like glowing fires coming down from the highest heaven; they had rapidly flown in a northeasterly direction and then descended. Our author conjectured that these might have been the remnants of the comet, but level-headedly added that he could not pass any definite judgment on the subject, since he had not observed the alleged "coals" himself. Yet, the story had been lent credence by Bazelius' own observations on the night of 6 January, when he had hardly been able to discern the comet at all, seeing only a small white stripe in the spot where formerly the hairy star had shone; on his second observation, an hour later, the comet had even vanished completely, however closely Bazelius scrutinized the sky.[174]

In a second short acknowledgment of the current debate on the nature and aetiology of comets, our author added that he had expounded the story of the fiery coals at such great length, "in order that scholars might the better ponder the material cause of comets."[175] Yet, he himself did not accord the slightest attention to any of the non-Aristotelian alternatives to the theory of comets as fiery, sublunary meteors. From his phrasing it becomes clear that apparently, he did not count himself among the "scholars"; see also his modest remark, quoted above, about the "more diligent observations of more learned men". Rather, Bazelius regarded himself as a simple

---

[174] Ibidem: "...waer het waerachtich dat diuersche persoonen my verhaelt hebben dat sy gesien hebben op den vi. dach Januarij ... eenen hoop gloeyende colen zom cleyne zom groot ende lanc al oft gloeyende vierbranden waren, wt den hoochsten vanden hemel commende, ende subytelijc naer het Noordooste vliegende ende zoo nederdalende. Het welcke zouden meugen wesen de Reliquien vander Comete, maer midts dat wy dat selue niet gesien hebben, en connen daer af niet vastelijc iudiceren, anders dan dat waerachtich is, dat wy... op den voornoemden vi. dach Januarij lestleden bemerct hebben seer luttel bescheets vande voornoemde Comete als maer een cleyn wit strepken daer die gheseten hadde, ende binnen een ure daer naer wederom aensiende het firmament en hebben het zelue (hoe nouwe ende lange wy daer op merckende waren) niet connen ghesien."

[175] Ibidem: "Dits hier zoo verhaelt omme dat de gheleerde te beter souden moghen duncken op de materiale cause vanden Cometen."

town physician, whose task it was, not to hold any learned discourses about the aetiology of comets, but rather to warn his fellow countrymen of impending calamities announced by the hairy stars, and counsel them on how to avoid or mitigate these disasters. He therefore devoted the larger part of his pamphlet to a 'calculation' of the consequences of the present comet, on the basis of its colour, size, brightness, motion, tail and other such determinants originating from the Arabo-Ptolemaic astrological tradition.

These matters will concern us in our next chapter; suffice it to repeat here that Bazelius' world view, as apparent from his 1578 pamphlet, was still entirely Aristotelian. Although he knew about the learned debate on the nature of comets, he did not himself renounce any of the Peripatetic tenets, nor did he even bother to mention a single one of the alternative views. He did not call into question the dogmas of the incorruptibility of the heavens, the central position of the earth, or the existence of solid celestial spheres, and even failed to raise these issues at all; unconditionally accepting the motion of the Peripatetic *primum mobile*[176], he seems to have entertained a completely traditional world view.

e. *Van Mauden and Portantius: Aflame and Hatched by Saturn*

Even less interested in the comet's physical constitution or its exact location than Bazelius was his Antwerp colleague David van Mauden. Unlike Bazelius, in whose pamphlet the comet had to share the readers' attention with the eclipses and the other ominous constellations foreshadowing miseries for the year 1578, van Mauden devoted a broadside entirely and solely to "the history of the new Comet of the year 1577, very diligently observed from the moment it first shone forth until its ending".[177] Apart from the introductory "Admonition to the Reader", the pamphlet consisted of eleven short sections, devoted respectively to "What I have, after the teaching of Ptolemy, noticed in this comet from its beginning to its end"; "Of the course and progress of this comet"; "Of the condition of the Heaven

[176] Ibidem: "... het roeren vanden Firmamente door het welcke die binnen xxiiii. uren rondsomme de geheele weerelt lopende is...".

[177] David van Mauden, *De geschiedenisse van de nieuwe Comete des Jaers 1577 seer neerstelick g'obserweert van datse begonst te schynen tot den eynde toe. Door M. David van Mauden Doctor inder Medicijnen, Lecteur der Chirurgijnen en gesworen Stadt Medecijn deser Stadt van Antwerpen* (Antwerp 1577).

over our horizon"; "The significance and power of this comet"; "The countries these plagues may affect", "These are the cities that are threatened"; "The persons who will be most afflicted"; "Of those born during the appearance of this comet"; "How long the operation of this comet will last"; "How these workings may end"; and "Of the operation of this comet in matters of religion". Once again, then, we are dealing with a work primarily devoted to astrological prediction and ethical exhortation, rather than astronomical investigation.

The few scattered clues about the author's views on comets' generation, constitution, and location, that the pamphlet does provide us with, are to be found in the first three sections. The initial one, which van Mauden self-professedly based on the teaching of Ptolemy, dealt with the two occasions on which our author himself observed the phenomenon: on 14 and 19 November. Although the comet had been spotted on 11 November, van Mauden himself first observed it on the fourteenth, when it found itself in the sign of Capricorn, displaying a very long and wide hairy tail that ended in a point. The tip of the tail had been red, but the closer it came to the comet's head, the brighter its colour; our author regarded this circumstance as evidence of the phenomenon's Martial and Mercurial nature (Mars being red and Mercury very bright). Undoubtedly, van Mauden had done his best to detect in the comet some characteristics that matched those of Mars and Mercury; after all, Ptolemy, on whom his account was based, had taught that all comets naturally produced the effects peculiar to those two planets.[178] Five days later, on 19 November, our author found the comet in the 27th degree of Capricorn, not far from the ecliptic. Its form and aspect had changed considerably: its complexion had become sad and murky, and its tail now pointed in an easterly direction, displaying a slight curvature and a reddish-dark colour. From this information, van Mauden concluded that, presently, the phenomenon was of a Saturnine and slightly Mercurial nature.

In his second section van Mauden included a brief and rather sketchy description of the comet's course through the constellations. From the sign of Capricorn, where it had originated, it moved north and passed under the Eagle, after which, moving in a northeasterly direction, it arrived in Aquarius, passed the equinoctial circle on 28 November under the constellation of the Dolphin, and moved

[178] See supra, chapter III, p. 57.

through Equiculus towards Pegasus, in whose chest it found itself on 27 December, close to the tropic of Cancer, so small that it could hardly be spotted. Thus, having passed through the ninth and tenth houses of heaven, it finally arrived in the eleventh, where, van Mauden confidently stated, it was to perish.[179] The comet's velocity had been very small: it described only one degree and twelve minutes a day, following the motion of Venus—or so van Mauden believed. Obviously, this statement was not at all accurate: we need only call to mind the much more precise observations of Gemma, who had detected a notable diminution of speed during the several months the comet shone.

Like Bazelius, van Mauden, despite first observing it as late as 14 November, thought the comet to have taken its origin during the night of the previous new moon, 9 November, when, extrapolating from the known daily movement of the comet during its visibility, it must have stood in the fifteenth degree of Capricorn.[180] Apparently, Bazelius and van Mauden (the latter self-professedly following the renowned Arabic astrologer Haly on the subject) both saw a connection between the comet's engendering and the phases of the moon, believing that the moment of new moon was most propitious for the birth of a hairy star. Van Mauden went on to describe in detail the aspects the comet had, during its course, displayed with respect to several of the planets: thus, it had started its life in conjunction with Saturn, and had throughout its duration remained in quadrature (an unlucky aspect) with Venus, which planet constantly "pursued" the hairy star "with a half-hostile face". The comet had at first been preceded by Mars, but around the middle of December, it had overtaken the fiery planet.[181]

Van Mauden's reasons for describing in such detail the move-

---

[179] Van Mauden, *Geschiedenisse:* "...ende vont haer den xxvii Decembris inden xii graet Piscium, loopende aende borst van Pegasus, ontrent Tropicum Cancri, ende daer wasse soo clyn ende soo nae by vergaen, datse niet wel gesien ende conde geworden: immers doorlopen hebbende het ix en x huys, isse comen tot int elfde daer sy in vergaen sal."

[180] Ibidem: "Volgende Haly (om dat wy haer recht beghinsel niet en hebben gesien) soo sal ick vande nieu Mane beginnen, de welcke nieu wert den ix dach Novembris des nachts ten xii uren ende xvi minuten, ... De mane ende de sonne in den xxvii graet en xv minuten van Scorpio int derde huys: de Comete int vijfde huys in Capricorno, mogelijc inden xv graet tellende soo veel achterwaerts tot aende nieu Mane, gelijc wise dagelicx hebben vonden voorwaerts gaende...".

[181] Ibidem: "Dese Comete heeft van het beginsel tot den eynde tijt gestaen inde Quadratuere van Venus, die doen was inde xvii graet van Scorpio, coniunct met

ments of the heavenly bodies with regard to the comet remained rather unclear. He undoubtedly believed them to have been of significance, either in connection with its meaning or in connection with its generation; yet, he did not draw any conclusions from his planetary observations. The only constellation he more or less explicitly connected with the comet's birth was the quadrature of Saturn, Mars, and Venus that had occurred "already before the revelation of this comet"; this quadratic aspect had been observed on 4 November, "when Saturn was near the Milky Way, where this comet has been hatched by him."[182] This formulation and the context in which it occurs make it likely that, in van Mauden's view, comets were formed by certain planets, at times when the aspects of the planets towards one another were favourable for such a generation; also, apparently the region of the Milky Way was best suited for the birth of comets. Our author did not express any opinion as to the phenomenon's distance from the earth, or on whether he believed it to be a sublunary meteor; yet, the tone of his entire pamphlet was so completely embedded in the Arabo-Ptolemaic tradition that it is hardly to be doubted that, to him, a comet was a burning terrestrial exhalation called into being by special planetary motions.

Even less outspoken was Johannes Portantius, the author of our last, very brief, pamphlet on the comet of 1577.[183] He included a very rough description of the phenomenon's motion, reporting that it had first appeared on 11 November in Capricorn, with Gemini as its horoscope, that it had originated near the ecliptic in the Milky Way, and that it had moved in the direction of the equator, past Aquila and Delphinus, in a northerly direction. Like Bazelius and van Mauden, Portantius believed the comet to have started burning in the night of the new moon. His use of the word "burning" moreover,

---

Mars Platice, welcke Venus dese comete altijt geselschap heeft gehouden met gelijcke graden in Quadrature vanden beghin tot den eynde toe. Mars die heeft altijt wat voor de Comete in graden geloopen ex Quadrato, tot den eersten December toe, doen Venus hem achterhaelde ende met hem coniunct was. Voorts syn sy beyde Venus ende Comete voorgeloopen ghelijckelijk ad gradum, tot malcanderen altijt in een vierde aspect, ende hebben Mars achtergelaten ende ontloopen, principalijc den xvi December. Maer Venus en heeft niet opgehouden dees Comete altijt te vervolgen met een gesichte van half vyantschap tot den lesten toe."

[182] Ibidem: ". . . al voor de openbaringe deser Comete, wasser een quadraet van Saturnus en Mars en van Venus ooc den iiii Novembris en Saturnus was by viam lacteam, daer dese Comete van hem gebroeyt is."

[183] Johannes Portantius, *Beschrijvinghe der nieuwer Cometen, Met aenwijsinge ende vermaninge wat zy bedieden ende voortbrenghen sal* (Antwerp, s.d.).

provides us with our only clue as to Portantius' cometology: "It started burning with the new moon, in the figure of a beautiful star of the nature of Mercury, with a burning and long tail".[184] Clearly, our author believed both the comet itself and its tail to be aflame, a conviction precisely and exclusively compatible with the Aristotelian view of comets as sublunary, fiery meteors. Furthermore, he also thought the nova of 1572 to have been a "great comet seen burning in the North".[185]

Such, then, were the Dutch writings on the comet of 1577. They clearly separate into two different categories, with Gemma and Heurnius on one side of the divide and Bazelius, van Mauden and Portantius on the other. The latter three authors, not connected to any university and writing for a mostly popular audience, made no attempts at incorporating into their pamphlets any cometary theory even slightly deviating from the widespread Aristotelian-Ptolemaic amalgam. As far as can be ascertained, they all thought of comets as burning terrestrial vapours attracted by, and assimilated to, the power of the stars. In spite of the fact that they most certainly knew about the existence of alternative views (be they the ancient cometological ideas of the Presocratics and the Stoics, or the more recent adjustments to comet lore proposed by such authors as Pena, Cardano, or other champions of the optical theory), and Bazelius even mentioned the current scholarly debate on the nature of comets, they based their explanations of the present phenomenon exclusively on medieval astrological sources such as Haly and Ptolemy. Their outlooks were firmly geocentric, and there is nothing in their writings to suggest the slightest deviation from the Aristotelian world view.

By contrast, both Gemma and Heurnius did try to come to grips with the various anomalies the appearance of the comet faced them with. Gemma, the only one of the authors of 1577 to address an exclusively scholarly audience, conceived of an amendment to the Aristotelian arrangement of the elements in their respective spheres, proposing to reduce the width of the sphere of air, and to replace the traditional sphere of fire with a region, just below the lunar orb, filled

[184] Ibidem: "...ende is met der nieuwer Mane brandende gheworden, in vorme van een schone sterre vander natueren Mercurij, met eenen brandenden ende langhen steert...".

[185] Ibidem: "...int iaer 1572. is een groote Comete int Noorden gesien branden...".

with an intermediary substance partaking of both celestial and terrestrial properties. Most comets, having been mathematically proven sublunary, Gemma stated, could be accounted for by according them a place in this intermediary region; the superlunary comet of 1577 was an extremely exceptional phenomenon, probably formed out of the divine quintessence itself. Yet, whereas three years earlier he had still counted superlunary comets among the divine and supernatural prodigies, in 1577 he admitted the possibility of a certain measure of change in the heavens. More specifically, he thought it not impossible that, prompted by certain constellations, the planets might conceive "offspring" that remained hidden under their rays until they shone forth at times of great significance. In support of this view he quoted the Pythagoreans, Hippocrates of Chios, and the Church Fathers.

Heurnius, too, framed a theory that allowed for superlunary comets. He conceived of a region similar to Gemma's 'ethereal orb', filled with a like intermediary substance; the main difference being that his 'astral element' was located just above rather than just below the moon. But the principle remained the same: by slightly amending the Aristotelian arrangement of the elements, superlunary comets (endowed with an intermediary nature because of their function of keeping the macro- and the microcosmos in touch with one another), could be accounted for without having to abandon any major tenet of Aristotelian physics. Especially the Stoic and Presocratic idea that comets might not even be newly formed celestial bodies, but rather entities coeval with the planets yet mostly invisible to earthly observers, was proving its fruitfulness as a compromise between the radical options of completely abandoning the Aristotelian dichotomy between heavenly and earthly regions, with all the disastrous consequences this would entail for Peripatetic physics, or, conversely, of denying the existence of superlunary comets despite convincing demonstrations to the contrary. Thus, Gemma and Heurnius were able to run with the hare of Aristotle and his physical dichotomy, and at the same time hunt with the hounds of the Stoics and Church Fathers with their conception of celestial alterability, and of the Presocratics and Seneca with their theory of suddenly appearing, yet not necessarily newly generated, 'novel' celestial bodies. A return to the ancients was needed in order to reconcile the Aristotelian world view, as coined in the Middle Ages, with modern observations and measurements.

3. *Forty Years of Development: 1577 to 1618*

Little did Gemma and Heurnius suspect, when they wrote their tracts early in 1578, how greatly the topicality of their 'intermediate' solutions to the problem of superlunary change was to be enhanced during the next few decades. When they published their writings, Tycho Brahe was only just beginning to ponder the consequences of his observations of the comet of 1577; he was not to start making public his findings, including his geoheliocentric world system, until 1588 (with the first official edition of his book appearing only in 1603). Furthermore, the reception of the Copernican system, with all its potential threats to the reigning Aristotelian world view, had as yet been largely restricted to the cautious compromise of the Wittenberg Interpretation, allowing astronomers to avoid the necessity of reconsidering their entire understanding of physics. And, finally, the weakness of human sense perception posed an as yet insurmountable limit to the accuracy of celestial observations: no matter how advanced their instruments, astronomers were still dependent on their naked eyes to the same extent as had been their Greek colleagues, some twenty centuries earlier.

Yet, the forty years that divided the comet of 1577 from its 1618 congener saw substantial changes in the way the heavens were perceived. Although these changes did not come out of the blue, and did not bring in their wake anything like the radical scientific turn that has often been postulated for the early seventeenth century by nineteenth-century historians[186], they did serve to make the tracts devoted to the comet of 1618 breathe an atmosphere relatively distinct from that to be found in the writings we have just left behind. A small avalanche of new discoveries and new ideas, or, to put it more correctly, new uses made of old ideas, descended on the scholarly community between the two comet years of 1577 and 1618. Still, the landslide did not take scholars completely by surprise; as we have seen, the ground had been prepared, slowly but surely, by the optical theory of comets, the revival of Stoic cosmology, the appearance of the nova, and the increasing number of accurate parallax measurements. The comet of 1577 had, in the eyes of those who recognized its superlunary position, already called for an adaptation of traditional cometary theory—preferably one that need not, or hardly,

---

[186] See supra, chapter II.

upset the core tenets of Aristotelian physics; the comet of 1618, appearing as it did in the wake of startling new evidence, was to scream for one.

a. *Dutch Pamphlets on the Comets of 1596 and 1607: Heymensz and Mulerius*

The strong emphasis here laid on the comets of 1577 and 1618, should not make the reader oblivious to the fact that these were by no means the only tailed stars to have appeared during the time span between 1577 and 1618. Indeed, the final decades of the sixteenth century and the opening ones of the seventeenth saw an unusually high number of novel celestial appearances: the years 1577, 1580, 1582, 1585, 1590, 1596, 1607 and 1618 were all marked by the advent of comets visible to the naked eye, while 1604 was honoured with the ignition of a second supernova. The exceptional frequency with which previously unseen citizens of heaven visited the earth's neighbourhood during precisely the period when astronomy and cosmology found themselves confronted with many an unorthodox theory, may be wondered at as a remarkable coincidence, and might entice the historian into tentatively postulating a connection between the fire of curiosity that must have been not only kindled, but also kept alive by so many novelties, and the bustling observational activity displayed by astronomers during this period. Add to this the life spans, during precisely this era, of such remarkable figures as Tycho Brahe, Johannes Kepler, and Galileo Galilei, and the stage is set for some genuine fireworks.

Curiously enough, though, the comets of 1580, 1582, 1585, 1590, 1596, and even the 1607 appearance of Halley's comet, generally considered to be a highly impressive sight, were accorded but very slight attention by the inhabitants of the Low Countries (who, it seems, had other things on their minds during those years than to scan the sky for new heavenly bodies). Only two very concise cometary pamphlets were published in the Netherlands between 1577 and 1618, one dealing with the comet of 1596[187], and one in which the

[187] Pieter Heymensz, *Een Prognosticatie, ofte voorsegginghe op de Komeet-Sterre met de steert, welcke haer verthoont ende gheopenbaert heeft den xvi. Julij na de nieuwe stijl, in dit teghenwoordige Jaer M. D. XCVI. Ghepractiseert door een Liefhebber van de edele Conste van Astrologie, M. Pieter Heymensz. Doctoor inde Medecijnen tot Delff* (Delft, s.d.).

comet of 1607 had to share the readers' attention with a prognostication for the year 1608 and a farcical parody on the peace negotiations between Spain and the United Provinces which had commenced in 1607.[188] Neither one of these works displayed more than an extremely superficial interest in the comets' physical and astronomical characteristics; the authors provided their readerships with vivid accounts of their respective comets' significance, without bothering to underpin their assertions with any but the most basic observations of the hairy stars' movements or physiques.

Pieter Heymensz, who chose to discuss the comet of 1596, according to the title page of his pamphlet occupied the position of physician in the town of Delft in the south of the province of Holland. In the same sentence he also referred to himself as "an amateur of the noble art of astrology", which was the quality in which he believed himself competent to discuss the present comet. Quoting his near-contemporaries Antonius Mizaldus and Cyprianus Leovitius, who had both published extensive compilations of astrological lore (Mizaldus' book, the *Cometographia,* even dealt exclusively with comets), Heymensz ventured to attribute to the comet a host of unpleasant consequences. On Mizaldus' authority, a comet with a tail in the sign of Cancer was said to foretell war, strife, the death of a prince, inundations, pestilence, and scarcity, while following Leovitius, our author felt inclined to ascribe to the present phenomenon "a year full of mischief, deceit, falsehood, and mendacity, fornication and adultery". Moreover, he did not even need the authority of Leovitius to underpin this prediction: "I, myself, can also see this by the look of this star".[189]

Not all of Heymensz' vaticinations were so vague. In fact, he belonged to those authors who were prepared to go to quite a high level of concreteness and practical applicability in their political predictions. Such forecasts, however, provide the material for our following chapter. For our present astronomical concerns, the harvest yielded by Heymensz' pamphlet is very poor. He summarily de-

---

[188] Nicolaus Mulerius, *D.D. Nicolaum Mulerium, van Brugghe, Doctor inde Medicynen van weghen der stadt Groeningen ende omme-landen. Zijne Practica. Met de beschryvinghe vande Comeet. Als oock mede vande Pasquil ende ghenoechlijck discours, hoe dat wy door Gods genade aen den Nederlandtschen Vrede-handel zijn ghecomen* (Groningen 1608).

[189] Heymensz, *Prognosticatie:* "Ende also Siprianus Leovitius seydt, ende ick oock can sien wt de Sterre, So wil dit wesen een Jaer vol schelmerije, bedroch, valscheyt ende leughentael, Hoererije ende Overspel. . .".

scribed the position of the comet on the night of its first appearance, on 16 July, when it stood high in the north, behind Ursa Maior, in the sign of Cancer, with a pale and sad, Saturnine complexion. Mars had been lord of the ascendant and had stood in Leo, in conjunction with Mercury, while the comet had been in conjunction with the Sun, in sextile aspect with Jupiter, and in quadrature with the Moon. This was all the information Heymensz bothered to enlighten his readers with; apparently, he deemed one observation enough for basing his far-reaching astrological predictions on, so that he did not make an effort to establish the proper course the comet must have followed through the constellations on the subsequent nights. The only meagre insight into the way in which Heymensz perceived the cosmos was provided by his assertion that the comet "whirled about along with the daily revolution"[190], testifying to the fact that, not surprisingly, our author was committed to the twenty-four-hour movement of the Aristotelian *primum mobile,* and consequently to a geostatic universe.

Only slightly more communicative was Nicolaus Mulerius, the author of a *practica* (almanac with prognostication) for 1608 that included a brief discussion of the comet of the previous year. As we shall be able to recount at greater length when we come to discuss this author's work on the comet of 1618, Mulerius in 1607 occupied the position of *medicus provincialis* in the northern province of Groningen, to which post he had been appointed in 1603. From about that time onwards, he began to issue annual almanacs that probably provided him with a welcome addition to his income. According to the *Effigies & vitae Professorum Academiae Groningae & Omlandiae,* Mulerius wrote such ephemerides for the years 1609 to 1626; however, recently, earlier almanacs of his have been discovered, testifying to the fact that he probably started his annual compilation of yearbooks as early as 1604. He undoubtedly did quite well out of this sideline: in a 1619 letter to the town secretary of Leeuwarden Mulerius estimated that another ten thousand copies of his almanac were to be printed, in addition to the ones that had already been issued.[191] If we take into consideration Craig Harline's statement,

---

[190] Ibidem: ". . .omdraeyende met de daghelijcsche beroerte. . .".

[191] G. C. Huisman, "Boeken en brieven van Nicolaus Mulerius" in *Bibliotheek, wetenschap en cultuur. Opstellen aangeboden aan mr W. R. H. Koops bij zijn afscheid als bibliothecaris der Rijksuniversiteit te Groningen* (Groningen 1990) 283-296: 284-5.

that "about 1,000 to 1,250 copies per run was common"[192] for pamphlets, we can conclude that Mulerius' little yearbooks enjoyed a very wide circulation.

Mulerius, then, despite his considerable scholarly capacities (which, again, we shall be able to praise at somewhat greater length when we come to discuss the comet of 1618) was by no means averse to producing the occasional (or even more than occasional) vernacular pamphlet. One such unpretentious yet lucrative little work was his *Practica* for the year 1608. It commenced rather abruptly with the question, "When have ever two eclipses been seen on one and the same day?" Answer: on Friday, 3 April, of the year 33 A.D., the first 'Good Friday'; Mulerius had calculated that, in addition to the solar eclipse that had occurred supernaturally after Jesus had died on the cross, the astronomical records showed that in the evening of the same day a bloody lunar eclipse had occurred in the East. This observation served as an introduction to Mulerius' discussion of the astronomical characteristics of the year 1608, in which he briefly explained the positions of the planets on 10 March, the day the Sun was to enter Aries, heralding the start of the new astronomical year. Inevitably, he followed these rather factual observations with a cryptical astrological interpretation.

Mulerius then got round to a discussion of unusual celestial events: the year 1608 was to be marked by no less than three eclipses, two solar and one lunar—although none of them would be visible in the United Provinces. Nonetheless, our author expressed the wish that God would avert from the Dutch the harmful consequences of the last of the eclipses, adding the moralizing comment that even worse would be the effects of "a pernicious eclipse which sometimes obscures the Sun of our reason, which can be rendered so dark that we mistake the straight for the crooked and the crooked for the straight."[193] Whenever God wants to punish us for such a temporary loss of judgment, Mulerius went on, he sends warnings in the guise of

[192] Craig E. Harline, *Pamphlets, Printing and Political Culture in the Early Dutch Republic* (Dordrecht 1987) 21. See also H.J. Nalis & J.L. Salman, "'Wie sal ick dan wes goedes konnen prognosticeren?' Deventer almanakken en prognosticaties in roerige tijden (1555-1610)", *Deventer Jaarboek* (1994) 6-39: 10, who estimate that the yearly circulation of almanacs and prognostications at Deventer during the second half of the sixteenth century amounted to at least 6,000.

[193] Mulerius, *Practica:* "God behoede ons voor de schade die dese leste Eclipsis mede brenghen sal. Ende noch meer voor een schadelijcker Eclipsis, welcke somtijdts comt voor de Sonne onses verstants, welcke so duyster can worden datwy trecht voor crom aensien ende 'tcrom voor recht."

clearly visible heavenly signs, to incite mankind to do penitence; such was the comet that appeared last year in September and October. It was the largest comet seen since the great comet of 1577, displaying a menacing pale face and long outstretched rays.

Mulerius then went on to a very brief discussion of the comet's course; like Gemma, Heurnius, and the other authors of 1577, he was interested in its astronomical characteristics only insofar as these were capable of contributing towards the discovery of the phenomenon's significance. He therefore laid much emphasis on the fact that the present comet had moved from the north-west to the south-east, while, conversely, the comet of 1577 had, according to Tycho's description, taken the opposite route, beginning in the south-west and ending in the north-east. Furthermore, the comet of 1607 was kindled in the spot where the 1577 one had perished, and had been extinguished under the same meridian where, thirty years earlier, Tycho's comet had first shone forth, and where, moreover, the nova of 1604 had stood. "Haec non accidunt fortuito", commented Mulerius, and "an extraordinary secret lies hidden in these [circumstances]: we have seen two new stars and one comet in seven years' time . . . and will yet see and hear great wonders."[194] Unfortunately, our author did not relate to his readers what kind of wonders these would be; the rest of his pamphlet was filled with a rather obscure astrological allegory on the current peace negotiations, and with a waggish parody on the same subject.

As far as the astronomical world view of Dutch authors was concerned, then, the period between 1577 and 1618 yielded a very meagre harvest. Heymensz and Mulerius, writing in the vernacular for a popular audience agitated by a continuing state of war, chose to stress the moral and religious overtones, rather than the physical characteristics, of the comets of 1596 and 1607. Not once did they make mention of the ongoing debate on the nature of comets, that had gained momentum since the appearance of the comet of 1577 and was fast becoming one of the foremost topics of interest to the international scholarly community.

The year 1609 was to be marked by two events that pushed cosmological questions similar to those raised by cometary appearances

---

[194] Ibidem: "Haec non accidunt fortuito. Daer is een sonderlijck secreet onder verborgen. In een seven Jaer tijdts hebben wy twee nieuwe sterren gesien ende een Comeet. . . . men sal noch groote wonderen sien en hooren."

even more to the forefront: the publication of Johannes Kepler's *Astronomia nova*, and the first telescopic observations of Galileo Galilei. Little did Mulerius realize, when, at the beginning of the year 1608, he announced great wonders in his *Practica*, how wondrously his prediction was shortly to come true.

b. *Magnetic Forces*

At the time of the appearance of the comet of 1577, the disciplines of astronomy and physics were still largely isolated from one another. Despite some attempts at reintegrating Aristotelian physical theory, with its homocentric, solid spheres, and Ptolemaic astronomy, with its fictional eccentrics and epicycles, the general attitude was one of detachment: the phenomena-saving devices of astronomy were taught by mathematicians as part of the *quadrivium*, while the physical aspects of the cosmos were discussed by scholars of natural philosophy. In accordance with this division of labour, Copernicus' *De revolutionibus* was appropriated by the astronomers and regarded as an eminently convenient calculational tool, notwithstanding its author's realist intentions.[195]

Most of the efforts of the adherents of the so-called Wittenberg Interpretation were directed at 'inverting' the heliocentric system, trying to refer the planetary distances yielded by the Copernican hypothesis to a stable, central earth. As we have seen, Tycho Brahe initially joined in this activity: in his 1574 lectures at Copenhagen university, he demonstrated the possibility of reconciling the figures of Copernicus to a geocentric world system, so that the motions of Mercury, Venus, Mars, Jupiter, and Saturn were referred to the Sun, while the Sun circled the Earth. However, at this early stage, the Dane failed to show any interest in the question of the reality value of such an arrangement.[196]

Tycho was not the first to propose a geoheliocentric order for the solar system; rather, geoheliocentrism emerged as a logical consequence of the Wittenberg Interpretation, with its efforts at employing Copernicus' calculations to their best advantage while retaining the

---

[195] See supra, chapter IV.

[196] Christine Schofield, "The Tychonic and semi-Tychonic world systems" in René Taton & Curtis Wilson eds., *Planetary Astronomy from the Renaissance to the rise of astrophysics. Part A: Tycho Brahe to Newton* (Cambridge 1989) 33-44: 33.

immobile, central earth required by Peripatetic physics. Thus, the idea of having the Sun revolve around the earth with all or several of the other planets as its satellites occurred more or less simultaneously to a number of astronomers during the 1570s.[197] As Christine Schofield has shown, apart from Tycho himself, Erasmus Reinhold, Albert Lonicerus, Christopher Rothmann, and François Viète all proposed various versions of such models; yet, none of them had abandoned the sceptical stance of Wittenbergism. "Each suggested it as a possible way of utilizing the mathematics of the Copernican theory, without tackling the problem of its physical and philosophical implications."[198]

Thus, the conception that a system in which the sun went round the earth with some or all of the other planets as its satellites was capable of yielding predictive accuracy, was widely diffused among astronomers of the 1570s; the geoheliocentric arrangement was, as it were, 'in the air'. Tycho, however, was the first (or in any case by far the most influential) to have proposed such a system as representing the actual state of affairs in the heavens.[199] Convinced that the heavy, sluggish earth was unsuited for perpetual motion, and relying on the authority of the Scriptures, Tycho refused to grant the earth any movement whatsoever; yet, for aesthetic considerations, he shared Copernicus' aversion of the Ptolemaic equant point. Moreover, observations made it quite clear that the planets' motions were

---

[197] See Richard A. Jarrell, "The Contemporaries of Tycho Brahe" in Taton & Wilson eds., *Planetary Astronomy*, A, 22-32; see also Owen Gingerich and Robert S. Westman, *The Wittich Connection: Conflict and Priority in Late Sixteenth-Century Cosmology* (*Transactions of the American Philosophical Society*, lxxviii, Part 7; Philadelphia 1988); and Christine Jones Schofield, *Tychonic and semi-Tychonic World Systems* (New York 1981) esp. pp. 20-49.

[198] Jones Schofield, *Tychonic and semi-Tychonic World Systems*, 20-49: 22.

[199] The question of Tycho's priority has from the outset been hotly disputed; Tycho himself was involved in lengthy and often vituperative debates on the subject, and recently his claims to priority have once again been called into question; see Gingerich & Westman, *Wittich Connection*. Although the accounts of the various squabbles between Tycho and his alleged plagiarizers, most notably Nicholas Raimarus Ursus, make interesting reading (see Nicholas Jardine, *The Birth of History and Philosophy of Science. Kepler's "A Defence of Tycho against Ursus" with Essays on its Provenance and Significance* (Cambridge 1984) and Edward Rosen, *Three Imperial Mathematicians* (New York 1986)), for our purposes the question of whether Tycho actually was the first to propose a realist geoheliocentric system is irrelevant; what matters is, that Tycho managed to "appropriate" the system and was subsequently regarded as its author; see Nicholas Jardine, "How to Appropriate a World System" (essay review of Gingerich & Westman, *The Wittich Connection*), *Journal for the History of Astronomy* 21 (1990) 353-357.

to a large degree dependent on the motion of the Sun, a fact not to be explained by means of any completely geocentric arrangement. Thus, he grew convinced that the celestial reality was accurately rendered by means of a compromise between Ptolemy and Copernicus, literally retaining the best of two worlds.[200]

While Tycho was tenaciously pondering such a possibility, roughly during the decade 1578-1588, a number of other scholars preferred a different approach. As we have seen, for a long time after the appearance of *De Revolutionibus,* Copernicus himself and his disciple Rheticus had been the only ones to actually endorse the physical truth of the heliocentric system, with the vast majority of their colleagues preferring the fictionalist Wittenberg Interpretation.[201] However, during the 1570s, several scholars grew convinced that the heliocentric arrangement of the planets in fact represented the true system of the world. Thus, the English mathematician Thomas Digges in his 1573 *Alae Seu Scalae Mathematicae,* a work dealing with the nova of 1572, publicly advocated the truth of the Copernican hypothesis and six years later even added a revolutionary new ingredient of his own: the infinity of the universe.[202] Other realist Copernicans included Kepler's teacher, Michael Mästlin; Christopher Rothmann, astronomer to Landgrave William IV of Hesse-Cassel; the Italian Giovanni Battista Benedetti; and the Spanish Augustinian monk and scholar Diego de Zuñiga, who in 1584 published a *Commentary on Job* in which he defended the reality of the Copernican system on Scriptural grounds, and which was to enjoy the dubious honour of being placed on the Index in 1616 along with *De Revolutionibus* itself.[203]

As is to be expected in the absence of convincing proof in favour of the Copernican system[204], these scholars based their realist interpretations of the heliocentric theory to a large extent on the aesthetic

---

[200] Jones Schofield, *Tychonic World Systems,* 55-58.

[201] See supra, chapter IV, p. 111. Possibly, the Englishman Robert Recorde, who published a *Pathway to Knowledge* in 1551, was also a realist Copernican, according to J.L.E. Dreyer, *A History of Astronomy from Thales to Kepler* (2nd ed. Dover 1953) 346-347. See also John L. Russell, "The Copernican system in Great Britain" in Jerzy Dobrzycki ed., *The Reception of Copernicus' Heliocentric Theory. Proceedings of a symposium organized by the Nicolas Copernicus Committee of the international union of the history of science* (Dordrecht/ Boston 1972) 189-239: 189-190.

[202] In an appendix to the *Prognostication Everlastinge* of his father; see Russell, "Copernican system", 194.

[203] Dreyer, *History,* 347-353; Jones Schofield, *Tychonic World Systems,* 14-16.

[204] Such proof did not become available until the nineteenth century, when stellar parallax was observed.

appeal exercised by the harmonious nature of the Copernican arrangement of the planets; they tried to further enhance its credibility by pointing out that it could be regarded as the revival of an ancient and venerable theory which had been proposed by such reverenced authorities as the old Pythagoreans.[205] Yet, some of the arguments they adduced were of a rather more down-to-earth nature. As Copernicus himself had pointed out, his system was capable, not only of accounting for, but also of explaining observed phenomena such as bounded elongation and planetary retrogradation. To minds willing to mitigate the sharp division between astronomy and physics, and to a certain extent to admit physical considerations to be taken into account when making a choice between rival world systems, such physical advantages of the heliocentric arrangement might, and did, lead to a realist stance in favour of the Polish master's theory.

Thus, the final decades of the sixteenth century saw the realist interpretation of both geoheliocentric and completely heliocentric world systems by a limited but influential number of scholars. However, realism necessarily entailed the intrusion of physical considerations into the study of the world system, a terrain on which, traditionally, astronomers had been left undisturbed to wield their fictionalist hypotheses as they pleased. This shift in division of tasks did not come about unproblematically. Realists, be they Tychonians or Copernicans, soon ran into problems when trying to underpin their new outlooks with plausible physical arguments. Thus, Aristotle's physics could not account for the fact that, in both the Tychonic and the Copernican systems, the planets revolved freely around the sun instead of the earth.

To Copernicus himself, this had not posed much of a problem, since he had retained the solid spheres in which the planets were embedded, simply placing the earth with its satellite the moon in the sphere where, formerly, the Ptolemaic planet Sun had been. However, as we have seen, by the time Tycho had completed his new world system he had lost faith in the existence of solid spheres; thus, he had to come up with some other plausible explanation of how the planets were kept in their orbits. This problem was not quite as new as has often been suggested; as we have seen, a number of respect-

---

[205] On the authority of pseudo-Plutarch, *De placitis philosophorum*, the heliocentric system was generally believed to have been adhered to by the ancient Pythagoreans Philolaus and Ecphantus, as well as by Aristarchus of Samos.

able scholars had, much earlier than Tycho, professed their belief in fluid, orbless heavens in accordance with the tenets of Stoicism. Thus, Robert Bellarmine in his Louvain lectures had argued in favour of corruptible, fiery heavens, in which the planets moved about "like the birds of the air or the fish of the waters." This phrase was often quoted by advocates of Stoic physics; however, to astronomers concerned with framing an accurate world system the analogy was not very convincing.

Both Tychonians and Copernicans, then, in the absence of solid spheres to carry the planets along, had to come up with some other motive mechanism. Tycho put forward the Platonic idea that the heavenly regions were animate, with the celestial bodies imbued with a vital spirit analogous to the one controlling the body of man. This spirit, he maintained, was bestowed on them at creation, and kept them moving in their respective orbits, so that they needed neither solid orbs nor angelic movers to propel them. Moreover, like Copernicus, Tycho possessed a Platonically inspired reverence for the Sun, suggesting that it played an important part in directing the other planets' motions, on whom it exercised an influence analogous to that of magnets on iron.[206] Such ideas, however, carried hardly any greater power of persuasion than Bellarmine's "birds and fish" analogy; not surprisingly, when Tycho asked his correspondent Rothmann's opinion on the geoheliocentric system, his most weighty criticism was directed at its inability to give a satisfactory physical explanation of the various planetary motions.[207]

An even more daunting task was faced by those realists who were of the Copernican persuasion. Not only did they have to come up with a physically plausible explanation of the motions of the planets in the absence of solid spheres; their removal of the earth from the centre of the cosmos also rendered invalid such core tenets of Aristotelianism as the dichotomy of celestial and terrestrial realms, or the rectilinear character of elementary motion. An entirely new science of physics was required, if the heliocentric arrangement of the planets was to carry any lasting conviction.

The man to devote a large part of his life to the contriving of such a science in defence of the Copernican system, was Johannes Kepler.

---

[206] Jones Schofield, *Tychonic World Systems*, 77-79.

[207] Ibidem, 100, and Kristian Peder Moesgaard, "Copernican influence on Tycho Brahe" in Dobrzycki ed., *Reception of Copernicus' Heliocentric Theory*, 31-55: 48-49.

Born in 1571, Kepler had never known the science of astronomy without the presence of *De Revolutionibus*; moreover, he had, more or less by chance, received his education from Michael Mästlin, one of the few men who were already realist Copernicans. Thus, from the onset of his career, Kepler had been firmly convinced of the truth of the heliocentric system, harmonious and beautiful, and capable of explanation where Ptolemy had provided mere description. Yet, from an equally early age, he had been haunted by the question after the physical underpinnings he felt must exist in order to keep such a perfectly ordered system going. Tycho's findings had rendered obsolete the belief in solid orbs; yet postulating angelic intelligences to replace them did not appeal to Kepler: "If there are no orbs, the conditions under which the intelligences and moving souls must operate are made very difficult, since they have to attend to so many things to introduce to the planet two intermingled motions. They would at least have to attend at one and the same time to the principles, centres, and periods of the two motions."[208] Such a heavy workload to Kepler seemed improbable, even for angelic beings. He therefore proposed a different solution, in keeping with his belief that Nature loved simplicity and unity: "But if the earth is moved, I show that most of this can be done with physical rather than animate faculties, namely magnetic ones."[209]

The resemblance between the driving power of the sun and the terrestrial phenomenon of magnetism had already been noted, by way of an analogy, by Tycho; Kepler, having explicitly set himself the task of "inquir[ing] into celestial physics and the natural causes of the motions"[210], went one step further and ventured to explain the heavenly power play by means of *real* magnetic forces. It was obvious that a connection existed between the sun on the one hand and the planets on the other; this was demonstrated most clearly from the observed fact that the more distant a planet was from the sun, the longer was its period. Kepler, with his mind set on elucidating the physical causes of the planetary motions, was therefore unable to subscribe to Copernicus's postulate, that the centre of the planets' orbits was not the real sun, but a mathematical point in its vicinity:

---

208 Johannes Kepler, *New Astronomy*, "Introduction to this work" (Prague 1609; translated by William H. Donahue, Cambridge 1992) 52.

209 Ibidem.

210 Ibidem, 48.

"A mathematical point, whether or not it is the centre of the world, can neither effect the motion of heavy bodies nor act as an object towards which they tend. . . . Let the physicists prove that natural things have a sympathy for that which is nothing."[211]

Rather, Kepler believed, the real sun should be regarded as the centre of planetary motions. From this vantage position, it was able to direct the planets' movements by exercising on them a celestial power, analogous to that of the magnet on earth. Many of Kepler's ideas on this magnetic sun-power were borrowed from William Gilbert, who in 1600 had published his book *De magnete* in which he had put forward his ideas on the subject of the earth's magnetism. According to Gilbert, the earth was a giant loadstone; by means of his experiments with *terrellae,* small magnetized globes, he demonstrated how the earth was fitted for circular motion, and how magnetic attraction was capable of acting without any material contact.[212] Although Gilbert himself had gone no further than to employ his magnetic theories to show the plausibility of the earth's daily motion around its own axis, his ideas were clearly capable of acting as the building blocks for the physical mechanism of the Copernican universe, for which Kepler was so restlessly searching.

Around 1602, while investigating the characteristics of the orbit of Mars, Kepler acquired a copy of *De magnete*; and sure enough, when seven years later he published the results of his research into Martial motion in one of the greatest astronomical works ever written, the *Astronomia nova,* the influence of Gilbert's magnetic philosophy was much in evidence in Kepler's attempt at elucidating the force that kept the planets in their orbits. The sun, placed in the true centre of the universe, was endowed with one kind of magnetic polarity, diffused over its surface. Because it rotated rapidly on its own axis, it continuously emitted emanations of magnetic force in all directions; the planets, possessing two Gilbertian poles of opposite polarities that strove to maintain their direction in the heavens, were, during the half of their rotations when their unlike poles faced the sun, attracted towards it, while during the other half, when their like poles faced the sun, they were repelled from it.[213] Thus, eccentric orbits were

[211] Ibidem, 54.

[212] Stephen Pumfrey, "Magnetical philosophy and astronomy, 1600-1650" in Taton & Wilson eds., *Planetary Astronomy,* A, 45-53: 47.

[213] Ibidem, 49.

generated. Moreover, the strength of the solar emanation decreased in inverse proportion to distance, thus explaining the varying speeds of the planets.[214]

Thus, Kepler to his own satisfaction succeeded in framing a convincing physical mechanism to account for planetary motions. There was no more need for solid orbs, nor for souls or intelligences: "As a result, every detail of the celestial motions is caused and regulated by faculties of a purely corporeal nature, that is, magnetic, with the sole exception of the whirling of the solar body as it remains fixed in space."[215] However, his theory was not greeted with much approval during the first quarter of the seventeenth century. According to Gingerich, the bulk of Kepler's ideas only came to the forefront as a result of the success of his *Rudolphine Tables,* published in 1627.[216] Those Peripatetics who accepted the fluidity of the heavens continued to ascribe the motions of the planets to the agency of souls or intelligences[217], as did the steadily growing ranks of adherents of the Tychonic system[218], some of whom even ventured to invalidate Kepler's celestial physics by demonstrating that his physical arguments ran counter to the entire science of magnetism.[219] However, magnetic forces did not constitute the only reinforcements to come to the aid of realist Copernicanism between the appearance of the comet of 1577 and that of 1618. While, in the year 1609, Kepler was seeing his new astronomy through the press, his colleague Galileo saw a new universe through the recently invented "optic glass".

c. *"Through Optic Glass the Tuscan Artist Views"*

According to Albert Van Helden, the discoveries made by several observers, most notably Galileo, when they pointed the newly invented telescope skyward in the years 1609-1610 were "bombshells indeed".[220] Galileo, Van Helden contends, in publishing the *Sidereus*

---

214 Owen Gingerich, "Johannes Kepler" in Taton & Wilson eds., *Planetary Astronomy*, A, 54-78: 64.

215 Kepler, *New Astronomy,* "Introduction", 68.

216 Gingerich, "Johannes Kepler", 77.

217 Edward Grant, *Planets, Stars, & Orbs. The Medieval Cosmos, 1200-1687* (Cambridge 1994) 369-370.

218 Jones Schofield, *Tychonic World Systems,* 227-229.

219 Martha R. Baldwin, "Magnetism and the anti-Copernican polemic", *Journal for the History of Astronomy* 16 (1985) 155-174.

220 Albert Van Helden, "The Telescope in the Seventeenth Century", *Isis* 65 (1974) 38-58: 51.

Nuncius "single-handedly took on the scholastic establishment which was in many cases part of the Church"; the booklet allegedly "struck at the roots of Aristotelian physics and metaphysics, which was the official natural philosophy of the Church."[221] "Yet", Van Helden continues in a tone of surprise, "only one year later, he [Galileo] had convinced the very men who were the guardians of the faith in matters of astronomy - the mathematicians of the Collegio Romano, as well as the head of the college, Cardinal Bellarmine... It seems therefore fair to say that except for a few cranks, scientists acted like reasonable men, no matter how conservative they were on the issue of the Great Debate, and it is in fact surprising how little resistance they mounted."[222]

Precisely in view of the relative lack of opposition to the telescopic discoveries, however, one may well wonder whether Van Helden's qualification of them as "bombshells" is entirely justified. By considering the events surrounding the invention and early application of the telescope in connection with the intellectual background around 1600, as described at length in our previous chapter, we may be able to somewhat alleviate Van Helden's surprise, and to put the telescopic discoveries of the early seventeenth century in their proper perspective.

The question of who can be regarded as the inventor of the telescope has received much attention, but has never yet been decisively answered. Spurred on by Galileo's statement, in his *Sidereus Nuncius,* that he had been induced to fabricate his first telescope by reports that "a certain Fleming had constructed a spyglass by means of which visible objects, though very distant from the eye of the observer, were distinctly seen as if nearby"[223], historians have investigated with great care the claims of Hans Lipperhey and Sacharias Janssen, lense grinders from the Zealand town of Middelburg, and of Jacob Metius from Alkmaar in northern Holland. Both Lipperhey and Metius applied to the Estates General of the United Provinces for a patent on the invention in the autumn of 1608, while the Committee of Councillors of Zealand on 14 October of the same year interviewed a young man who claimed also to know the art of mak-

221 Ibidem.

222 Ibidem, 51-52.

223 Galileo Galilei, *Sidereus Nuncius,* translated by Stillman Drake in *Discoveries and Opinions of Galileo* (Garden City 1957) 28.

ing spyglasses.[224] It has proved well-nigh impossible to ascertain which one of these men could rightfully be called the inventor of the telescope. The most recent research on the subject has even made it appear that none of them can in fact be credited with the invention; rather, the combination of a convex and a concave lens as a device for magnifying distant objects had been employed for various purposes as early as the final decades of the sixteenth century, and the three Dutchmen (or one of them) can at most be said to have first realized the full potential of the instrument.[225]

As in the case of the invention of the realistically interpreted geoheliocentric world system, however, the question of who was the "real" inventor of the telescope does not really carry much weight; the important point to establish is, who in the end successfully appropriated the invention. In neither case can there exist any doubts on that score: as the geoheliocentric world system came to bear the epithet 'Tychonic', so the first telescopic discoveries became 'Galilean'. Galileo may not have been the first to point a telescope skyward (the Franeker professor of mathematics, Adriaan Metius, in 1614 claimed that he had seen the moons of Jupiter, the sunspots, the mountainous nature of the moon, and the phases of Venus as early as 1608, through the telescope constructed by his brother Jacob[226]), but the priority of publication definitely belonged to the Italian. Whereas Jacob Metius, who seems to have been rather a withdrawn figure, was very secretive about his telescope and would hardly allow anyone to see it, Galileo, immediately after assembling his own spyglass and making his first observations, set up an effective publicity campaign, going around with his telescope to show his discoveries to anyone who was willing to see, dedicating his *Sidereus Nuncius* to the powerful Medici family, and having his booklet and telescopes distributed to foreign nobles through the Medici ambassadorial network.[227] If the telescopic discoveries were not exactly bombshells, the way in which copies of the *Sidereus Nuncius* descended

[224] Albert Van Helden, *The invention of the telescope* (*Transactions of the American Philosophical Society* 67, 4 (Philadelphia 1977)) 20-21.

[225] Ibidem, passim.

[226] In his *Institutiones Astronomicae;* see H.A.M. Snelders, "Alkmaarse natuurwetenschappers uit de zestiende en zeventiende eeuw" in *Van Spaans beleg tot Bataafse tijd. Alkmaars stedelijk leven in de zeventiende en achttiende eeuw* (Alkmaarse historische reeks 4, Zutphen 1980) 101-122: 108.

[227] Mario Biagioli, *Galileo Courtier: The Practice of Science in the Culture of Absolutism* (Chicago 1993) 133-135.

on scholarly and courtly Europe might perhaps be compared to a fall-out.

The *Sidereus Nuncius,* in which, on 12 March 1610, Galileo published his findings after about eight months of telescopic observation, immediately attained widespread renown. In it, he presented three fundamental discoveries that were to become the subject of considerable excitement with the intelligentsia throughout Europe. First of all, the 'Tuscan artist' had focused his telescope on the nearest of all celestial objects, the Moon, to find that, next to what he called the "large or ancient spots", that had been visible with the naked eye throughout the ages[228], there could be discerned a great number of smaller spots all over the lunar surface, especially in the lighter part. In order to account for these 'new' spots, Galileo included a demonstration to prove that the Moon's surface was "not smooth, uniform, and exactly spherical, as the majority of philosophers believe it (and all other heavenly bodies) to be, but uneven, rough, and full of cavities and prominences, it being (not unlike the earth's surface) in relief with mountain chains and deep valleys."[229]

Next, Galileo went on to discuss the stars and planets. He stressed that the stars were not seen enlarged by the telescope in the same ratio as other objects, because the instrument removed from them their adventitious and accidental rays (which made them seem larger to the naked eye than they really were) and then enlarged just their simple globes. A further interesting phenomenon was the difference in telescopic appearance between stars and planets: whereas, through the telescope, planets were seen "perfecly round and definitely bounded, looking like little moons, spherical and flooded with light all over", the fixed stars showed "[no] circular periphery, but have rather the aspect of blazes whose rays vibrate about them, and they scintillate a very great deal."[230] However, by far the most amazing discovery Galileo made concerning the fixed stars, was that there existed innumerously many more of them than had ever been seen with the naked eye: "These are so numerous almost to defy belief."[231] Through the telescope, the largest of these previously unseen

[228] Galileo, *Sidereus Nuncius,* translated by Stillman Drake in *Telescopes, Tides, & Tactics. A Galilean Dialogue about the "Starry Messenger" and Systems of the World* (Chicago/London 1983) 23.

[229] Ibidem, 24.

[230] Ibidem, 49.

[231] Ibidem.

stars appeared larger and brighter than stars of the second magnitude seen with the naked eye, and it was virtually impossible to depict all the 'new' stars on any star map, so many were they. Furthermore, the Milky Way, too, was "nothing but a congeries of countless stars, grouped in clusters", as were the various celestial "nebulae".[232]

Still, Galileo saved what he considered his most momentous discovery for the end of his treatise: "There remains the matter which in my opinion deserves to be considered most important of all - disclosure of four PLANETS never seen from the creation of the world up to our time...".[233] In the course of January, 1610, our author observed three (later four) small stars in the vicinity of Jupiter, which lay in an exact straight line parallel to the ecliptic and appeared more splendid than other stars their size.[234] After a number of observations, he realized that they had to be Jovian satellites: "I now decided beyond all doubt that there existed in the heavens three stars wandering about Jupiter as do Venus and Mercury about the sun... Nor were there just three such stars...: four planets complete their revolutions around Jupiter...".[235] The existence of such 'moons' of Jupiter, as they were often called, assumed special significance for Galileo in that he saw their discovery as an opportunity for dedicating them (and in so doing recommending himself) to the Grand Duke Cosimo de' Medici, christening them the "Medicean Stars".

Galileo by no means abandoned his telescopic observational activity after the publication of the *Sidereus Nuncius*. In the summer of 1610, he discovered two "ears" or "handles" on either side of the planet Saturn, which he supposed to be stationary objects accompanying the planet, while in the autumn of the same year, he noticed that Venus no longer looked round, but rather gibbous. At the end of December, he was able to affirm for a fact that Venus displayed phases like the moon's—or, as he had cryptically formulated his hunch in an anagram he had sent to Kepler on 11 December: "Shapes of the moon are imitated by the mother of loves".[236] The latter observation could lead to only one conclusion: Venus revolved around the sun. The sun itself, moreover, was the object of yet an-

---

[232] Ibidem, 52.
[233] Ibidem.
[234] Ibidem, 59.
[235] Ibidem, 61.
[236] Stillman Drake, *Galileo: Pioneer Scientist* (Toronto 1990) 137-138; *Haec immatura a me iam frustra leguntur o.y.*, to be solved as *Cynthiae figuras aemulatur mater amorum.*

other telescopic discovery, first made in 1610 by Thomas Harriot, and repeated in 1611 by the Jesuit Christopher Scheiner of Ingolstadt: it was covered with dark spots, appearing to move across its surface.

The discoveries laid down in the *Sidereus Nuncius* and in later publications concerning the phases of Venus and the sunspots were greeted with great enthusiasm throughout Europe. Johannes Kepler, who was one of the first to hear about the novelties, was characteristically carried away with enthusiasm and immediately sat down to compose a *Conversation with the Sidereal Messenger,* in which he confessed that, after he had received Galileo's booklet, "For a long time I had stayed at home to rest, thinking of nothing but you [Galileo]".[237] Nor did one have to be a convinced Copernican like Kepler to appreciate Galileo's achievement. Observing the sky by telescope soon became a very popular activity, and by the autumn of 1609, even before the *Sidereus Nuncius* was published, spyglasses were commonly for sale in the major cities of Europe.[238] Despite the fact that most of these were of a distinctly inferior quality to the one employed by Galileo (as late as 1637, the Dutch astronomer Martinus Hortensius wrote to the Italian master complaining that "we have found none in Holland up to now which can be polished to the precision which is required for these observations [of Jupiter's satellites, for longitude determination]. For the best usually show the disk of Jupiter hairy and poorly defined, whence the Jovian stars in its vicinity are not perceived well"[239]), at least the numbers in which telescopes were offered for sale testify to the general public's interest in the new discoveries.

Traditionally, much has been made of the damaging evidence against the Aristotelian world view, supposed to have been provided by Galileo's telescopic findings. Thus, the discovery of the Jovian satellites demonstrated that there existed more than one centre of motion in the universe; apparently, not all celestial motions were caused by the *primum mobile.* Also, the sunspots and the lunar mountains could allegedly be used as arguments against the perfection of the heavens, and might moreover enhance the credibility of the Copernican view of the earth as 'just another planet', since, clearly, the difference between the corruptible earth and the incorruptible plan-

---

[237] For a translation of this work, see Edward Rosen, *Kepler's Conversation with Galileo's Sidereal Messenger* (New York/ London 1965).

[238] Van Helden, *Invention of the telescope,* 21.

[239] Van Helden, "Telescope in seventeenth century", 43.

ets was by no means as great as had been assumed by Aristotle. And, finally, the fact that Venus went through phases similar to those of the moon provided undeniable proof that the planet revolved round the sun rather than the earth.

In accordance with the traditional view of the supposed injurious nature of the telescopic novelties to the Aristotelian world view, it had often been stressed that orthodox Aristotelian philosophers refused to look through Galileo's telescope, or, if they did, denied that what they saw was the truth, denouncing the telescopic sights as optical illusions. Yet, although there were undoubtedly some Peripatetics who would have nothing to do with Galileo's findings, or tried to enter into vituperative disputes with him (like Galileo's ardent opponent, Lodovico delle Colombe, who ventured to explain away the lunar mountains by suggesting that the apparent valleys on the moon were filled with a perfectly smooth and spherical layer of invisible crystalline material), the most frequently adduced case of reluctance to look through the telescope, that of the Paduan Peripatetic Cesare Cremonini, is probably spurious, having been made up years later[240], while, though the eminent Jesuit mathematician Christopher Clavius's first reaction to the news of Galileo's discoveries may have been "that in order for Galileo to see the things he claimed to see through the telescope, he first had to put them in the telescope"[241], yet a year later he feasted the Tuscan at the Collegio Romano. The Jesuits had obtained a telescope in November 1610, through which they soon observed the phases of Venus as well as the novelties of the *Sidereus Nuncius*. At this point, their incredulity gave way to admiration; early in 1611, they informed Cardinal Bellarmine that the new phenomena were indeed to be observed in the heavens, and they invited Galileo to Rome. When he arrived there, in the spring of 1611, he was honoured and feted by the Jesuits of the Collegio Romano as well as by the members of the young Accademia dei Lincei, to which he was elected, and whose fellows gave the new instrument the name 'telescope'.

Thus, the question presents itself of why, if Galileo's discoveries were as lethal to the traditional world view as has often been maintained, his *Sidereus Nuncius* met with such overwhelming success, espe-

---

240 Charles B. Schmitt, *Cesare Cremonini, un Aristotelico al tempo di Galilei* (Venice 1980) 14.

241 Van Helden, "Telescope in the seventeenth century", 51-52.

cially among members of a society whose task it was to guard this world view. The answer to this question is twofold: firstly, the telescopic novelties were not as menacing to the Aristotelian world view as they might seem at first glance; and, insofar as they did entail an outright contradiction of Aristotelian tenets, these tenets had for the most part been challenged before, and their denial could be accommodated to a world view that, as we have seen, had from the outset borne a chameleonic character.

First of all, those who were unwilling to part with the centrality of the earth or with Aristotle's physics, had at their disposal, by the time Galileo's discoveries became known in 1610, a ready-made world system that was capable of combining the telescopic novelties with the time-honoured Peripatetic physics, as well as with Scriptural texts upholding the centrality of the earth: the system contrived by Tycho. From 1602 onwards, the Tychonic arrangement of the planets had become well known and respected in scholarly circles, although it did not immediately gain widespread acceptance. However, the forceful impact made by the *Sidereus Nuncius* in 1610 did the geoheliocentric arrangement a world of good. Not surprisingly so: while retaining the stable earth, it was able to account for both the phases of Venus and the satellites of Jupiter in a completely convincing way. Thus, in the Tychonic arrangement, Venus circled the sun, not the earth; therefore, her phases might be explained just as soundly in the geoheliocentric system as in the Copernican. Also, the Jovian satellites were no more problematical in a universe in which the sun circled the earth with the other planets as its satellites than in a universe in which all planets circled the sun. Rather, they constituted less of a problem in the Tychonic system, in which satellites in fact figured much more prominently than in the Copernican; the heliocentric arrangement knew just the one satellite (the moon), whereas, to the Tychonic, all the planets except the earth, the sun and the moon could be regarded as satellites.

Thus, both the phases of Venus and the so-called 'Medicean stars' could easily be fitted into the Tychonic system, which many astronomers could (and did) adopt without having to abandon traditional, earth-centred physics, and without upsetting orthodox biblical exegesis. After 1610, those astronomers who had not yet adopted the geoheliocentric arrangement of Tycho soon did.[242] Even Galileo

[242] Ibidem, 53; Peter Dear, "The Church and the new philosophy" in Stephen

himself did not draw any revolutionary conclusions in the *Sidereus Nuncius;* although he had been of the Copernican conviction in private for several years, he did not speak out on the issue of heliocentrism in his 1610 booklet, probably because he realized that all the evidence it contained was as easily reconcilable with the Tychonic as with the Copernican system.[243]

What the telescopic discoveries *did* establish beyond doubt, was the untenability of the Aristotelian postulate of celestial perfection, as well as the non-existence of any solid celestial spheres. After all, the moon was clearly not perfectly spherical, but rather pockmarked, while the sun was tainted with various black spots, floating across its surface and constantly changing shape. Moreover, the fact that no less than four bright stars made their revolutions around Jupiter rendered the concept of solid spheres in which all celestial bodies were supposed to be embedded, completely untenable.

However, both the idea that the heavens were incorruptible and the notion that they were full of crystalline orbs had been dealt very serious blows long before the advent of the telescope. As we have seen, the case for celestial corruptibility had been extensively argued in the course of the sixteenth century, inspired by the revival of Stoic cosmology and the teachings of the Church Fathers. Thus, we have seen Robert Bellarmine, the future Cardinal, defend the fiery and elemental nature of the heavens in his 1570-2 Louvain Lectures, on the basis of the Church Fathers as well as the ancient non-Aristotelians.[244] Moreover, the Stoic view of the heavens as one fluid continuum, obviously incompatible with the notion of solid spheres, had also found its adherents, Bellarmine among them, during the final quarter of the sixteenth century. Tycho's observations of the comet of 1577, and his orbless system of the world, published in 1602, undoubtedly further discredited the belief in solid spheres.

As for the astounding number of previously unseen fixed stars, this, too, was a phenomenon that had so to say been 'predicted' by several authors from antiquity. Thus, Democritus had proposed an infinity of worlds in infinite space, while, in his discussion of comets, the great Stoic Seneca had suggested that there might be very many

---

Pumfrey, Paolo L. Rossi & Maurice Slawinski eds., *Science, culture and popular belief in Renaissance Europe* (Manchester 1991) 119-139: 135.

[243] Drake, *Galileo: Pioneer Scientist,* 134.

[244] See supra, chapter IV, p. 103.

more celestial bodies than the ones seen with the naked eye, and that a future age might make noteworthy discoveries concerning them. Statements such as these may have induced the Jesuit Christopher Scheiner to suggest that the sunspots might be due to small planet-like bodies, circling the sun and partly eclipsing it.[245] Also, the idea of the resemblance of the moon's uneven surface to that of the earth had been advanced by, once again, Democritus, as well as Anaxagoras, who both believed that the moon was like the earth in having mountains and glens and plains.[246] The Pythagoreans, too, were favourably disposed towards this notion, a fact Galileo did not fail to mention in his *Sidereus Nuncius* (to facilitate acceptance of the idea with his readers, as well as, probably, because the same Pythagoreans were also credited with endorsement of a heliocentric world system).

The telescopic discoveries, then, were not as revolutionary as they might seem at first glance. Many of them had already had their way prepared by the Renaissance passion for unearthing the most ancient wisdom, and by the concomitant popularity the cosmology of the Stoics had come to enjoy during the sixteenth century; moreover, astronomically, the phenomena brought to light by Galileo's telescope could easily be accounted for by means of the, already existing, Tychonic system. The greatest immediate effect of the celestial novelties was probably the enhancement of the popularity of the geo-heliocentric arrangement, which, with its fluid heavens, solar magnetism, numerous satellites, and centrality of the earth, was perfectly suited to accommodate the celestial novelties to the framework of Aristotelian physics and Scriptural authority.

Robert Bellarmine, who in 1610 occupied the important position of personal theological consultant to pope Paul V, was one of the first persons to have been persuaded of the reality of Galileo's discoveries.[247] Sure enough, they cannot have contained anything very offending to the Cardinal: he had long been convinced of the fluidity and corruptibility of the heavens, and Galileo's book did not include any open allusion to realist heliocentrism. However, a few years later, the situation had changed somewhat. From 1611 onwards, Galileo

---

[245] For a discussion of Scheiner's arguments and Galileo's refutation of them, see William R. Shea, "Galileo, Scheiner, and the Interpretation of Sunspots", *Isis* 61 (1970) 498-519.

[246] See D.R. Dicks, *Early Greek Astronomy to Aristotle* (Cornell 1970) 82.

[247] Drake, *Galileo: Pioneer Scientist*, 144.

had become engaged in a debate on sunspots with the Jesuit Scheiner. The dispute was fought by means of a series of letters on the subject, which were published and thus became accessible to the general public. In an appendix to one of these letters, printed at Rome in 1613, Galileo had finally made public his real conviction: that the Copernican arrangement was to be regarded as the true system of the universe, and had tried to underpin it with undeniable proof.[248] Two years later, in 1615, the Carmelite Foscarini, too, openly defended the reality of the Copernican system, while Galileo published a *Letter to the Grand Duchess Christina* in which he upheld that the Bible and Nature spoke two different languages.

These allegations could not be left unanswered. Bellarmine notified Foscarini and Galileo of the fact that, in the absence of incontrovertible proof of the truth of the heliocentric system, it should be treated as a mathematical hypothesis only, in accordance with the time-honoured practice that severed astronomy from physical and cosmological considerations, since the endorsement of the reality of Copernicus's arrangement conflicted with many biblical passages, as well as with the dicta of the Church Fathers. Bellarmine undoubtedly considered the latter by far the most important. After all, against almost every Scriptural passage in favour of geocentrism could be quoted another one that favoured a moving earth, so that the literal interpretation of biblical passages did not provide a very stable guideline. The Church Fathers, however, whose authority had been enormously reinforced by the Council of Trent, where it had been decided that it was absolutely forbidden to go against their unanimous interpretation, had, on the issue of geocentrism, all concurred in favour of Aristotle.

As a result of Galileo's and Foscarini's publications, in February 1616 the Pope asked the Holy Office for its judgment on Copernicanism. After consulting with Bellarmine, he submitted two propositions to the theologians known as the Committee of Qualifiers: first, that the sun was fixed; and second, that the earth moved. Both these notions were duly condemned; although they were not formally denounced as heresies, they could no longer be held or defended by Catholics, and Foscarini's book was placed on the Index along with *De Revolutionibus* itself, at least until properly corrected. Bellarmine, as

[248] Ibidem, 145.

head of the Holy Office, instructed Galileo on the decree's contents, which, however, did not mention the Tuscan's name.[249]

The effect of the Decree of 1616 was not at all dramatic. In 1620, a list of corrections to *De Revolutionibus* was ordered, but, outside Italy, hardly anyone actually introduced them into their copies of Copernicus' work. All in all, the decree did not cause a great stir, not having established anything as a heresy. Copernicanism still was not much of an issue; for the time being, by far the most common stance was the adoption of the Tychonic system, which reconciled the telescopic discoveries with Aristotelian physics, and did not offend either Scripture or the authority of the Church Fathers that had so prominently come to the fore as a result of the Counter-Reformation.

4. *Dutch Reactions to the Comet of 1618*

a. *Counter-Reformation at Louvain*

One of the most northerly bulwarks of the Counter-Reformation was the university of Louvain, in the province of Brabant in the Southern Netherlands. Founded in 1425, the institution had always displayed a marked loyalty to the Holy See. Thus, it had never accepted conciliarism, a theory that derogated from the power of the popes[250]; and, at the very beginning of the Reformation, the Louvain theological faculty had been the first institution to denounce Luther's works, issuing a condemnation on 7 November 1519, several months before Leo X's bull against the Wittenberg reformer was promulgated in the summer of 1520.[251] In the decades that followed, Louvain university revealed itself a staunch combatant of heresy, publishing the first ever Index of Prohibited Books in 1546, issuing a new Latin edition of the bible, the so-called 'Louvain Bible', in 1547 by way of an antidote to the many heretical bibles in circulation, and producing an edition of the complete works of Saint Augustine in 1577.[252]

Louvain professors also made a considerable contribution to the

---

[249] See Drake, *Galileo: Pioneer Scientist*, 173-177, and Albert Van Helden, "Galileo, telescopic astronomy, and the Copernican system" in Taton & Wilson eds., *Planetary Astronomy*, A, 81-105: 94-97.

[250] Léon van der Essen, *L'Université de Louvain (1425-1940)* (Brussels 1945) 16.

[251] Ibidem, 24-25.

[252] *De Universiteit te Leuven 1425-1985* (Louvain 1986) 110-112.

Council of Trent. At all three sessions of the Council, delegations from Louvain were present, whose opinions were held in high esteem. Especially the theologian Ruard Tapper, chancellor of the university, who attended the 1551-2 session, seems to have played quite an important role: at the Council meetings, he was asked for his opinion immediately after the Pope's own theologians had given theirs.[253] In 1565, the Tridentine decrees were published at Louvain university with great pomp, and all members of the academic community were obliged to abjure Luther and his doctrines.[254]

By the time the last session of the Council of Trent had been concluded, the Netherlands found themselves on the eve of the great conflict known as the Dutch Revolt. As we shall see in a later chapter, this episode, which has long been portrayed as a struggle for political and religious freedom opposing the Protestant Dutch to the Catholic Spanish ruling bureaucracy, in fact rather bore the characteristics of a civil war, with religious and national demarcations by no means as clear-cut as has been suggested in the past. However, despite considerable confusion and conflicts of loyalties the official policy of Louvain university had always been to profess unconditional loyalty to the King of Spain. Obviously, this did not safeguard the institution from suffering the consequences of the horrible war fought on Flemish soil; the seemingly unending series of deprivations and lootings both the town and the university of Louvain underwent during the final decades of the sixteenth century has been described in detail by Léon van der Essen on several occasions.[255]

Yet, when, in 1618, the second of our great comets appeared, the ancient institution had recovered much of her former strength. The greater part of the Southern Netherlands had been recaptured by the Duke of Parma, the overwhelming majority of Protestants had fled,

---

[253] F. Claeys Bouuaert, *L'ancienne université de Louvain. Études et Documents* (*Bibliothèque de la revue de l'histoire ecclésiastique*, fasc. 28, Louvain 1956) 59-61; *Universiteit te Leuven*, 110.

[254] Van der Essen, *Université de Louvain*, 29; Claeys Bouuaert, *Ancienne université*, 69-70.

[255] Van der Essen, *Université de Louvain*, 30-58; Léon van der Essen, "Les tribulations de l'Université de Louvain pendant le dernier quart du XVIe siècle", *Bulletin de l'Institut historique belge de Rome* 2 (1922) 61-86. See also J. Grisar, "Die Universität Löwen zur Zeit der Gesandschaft des P. Franciscus Toletus (1580) nach bisher unbenützten Quellen des Vatikan. Archivs" in *Miscellanea Historica in honorem Alberti De Meyer* (*Universiteit te Leuven, Publicaties op het gebied der geschiedenis en der philologie*, 3e reeks, 23e deel, Louvain/Brussels 1946) 941-968.

and, in 1609, a Twelve Years Truce had been concluded with the northern rebels. Although, occasionally, Dutch and French troops still laid siege to the town of Louvain (as they did in 1635, when, after the beleaguerment was lifted, both Erycius Puteanus and Libertus Fromondus wrote small works in praise of the heroism displayed by the *Lovanienses*)[256], on the whole, peace and quiet had returned to the Spanish Netherlands. Under the benevolent, ardently Catholic reign of the Archdukes Albert and Isabella, Louvain university regained its former splendour, and its professors were able once again to devote themselves wholeheartedly to their studies.

Such was the scholarly climate in the Southern Netherlands, when Galileo's *Sidereus Nuncius* appeared in 1610. The booklet at once raised a great deal of interest, and, as elsewhere in Europe, telescopic observations soon came into vogue. The Prince-Bishop of Liège was one of those potentates on whom Galileo generously bestowed the gift of a telescope and several copies of the *Sidereus Nuncius;* and Archduke Albert, too, displayed a marked sympathy for the Italian scientist and showed much interest in the scientific novelties, ordering one of Galileo's former pupils, Daniel Antonini, who had taken up residence in Brussels, to show him every letter he received from his erstwhile teacher.[257] The art of constructing telescopes was from the outset taken up diligently by Netherlandish artisans; thus, the Prince-Bishop was, by the time he received Galileo's telescope, already in possession of several spyglasses of superior quality.[258] Also, the Antwerp Jesuit Odo Van Maelcote, who at the time was one of the four astronomers attached to the Collegio Romano, had been given a telescope fabricated in the Netherlands by a fellow citizen of Antwerp; this spyglass was of such good quality that it enabled the Jesuits at Rome to observe the strange appearances of Saturn as early as 1609.[259] Both Van Maelcote and his fellow-Jesuit Gregorius a Sancto Vincentio from Bruges in western Flanders spent long nights observing the Galilean discoveries at the Collegio Romano, and when they

---

[256] Erycius Puteanus, *Historiae Belgicae liber singularis de obsidione Lovaniensi;* Fromondus' work, unfortunately, was never published; see L. Ceyssens, "Le Janséniste Libert Froidmont (1587-1653)", *Bulletin de la Société d'Art et d'Histoire du Diocèse de Liège* 43 (1963) 1-46: 13.

[257] Georges Monchamp, *Galilée et la Belgique. Essai historique sur les vicissitudes du système de Copernic en Belgique* (Saint-Trond/Brussels/Paris 1892) 16, 26.

[258] Ibidem, 16.

[259] Ibidem, 23.

returned to their home country, both taking up residence in Louvain in 1612, they must have brought with them considerable skill and expertise at telescopic observation of the sky.[260]

The same Gregorius a Sancto Vincentio, who became a professor of astronomy at the Jesuit college of Antwerp, in 1619 wrote a series of theses on comets, undoubtedly inspired by the appearance of the comet of 1618. He defended them at Antwerp in front of his pupils at the college; although, sadly, the theses themselves, which were printed at Louvain in 1619[261], appear to have been lost[262], Henri Bosmans has been able to unearth a report of the occasion, preserved in the manuscripts of the Antwerp Jesuit college. According to the anonymous Jesuit who wrote the account, the members of the Antwerp college had on many occasions observed the comet's vicissitudes and its entire course by means of telescopes. Afterwards, they had learned that expert observers from abroad preferred the Antwerp observations to all others. When, after the comet's expiry, A Sancto Vincentio had publicly defended his theses, his audience had been so numerous that the occasion resembled a play rather than a discussion of a cometary appearance.[263]

Three years before the comet's appearance, in December 1615, Libertus Fromondus, a young professor of philosophy at the Louvain Arts Faculty, was called upon to preside over the *Saturnalia,* a festival held annually at Louvain university around the feast of Saint Lucia,

---

[260] Ibidem, 23-24.

[261] Gregorius a Sancto Vincentio, *Theses de Cometis* (Louvain 1619). Erycius Puteanus mentions them at the end of his own treatise on the comet of 1618, *De Cometa anni 1618 novo Mundi Spectaculo Libri Duo Paradoxologia* (Cologne 1619) 168: "Inter Mathematicorum hodie primos Reverendum, & rarissimae subtilitatis Virum *Gregorium A S. VINCENTIO* habeo: quippe magni *Clavii* e Societate *Iesu* Auditorem: is in *Thesibus de Cometis,* peculiarem de admirando hoc spectaculo opinionem exprimens, & qua Paradoxa quaedam nostra confirmentur, esse eiusmodi in coelo Globos, eiusmodi Corpora pariter profitetur."

[262] Sancto Vincentio's assiduous biographer, Henri Bosmans, has been unable to locate any copy of the work; see *Biographie Nationale de Belgique* XXI, col. 141-171 s.v. "Grégoire de Saint-Vincent".

[263] Report published by Henri Bosmans in his *Documents inédits sur Grégoire de Saint-Vincent* (Brussels 1903) 12: "Anno 1619. . . Mathematici quoque magnam sibi apud peritiores opinionem doctrinae compararunt. Nam visus hoc anno Cometes eorum industriam plurimum acuit. Saepe enim dioptrices machinis usi, omnes eius varietates omnemque cursum observarunt. Et vero quas nostri Antverpiae collegerunt observationes, prae caeteris alibi collectis, intellectum est placuisse viris ea aetate peritissimis. Cum vero typis evulgatae observationes illae publice defenderentur, tantus fuit auditorium ex ordine concursus, ut non problemata de Cometis sed Comoediam spectare viderentur."

at which a master who had been elected president had to answer quodlibetical questions posed to him by the audience. Fromondus later had his contribution to the feast of 1615 published under the title of *Saturnalitiae Caenae, Variatae Somnio, sive Peregrinatione Caelesti*; the work seems to have enjoyed some popularity, since it was reprinted as late as 1665. This acclaim may well have been due in large part to the intermezzo Fromondus inserted between Questions XIV and XV, in the form of a 'Celestial Peregrination' in which the author described how, in a dream, he was carried up on high on the back of the winged horse Pegasus and thus got a chance to see at close range all the novelties that, in waking life, were only visible to earthlings by means of the telescope.

With regard to its form, the *Peregrinatio Caelestis* bears some resemblance to Galileo's later *Dialogo sopra i due massimi sistemi del mondo*. The greater part of the piece is taken up by a conversation between Fromondus on the one hand and his *Genius*, a sort of omniscient guardian angel, on the other. The Genius, driving the horse Pegasus, has captured the sleeping Fromondus and installed him on the flying steed's back, holding onto him by his belt to prevent him from falling. In this manner, the demurring Fromondus is willy-nilly carried up into the highest heaven, continually uttering Peripatetically-inspired objections, thus playing a role similar to that of Simplicio in the *Dialogo*. The Genius, conversely, displays the confidence and common sense of Salviati, invalidating every one of his protégé's objections and in the process expounding many of the new theories and telescopic discoveries of the recent decades.

Thus, in response to Fromondus' fear of being burned to ashes on traversing the Aristotelian sphere of fire, the Genius proposes a new arrangement of the elements: adducing several optical arguments, he states that there exists no such thing as a sphere filled with fire, at least, not with fire of the kind to be found on earth. After all, it has been shown that the planets do not scintillate, but shine with a calm and undivided lustre; this could not be, if their rays had to penetrate a fiery sphere of the kind postulated by the Peripatetics.[264] The tradi-

[264] Libertus Fromondus, *Saturnalitiae Caenae, Variatae Somnio, sive Peregrinatione Caelesti* (Louvain 1616) 69: "Ideo, inquit [Genius], ut tam securus, quam in aqua globus hic esses, Natura tale [vorax Elementum ignis] nullum caelis affudit. neque hodie apud vos faciunt Opticorum doctissimi: quorum fidem aestimare, ut in re quae propitim ad ipsos, deberetis. Planetas, inquiunt, etiam tranquillos illos, & quietae lucis, coruscos e terra spectabitis, si per aetherium aliquem ignem dejiciunt radios." On

tional conception of the air, too, needs to be reconsidered: it should not be thought of as one of the elements, endowed with the primary qualities of heat and moisture. Rather, the air should be regarded as devoid of any qualities of its own, but providing a meeting ground for the qualities of, on the one hand, the inferior terraqueous globe, and, on the other, the superior heavenly region. The traditional primary qualities are sufficiently encompassed in the elements of earth and water, that are cold, moist, and dry, and in the heavens, that provide all the heat to be found on earth.[265]

Leaving the so-called sphere of fire behind, Fromondus and his Genius then reach the truly celestial regions. Our author expresses his amazement when he is able to see for himself that the heavens are by no means filled with any of the traditional eccentric and epicyclic spheres of Ptolemaic astronomy, but rather with a continuous, fluid, ethereal substance. He explains to the Genius, that, in the science of astronomy, solid spheres have always been deemed absolutely necessary to save the planets' appearances; however, after hearing out this whole complicated exposé, Fromondus' heavenly mentor expresses his impatience at such nonsense: "If I were capable of sleeping, you would make me doze off with your old wives' tales. Nature is incapable of fabricating so many perplexities."[266] Optical arguments, as well as the appearances of superlunary comets, have demonstrated the absence of the solid spheres; the heavenly bodies do not need solid orbs to keep them in motion any more than do birds or insects.[267] Nor is there any need for angelic movers; the starry fires possess an inborn virtue that enables them to follow their proper paths. The Genius compares this virtue to the power that sets elementary bodies in motion on perceiving the presence of a vacuum.[268]

On his celestial journey, Fromondus' greatest wonder is directed

---

Fromondus' *Peregrinatio Caelestis*, see Monchamp, *Galilée et la Belgique*, 35-44.

[265] Fromondus, *Peregrinatio Caelestis*, 70-72: "Gravia duo Elementa Terram & Aquam in centrum, & habitationem animalium abiecisse. Unde corporibus mixtis plaeraque omnis etiam materies: frigus quoque multum, humor & siccitas, quae Elementa illa habent a se. Principium vero caloris . . . & fontem in caelo posui, quem largiter astra fundunt, aut iuvat etiam vicemque subit terrenus ac ministratorius vester ignis... Ratio quoque adsidet, instatque aerem veluti quoddam aequor, caelum Terramque interfusum esse, in quo qualitates primae superne inferne repugnarent."

[266] Ibidem, 75: "Genius ibi velut expergitus. Somnium, inquit, mihi, si dormire possim, concinnes istis fabellis. Tot tricas tricari Natura nescit."

[267] Ibidem, 75-77.

at the novel appearances of the planets, which, having seen them only through the telescope, he is now able to scrutinize at close range with his naked eyes. On reaching the moon, our author triumphantly tells the Genius that, when on earth he had heard of the recent telescopic discoveries, he had within a few hours been able to come up with a theory reconciling the moon's pockmarked appearance with traditional cosmology. The Genius dismisses this theory, which consisted of postulating a layer of perfectly smooth, translucent and therefore invisible material around the rough lunar surface, without, however, underpinning his disapproval with any sound reasons.[269]

Travelling on through the fluid aether, Fromondus and his Genius pass many small planets, unknown to earthlings and so tiny that even a telescope cannot render them visible.[270] They reach Venus, whose phases Fromondus is able to discern, on which evidence he reaches the conclusion that the planet must revolve around the sun. Our author tries to observe the sun, too, but is only able to do so with screwed-up eyes; he therefore attributes the spots he can discern on the solar surface to the blinding of his eyesight by the sun's excessive light, rather than to any small planets circling it (thus dismissing the theory that had been proposed by Scheiner, without, however, naming the Ingolstadt Jesuit).[271] Fromondus greets the four satellites of Jupiter with great enthusiasm; and, on reaching Saturn, who is "no longer Saturn, but the tricorporal Geryon", he quotes a passage in Italian from Galileo's letter to Giuliano de' Medici, in which the "Paduan Argus" had expounded his observations of Saturn's strange appearances.[272] On the basis of the constant image as yet displayed by the planet and its two "handles", Fromondus speculates that pos-

---

[268] Ibidem, 88: "Volantes, inquam ego, cum sideribus istis e gente vestra aurigas suspectaveram, solitudinem iam miror: & merito Errat audax & effraene sine Rectoribus agmen. Peream, reponit, si Phaëton hic ullus, aut caeleste veterinum fuit. Labyrinthos suos norunt. . . astrici ignes, & ingenua suaque virtute Elementorum circumvolvuntur Orbem. Ut enim vis naturalis omnifarii motus efficax Elementariis corporibus, probabili Philosophantium mente, adsita ad replendum vacuum, ubi dehiscere coepit: ita talis aliqua est, quae caelestia publico Universi bono circumagit. . .".

[269] Ibidem, 80-81.

[270] Ibidem, 81: "Abhinc mille gyris inter agmina Planetarum, qui parvitate nobis ignari aetherium illud aequor frequentant, cautabundi ad sideream sphaeram evasimus."

[271] Ibidem, 85: "Solem nisi collecta & conniventi palpebra inspicere non valui. & suberrantes maculas palante aliquotfariam nigra per aureum sideris corpus eclipsi, oculorum ab acri scintillatione magis caligines inverte divinabam, quam obscurum submeantemque celatim stellarum gregem."

[272] Ibidem, 86-87; see Monchamp, *Galilée et la Belgique*, 42 n. 3.

sibly, like the moon, Saturn is a giant globe whose various parts possess different densities.[273]

On several occasions in the course of his celestial peregrination, Fromondus expresses a highly favourable attitude towards a realistically interpreted heliocentric conception of the universe. Thus, on seeing at close range the lunar mountains, our author exclaims that "I may die if I did not almost join the ranks of the Copernicans, when I saw very high mountains on the moon". He even ventures to suggest that the moon may be inhabited, "just like the other planets, one of which we live on (according to the Copernicans)."[274] There might even be seas on the moon, Fromondus adds; after all, the Copernicans hold that any lunar waters would tend towards the centre of the moon, and would therefore not be in danger of dropping off its surface and onto the earth.[275] Later, when discussing the Jovian satellites, our author repeats the Copernican view that each planet constitutes a little world and has its own inhabitants, and that thus the Jovian moons might serve the inhabitants of Jupiter in the same way that our moon serves us.[276] Finally, Fromondus even suggests that the Sun, which, according to the Copernicans, is placed in the centre of the universe, is especially well suited to contain within its fiery bowels Hell and demons; after all, the centre of the universe is a most appropriate location for the eternally damned, since it is as far removed as possible from the Empyrean heaven, the abode of the Blessed.[277]

Although Fromondus nowhere in his Celestial Peregrination explicitly counted himself among "the Copernicans" whose views he so favourably rendered, it seems fair to state that, at the very least, as a

---

273 Fromondus, *Peregrinatio Caelestis*, 87: "Ex constanti vero illa & perenni trium astrorum imagine, Saturnum ingentem esse globum multimodae, ceu Luna, opacitatis affirmatissime (nisi labente postea tamen aevo posituram migrent) decerno."

274 Ibidem, ". . . & dispeream, nisi paene eram ut Copernicanis adessem, ubi altissimos, qua maculosa est & illunis, montes vidi; & incolam inibi aliquem suspectarem, ut in Planetarum reliquis, quorum etiam unum (Terram) [Marginal gloss: "Copernicanorum mente"] nos inhabitaremus."

275 Ibidem, 84.

276 Ibidem, 85-86: ". . .& nescio. . . cui bono aut fini siderum ista lascivia, nisi Iovialis aliquis ibi Mundus erit; & regalem Planetam caelestis creatura habet ignorata mihi."

277 Ibidem, 86: "Daemones volo, cum furvo toto Infernorum grege, centrali eius cavitati instruant, ubi [Marginal gloss: "Nam ibi ipsis Centrum Universi: in cuius ora & abside Empyreum, Beatorum sedes.] longissime, uti debet, a Beatorum gaudiis, & candenti hoc Anaxagorae ferro clusi, acrius quam in Perilli bove torreantur."

twenty-eight-year-old in 1615 he was not at all averse to the idea of realist heliocentrism. Even taking into account the fact that the work was written on the occasion of a none too serious feast, and that, whenever the text mentioned the planetary nature of the earth or the central position of the sun, printed marginal notes were carefully added to the effect that it was just "the Copernicans" who adhered to these opinions, Fromondus' phrasing leaves no room for doubt that, to him, the Copernican view seemed plausible. However, significantly, our author did not ascribe any of the pro-Copernican utterances to the Genius; Fromondus' omniscient mentor prudently kept silent on the issue of heliocentrism. This procedure suggests that, to the Fromondus of 1615, the absence of solid spheres, the non-existence of a sphere of fire, and all the other issues on which the Genius pronounced his indisputable verdict, had been demonstrated beyond doubt; the Copernican question, however, being still undecided, could not be broached by an all-perceiving being such as the Genius, and utterances in favour of heliocentrism were therefore systematically ascribed to the ignorant figure of Fromondus.

Despite his early sympathies for the heliocentric system, in later life Fromondus gained renown as an assiduous opponent of realist Copernicanism. Between 1629 and 1634, he engaged in a fierce polemic with Philip van Lansbergen, one of the first Dutch Copernicans, and later with his son Jacob van Lansbergen.[278] In his two contributions to this debate, the *Ant-Aristarchus* of 1631 and the *Vesta* of 1634, Fromondus maintained that a heliocentric universe, realistically interpreted, was absolutely untenable. He was prepared to concede to the Copernicans the physical resemblance between the earth and the planets; nonetheless, to him, this concession did not entail the necessity of removing the earth from the centre of the universe. He agreed that the planets circled the sun; only, the earth was not one of them: it remained fixed at the centre of the universe, as in the

---

[278] In 1629, Philip van Lansbergen published his *Bedenckingen op den Dagelyckschen, ende Jaerlyckschen loop van den Aerdt-Cloot, Mitsgaeders op de ware af-beeldinghe des sienlijcken Hemels* (*Thoughts on the Diurnal and Annual Course of the Terrestrial Globe, and on the True Depiction of the Visible Sky*), in which he defended the heliocentric universe; Fromondus' dismissive response appeared in 1631 in his *Ant-Aristarchus sive Orbis terrae immobilis Liber I*; Philip's son Jacob defended his deceased father in an *Apologia pro Commentationibus Philippi Lansbergii in Motum Terrae Diurnum et Annuum, Adversus Libertum Fromondum Theologum Lovaniensem et Joan. Baptista Morinum* (1633), to which Fromondus replied in his *Vesta, sive Ant-Aristarchi Vindex* (1634).

Tychonic system.[279] Fromondus expounded the traditional arguments against the motion of the earth with such persuasive force, that, after having read the *Ant-Aristarchus,* Galileo wrote to Diodati and Gassendi that "among all Copernicus' adversaries that have hitherto come to my attention, Fromondus seems to me the most sensible and most capable".[280]

It is not difficult to discover the reason of Fromondus' change of attitude on the issue of Copernicanism. After all, in 1615, no ecclesiastical censure had yet been passed on the subject of heliocentrism. The year 1616, however, had seen the pronouncement of a papal decree forbidding Catholics to hold or defend the doctrine of Copernicus; moreover, in 1633, Galileo himself had been forced to abjure his Copernican convictions and the Copernican system had been declared false. Fromondus was one of the first to make known the decree of 1633 to the scholars of the Southern Netherlands, including the entire text of the letter, by means of which the apostolic vicar in Brussels had communicated the inquisitorial decision to Louvain university, in the preface of his *Vesta.*[281] In the same preface, our author indignantly recounted how it was rumoured that he, Fromondus, had once been a Copernican, but, alarmed by the inquisitorial decree, had fled back into the Ptolemaic and Aristotelian camp. Fromondus vehemently denied this misrepresentation of reality: in fact, he had always admired Aristotle, and had at most on occasion praised the ingeniousness of the Copernican system in front of his students.[282]

On the basis of the evidence presented by the *Peregrinatio Caelestis*

---

[279] For a detailed discussion of Fromondus' anti-Copernican arguments, see Monchamp, *Galilée et la Belgique,* 82-93.

[280] Monchamp, *Galilée et la Belgique,* 95.

[281] Libertus Fromondus, *Vesta, sive Ant-Aristarchi Vindex, Adversus Iac. Lansbergium Philippi F. Medicum Middelburgensem. In quo Decretum S. Congregationis S.R.E. Cardinalium anno MDCXVI & alterum anno MDCXXXIII adversus Copernicanos terrae motores editum, iterum defenditur* (Antwerp 1634), "Ad lectorem".

[282] Ibidem: "Fama etiam ibi (sic audio) apud quosdam temere vulgavit, me olim etiam in Copernici sensu & Philosophia fuisse, sed Decreto Sacrae Congregationis Cardinalium exterritum, in Aristotelis & Ptolemaei castra refugisse. Quam errant! Nimis enim magna semper admiratione in Aristotele defixus fui, quam ut tralalitiis & ridiculis argumentis, levitateque desultoria potuerim unquam alio transfugisse. Sed hoc, opinor, volunt dicere: Me inter discipulorum scholas, ingenium Copernici & terrae motus subtilitatem (sed ut veteres quidam malas feminas aut febrim quartanam) aliquando laudavisse, & argumenta Aristotelis ac Ptolemaei, ad exercitationem ingenii, conatum in speciem solum & dicis ergo dissolvere."

and the *Vesta,* Fromondus might seem to have been guilty of a certain amount of hypocrisy. After all, in 1615 he had but thinly disguised his sympathy for the realist heliocentric system, while in 1634, he denied having ever considered it as anything more than an ingenious fiction. Moreover, his hasty reconsideration of his world view as a result of the Roman decrees may seem all the more surprising in the face of his unwavering attitude on the issue of Jansenism.

While a student at the Louvain faculty of arts, Fromondus had befriended Cornelius Jansenius, his elder by a few years; when Fromondus arrived at Louvain in 1604, aged seventeen, Jansenius had just finished his courses in the arts faculty and embarked on the study of theology. After a ten-year intermezzo, during which time Jansenius arduously studied the works of Saint Augustine in France, while Fromondus taught philosophy at the Premonstratensian college in Antwerp, the two friends met once again in 1617, when Jansenius returned to Louvain to occupy one of the chairs of theology there. Fromondus became one of his first students, while at the same time himself teaching rhetoric and philosophy in the arts faculty. For a time, the two friends even shared a house, ambitiously named the 'New Sorbonne' by Jansenius.[283]

In 1630, Fromondus succeeded his friend as regius professor of Divinity, Jansenius leaving for Ypres, to which diocese he had been elected bishop. While, during the decade that followed, Fromondus devoted a share of his time to combating cosmological heterodoxy in his polemic with the Lansbergens, Jansenius spent most of his giving birth to one of the most controversial and allegedly heretical books ever written, his monumental *Augustinus.* Applying the method of positive, as opposed to speculative, theology, he scrupulously expounded what, in his view, could be regarded as the true doctrine of the historical bishop of Hippo concerning divine grace and human free will. He hoped and trusted that his book would make a decisive contribution towards solving, in a historically accurate manner, what had long been the hottest issue in Counter-Reformation theology: the question of grace.[284] Thus, he regarded the *Augustinus* not so much as a vehicle for his own personal views on the subject of grace,

---

283 Ceyssens, "Janséniste Libert Froidmont", 3-9.

284 On the papal concern over this question, and the prehistory of Jansenius' involvement with Saint Augustine, see L. Ceyssens, "Diepere gronden van het Jansenisme", *Tijdschrift voor theologie* 6 (1966) 395-420.

but rather as an exposition of the pure doctrine of Saint Augustine, stripped of any subsequent accretions.

Jansenius did not live to see his magnum opus published. When he died in 1638, the book was completed in manuscript form, ready to be printed, and paper and even an entire printing press had been purchased. On his death bed, the bishop had as a final wish asked Fromondus and Henri Calenus, another friend of his student days, to see the *Augustinus* through the press. Both men, but especially Fromondus, being the better versed in theology, scrupulously acquitted themselves of their task, with Fromondus correcting all the proofs. Printing started, in secret, with the Louvain printer Jacques Zegers; however, several months before the book appeared, the Jesuits got wind of the affair. Being staunch opponents of the Augustinian current in theology, they informed the papal internuntius and demanded that printing be discontinued, calling upon the papal decree of 1625 by which it had been forbidden to discuss the question of grace.[285]

Fromondus mounted a firm resistance in the face of his Jesuit opponents, who enjoyed the support of the Holy See.[286] He maintained that the decree of 1625 had never been promulgated in the Netherlands, and that he did not, therefore, consider himself bound by it. Even after the papal condemnation, in 1641, of both the *Augustinus* itself and all the writings that had appeared attacking or defending it, which included all of the apologies Fromondus himself had written for Jansenius, he remained an ardent defender of the irreproachability of the contents of his friend's book. All in all, Fromondus' name came to be mentioned eight times on the Index of Prohibited Books; several times, attempts were made to remove him from his Louvain chair of theology; and his candidature to the bishopric of Tournai was successfully suppressed.[287] Yet, in spite of these hardships, and in spite of the papal condemnation of Jansenism on two occasions during his lifetime (in the 1643 bull *In eminenti* and the 1653 bull *Cum occasione*) Fromondus never wavered from his standpoint, that the *Augustinus* did not contain anything offensive, let alone heretical.

---

[285] Ceyssens, "Janséniste Libert Froidmont", 19-28.

[286] For the power politics of the Jesuits in Rome, see Ceyssens, "Diepere gronden", 408-416.

[287] Ceyssens, "Janséniste Libert Froidmont", 35-36.

Why did Fromondus so stubbornly disobey the papal decisions on the issue of Jansenism, while, a decade earlier, he had so hastily and obediently reconsidered his views on Copernicanism in the face of, if anything, doctrinally less binding bulls condemning Galileo without actually declaring heliocentrism heretical? The solution to this question is probably to be found in Fromondus' interpretation of papal authority, which came to the fore quite clearly in his polemic with the calvinist Martin Schoock. This debate, on the question of which church was the true Church of Christ, had originally been conducted by Jansenius on the Catholic side and Gisbertus Voetius on the Protestant; during the years 1636-1645, however, the places of the two initial protagonists had been taken by their most promising pupils, Fromondus and Schoock.

In the course of the polemic, Fromondus developed the view that the whole essence of the Christian faith was contained in Holy Writ. Neither the Popes, nor the Council of Nicea had created any new articles of faith; they had only served to make explicit, by means of dogmatic definitions, what had previously been contained implicitly in Sacred Scripture. Only after such an official 'explicitation' of tenets that had hitherto lain dormant in the Bible, did the denial of the dogmas contained in a papal decree explicitizing them amount to heresy.[288] Thus, to Fromondus, the issue of Copernicanism had been decided with the promulgation of the decrees of 1616 and 1633, by means of which papal authority had rendered explicit, and therefore binding, the biblical doctrine of geocentrism. Whereas, before the decrees, there had been room for doubt, with several biblical passages seemingly favouring heliocentrism rather than geocentrism, after their publication the issue had been decided. Not so, however, in the case of Jansenism; to Fromondus, as to Jansenius, the *Augustinus* contained the historically accurate doctrine of a Father of the Church, and therefore deserved more loyalty than the papal bulls condemning it.[289]

Despite his alleged heterodoxy with regard to Jansenism, then, taken all in all, Fromondus was quite a conservative figure, at least during the latter part of his life. During the 1620s and 1630s, he

---

[288] Achiel Roets, "Libertus Fromondus (1587-1653) en het kerkelijk gezag", *Sacris erudiri* 2 (1949) 335-366: 352-355.

[289] See especially L. Ceyssens, "Le drame de conscience augustinien des premiers jansénistes" in: *Augustinus magister. Communications du congrès international augustinien* (Paris 1954) Part II, 1069-1076: 1073.

vehemently attacked realist heliocentrism with the aid of an arsenal of arguments derived from Peripatetic physics; and, when in 1637 Descartes asked him for his opinion on the newly written *Discours de la méthode,* Fromondus made no secret of his disapproval, chiding the Frenchman for relying on the reprehensible philosophy of Epicurus.[290] Fromondus thus formed one of the very few exceptions to the rule that, in the majority of cases, Jansenists were favourably disposed towards Cartesianism.

While Fromondus, the philosopher and theologian, thus showed himself generally cautious when it came to the abandonment of Peripatetic tenets, his colleague Erycius Puteanus, professor of Latin at the Collegium Trilingue, displayed more enthusiasm for non-Aristotelian schools of thought. The humanistic education he had received, above all, from the great Justus Lipsius, had provided Puteanus with an intellectual background of a markedly Stoic orientation. As mentioned above in chapter IV, Lipsius had been the first to systematically expound the philosophy of Stoicism, editing the works of Seneca and supplementing these with two Neostoic handbooks, the *Manuductio ad Stoicam philosophiam* and the *Physiologia Stoicorum*, in which the great humanist demonstrated the compatibility of ancient Stoicism with the Christian faith.[291]

However, despite the fact that Lipsius, following the ancient Stoics, had emphasized the necessity of familiarizing oneself with Stoic cosmology as a prerequisite for leading an ethically sound life and attaining the much-praised Stoic wisdom, his pupil Puteanus did not display much proclivity towards natural philosophy. During the early part of his career, comprising roughly the period 1597-1615, his main interest lay with rhetoric and eloquence. After having studied law and history at Louvain for two years, Puteanus left for Italy in 1597, with letters of recommendation from his former teacher, Lipsius, by means of which he was easily able to procure a series of respectable teaching positions at humanist colleges in Milan. In 1603, he was even appointed royal historiographer of the Milanais, in which quality he produced several historical works dealing with the glorious past of the

---

[290] Ceyssens, "Janséniste Libert Froidmont", 16.

[291] On Lipsius' Stoicism, especially with regard to natural philosophy, see Jason Lewis Saunders, *Justus Lipsius. The Philosophy of Renaissance Stoicism* (New York 1955).

region. In Italy, Puteanus also became involved in the stylistic debates opposing Lipsianists to Ciceronians.[292]

When, in 1606, Puteanus returned to Louvain to succeed the deceased Lipsius as professor of Latin at the Collegium Trilingue, his interests initially remained mostly confined to rhetoric and historiography. The philosophy underlying his works was practical and Stoically-inspired: in order to acquire happiness, man had to follow the guidance of his reason and live in accordance with nature. Monchamp believed Puteanus to have attempted to revive Epicureanism at a time when Gassendi, whose name was later to be firmly attached to the rehabilitation of ancient atomism, was only seventeen[293]; however, Simar has shown that our author was in fact merely interested in Epicurean moral philosophy, ignoring Epicurus' physiological and atomistic theories.[294] Nonetheless, Puteanus did count Arnold Geulincx, later to become one of the foremost Flemish Cartesians, among his pupils.

On the political level, Puteanus emphasized the importance of peace and quiet, and the absence of differences of opinion within society. These ideas at one point got him into considerable trouble: when, in 1633, he published his *Statera belli ac pacis,* in which he advised his government to negotiate a definitive peace with the Dutch rebels and to renounce all Spanish claims to sovereignty over the northern provinces, he was questioned by the Privy Council and forbidden to leave Brussels. Only with the aid of his many influential friends did he manage to clear his name.[295]

Only during the later stages of his career did Puteanus begin to display some attention for the natural sciences. He became interested in the science of chronology, on which subject he published several works; and he befriended the astronomers Godfridus Wendelinus, who was a Copernican, and Florent Van Langren. With the latter, he conducted an extensive correspondence in the Dutch language.[296]

---

[292] Th. Simar, *Étude sur Erycius Puteanus (1574-1646) considéré spécialement dans l'histoire de la philologie belge et dans son enseignement à l'université de Louvain* (Louvain/Paris/Brussels 1909) 1-6.

[293] Monchamp, *Galilée et la Belgique,* 67.

[294] Simar, *Puteanus,* 41-42.

[295] Ibidem, 25-26.

[296] Published with an introduction by J.J. Moreau, *Honderd veertien Nederlandse brieven van Erycius Puteanus aan de astronoom Michael Florent Van Langren* (Antwerp/Amsterdam 1957).

Although Van Langren was a very capable astronomer, who competed with Galileo over the invention of a method for determining longitude at sea, and who produced the first true map of the moon[297], this collection of letters contains very little discussion of the issues current in early seventeenth-century astronomy. Characteristically, Puteanus' letters to Van Langren dealt at length with questions of patronage, and with the nomenclature to be used in selenography, with the vain humanist displaying a childlike enthusiasm when his correspondent decided to honour one of the highest mountain-tops on the moon with the name of 'Puteanus'. Thus, the contribution made by Lipsius' successor to natural philosophy remained rather meagre; in fact, the only one of his works on scientific subjects to have found favour in the eyes of historians of science was his work on the comet of 1618.[298]

The appearance of the last and brightest of the three comets of 1618 did not pass by unnoticed to the scholars of the Southern Netherlands. We have already seen how the Antwerp Jesuits, equipped with a number of good telescopes, carefully observed the phenomenon, and how, after its disappearance, their star mathematician Gregorius a Sancto Vincentio discussed the hairy star in front of a packed hall. Meanwhile, in Louvain, Erycius Puteanus with characteristic enthusiasm subjected the comet to nightly scrutiny, observing it from the ramparts of the somewhat dilapidated Imperial Castle of which he was the proud tenant. Nor did the spectacular arrival fail to elicit the interest of thirty-one-year-old Libertus Fromondus, professor of philosophy at Louvain. He had from the outset been very interested in telescopic observation, and the novelties discovered by Galileo had occupied his thoughts to such an extent that he had devoted a large part of his contribution to the 1615 *Saturnalia* to a discussion of the previously unthought-of characteristics of the heavens. While the comet was still visible, on 10 January 1619, Fromondus wrote to his older colleague, Thomas Fienus, who at the time occupied the chair of medicine that had once been held by Cornelius Gemma, to ask him for his opinion on the subject of the tailed star. Fienus promptly responded to his young colleague's request, and just a week later, on

---

[297] See Ewen A. Whitaker, "Selenography in the seventeenth century" in Taton & Wilson eds., *Planetary astronomy*, A, 119-143: 127-129.

[298] Simar, *Puteanus*, 221; Monchamp, *Galilée et la Belgique*, 67.

18 January, he sent Fromondus a dissertation on the comet. The philosopher immediately sat down to compose a reply to Fienus' arguments, and subsequently had both treatises printed at Antwerp.[299]

It is to the three contributions to cometology of Fienus, Fromondus, and Puteanus, that we must now turn. As, in the present chapter, our concern is primarily with the cosmological side of the cometary question, we may once again ask ourselves, as we did when discussing the Dutch reactions to the comet of 1577: to what extent had the erosion of the Aristotelian world view progressed in the Netherlands of 1618? How did the three basic Peripatetic tenets of geocentricity, dichotomy between celestial and terrestrial regions, and uniformity and circularity of celestial motions fare in an age that had seen the publication of the Tychonic world system, the new physical astronomy of Kepler, and the telescopic discoveries of Galileo? What position did our authors occupy vis-à-vis Copernicanism and Stoic cosmology? In short, what location and what nature did they assign to the comet, and what consequences did this entail for their world views?

b. *Thomas Fienus: "Esse corpus caeleste, atque esse in caelo"*

Both with regard to its form and with regard to its tone, the tract Thomas Fienus devoted to the comet of 1618 greatly differred from that written on the occasion of the tailed star of 1577 by his predecessor in the Louvain medical faculty, Cornelius Gemma. Whereas Gemma had been primarily interested in discovering the divine message that was contained in the miraculous phenomenon, and in the moral and ethical exhortations it was undoubtedly meant to direct at the warring peoples of Europe, his successor Fienus concentrated rather on the question of the comet's location, without attaching any moral connotations to the outcome of his investigation. To him, the main question to be answered was, whether the comet was a heavenly body and some kind of star, or, conversely, a meteor or burning exhalation; or, which amounted to the same thing, whether it was

---

[299] *De Cometa anni 1618. dissertationes Thomae Fieni in Academia Lovaniensi Medicinae et Liberti Fromondi Philosophiae Professorum. In quibus tum istius motus, tum aliorum omnium essentia, effectus, & praesagiendi facultas declarantur* (Antwerp 1619). On the three Louvain tracts on the comets of 1618, see Monchamp, *Galilée et la Belgique*, ch. IV and V.

located in the heavens or in the air.[300] By far the greater part of his tract was devoted to answering this question.

In order to do so, Fienus employed an unmistakably Aristotelian method. With regard to its form, his tract closely resembled the many seventeenth-century treatises that were still based on the scholastic way of dealing with scientific problems: the *quaestiones*. Although, unlike his medieval predecessors' contributions to cometology, Fienus' tract did not constitute a commentary on Aristotle's *Meteorology*, question by question elucidating the Philosopher's text, nonetheless it breathed an overwhelmingly Aristotelian atmosphere. After a concise passage in which he set forth his own observations of the comet of 1618, our author went on to expound eight arguments in favour of the Aristotelian theory of comets as burning terrestrial vapours. Next, he reminded his readers of the fact that many authoritative scholars of antiquity had, on the other hand, believed comets to be celestial bodies, and went on to put forward eleven arguments intended to demonstrate the non-viability of the idea of sublunary comets, as well as four further grounds for believing comets to be heavenly bodies. Finally, he answered the eight Aristotelian arguments set forth at the beginning of his tract by means of eight counter-arguments, showing that the Peripatetic theory of comets was untenable. Thus, like his medieval predecessors, Fienus commenced his reasoning by expounding all possible arguments against the conclusion he wished eventually to defend, after which he systematically torpedoed every one of them.

Another symptom of Fienus' traditional bent of mind was the completely qualitative way in which he dealt with the description of the comet's course through the heavens. His observations were even less accurate than Gemma's, and his parallax determinations even cruder. Our author himself was the first to admit that his investigation of the comet's path was not as precise as it might have been: "When I said farewell to the comet, I promised that, on its, or one of its congeners', next return, I would observe its motion more accurately. For, this time, I could not precisely observe its altitude above the horizon, its exact distance from the other stars, or the correct

---

[300] Thomas Fienus, *De cometa anni 1618 dissertatio* (Antwerp 1619) 17: "Mihi omnium opiniones examinare propositum non est: sed hoc solum, An Cometa sit corpus caeleste, & sidus aliquod; an vero Meteorum, & exhalatio ignita; & quod in idem recedit; an sit in caelo, an in aere."

proportion of the irregularity of its movement, because I did not have at my disposal an astronomical radius, an instrument principally used for determining the dimensions of celestial bodies."[301]

Despite his lack of reliable instruments, however, Fienus did go to great pains to ascertain, as accurately as possible, the comet's trajectory with regard to the constellations. He observed it as often as he could, usually three times a night, at six o'clock in the evening, midnight, and six o'clock in the morning.[302] He did not remember the exact day on which he first saw the comet, somewhere in the beginning of December. Certain farmers maintained that they had seen it as early as September or October; however, Fienus invalidated this statement by pointing out that, if the farmers had been right, the comet would at the beginning of October have stood in the fourteenth or fifteenth degree of Scorpio. Yet, at the time, the Sun had stood in the sign of Scorpio, so that its rays would undoubtedly have rendered the nearby comet invisible.[303]

When Fienus first saw the comet, it stood below the bright star Arcturus in the constellation of Bootes. Because, from there, it had moved north in a straight line, our author conjectured by extrapolation from the rest of its path that it had arrived there via Scorpio and the right knee of Bootes. From Arcturus, it moved northward, through the signs of Bootes and Ursa Maior, and Draco's tail, until it arrived near the Pole Star, where it disappeared.[304] After this summary rendition of the comet's path, Fienus went on to describe in detail its position on the nights of 11, 14, 17, 22 and 24 December and 11 and 18 January, for each of these days stating its location with regard to the nearest of the fixed stars, its altitude on the meridian, and its distance from the zenith.[305] From the fact that, to de-

[301] Ibidem, 14: "Quapropter ei valedixi: promittens, quod proximo, vel ipsius, vel confratrum suorum, reditu omnem eius motum exactius essem observaturus. Nunc enim, quia radio Astronomico carui, cuius praecipuus in caelestium syderum dimensionibus est usus: neque eius altitudinem supra Horizontem singulis inspectionis horis: neque exactam a reliquis stellis distantiam: neque certum modum irregularitatis motus eius, ita praecise potui observare."

[302] Ibidem, 9: "Itaque aspexi & considerate frequenter, & fere toties, quoties per aurae serenitatem licuit: plerumque ter una nocte, hora sexta vespertina, duodecima noctis, & sexta matutina."

[303] Ibidem, 10. The phenomenon seen by the farmers was probably the first of the three comets of 1618.

[304] Fienus, *Dissertatio*, 11.

[305] Ibidem, 11-14.

scribe the comet's location to nearby stars, Fienus invariably used the phrase "on a straight line between" (as, for example, on 22 December, when it stood "on a straight line between the last star in the tail of Ursa Maior and the inferior of the two stars in the left hand of Bootes"[306]) one may conclude that, presumably, lacking a Jacob's staff, he used a piece of thread to determine the comet's position.

The comet's tail had been of a fearsome size: according to Fienus' estimate, when the comet was near Arcturus, the tail occupied a space of forty to fifty degrees. However, the further the comet moved towards the Pole Star and away from the sun, the more the tail diminished, just as the head grew less luminous. The tail's brightness also depended to some degree on the clearness of the sky. It had been impossible for Fienus to determine the ratio with which the comet's light grew dimmer, because of the many cloudy nights that had occurred; on occasion, he had been unable to observe the phenomenon for as much as a fortnight due to bad weather. Still, its movement was undoubtedly irregular: at first, it had covered 2 1/2 to 3 degrees a day, which number had grown progressively less as the comet diminished in size, so that, just before its disappearance, its daily advancement amounted to scarcely one degree. During the period of its visibility, the comet covered a total distance of at least seventy degrees; its motion had been in a straight line from which it had not once deflected, from the tropical circle towards the pole; from these data it could be deduced that it had described part of a great circle, dividing the sphere of the heavens into two equal halves.[307]

After this description of the comet's trajectory and its external appearance, Fienus embarked on the investigation of the question posed to him by Fromondus: what was the phenomenon's nature and essence, and where was it located? In accordance with a time-honoured tradition, he first of all set forth the many different opinions that had in the course of time been advanced on the subject. On the basis of pseudo-Plutarch's *De placitis philosophorum,* Aristotle's *Meteorologica* and Seneca's *Naturales Quaestiones,* our author distinguished three different positions among the ancients: the Pythagoreans, Hippocrates of Chios, Aeschylus, and Apollonius Myndius

---

[306] Ibidem, 12: "... in recta linea inter ultimam stellam caudae Ursae Maioris, & proximam ac infimam stellarum illarum, quae sunt in sinistra manu Bootis...".

[307] Ibidem, 15.

thought a comet to be a celestial body; Anaxagoras and Democritus believed it to be an optical illusion arising from a conjunction of planets; and Aristotle and most Peripatetics regarded it as a hot, smoky, fatty exhalation burning in the upper air. Furthermore, Fienus mentioned two more recent cometary theories: that of Cardano, who thought of comets as globes formed out of celestial matter, that were at times illuminated by the sun and thus rendered visible, and that of Scaliger, according to which a comet was an exhalation imbued with sunlight.[308]

Despite the fact that he was well aware of the existence of all these theories, Fienus announced that he did not intend to examine every one of them. To him, there was only one question that really mattered: to determine whether the comet was a star in the heavens or a meteor in the elemental air below the moon. For, if it were a celestial body, it would necessarily be located in the heavens, while, if it were a meteor, it had to be below the moon, since it was unthinkable for heavenly bodies to descend below the moon and attain the airy region, or, conversely, for an elemental exhalation to rise up into the celestial regions.[309] From this remark it becomes clear that Fienus did not doubt the existence of the insurmountable Aristotelian barrier constituted by the moon's sphere, separating terrestrial and celestial regions.

In an equally Aristotelian vein, our author commenced his investigation of the comet's location by enumerating all the authorities in favour of the Peripatetic cometary theory, according to which comets were burning exhalations. In addition to Aristotle, this opinion had been held by Xenophanes, Epigenes, Alexander[310], Albertus Magnus, Averroes, Saint Thomas Aquinas, Agostino Nifo, and many others. Fienus cautiously added that, among the moderns, Regiomontanus had *tried* to prove that the comet of 1475 stood in the air below

---

[308] Ibidem, 16.

[309] Ibidem, 17: "Mihi omnium opiniones examinare propositum non est: sed hoc solum, An Cometa sit corpus coeleste, & sidus aliquod; an vero Meteorum, & exhalatio ignita; & quod in idem recidit; an sit in coelo, an in aere. Haec enim se mutuo sequuntur, & conversive dicuntur. Nam si est sidus, & corpus aliquod coeleste, ergo est in coelo, & e contra: si est Meteorum, & exhalatio incensa, ergo est in aere, & e contra: nemo enim dicet corpora coelestia, vel extra coelum in aerem descendere, vel exhalationes elementares in coelum adscendere."

[310] Fienus probably means Alexander of Aphrodisias, a second century A.D. commentator of Aristotle's works, who among other works produced a commentary on the *Meteorology*.

the moon, and *it was said* that Johannes Vögelin had done the same thing for the comet of 1532. Apparently, these two attempts did not seem as convincing to Fienus as they had seemed to Gemma forty years earlier; as we have seen, Gemma in 1577 did not doubt the fact that many comets, among them those of 1475 and 1532, had been proved sublunar. Despite his hesitating stance, however, Fienus went on to carefully expound eight Peripatetic arguments against the view of comets as celestial bodies, reminding his readers that, once this option had been disproved, it would necessarily follow that comets were, indeed, burning meteors, in accordance with Aristotle's dicta.[311]

The arguments Fienus adduced against the view of comets as celestial bodies, were of two kinds: some of them, based on common sense experience, were intended to show that the appearances displayed by comets were much more easily to be reconciled with burning exhalations than with heavenly bodies; while some of them, conversely, were based on the dicta of ancient authorities. To the first category belonged the observation that, contrary to stars, which always remained the same size, comets grew larger and smaller in the course of their existence. Such a variation in size could easily be explained by assuming them to consist of burning exhalations, which would obviously expand or diminish in accordance with the amount of fuel available to keep them on fire.[312] In addition, their character of fiery vapours explained why comets were oblong in form, while stars were globular. After all, terrestrial exhalations always rose up from the surface of the earth perpendicularly, so that, once they caught fire, it was natural for them to display an outstretched shape, with the flame following its fuel in a straight line. Also, stars were nontransparent, while, conversely, comets were pellucid: sense experience showed that it was possible to see the fixed stars through cometary tails.[313] Even their movement could be satisfactorily accounted for by assuming them to be exhalations: the fire with which they burned needed to follow its fuel to remain kindled, and this

---

311 Fienus, *Dissertatio*, 17-18: "Porro ipsum esse exhalationem in aere accensam hoc argumento videtur probari. Vel Cometa est talis exhalatio accensa, & Meteorum; vel est corpus coeleste: non est corpus coeleste; ergo. Maior est vera; sequela est bona. Minor videtur posse probari variis rationibus."

312 Ibidem, 19-20.

313 This argument is to be found in Seneca, *Naturales Quaestiones*, VII.18.

explained the inconstancy and irregularity of their motion, a feature completely irreconcilable with the nature of celestial bodies.[314]

The second category of arguments Fienus adduced in favour of sublunary comets was based almost entirely on the time-honoured postulates of Aristotelian cosmology. Thus, he stated that if comets were celestial bodies, either they had been in existence from the creation of the world, or they were newly generated entities. The first assumption, however, was untenable, because if they were eternal bodies, it was impossible to come up with any sound reason why they should not, then, appear every night, along with all the other stars. The reasons given by the Pythagoreans, that most of the time they were rendered invisible by the rays of the sun, and by Apollonius of Myndus, that they usually lay hidden in the depths of the universe, had been sufficiently refuted by Aristotle and Seneca. But the second possibility, that comets were newly generated at the moment of their first appearance, was equally false: after all, Aristotle had taught that the heavens were unalterable, and that no change had ever been observed to have taken place there.[315] Moreover, comets could not belong to the category of fixed stars, because they moved; but nor could they be planets, because only seven of those existed, none of which ever moved outside the bounds of the zodiac, while cometary motions were completely erratic and could take place anywhere in the skies.[316] Finally, if comets were celestial bodies, this would entail the necessity of assuming the heavens to be fluid, and devoid of any solid spheres. Such an assumption, however, ran counter to Aristotle's dicta, the generally accepted Peripatetic philosophy, and Holy Writ, which on many occasions spoke of "the heavens" in the plural.[317]

In the course of this enumeration of completely traditional argu-

---

[314] Fienus, *Dissertatio*, 20-21.

[315] Ibidem, 19: "Non generantur etiam de novo: quia in coelo non fiant generationes, aut corruptiones; neque unquam in coelo ulla percipitur mutatio. Coelum enim, ut inquit Aristoteles 1. De Coelo, capite 3. textu 20. est ingenerabile, & incorruptibile."

[316] Ibidem, 18: "Non enim est stella fixa, quia movetur; non erratica, quae Planetae appellantur: quia tales ab omni aevo notatae sunt tantum septem: Cometa non est ullus ex illis. Insuper Planetae omnes feruntur per Zodiacum, & ab Occidente in Orientem; cometa non; sed movetur in omnem situm. . .". This argument was taken directly from Aristotle; see Aristotle, *Meteorology* I.vi.343a.

[317] Fienus, *Dissertatio*, 22: "Denique sequeretur, quod etiam eodem modo Planetae per coelum moverentur; & consequenter quod nulli essent caeli; sed quod esset unum continuum liquidum corpus, a terra usque ad inania spacia, per quod sidera

ments in favour of Aristotle's cometary theory, Fienus already made it clear to the careful reader that, in the end, he himself was not going to side with the adherents of the 'burning exhalation' theory. As in his account of the parallax measurements of Regiomontanus and Vögelin, where as we have seen he used phrases such as "he tried to prove" or "it is said that" the comets of 1475 and 1532 were below the moon, again in his exposition of Aristotelian arguments our author several times hinted at their lack of persuasive force. Thus, the comets' oblong figure "*seems* to be an argument" in favour of their sublunary nature, and the absence of solid spheres "*seems* to run counter to Aristotelian doctrine, common Philosophy, and Sacred Scripture".[318] Thus, to the advertent reader, it cannot have come as a complete surprise when, after having exhaustively enumerated all arguments in favour of the burning exhalation theory, Fienus abruptly announced that, nonetheless, in his own opinion comets were by no means meteors or exhalations burning in the air, but celestial bodies shining in the heavens.[319]

Significantly, the very first tactic Fienus employed to strengthen this point of view was to spell out the names of all those past scholars who had been of the same opinion. Many great men had believed comets to be celestial bodies: the ancients Anaxagoras, Democritus, Pythagoras, Diogenes, Apollonius of Myndos, Hipparchus, Hippocrates of Chios, Aeschylus, and Seneca, the Arabs Mashallah, Haly, and Albumasar, and the moderns Tycho Brahe, Cornelius Gemma, Michael Mästlin, Girolamo Cardano, Marcellus Squarcialupus and Thaddaeus Hagecius had all adhered to some version or other of the hypothesis that placed the tailed stars beyond the moon's sphere.[320] This constituted a very impressive list of authorities, the esteem for some of whom, such as Pythagoras and Seneca, had risen to great heights during the sixteenth century; furthermore, they were eighteen in number, and thus twice as numerous as the nine adherents of the Aristotelian theory Fienus had been able to come up with.

With so many eminent scholars in its favour, the 'celestial body'

---

vagarentur. Quod doctrinae Aristotelicae, & communi Philosophiae, immo etiam sacris litteris, quae frequenter coelos in plurali numero nominant, videtur repugnare."

[318] Ibidem, 20: "Imo illa longitudo videtur esse argumentum..."; 22: "Quod... videtur repugnare."

[319] Ibidem, 23: "Hisce tamen non obstantibus, opinio mea est, Cometas esse in coelo, & esse corpora coelestia, & minime meteora aut exhalationes in aere accensas."

[320] Ibidem, 23.

theory made an auspicious entry into Fienus' tract. Furthermore, our author managed to advance no less than eleven arguments to the effect that comets could not possibly be held to be sublunary objects located in the unstable airy region. The majority of these were drawn from Cardano's *De subtilitate* and *De rerum varietate,* and from Seneca's *Naturales Quaestiones*; Fienus' own contribution, though not quite non-existent, was in any case not very significant. From Cardano, he borrowed the idea that comets could not consist of terrestrial exhalations because these could not possibly ascend to such great altitudes as those that had been observed for many of the recent tailed stars. Fienus illustrated this idea with the example of the comet of 1618 itself, which had accurately followed the motion of the *primum mobile,* moving due north without in the slightest way lagging behind the daily revolution. This meant that it was situated very close to the heavens, for objects situated below the moon at some distance from the celestial region could not possibly follow the daily movement of the heavens as effortlessly as had the present comet. But terrestrial exhalations never rose to such great heights, not even if they were attracted by one of the stars or planets. For the comet observed by Fienus had not moved towards the zodiac, within whose bounds the planets made their revolutions, but away from it, so that it could not possibly have been attracted by a planet; moreover, if it had been drawn by one of the fixed stars, it would have ceased moving on reaching that star, which had not been observed to happen.[321] Also, Fienus added, it seemed quite impossible for hot exhalations to pass through the cold middle region of the air without being transmuted into meteors such as thunder and lightning.[322]

In another argument borrowed partly from Cardano, Fienus pointed out that the comet's tail, covering more than 45 degrees of the heavens, must have been immensely outstretched. However, if this vast object were a burning exhalation, it would require such an enormous amount of fuel to keep its flame alive, that, even if the

---

321 Ibidem, 24-25. See Hieronymus Cardanus, *De Subtilitate* (Basle 1554) 117, where he states that many comets are observed at much greater altitudes than the maximum height to which terrestrial vapours can rise.

322 Fienus, *Dissertatio,* 25-26: "Accedit quod non videatur, quomodo exhalationes calidae transire possint mediam ac frigidam regionem aeris sine eo quod illic alterentur, destruantur, includantur, & in fulmina tonitruaque convertantur." See Aristotle, *Meteorology,* II.ix for the Aristotelian explanation of thunder and lightning, which according to the Stagyrite were meteors generated when dry exhalations were trapped in condensing clouds and forcibly ejected from them.

whole earth were converted into exhalations, this would not suffice to allow the comet to remain kindled for several months on end.[323] Moreover, Fienus added, such an amount of terrestrial fumes located at a great height would have to be visible to us in the form of a huge, black cloud, often eclipsing the stars and the comet's head with its opaque mass; however, this ran counter to the actual observed facts, which showed that, on the contrary, in clear weather all the stars in the vicinity of the comet were very well visible.[324]

Many of Fienus' anti-exhalation arguments were drawn from the sections of book VII of Seneca's *Naturales Quaestiones* in which the great Stoic set forth his own view of the nature of comets. Thus, Fienus stated that if a comet were an exhalation, it would not remain constant in shape, but would continually change its appearance, following its fuel in any direction where this might happen to be located.[325] Also, exhalations, being light and agile, and located in the naturally liquid and unstable element of air, tended to move unpredictably in all directions, while the comet of 1618 constantly proceeded with a fixed motion.[326] Furthermore, fiery meteors never lasted for very long; instead, they were burnt up all at once and were quickly extinguished, as was the case with lightning, shooting stars, and other flames in the sky. Yet, comets could remain visible for as much as an entire year, as was testified by Josephus with regard to the one that shone over Jerusalem before its destruction.[327] And, if a

---

[323] Fienus, *Dissertatio*, 26: "At non videtur possibile, tantam exhalationis quantitatem, e terra in altum potuisse attolli, quae non solum vastum illud corpus constituerit, sed etiam ei tanto tempore pabulum suppeditarit: ne quidem etsi tota terra in exhalationes converta fuisset; ergo." See Cardano, *De Subtilitate*, 117.

[324] Fienus, *Dissertatio*, 26-27: "Porro tanta quantitas fumidarum, ac terrestrium, & viscosarum exhalationum loco illo supremo constituta deberet ibi esse visibilis, & deberet apparere in modum nigrae, & opacae nubis, & deberet sidera quibus subjicitur, & ipsum etiam Cometam frequenter obscurare, & offuscare, sicut a nobis videri non possent... At hoc nunquam factum fuit... Quin e contra manifestum est, quotiescunque dies est serenus... vicinas omnes, ac circumiectas stellas perfectissime apparere."

[325] Ibidem, 27-28: "Denique, si esset exhalatio accensa, non semper servaret eandam figuram; sed prout ignis modo hac, modo illac excurrit ad suum alimentum, deberet exhalatio illa interdum magis ardere ad dextram, interdum magis ad sinistram; interdum magis in hoc extremo, interdum magis in opposito. At ita non est; sed semper Cometa sese habet eodem modo, & eadem figura semper in uno extremo est globus instar stellae, & nunquam in alio." See Seneca, *Naturales Quaestiones*, VII.23.1.

[326] Fienus, *Dissertatio*, 28-29; see Seneca, *Naturales Quaestiones*, VII.22.1.

[327] Fienus, *Dissertatio*, 29: "Ignita meteora non diu durant, sed statim tota simul

comet consisted of fire, it would be observed to grow larger and smaller on alternate days, because the amount of fuel ascending from the surface of the earth to feed them would vary on different days. Yet, on the contrary, comets were observed to increase and decrease very gradually, without any sudden variations.[328]

Moreover, it was natural for anything consisting of fire to move upwards in a straight line; the proper motion of comets, however, was invariably observed to be circular. Whereas Seneca had still been somewhat wavering on this point, not venturing to generalize beyond the appearances of the two comets he had himself observed, Fienus was prepared to affirm without qualification that "A comet moves with circular motion."[329] Also, for a comet to be a burning exhalation, it would necessarily have to be observed moving downwards in the direction of the earth, since, according to Peripatetic cometology, the vapours constituting comets were first set alight on reaching the sphere of fire; therefore, their highest extremity was first burnt up, after which they would be forced to descend towards the lower regions to find more inflammable material to feed on. Thus, the comet would have to be seen to move all the way towards the earth's surface; yet, this never happened, as Seneca had pointed out. On the contrary, added Fienus, all comets were seen to be carried up on high just before disappearing.[330] To this Senecan argument, Fienus himself adjoined another anomaly of the Peripatetic theory: why, if comets consisted of burning exhalations, did they never occur in the middle regions of the air? After all, these layers of the atmos-

---

accenduntur, ut inquit Seneca: & brevissimo tempore extinguuntur. . . . At Cometae diu durant. . . .". See Seneca, *Naturales Quaestiones*, VII.23.1.

[328] Fienus, *Dissertatio*, 31-32; see Seneca, *Naturales Quaestiones*, VII.23.2.

[329] Fienus, *Dissertatio*, 30: "Ignis motu proprio movetur per lineam rectam: Cometa, motu proprio . . . movetur circulariter. . . . Affert hoc argumentum Seneca libro 7. naturalium quaestionum capite 23 & quidquid de aliis sit, duo saltem, inquit, tempore nostro sic motum suum in orbem duxerunt." See Seneca, *Naturales Quaestiones*, VII.23.1: "Etiamnunc ignis aut it quo illum natura sua ducit, id est sursum, aut eo quo trahit materia cui adhaesit et quam depascitur; nullis ignibus ordinariis et caelestibus iter flexum est. Sideris proprium est ducere orbem. Atqui hoc an cometae alii fecerunt? Nescio; duo nostra aetate fecerunt."

[330] Fienus, *Dissertatio*, 30-31: "Exhalationes debent prius accendi parte superna, quam inferna; eo quod illa parte sint viciniores coelo, aut elemento ignis, a quo secundum Peripateticos accenduntur; unde, si Cometa esset ignis, deberet semper descendere deorsum, ac versus terram; prosequendo pabulum suum quod est ab infra. . . deberet Cometa, ut inquit Seneca, aliquando ad terram usque descendere. . . . At hoc non fit; quin potius cum pereunt, omnes videntur in altum ferri; ergo." See Seneca, *Naturales Quaestiones*, VII.22.2.

phere were often full of suitable, hot and dry exhalations, as was apparent from the fact that it was the region in which other fiery meteors, such as lightning, were generated. Thus, the fact that comets never occurred in the middle region, but always just below the sphere of fire in the very highest layers of the air, was inexplicable by means of the Stagyrite's cometary theory.[331]

Fienus added several other arguments of his own that were not to be found in Seneca's refutation of the Aristotelian theory. Thus, he stated that, if a comet were an exhalation, which was by definition a volatile and easily dissipated substance, it would in an instant be torn apart by the vertiginous motion of the *primum mobile*.[332] Also, exhalations always tended to cluster together in the form of a globe, as was observed, for example, in the case of clouds, while comets displayed an oblong figure.[333] And, while fire tended to scintillate, because not all parts of the alimentary material were set alight at precisely the same moment, comets, on the other hand, were never observed to scintillate, and were therefore not fires.[334] Finally, burning exhalations, being dependent on the presence of suitable fuel to keep them alive, would never be able to move over such vast distances as comets were observed to do: after all, it was extremely unlikely that in every one of the many regions over which the comet of 1618 had passed, the appropriate kind of hot and dry exhalation had risen up at precisely the right time and to precisely the right height to keep the comet's flame going. In any case, the comet would not have been able to pass over the ocean, since the only kind of exhalation watery surfaces were able to produce was cold and moist, not hot and dry.[335]

To Fienus, this profusion of arguments sufficed to satisfactorily resolve the question posed at the beginning of the tract: namely, whether the comet was a meteor or a celestial body. After all, they

---

331 Fienus, *Dissertatio*, 31: "Cometae deberent aliquando generari prope terram in media aeris regione. Nam cum ibi frequenter plurimae exhalationes congregentur, & accendantur, ut patet in fulmine, atque aliis meteoris ignitis; sequitur quod ibi accensae exhalationes deberent tam bene Cometae figuram, ac naturam referre, quam accensae in suprema aeris regione."

332 Ibidem, 27.

333 Ibidem.

334 Ibidem, 29-30.

335 Ibidem, 32-34: "Maior probatur, quia quando ab illa regione [super quam habuerunt originem] recederent, & decurrerent super vertice aliarum regionum, longe distantium, non reperirent ibi neque praecise in illa altitudine, qua decurrunt, paratas exhalationes, a quibus possent foveri, & sustentari; proinde deberent exstingui atque interire."

conclusively demonstrated that it was not a burning exhalation, leaving open only the possibility that it was some kind of heavenly object.[336] In addition, the view of comets as celestial bodies could be proved by means of a separate set of arguments, intended to show that the appearances of the hairy stars were best explained by assuming them to be located beyond the moon. These arguments, four in number, were all based on the characteristics of the comet's motion.

First of all, Fienus stated, movement in a straight line was proper to sublunary bodies, while the nature of heavenly objects was to move circularly; comets were observed to move in circles; therefore, comets belonged to the genus of heavenly bodies.[337] Secondly, both the present comet and the comet observed in 1577 by Tycho Brahe had described sections of perfect 'great circles', passing through the centre of the universe and dividing the heavenly globe into two equal halves; this was a kind of movement that only celestial bodies were capable of describing.[338] Also, as mentioned above, the comet of 1618 had followed the daily motion of the *primum mobile* without lagging behind even the slightest bit. This meant that it had to be located in the heavens; after all, it was natural for celestial bodies to be transported along with the daily revolution, and that which is moved *per se,* out of a natural inclination, is necessarily moved with greater speed than that which is moved *per vim,* being carried along violently merely as a result of its vicinity to the mover (as was the case of comets according to the Aristotelian theory).[339] By the same argument, the comet could conclusively be demonstrated to be located beyond the moon; for even the moon itself, despite being a celestial body, showed a divergence from the motion of the *primum mobile,* lagging behind some thirteen degrees a day. The comet of 1618, on the other hand, had shown no such deviation, moving due north

[336] Ibidem, 35: "Atque his argumentis videtur evidenter probatum Cometam non esse exhalationem accensam, aut meteorum; proinde sequitur esse corpus caeleste, atque esse in caelo."

[337] Ibidem: "Primo, motus rectus est proprius corporum sublunarium; circularis caelestium; at Cometa movetur motu circulari, ut supra est ostensum; ergo."

[338] Ibidem, 35-36.

[339] Ibidem, 36: "Hic Cometa exacte assecutus est motum primi mobilis; ergo est in coelo. Sequela probatur; quia si esset in regione elementari, non potuisset tam celeriter ferri, atque ipsum coelum. Quia, quod per raptum fertur, propter viciniam, & per vim; non potest tam celeriter ferri, sicut illud quod fertur primo et per se, & ex proprietate naturali. . ."

without deflecting towards the east.[340] Finally, Fienus added, despite having observed the comet several times a night, I have not been able to detect any parallax or diversity of aspect for it with regard to the fixed stars (apart, of course, from that caused by its proper northward motion). Therefore, it was beyond doubt a heavenly phenomenon.[341]

Thus, Fienus, by means of an abundance of arguments based in large part on Seneca's cometary theory, owing something to Cardano's views on the subject of tailed stars, and supplemented with his own thoughts and observations occasioned by the comet of 1618, reached a conclusion that was quite incompatible with Peripatetic cometology. Yet, on close scrutinity, not a single one of the arguments discussed so far constituted a serious break with Aristotelian physics. On the contrary, most of them were in fact unthinkable without the presupposition of a solidly Peripatetic outlook on the ways in which nature was wont to operate. Thus, underlying many of the arguments against the view of comets as burning exhalations were the tenets of Aristotelian meteorology: terrestrial exhalations are hot and dry; they are easily dissipated; they rise up from the earth's surface through the several different layers of air; they are capable of catching fire; they constitute the basic ingredients of meteors such as thunder, lightning, and shooting stars. These were all statements to which Fienus wholeheartedly subscribed; he merely advocated the exclusion of one kind of Aristotelian meteors, comets, from the laws of Peripatetic meteorology.

The contents of his arguments also made it clear that he agreed with Aristotelian elemental theory (at least, as far as the sublunary region was concerned). Earth, the heaviest of the elements, to Fienus was at rest at the centre of the universe, as can be deduced from the fact that he on several occasions professed an unquestioning belief in the daily rotation of the *primum mobile*. Its surface was partly covered with oceans, and around it came several concentric layers of air (of which, as we have seen, Fienus mentioned the middle and the superior one); next came the sphere of fire. The moon's sphere constituted an impassable barrier between sublunary and superlunary regions, each of which obeyed its own kind of laws: heavenly bodies

[340] Ibidem, 36-37. Fienus borrowed the basic idea of this argument from Cardano, *De Subtilitate*, Book IV, 118.

[341] Fienus, *Dissertatio*, 38.

described circular motions, while sublunary objects tended to move in straight lines.

Thus, Fienus was able to argue a position on the nature of comets that was completely at odds with Aristotle's views on the subject, without having to abandon any of the basic tenets of Peripatetic cosmology. Doing so was undoubtedly facilitated by the availability of the point of view of Seneca, on whom Fienus heavily relied. As described above, most of our author's arguments discussed so far were drawn from the arsenal supplied by the eminent Stoic. This explains why Fienus managed so effortlessly to reconcile the idea of superlunary comets with the basics of the Aristotelian world view: after all, these were tenets that were common to Aristotelian and Stoic cosmology alike. In the Stoic universe, as in the Aristotelian, there existed two distinct regions, the celestial realm above and the terrestrial realm below the moon. In both world views, these regions were governed by sets of different laws, among them the rule that, to both Stoics and Aristotelians, celestial motions were invariably circular, while elementary motions could not be other than rectilinear. Finally, to Seneca as to Aristotle, the earth was at rest at the centre of the universe, while the daily movement of the stars and planets was caused by the revolution of the *primum mobile*.[342]

Yet, Fienus did not rest content at having demonstrated that comets were celestial bodies: "Truly it does not suffice to have proved this opinion by means of various arguments, unless we also invalidate the arguments in favour of the opposite point of view."[343] He accordingly went on to disprove the eight arguments in favour of the Aristotelian cometary theory, which he had set forth at the beginning of his tract, by means of eight counter-arguments, based on three main

342 Fienus explicitly addressed the question of whether or not the earth was at rest at the centre of the universe, in a small tract entitled *Thomae Fieni Disputatio An Caelum moveatur & Terra quiescat, ad generosos et nobiles viros DD. Thobiam Matthaei et Georgium Gays*, which was printed along with the dissertations on the comet of 1618 by Fienus and Fromondus. It contained a completely traditional defense of Aristotelian geocentricity, with the author quoting scriptural texts against the movement of the earth, and expounding such arguments as the alleged fact that, if the earth moved, a stone dropped from the top of a tower would never reach the ground next to the tower's base, or arrows shot in a westerly direction would cover a much smaller distance than arrows shot in an easterly direction, on the assumption that the earth moved from East to West.

343 Fienus, *Dissertatio*: "Verum non sufficit variis argumentis hanc sententiam comprobasse, nisi contrariae quoque opinionis argumenta solvantur."

sources: Seneca, the telescopic discoveries of Galileo, and the dicta of the Church Fathers. It was in the course of this direct refutation of Peripatetic cometary theory that a somewhat less orthodox side to Fienus' outlook came to the fore.

Throughout this section of his tract, Fienus' tone was one of marked scepticism. In response to the Aristotelian question of what kind of celestial body a comet could possibly be, given the fact that it was neither a fixed star nor a planet, he wrote down a page-long exposé that almost reads as a manifesto of the sceptical stance in natural philosophy which was enjoying a revival during the early half of the seventeenth century. "I cannot say what kind of celestial body is a comet, any more than could those many great men who lived before me... Of God, the soul, and all the stars, we know what they are not; yet we do not know what they are. It is truly a miracle of nature, suitable for us to exercise our minds on; yet we have to admire it, without ever being able to actually comprehend it. Who knows what goes on in the heavens, and what kind of bodies and substances exist there, as yet invisible to us, and what changes take place there, perhaps even daily? Our entire perception of the substance, virtues, and permutations of the stars and the heavens (apart from their motions, and from the effects of the sun and moon, which are perceptible to the senses): it is all nothing but conjecture."[344]

Because of the limitations of the human senses and man's intellectual capabilities, Fienus believed, human beings would never be able to attain certain knowledge as to what kind of body was a comet. He quoted several possibilities, which all seemed equally probable: comets might be new stars, miraculously created by God to inspire terror and compel admiration for his majesty; bodies naturally generated in the heavens from time to time, from material and efficient causes unknown to us; very distant fixed stars that for unknown reasons occasionally left their place of residence and moved to celestial regions in the vicinity of the earth; or even a special kind of planets

[344] Ibidem, 39: "Quale autem sit, ego quidem dicere non possum, non magis, quam tot magni viri, qui ante me fuerunt. ... De Deo, de anima, de sideribus omnibus, scitur quid non sint; sed non scitur quid sint. Miraculum vere est naturae, idoneum ingeniis nostris torquendis, quod admirari debemus, cognoscere nunquam poterimus. Quis scit, quid in coelo agatur, quae ibi corpora, quae substantiae nobis invisibiles adhuc sint, & quae ibi quotidie fiant permutationes? Omnis notitia nostra de siderum coelorumque substantia & virtutibus, ac permutationibus, excepto motu, & Solis, Lunaeque effectibus, qui sub sensum cadunt; non est nisi coniectura."

describing orbits eccentric to the earth, and rendered visible to earth-dwellers only in the very lowest portions of their trajectories.[345]

Despite his scepticism, however, Fienus did take pains to argue at somewhat greater length in favour of the last view mentioned, which was basically that advocated by Apollonius of Myndus, who thought of comets as planetlike bodies describing extremely eccentric orbits.[346] This view is not as absurd as it may seem at first glance, our author maintained. Why should not more than seven planets be found in the heavens? We conceive of just seven of them merely due to the limitations of our senses: our naked eye can see only a fraction of celestial reality. It is well possible that there exist very many objects in the heavens that we cannot see with our feeble eyesight; this view has recently been corroborated by Galileo, who with the aid of the telescope discovered three small new planets making their revolutions around Jupiter.[347]

Furthermore, Fienus added, the irregular motion of comets seemed also to argue in favour of the view of Apollonius. For the comet of 1618 had moved fastest when its size was at its maximum; when it grew smaller, its movement also slackened. These observed facts seemed to be explicable by assuming that the comet, when close to the earth, moved in a curved line, while later on, when it returned to the depths of the sky, its motion was more rectilinear; after all, experience showed that objects moving towards or away from the observer in a straight line seemed to move more slowly, or even to stand still, while objects describing a curved path seemed to cover more ground.[348] Fienus declared groundless Seneca's objection, that for Apollonius' theory to be valid comets would have to increase in

---

[345] Ibidem, 39-40.

[346] See Seneca, *Naturales Quaestiones*, VII.17.

[347] Fienus, *Dissertatio*, 40-41: "Plures enim quam septem planetas in coelo reperiri, nihil miri est. Non constituti sunt nisi septem, quia multo plurium motum oculus noster advertere hactenus non potuit. Sed quid impedit adhuc multo plures eosque parvos esse, qui videri a nobis, quorumque motus observari minime potuerit. Ita esse tandem quoque seculo nostro docuit Galilaeus Galileius, qui beneficio dioptrae Batavicae, adhuc tres parvos, ac novos Planetas adinvenit, perpetuo circa Iovem decurrentes. . .".

[348] Ibidem, 41: "Num cum Cometa noster maximus apparuit, eius motus fuit celerrimus; ut minor evasit, factus quoque fuit valde lentus. An hoc est, quod quando terrae propinquus fuit, motum habuit transversum; quando vero in altum rediit, magis rectum? Etenim quae in transversum moventur, celeriter moveri, & magnum spacium conficere; quae vero in rectum a nobis recedunt, vel ad nos accedunt vix moveri, exiguumque spacium pertransire; immo quodammodo stare videntur."

size the closer they approached the earth, which was not the case; after all, the fact that most comets were biggest in size on the day of their first appearance could be due to other causes. Thus, the postulated increase in size might take place during periods when our vision was impeded by interjacent clouds or sunrays. Moreover, often enough comets actually had been observed to grow during the first days of their visibility.[349]

Despite this sympathetic defense of the Apollonian theory, however, Fienus once again stressed his ignorance of the essence and nature of comets, and declared himself satisfied with the result already obtained, namely, the demonstration that they were celestial bodies (of whatever kind) instead of burning meteors. Further on in his refutation of the arguments in favour of the Aristotelian theory, too, his sceptical stance repeatedly came to the fore. Thus, in response to the Peripatetic objection that comets increased and decreased in size, while celestial bodies never did, our author retorted that, perhaps, they did not really grow bigger and smaller at all, but only seemed to do so to human eyes, owing to the fact that they advanced towards and receded from the earth. He was prepared to admit that comets displayed many characteristics uncommon to the other stars; yet, to him, this did not prevent them from belonging to the class of celestial bodies. The only thing it proved was that the tailed stars were to be thought of as miracles of the heavens, analogous to the many miraculous occurrences to be encountered on earth.[350] Similarly, Fienus granted the Peripatetics the fact that cometary motions were irregular; yet once again, he was not prepared to conclude, on the basis of this observation, that they could not be celestial objects, referring his readers back to the passage described above, in which he had explained that it was well possible that the hairy stars only seemed to accelerate and decelerate.[351]

Several of Fienus' responses to the Aristotelian arguments were, once again, borrowed from Seneca. Thus, he maintained that, contrary to the belief of the Peripatetics, it was not necessary for bodies of the same species to possess exactly the same motions. After all,

---

[349] Ibidem, 42. For Seneca's objection, see *Naturales Quaestiones* VII.18.

[350] Fienus, *Dissertatio*, 44: "Ut ut sit, fateor in Cometis multas reperiri conditiones, ac proprietates aliis sideribus non communes; sed ex quibus tamen non conficitur, ipsos caelestia corpora non esse, aut esse Meteora; sed hoc solum: esse coeli miracula, qualia & in terris non pauca reperiuntur."

[351] Ibidem, 51.

Venus, Mars, and Saturn were all planets; yet their movements were quite diverse; and, while all men were of the same species, yet one of them might move towards the right while another one moved to the left. It was sufficient for objects of the same species to possess motions of the same genus, and in the case of celestial bodies, this postulate was fulfilled: all of them, including comets, moved in circles.[352] Furthermore, following Seneca, Fienus also pointed out that comets were not in fact transparent, as the Aristotelians believed; it was only the cometary tails through which the fixed stars could be observed, these being nothing more than rays of light.[353] Also, our author joined his Stoic predecessor in observing that, contrary to the Peripatetic view, comets were not oblong in shape, but globular, just like the other stars; they only appeared to be outstretched because of their tails, which, however, were not part of the comet proper but were rather constituted by sunrays diffused through the comet's head.[354]

This observation inspired Fienus to include an extensive defence of the optical theory of comets, according to which cometary tails were luminous rays of sunlight instead of outstretched burning exhalations. He tried to render convincing this point of view by means of a variety of arguments, most of them based on the experience of the senses. Thus, one needed only to take a glance at the tails to see that they did not look like fire, but rather resembled the rays of the sun; just as the eye could distinguish differences in intensity in sunlight, so also the cometary tails were observed to be composed of rays of unequal luminosity.[355] Also, it was possible to see the fixed stars through the lustre of the tails, while the flames of burning objects were invariably nontransparent; and, if the tails were burning exhalations, they would display a very irregular variation in luminosity, while in actual fact they were seen to decrease in intensity very gradually.[356] The most conclusive argument in favour of the optical theory, however, was the observed fact that the tails invariably pointed in an antisolar direction; Fienus quoted Cardano, Apian,

352 Ibidem. See Seneca, *Naturales Quaestiones*, VII.27, where he elaborates on the theme of Nature's variety, and on the fact that all the stars are celestial bodies, even though they may look very different and behave in very diverse ways.

353 Fienus, *Dissertatio*, 51; see Seneca, *Naturales Quaestiones*, VII.26.1.

354 Fienus, *Dissertatio*, 44; see Seneca, *Naturales Quaestiones*, VII.26.2.

355 Fienus, *Dissertatio*, 45.

356 Ibidem, 46.

and Gemma Frisius for this observation, and added that he, too, had discerned this permanent antisolarity for the tail of the comet of 1618.[357] In response to the possible objection, that if cometary tails were indeed formed by sunlight, it was inexplicable why the other stars, or at least some of them, did not appear equally hairy, Fienus replied, in an argument borrowed from Cardano[358], that comets were formed out of a substance different from that of the other stars; they were of an intermediate nature, between density and rarity, so that they were capable of both receiving and transmitting the sunlight, which other stars could not due to their greater density.[359]

Fienus' most scathing criticism of certain tenets of Aristotelian cosmology was contained in his refutation of two Peripatetic arguments in favour of the burning exhalation theory that were based on nothing but the authority of Aristotle and the Scriptures. Thus, in his exposition of Peripatetic cometology at the beginning of his tract, he had put forward the argument that superlunary comets were unthinkable because the Stagyrite had declared the heavens to be unchangeable. To this statement, Fienus now responded that "The idea that the heavens are not susceptible to generation and corruption is not an article of faith, but a highy uncertain axiom, which can be as easily denied as Aristotle affirmed it, because it cannot be proved by any valid reasoning."[360] First of all, our author stated, it was simply not true that no change had ever been discerned in the heavens. Comets and novae had been observed through the ages: according to Pliny a new star had been seen in the time of Hipparchus; the Arab Haly professed also to have watched a nova located above the sphere of Venus; and, of course, recently, a new star had shone forth in the

[357] Ibidem, 46-47: "Probatur Septimo, & quidem caudam esse splendorem luminis Solis; ex eo, quod cauda semper sit Soli aversa; quod ut Cardanus lib. 4 de subtilitate, & Appianus, & Gemma Frisius animadverterunt; ita in hoc nostro Cometa semper fuit animadversum."

[358] See Cardano, *De rerum varietate* (Basle 1557) Book I, Chapter I, where he argues that comets are of a middle nature between the stars, which are capable of absorbing sunlight, and the moon, which can only reflect it.

[359] Fienus, *Dissertatio*, 49: "...diversitas autem ex eo necessario esse debet; quod Cometae habeant alium modum substantiae, quam aliae stellae; & mediam dispositionem inter raritatem, & densitatem; sic ut & lumen Solis recipere, & etiam transmittere possint; quod aliis stellis propter maiorem densitatem non contingit."

[360] Ibidem, 42-43: "Caelum esse ingenerabile, & incorruptibile non est articulus fidei, sed axioma valde incertum, quod ea facilitate potest negari, qua ab Aristotele est assertum; eo quod nulla valida ratione possit demonstrari."

constellation of Cassiopeia, studied among others by Fienus' predecessor, Cornelius Gemma.[361]

Moreover, even if Aristotle were right, and no human eye had ever perceived any alterations in the heavens, Fienus went on, this still would not prove the absence of celestial change. After all, it was well possible that enormous generations and corruptions took place in the heavens, unseen by us humans due to the weakness of our eyesight and the vast dimensions of the cosmos. If we were located in the sphere of the moon, our author added, we would undoubtedly not be able to discern any of the changes that we knew continually to take place on earth.[362] Thus, once again, Fienus stressed the inadequacy of human sense perception and the impossibility for us to attain certain knowledge of the heavens; Aristotle's postulate of celestial incorruptibility was therefore completely unfounded and boldly over-confident.

The second Peripatetic tenet that, to Fienus, had been rendered entirely groundless, was the assumption of the existence of solid planetary spheres. First of all, once again, the appearances of comets, which had so painstakingly been demonstrated to be superlunary in the earlier portion of his tract, had shown quite convincingly that the heavens were not at all solid, but fluid.[363] The biblical texts quoted by the Peripatetics in defence of solid heavens, to Fienus lacked persuasive force: where, in the Book of Job, it was written that the heavens were "most solid", this did not necessarily mean that they were hard and impenetrable; it might mean merely that they were corporeal, possessed three dimensions, and were no imaginary, empty spaces; or that, notwithstanding their fluidity, they stood firm and unwavering at God's command; or again, that only the firmament was said to be solid because it carried the fixed stars firmly attached to it.[364]

Furthermore, many loci from the Scriptures and the Church Fa-

---

361 Ibidem, 43-44.

362 Ibidem, 43: "Possunt enim maximae generationes, & corruptiones, maximaeque permutationes in coelo fieri; etiamsi a nobis non conspiciantur. Hoc visus nostri debilitas, & immensa coeli distantia faciunt, nec magis permutationes, aut generationes ullas in hoc mundo inferiore videremus, si in coelo Lunae constituti essemus . . . quam modo in coelo videmus."

363 Ibidem, 52: "Etenim coelum esse liquidum, & pertransibile vel soli hi tam diversi Cometarum motus, quos tam variis argumentis, atque efficacibus, coelestes esse, ac per coelum vehi probatum est, evidentissime ostendunt. . .".

364 Ibidem, 52-53: "Respondeo: non certo posse dici, quid ibi per verbum (solidis-

thers could be adduced in support of the fluidity of the heavens.[365] From Holy Writ, Fienus cited, among other passages, Isaiah 51:6 ("for the heavens shall vanish away like smoke") in favour of fluid heavens, and Ecclesiastes 1:5 ("The sun also ariseth, and the sun goeth down, and hasteth to his place where he arose") and Joshua 10:13 ("And the sun stood still, and the moon stayed") to show that the Bible tended to ascribe movement to the heavenly bodies themselves, not their orbs.[366] In addition, many of the Church Fathers could be adduced in support of the same point of view; Fienus cited Justin Martyr, Eusebius, Diodorus Tharsensis, Saint Augustine, Saint Basil, and John Chrysostom, as well as the famous passage in which Origen had likened the motions of the heavenly bodies to those of animals on earth, fish in the waters, and birds in the sky.[367]

The Peripatetic objection, that one single fluid heaven would continue *ad infinitum*, which ran counter to Scripture, Fienus retorted that this was not the case, since in his view the firmament, to which the fixed stars were attached, was indeed solid; if not, it would be unable to carry so many motionless stars. The fact that, in the Holy Writ, the heavens were often referred to in the plural, was due to an error of translation: John Chrysostom (and, following him, also Thomas Aquinas and Robert Bellarmine[368]) had pointed out that the Hebrew word for "heaven" (*shamayim*) exists only in the plural. Nor did the story in 2 Corinthians about Saint Paul being carried up towards the "third heaven" prove the existence of a plurality of heavens: the passage should be understood to mean by the "first heaven", the

---

simi) intelligatur, & quae sit proprietas verbi Hebraici, quod in verbum (solidissimi) est translatum. Possunt enim coeli dici esse solidi, non quod sint duri & impervii, sed quod sint corporei; constantes trina dimensione, sicut alia corpora, & sicut quaevis corpora tangibilia ... quodque non sint spacia imaginaria, omni corpore vacua. Vel potest dici quod dicantur solidi; quod etsi liquidi sint, stent tamen suo loco firmiter librati, solo nutu, & imperio Dei; vel potest etiam dici, quod ibi nomine coelorum veniat intelligendum solum firmamentum, quod quia stellas fixas firmiter continet, firmamentum ac solidum potest vel appellari, vel revera est."

365 On the early seventeenth-century practice of quoting biblical passages in favour of either celestial solidity or celestial fluidity, see Grant, *Planets, Stars, & Orbs*, 351-353.

366 Fienus, *Dissertatio*, 53-54: "... variis locis apparet, quod sacra scriptura apertis verbis, motum non coelis, aut orbibus, sed ipsis astris adscribat."

367 Ibidem, 54.

368 See Robert Bellarmine, *Louvain Lectures*, in Ugo Baldini & George V. Coyne eds., *The Louvain Lectures (Lectiones Lovanienses) of Bellarmine and the Autograph Copy of his 1616 Declaration to Galileo* (*Vatican Observatory Publications, Studi Galileiani*, 1 (2)) (Vatican City 1984) 16.

elementary region, while the "second heaven" referred to the location of the stars and the "third heaven" to the Empyrean, the abode of the Blessed.[369]

On the whole, we can conclude that Fienus' world view was still overwhelmingly Aristotelian in nature. We have seen how he managed to reconcile his non-Aristotelian belief in the superlunary nature of comets with a continued adherence to the three basic postulates of Peripatetic cosmology: geocentricity, fundamental dichotomy between sublunary and superlunary regions, and circularity of celestial motions. This was rendered all the more easy for him, since he could rely on Seneca, who had argued for superlunary comets while at the same time subscribing to the basic tenets of the Aristotelian world view, which happened also to be fundamental constituents of the Stoic cosmos. However, when it came to answering two questions to which the Peripatetics and the Stoics had proposed different solutions (namely, whether the heavens were unalterable, and whether they were solid or fluid) Fienus unhesitatingly chose to defend the Stoic point of view: to him, the heavens were corruptible and fluid.

At this point we may well wonder what had induced Fienus to abandon celestial incorruptibility and solidity. On both issues, he appealed to the observations of comets and novae, whose appearances could not be reconciled with Aristotle's views. Yet, even more important to him, judging by the amount of attention he devoted to it, was the appeal to ancient authorities. The solidity of the spheres had been disproved by the motions of comets, but was also incompatible with the dicta of the majority of the Church Fathers (who, especially since the Council of Trent, enjoyed great authority). And the heavens were changeable, not only because of the evidence provided by comets and novae, but also because the limitations of sense perception made it inviable to categorically maintain that the heavens were incorruptible.

The dominant mood of Fienus' tract was one of scepticism: he realized only too well that to disprove Aristotle's exhalation theory was one thing, but to state what kind of bodies comets actually were was quite another. He drew much of his intellectual modesty from Seneca: on several occasions, the Stoic had stressed how little man as yet knew about the universe, and how unwise it was to rule out any

---

369 Fienus, *Dissertatio*, 56.

possibility on the flimsy basis of our weak human senses. Seneca's discussion of comets was pervaded with remarks like "How many other bodies besides these comets move in secret, never rising before the eyes of men!"; "These are matters pertaining to comets which have impressed me and others. Whether or not they are true only the gods know, who have knowledge of the truth"; "Anyone who thinks that nature is not occasionally able to do something she had not done frequently, simply does not understand the power of nature"; or "There are many things that we concede exist; what their qualities are we do not know."[370]

To Fienus, Seneca's guidance was indispensable; not only were the vast majority of his arguments in favour of celestial comets drawn from the Stoic's work, but he also copied his Roman predecessor's attitude of scepticism with regard to the human capability of grasping the laws of nature. If anything, this moderate pessimism had been aggravated by Galileo's telescopic discoveries: although Fienus mentioned them only once, and even misquoted Galileo, speaking of only three Jovian satellites whereas the Italian had discovered four of them, they do seem to have influenced his cosmological outlook. However, this influence was of a different kind from the impact historians have usually postulated for Galileo's discoveries, claiming them to have rendered more viable the Copernican, or at least the Tychonic, world view, and to have constituted an important step on the road to the new science[371]; on Fienus, they seem to have had almost the opposite effect, corroborating his Senecan sense of inadequacy of human sense perception and unattainability of knowledge of the natural world. On the basis of Seneca, Galileo, and the Church Fathers, only one conclusion could be reached: comets were celestial bodies, moving about in fluid and changeable heavens about which not a single more definite statement could be made.

c. *Libertus Fromondus: "Peregrinum in Regione Planetarum Sidus"*

On receiving his older colleague's tract, Fromondus professedly did not read the work, but rather 'devoured' it. Impressed with the way in which Fienus had invalidated the cometary theory of the Stagyrite,

---

[370] Seneca, *Naturales Quaestiones*, VII.30.3; VII.29.3; VII.27.5; VII.25.1; translated by T.H. Corcoran, Loeb Classical Library edition (Cambridge Mass./London 1972).

[371] See supra § V.3.c.

he claimed that the advent of the comet of 1618 had presaged the death of Aristotle, if nothing else: "Begone, you Peripatetics, and arrange an imaginary funeral for your Prince!", the young philosopher triumphantly exclaimed.[372]

Nonetheless, Fromondus from the outset made it clear that it was not his intention to do away with Peripatetic doctrine in any drastic way. First of all, the Greek master's teachings had been discredited only insofar as cometary theory was concerned; moreover, Aristotle himself had admitted that the investigation of comets was very difficult, and that it sufficed to produce a theory that was possible, since in these matters certainty was not to be attained. Thus, Fromondus said, the master had left open an opportunity for his followers to entertain deviating ideas on the subject of comets; moreover, even the most sworn Peripatetics were by now agreed that the Milky Way (out of whose material Aristotle believed comets to be formed) was not situated in the sublunary region, but in the firmament. Why, then, Fromondus pondered, should it not be possible for us to remain Peripatetics, and yet count comets among the planets? After all, he jokingly added, it was not to be supposed that the Stagyrite preferred his comets to milk, so that he would probably sooner part with the former than the latter.[373]

Thus, Fromondus explicitly associated himself with the Peripatetics and with Aristotelian doctrine; as long, that is, as the latter was allowed to be adapted to new opinions wherever necessary. Also, like that of Fienus, the structure of Fromondus' tract reflected a fundamentally Aristotelian way of dealing with scientific problems: after the obligatory description of the comet's appearance and motion, with which every cometary tract began, our author spent five sections attacking the Aristotelian theory of comets on several different

---

[372] Libertus Fromondus, *Dissertatio de Cometa anni 1618* (Antwerp 1619) 79-80: "Et Cometas nihil minari dicis? falsum est. Hic certe Aristoteli nostro, non minas solum, sed exitium tulit. . . . Ite Peripatetici, & imaginarium funus facite vestro Principi."

[373] Ibidem, 80-81: ". . . [Aristoteles] opinioni suae quasi diffidens, negotium hoc Cometarum difficile in primis ait, & abditum; & in quo sit satis docere quid fieri possit, non quid fiat. Libentissime id ab ipso audivi; quia Peripateticis suis videbatur fecisse gratiam seorsim aliquid a Magistro opinandi. . . . Viden' ut doctrinam suam Cometarum struxerit in altera Galaxiae? Haec tamen pridem a iuratissimis etiam Peripateticorum subruta est, & consensu translata in firmamentum. Cur non & nos Peripatetici perseverare, & Cometas Planetis saltem miscere possimus? Non existimo Cometas suos prae Lacte dilexisse nostrum Stagyritam: immo facilius illos, quam istud dimissurum e manibus."

grounds, after which, in the last and most important section of the treatise, he argued extensively in favour of celestial comets, and expounded his own views on the nature and essence of the hairy stars. Thus, like Fienus, Fromondus employed the framework of the standard scholastic *quaestiones*-method in order to arrive at a conclusion diametrically opposed to that of the Stagyrite himself.

Fromondus' approach also resembled that of his older colleague in that he interlaced his fundamentally Aristotelian world view with other, non-Aristotelian elements taken from Stoic cosmology and the recent telescopic observations. However, while Fienus had based himself primarily on Seneca, with Galileo's discoveries playing a very minor, and to a large extent negative, role, with Fromondus the balance decidedly swung the other way. Relying to a large extent on arguments based on common sense experience, with telescopic observation as its most refined mode, he displayed much less scepticism than his colleague, and was prepared to make far more confident assertions concerning the nature and essence of hairy stars.

Fromondus' observations of the path the comet of 1618 had followed through the sky were not much more accurate than those of Fienus. Although he had at his disposal a telescope, through which he observed the new phenomenon "fairly inquisitively"[374], he on several occasions complained of his lack of measuring instruments, which had among other things prevented him from accurately determining the comet's parallax.[375] Nonetheless, he recorded, as best he could, the comet's locations for 17 nights (out of a total of 52 nights on which the comet had been visible). Again like Fienus, Fromondus often determined the comet's position with regard to the constellations by stating that it stood "on a straight line" with two of the nearest stars, or that it formed a "triangle" with two such stars, so that it is probable that Fromondus, too, for want of any better instruments, used a piece of thread in order to conduct his observations.[376]

---

[374] Ibidem, 82: "Die Novemb. 30 mane calide ad spectaculum evolavi: & inspexi per dioptricum tubum satis curiose."

[375] Ibidem, 122: "De parallaxi, unde altitudo eius tenuius inquiri potuit, inopia instrumentorum, non possum certi aliquid."

[376] See for example ibidem, 86-87: "Vicesima tandem primo hora 3 matutina, in confinio Librae & Virginis clarissime visus, infra cubitum Bootis duos circiter gradus; ita ut Benenack in cuspide caudae Ursae Maioris, cubitus Bootis, & Cometa Meridionalior, tenerent angulos trianguli, cuius brevissimum latus, a Cometa in cubitum, longissimum, a Cometa in Benenack exiret."; or ibidem, 88-89: "Die 13 Ian. vesperi hora 11 iam linea recta ab ultima Draconis per Cometam in spacium inter radicem caudae Ursae Minoris, & sequentem in cauda medium exibat."

As for the exact distances of the comet to the most nearby of the fixed stars, our author could do no more than utter cautious guesses, such as: "The comet was removed about one degree and thirty minutes from the [star of the third magnitude in the] hip of Bootes, as far as I was able to conjecture."[377]

Fromondus showed himself especially interested in the comet's tail —or rather, its beard, since unlike the comet of 1577, this particular hairy star was preceded by its effluxion rather than followed by it. Whereas Fienus had included just one description of the beard, Fromondus paid it a great deal of attention, recording its size, its luminosity, and the direction in which it pointed, for almost every night on which he had been able to observe the comet. When Fromondus first saw the comet, on 30 November, the portion of the beard closest to the head was narrow, but dense; yet the further removed parts were more dispersed. On that day, the beard as a whole was curved, half-oval in shape; yet almost a week later, on 5 December, it had grown much straighter.[378] During the first half of December, the comet's beard was immense in size: on 6 December, it easily covered thirty degrees of the zodiac, and the following day, it even extended as far as fifty degrees, stretching out from the tail of Ursa Maior, across the equinoctial colure, under the southerly wheels of the Wain, to the left knee of the Bear.[379] By far the most interesting observation concerning the beard was made on 12 December, when Fromondus noticed that, whereas in the middle of the night not a trace of it had been visible, later on, when the sun began to rise, the beard slowly but surely began to appear, in the same shape as the day before. Our author compared this phenomenon to the morning twilight, which was also progressively illuminated by the sun as it slowly rose closer to the horizon.[380]

After his detailed description of the comet's path, Fromondus provided his readers with a summary description of its course, duration,

---

[377] Ibidem, 85: "Aberat vero a coxendice Cometa, quantum conjicere valui, gradu uno & 30 fortasse scrupulis."

[378] Ibidem, 83-84.

[379] Ibidem, 85.

[380] Ibidem, 85-86: "Die 12 Decemb. alta & concubia nocte miratus sum conspicuo toto plaustro, de barba ne pilum comparere. Tandem post medium primae & Sole inferum Meridiani arcum elapso, deformari paulatim coepit, loco eodem quo pridie & pari longitudine: ut plane matutina crepuscula solent, cum Sol radios quos praemiserit, magis magisque implet & intendit, quo propius ad Horizontem scanderit."

and motion. He believed that the phenomenon had lasted for 52 days, from its first sighting in Holland on 27 November 1618 until 18 January 1619, when Fienus had last discerned it. Nonetheless, Fromondus was convinced that it had in fact passed away completely only after a longer period of time, but that due to the weakness of human eyesight it had not been visible during the last leg of its journey.[381] During its visibility, it had traversed the signs of Scorpio, Libra, Virgo and Leo, and had disappeared in Cancer. Its velocity had been greatest during the earlier phase of its appearance; later, it had grown progressively slower. Finally, Fromondus included a rough estimate of the comet's speed on single days; thus, he maintained that between 6 and 10 December it had covered a distance of 17°, which amounted to 3° 24' a day; between 11 and 12 December, it had advanced 4°; and from 28 December to 18 January it had covered 20°. This last figure amounted to somewhat less than one degree a day; however, Fromondus added, "the number of degrees should not be divided equally among the different dates; rather, something should be added to the earlier days, on which it moved faster, and something should be subtracted from the later days."[382] However, he did not provide his readers with any fixed ratio to compute the exact distance covered by the comet on individual days.

After this rendition of the comet's appearance and motion, Fromondus got round to the real purpose of his tract: the invalidation of Aristotle's theory of comets as burning terrestrial vapours, and the defence of his own belief in superlunary, celestial comets. Like Fienus, he chose first of all to attack the various elements of the exhalation theory, to which purpose he composed five chapters in which he argued, from five different points of view, that comets could not possibly be fiery vapours: "Because of the height to which exhalations rise"; "Because of the volatile and dissoluble nature of fire"; "Because of scintillation"; "Because of their motion"; and "Because of their beards".

Before embarking on the refutation proper of Aristotle's theory,

---

[381] Ibidem, 90: "Duravit autem noster a 27 Novemb. an. 1618 quando primum in Hollandia conspectus, ad 18 Ian. an. 1619 id est dies 52. Exire tamen indubie longius, quia parvitas adhuc spirantis aliquamdiu fugere oculos nostros potuit."

[382] Ibidem: "A 28 autem Decemb. ad 18 Ian. an. 1619 id est, 21 diebus confecit 20 circiter grad., id est, non totum unum singulo quoque die. Non tamen aequaliter per omnes dies spargendus graduum numerus est: dies priores quibus incitatior fuit cumulandi, & aliquid detrahendum est posterioribus."

however, Fromondus thought it necessary to cursorily invalidate the theory of Anaxagoras and Democritus, according to which comets were formed out of the conjunction of several planets. This idea could not be reconciled with the experience of many ages, which showed that, unlike planets, comets often moved in the same direction as the *primum mobile*, and deflected from the ecliptic much further than planets ever did.[383] Thus, Fromondus agreed with Aristotle, that the 'planetary conjunction' theory was invalid. As for the other ancient Presocratics refuted by the Stagyrite in his *Meteorology*, Fromondus announced that he would treat their theory, that comets were celestial bodies, at greater length in his last chapter. Presently, he first of all turned his attention to the investigation of Aristotle's own theory, in which, as usual, the Stagyrite scorned the ancients; he had been one of the first to take comets down from the heavens and to place them in the element of air.[384] After Aristotle, almost all astronomers, even the most proficient ones, among them Ptolemy himself, had imagined themselves certain of the truth of the Aristotelian theory, until the new star of 1572 had induced most of them to return to the more ancient theory.[385] Thus, Fromondus explicitly designated the recent opinion, that comets were celestial bodies, as a return to the most ancient, pre-Aristotelian tradition, rather than as a break with the past occasioned by new inventions.

The first "engine" with which Aristotle could be "assaulted"[386], Fromondus went on, was the argument from the maximum height to which exhalations could possibly rise. Like Fienus, Fromondus pointed out that comets could not possibly consist of terrestrial exhalations, since these were unable to rise to the altitudes at which comets were usually observed. Even if it were admitted that earthy and watery vapours could rise up as high as fifteen German miles above the earth's surface, the Peripatetic theory could not be saved, since even the lowest estimates (those of Nolthius, Scultetus, and

---

383 Ibidem, 92-93.

384 Ibidem, 93: "Aristoteles spretis, ut solet, veteribus, Cometas fere inter primos caelo detraxit, & aerio Elemento miscuit."

385 Ibidem, 94: "Post Aristotelem universi fere, etiam Astronomiae Praesides, in quibus ipse Ptolemaeus, quasi inventionis iam & veritatis securi abierunt; donec tandem saeculo nuper lapso occasio maxime Cassiopaeanae stellae plerosque ad respectum veteris opinionis permovit."

386 Ibidem: "Quibus autem machinis petitur Aristoteles? Primo. Non elevantur a terra vapores, aut exhalationes conspicuae tam alte, quam est Cometae locus."

Buschius, who had judged the 1577 comet sublunary) placed comets far beyond fifteen German miles from the earth.[387]

Next, Fromondus ventured to play the devil's advocate, attempting to save the Aristotelian theory by postulating the existence of extremely subtle and rare terrestrial spirits, capable of ascending higher than their denser and fatter counterparts. After all, the exhalations visible to us, our author asserted, all possessed a certain gravity, density, and opacity, so that they absorbed and reflected the light of the sun. Thus, the maximum altitude measured for exhalations by the opticists was valid only for these gross and heavy vapours; however, it was not unthinkable that other, subtler spirits could rise up much higher, but would remain invisible to earth-dwellers, being incapable of reflecting light. A Peripatetic, then, might postulate that comets consisted of such extremely rare spirits, located as high up as the sphere of fire, and condensed into a globe.[388]

Not suprisingly, though, Fromondus set out to crush this theory as quickly as he had expounded it. After all, even if in themselves weightless, once they had coagulated together the spirits would undoubtedly gain a certain measure of gravity, so that the newly formed 'comet' would soon descend towards the altitude of the clouds. Moreover, even if it were upheld that the matter of which these comets were formed was pure fire of the kind with which the Aristotelian sphere of fire was filled, the argument would not be valid, since it was well known that the matter on which the burning globe fed consisted of earthy exhalations, which possessed gravity and would cause the whole comet to descend. Even if the subtle fire were to permeate all the comet's pores, this would still not serve to prevent the body from subsiding, as could be deduced from the analogy with a glowing coal, which, when thrown up into the sky, invariably fell down again.[389]

Nor would the well-worn postulate of celestial influence (the power of certain stars and planets to attract comets in the way that magnets

---

[387] Ibidem, 95. Nolthius had located the comet of 1577 at 6880 German miles, Scultetus at 7188, and Buschius even at 20.640 German miles from the earth's surface.

[388] Fromondus, *Dissertatio*, 96: "Spiritus vero alii subtiliores, & in confinium Elementi ignis subvecti, crespusculi radios non sentiunt, transmittuntque per suam raritatem irrefractos. Horum altitudinem nullis oculis videre possunt perspectivi; quia non possunt ipsos. Ex talibus autem excelsissimi illi Cometarum consistunt & densantur."

[389] Ibidem, 96-97.

attract iron) do the Peripatetic case any good. For, if all the stars, consisting of the same kind of matter, were indiscriminately capable of attracting comets, the hairy stars would be torn to pieces because all the stars in their vicinity would exercise their powers on them at the same time. Ptolemy's contention, that cometary matter was held together by the coldness of the planet Saturn, was invalid; after all, the comet of 1618 had disproved it, since when it was closest to Saturn it had suffered its greatest loss of material.[390]

Another way of invalidating the Peripatetic cometary theory was to argue "from the volatile and dissoluble nature of fire". If comets were fiery bodies, Fromondus stated, they would suffer all kinds of changes of shape and perhaps even instant expiry; after all, fire by its nature caused the material it fed on to dissipate. Furthermore, comets, resisting the movement of the *primum mobile* with their own wayward motions, would be liable to be torn apart especially easily if they consisted of volatile fire.[391] Aristotelians might defend their theory by pointing out that, possibly, the shape of the comet's head did, indeed, vary, as was to be expected of a burning body, but that we on earth were unable to see these changes due to the enormous distance. But this hypothesis did not hold true; Fromondus himself had clearly seen, through the telescope, that the comet's head was invariably round and well-defined.[392] Nor could it be maintained that the comet's matter was held together by means of a kind of extremely solid glue, or that its shape was invariable because any gaps in the comet's head were immediately repaired by fresh exhalations newly risen up. If this were so, Fromondus agreed with Fienus, the comet's body would suffer a continual eclipse because its lower side would be surrounded by a thick cloud of newly ascended vapours.[393]

Fromondus also concurred with the idea put forward by Fienus, that it was impossible to satisfactorily explain, by means of Peripatetic cometology, why no comets ever came into being in the lower

[390] Ibidem, 98-99.

[391] Ibidem, 100: "Argumentum hoc validum est, ubi praesertim Cometa non solo motu diurno cum primo mobili in Occasum fluit, sed alio proprio in Ortum renititur, vel abit in Septentrionis aut Meridiei latus. sic enim inter utrumque motum oritur rixa, quae facile discordem a se nec bene cohaerentem ignis materiam dissipabit, aut diducet. Hinc Cometa statim dissolutio, aut, ut minimum, mille materiae vicissim figurationes, & turbatio rotunditatis, quam perpetuo in capite, nisi iam plane expirantis, notamus."

[392] Ibidem, 100-101.

[393] Ibidem, 101.

regions of the air. After all, he argued, an exhalation, being a light body, will tend to fly away from the centre of the world in a straight line; therefore, each of the atoms of which the exhalation consists will move in a direction slightly different from that of the other atoms, causing the exhalation as a whole to progressively rarify during its ascent. Thus, exhalations located in the lower regions of the air were invariably denser than their more elevated counterparts, and better capable of coagulating into the globular shapes necessary for Peripatetic comet formation. On the basis of Aristotelian theory, therefore, we should expect comets to be located right over our heads, rather than at the immense distances at which they were actually observed.[394]

Moreover, if comets consisted of fire, they would scintillate, since experience showed that flames tended to move about restlessly and to follow their fuel. On the basis of this argument, Fromondus was inclined to include the hairy stars in the genus of planets; after all, like planets, they shone with a calm and unperturbed lustre, whereas the fixed stars scintillated and vibrated as if they were celestial fires.[395] This observation inspired Fromondus to include a long digression in which he assessed the merits of several different theories that had recently been put forward to account for the difference in splendour between stars and planets. Several modern authors, such as Tycho Brahe and François d'Aiguillon, had lately revived the ancient theory that the scintillation of the fixed stars was due to the fact that they continually revolved about their own axes. D'Aiguillon, undoubtedly inspired by the recent telescopic discoveries, had added that the stars' surfaces might well be spotted and dotted over with inequalities of light, so that by quickly turning around, they sometimes showed the earth their resplendent sides, and sometimes their mutilated and broken ones.[396] Conversely, none of the planets thus rotated about their own axes, as was demonstrated by the appearances of the moon, which constantly displayed only one of its sides.

Yet, Fromondus stated, the credibility of this opinion had recently

---

[394] Ibidem, 102-103.

[395] Ibidem, 103: "In genere ergo Planetarum Cometam ponemus; quia simili, id est, quieta & tranquilla luce. Stellae vero fixae, tanquam caelestes quidem ignes, scintillant & vibrantur."

[396] Ibidem, 104: "Addit vero Aguilonius maculosas forte & lucis inaequalitate aspersas fixarum superficies; ut rapidissime circa centra sua vertiginando, nunc aerem & plenae lucis faciem, nunc plagosam & fractae nobis obvertant; & inde interruptio illa luminis, & species iactationis."

been somewhat weakened by the observation that the sun carried around with it certain spots, constantly changing their position; this would seem to demonstrate that, unlike the moon, the sun in fact did turn around its own axis. Some people had tried to explain the sunspots by assuming small stars to revolve about the sun, by analogy with Jupiter's four Medicean satellites; yet Fromondus knew that Fienus favoured the former explanation: "You rather believe that the Sun is darkened by a number of birthmarks adhering to its surface and being carried round with it."[397]

If the sun's rotation were to be accepted, Fromondus proposed, perhaps we might explain the fact that Mercury, Venus, and Mars tend to scintillate as well, by assuming that these planets, too, revolved about their own centres. In any case, the fact that Venus displayed phases like those of the moon could, according to Fromondus, only be satisfactorily explained by assuming that that planet revolved not only about the sun, but also about its own axis. Otherwise, he queried, how could Galileo have seen Venus in semicircular shape when at aphelion? After all, if she did not constantly turn different sides, not all equally capable of reflecting light, towards the sun, she should, like the moon, have been full, not half-full, at aphelion.[398] Jupiter, on the other hand, seemed to be more tranquil; but Saturn was the calmest of all the planets, radiating with a most constant and unwavering lustre. Therefore, Fromondus did not believe that Saturn turned around its axis; the less so, since it had been observed that it consisted of three small orbs none of which ever changed place with regard to one another.[399]

From all these observations, Fromondus went on, it might be deduced that all the planets revolved about their own axes, like the fixed stars—except the moon and Saturn—because of the observed constancy of their light. However, when, on many occasions, our

---

397 Ibidem, 105: "Tu nativa macula & adhaerente lividum Solem mavis, & cum astro subiecto eam circumagi."

398 Ibidem: "Certe decretio illa & accretio Veneris, nec Lunari plane ritu, non eam circa Solem tantum, sed circa proprium etiam centrum volvi satis instant, partesque perpetuo alias aliasque Soli obvertere, quae non omnes ex aequo rapiant eius radium. Si enim id non sit, nescio quo modo Galilaeus Venerem in maxima a Sole digressione semicircularem viderit, cum alias plenum debuisset. sic enim Luna impletur, cum longissime a Sole discessit."

399 Ibidem, 106: "Saturnus ergo non videtur gyrari circa suum centrum: si motus iste lucem divaricat, & speciem facit coruscae nictationis. & aliud etiam argumentum habeo immobilitatis Saturni: discretio illa trium in ipso orbiculorum (de quibus in *Peregrinatione* nostra diximus) quos nunquam mutat & pervertit."

author had tried to reconcile this "enticing opinion" with the actual celestial occurrences, he had been unsuccessful. Most damaging to his hypothesis had been the observation that, although the fixed stars did indeed scintillate a great deal shortly after having risen above the horizon, later on, when they reached the meridian, they started to calm down and shine with an uninterrupted radiance.[400]

Thus, to Fromondus, the issue of the causes of the scintillation of the fixed stars was still quite undecided; as far as the planets were concerned, however, he was convinced that, apart from the moon and Saturn, they all revolved about their own axes. Nonetheless, the fact remained that the light emitted by the planets was of a much more constant and uninterrupted nature than that with which the fixed stars shone. Therefore, returning to the initial question of whether comets were fires, planets, or stars, Fromondus concluded that, judging by the nature of their light, they belonged neither to the genus of fixed stars nor to that of fiery meteors, but should be considered a special kind of planets.[401]

Fromondus next employed the argument based on the comet's motion, to show how wrong the Aristotelians were in supposing comets to be situated in the sublunary regions. After all, the comet of 1618 had, like that of 1577, followed a circular, and therefore clearly celestial path, that had constituted more than a quarter of a 'great circle' on the celestial globe. Contradicting Fienus' belief, that the reason that the comet had not lagged behind the movement of the *primum mobile* was to be sought in the fact that it had stood high in the heavens beyond the moon, Fromondus stated that the reason was rather to be found in the comet's proper motion. If it had simply been in the heavens and followed the *primum mobile* without adding any proper motion of its own, Fromondus asserted, the comet would have shown a deviation towards the east; the fact that it had not

[400] Ibidem, 106-107: "Tentavi saepe an haec prima fronte ita blandiens opinio experientiae & caelo responderet: & vidi difficulter posse. Inter omnia nil me sic alienavit, quam videre multis noctibus omnes fixas . . . acerrime scintillantes circa Horizontem, immo toto fere arcu ascensionis usque ad Meridianum. cum vero huc, aut circa pertigerint, crinem incipere componere, & molli solum nec frequenti nictu vibrare."

[401] Ibidem, 107: "Verumtamen sive hac ratione, sive alia quacumque fiat illa micatio & siderum lascivia videmur hoc nacti Cometam non de genere firmamento affixarum aut ignem fuisse, quia non lubricae & petulantis lucis, sed tranquillae & aemulae quasi Planetarum."

done so, was to be attributed to the effect of its own northward motion.[402]

Given the fact that both the comet of 1577 and the present one had not deviated from their segments of a great circle, even by a hair's breath, Fromondus went on to ask, who could possibly believe them to have been constituted out of a knot of fluctuating fire situated in the changeable, unstable airy regions?[403] The Peripatetics' case could not be saved by adducing the "nonsensical notion" put forward by certain people, who held that it was not in fact one and the same comet that moved from spot to spot, but, rather, several new comets took one another's place in a chain-like succession. After all, it was preposterous to believe that new fuel would become available for these successive comets at locations all situated on precisely the same great circle, and all exactly adjacent to one another. Furthermore, was it not totally implausible to suppose that each 'old' comet would expire at the very same moment its successor came into being?[404]

If, however, Aristotle's cometary theory were not to be rejected on other grounds (which of course, in Fromondus' opinion, it was) there did exist an opportunity for the Peripatetics to plausibly account for the characteristics of cometary motion: they might ascribe it to the agency of an angelic intelligence. After all, admitted Fromondus, it could not be denied that angels were capable of regulating any sublunary exhalation's motion to the effect that it described a segment of a perfect great circle, and of preventing the comet's material from falling apart, or even providing a new comet if the old one was in danger of expiring. If the comet's final cause was to terrify mortals and to induce them to repent of their sins, such might well explain why the conducting angel would, after a while, allow the comet to

[402] Ibidem, 107-108.

[403] Ibidem, 109: "Si vero, quod verissimum est, Cometes noster, ut plures alii, a segmento circuli maioris ne vel latum unguem motu proprio exciderit; quis prudenter id existimet in pendulum aliquem, & instabilissimo aeris Elemento fluctuantem ignis nodum posse cadere?"

[404] Ibidem, 109-110: "Quasi vero singulis momentis nova materia velut ex compacto sub eodem circuli magni filo perpetuo disponatur, eadem etiam quantitate & forma, ut Cometa eam inflammet, seque in filio per omnia simillimo centies millies amplius regeneret. Cur etiam semper a parte Cometae anteriori materia praeparatur? cur non retro? cur non ad latus? quae causa veterem Cometam praecise sic interficit, ut recte aboriatur & dispareat cum exoritur novus?"

grow smaller and weaker, and finally to die out once its sole purpose, to notify mankind of God's anger, was fulfilled.[405]

However, such an explanation did not appeal very strongly to Fromondus. First of all, it seemed "far-fetched and strained, and not sufficiently physical"[406]; and, moreover, if the agency of an intelligence had to be postulated to keep comets in motion, would it not be much more logical to locate them in the heavens, so that they could "rotate along with the eternal assembly of the stars and by means of a celestial intelligence"? Thus, their movement would be "solemn, but not as miraculous [as that of sublunary comets driven by angels]".[407] To Fromondus, then, the attribution of cometary movements to the agency of celestial intelligences seemed much the more likely explanation: it did not entail the necessity of positing any miraculous, angelic interventions, and accounted for the motions of comets in precisely the same way as for those of the planets. The possible counter-example of the star of the Magi could not make Fromondus in any way reconsider this opinion. The star of Bethlehem, he said, had not been an ordinary comet; instead, it had stood below the clouds, so close to the earth's surface as to be able to designate one single stable. Ordinary comets, on the contrary, were always observed at enormous distances from the earth. Therefore, the star of the Magi had been very exceptional and miraculous, and its path had been directed by an angel at God's command.[408]

The last argument Fromondus adduced against the Peripatetic exhalation theory was based on the characteristics of cometary tails. Our author went to great pains to show that, on the Aristotelian assumption that the tails consisted of burning exhalations slightly more subtle than those of which the heads were formed, the observed facts could not possibly be reconciled with the laws of Aristotelian physics. How, for instance, was it to be explained that cometary tails invariably pointed away from the sun, as had been demonstrated beyond doubt by Apian, Gemma Frisius, Cornelius Gemma, Fracastoro, Cardano, and Julius Caesar Scaliger? If they were exhal-

---

405 Ibidem, 110.

406 Ibidem, 111: "Verumtamen quaesitum hoc & violens forte videatur, nec satis Physicum."

407 Ibidem: "... & si intelligentiae opera placet, probabilius profecto in caelum Cometam evehemus, ut ibi in siderum aeternorum coetu & caelesti intelligentia rotetur, solemniter nec tam miraculose."

408 Ibidem, 111-112.

ations, they would, on the contrary, usually be seen directed straight at the sun, because the side of the comet nearest to the sun would be rarified by that body's heat, so that the kind of subtler exhalation of which the tails were supposed to be formed would be in more plentiful supply on the side nearest to the sun than in the antisolar direction. Furthermore, if the tails consisted of fire, they would, according to Peripatetic physics, point straight upwards because of their rarity, which was clearly not the case. Nor could it be maintained that the flame of the tail could only leave the comet's body through an airhole; according to Fromondus, this theory could not be reconciled with the rare and porous nature of comets' heads, which would allow the flames to leave on whichever side they pleased.[409] Also, according to the exhalation theory, cometary tails, being subtler and therefore lighter than the heads, should be seen pointing up into the sky in a straight line originating in the heads, which was not the case.[410]

But perhaps the most conclusive evidence of all could be gathered by simply observing the tail of the comet of 1618, Fromondus assured his readers. At a single glance, anyone could see that it was no earthy material being combusted by fire, but rather consisted of lightbeams passing through the head, through which even the smallest stars were easily visible.[411] Original and compelling evidence in favour of this 'optical' theory of cometary tails was furthermore provided by Fromondus' own observations of the comet of 1618. Several times, our author had noticed that during the early hours of the night, no trace of a tail had been visible; it had only begun to appear, slowly but surely, after midnight, when the sun once again entered the northern hemisphere. In the face of this evidence, Fromondus believed, it was no longer possible to deny the fact that the tail (or rather, the beard) had consisted of nothing else than sunrays passing through the comet's head, refracted, and spread out in the form of a beard.[412]

Thus, Fromondus explicitly subscribed to the optical theory of

---

[409] Ibidem, 114: "Si dicant non patere flammae exitium, nisi ubi materia spiraculum aperuit, ut videmus in fornacibus. Respondeo non habere id locum in materia Cometae, spongiosa & ab igne rara, quae quaelibet se potest expirare."

[410] Ibidem, 115.

[411] Ibidem: "Denique si quis unquam alius, hic noster Cometa exhalationes illas caudae refutavit. Non enim flagrantem in aliqua materia ignem referebat: sed radiorum per caput traiectorum claram imaginem: per quos etiam minimas stellulas facile saepe vidimus. . .".

[412] Ibidem, 115-116: "Id etiam clarissime deprehendi, ubi nocte hora octava,

cometary tails. Not only was it more easily to be reconciled with the observed facts than the exhalation theory; it was also capable of accounting for most of the vicissitudes displayed by the tails of the different comets that had recently been seen. He explained the fact that Tycho Brahe had observed the tail of the comet of 1577 pointing straight away from Venus instead of the sun, by means of an appeal to the varied effects refraction could have on lightbeams. He maintained that the substance of comets' heads did not possess a single, uniform density or opacity, but that, on the contrary, inside the heads patches of dense material alternated with fissures filled with much rarer stuff, rather like the human head was filled with brains and moisture. Therefore, the solar rays that passed through the head were all refracted into different directions, depending on the density of the part of the head they encountered.[413] This explained why the tail of Tycho's comet had pointed in the direction opposite to Venus instead of to the sun. It could also account for the fact that some comets displayed longer tails when closest to the sun, while, conversely, the tails of others grew larger the further they moved away from the sun; why some tails were straight while others were curved; and why the comet of 1618 had initially turned the convex side of its tail towards the north, and later its concave side. All these vicissitudes were due to the various ways in which the solar rays were refracted due to the unequal densities of the different parts of the comets' heads.[414]

Thus ended Fromondus' attack on Peripatetic cometology. In the course of his five anti-Aristotelian arguments, he unhesitatingly and conclusively exploded the theory of comets as burning terrestrial va-

---

nona, decima & porro nullum barbae vestigium circa Ursam ubi mane videram apparebat: quae solum post mediam noctem & Sole in Hemisphaerium nostrum resurgente, formari pallido quodam & tenui radiorum tractu incipiebat: qui magis tamen magisque ex vicinia Solis adventantis incendebatur. Hoc argumentum vix puto ulla specie eludi. Itaque barba illa planissime radii solares per caput Cometae transverberati erant, ibique ob materiae densitatem refracti, & sparsi deinde in barbam."

413 Ibidem, 117-118: "Itaque dico, etiamsi barba non directissime a Sole aversa spargeretur, radios tamen Solis esse, non alterius cuiusvis errantis aut fixi, qui per Cometae occiput cuneati mento aut ore vomebantur. Quod autem aliquando non recta, ob variam in cerebro, ut sic dicam, Cometae radiorum refractionem accidit. Substantia enim capitis Cometae non uniformis densitatis aut opacitatis est, nec per quemlibet circumferentiae partem aeque expergi possunt. Velut enim rimis quibusdam alicubi dehiscit, quas replet materia rarior, hic radii trajiciuntur facilius, quam ubi sese densat."

414 Ibidem, 119.

pours, refuting every counter-argument the Peripatetics might conceivably come up with, and demonstrating that not a single one of these objections could be held valid in the face of the evidence provided by the observed facts. Yet, like Fienus, Fromondus did not intend to break with Aristotelian physics on the basis of his rejection of the Stagyrite's cometary theory. Quite the contrary: in fact, the entire section of his tract discussed so far can be read as an attempt to reconcile the outcome of the most recent observations of comets (such as their superlunary location, the antisolarity of their tails, and their calm and unwavering luminosity) with the inexorable laws of Aristotelian physics, to most of which Fromondus unquestioningly subscribed.

Significantly, at the very beginning of his tract, he explicitly stated that he intended to argue for superlunary comets, yet at the same time to remain committed to Aristotelianism in general, believing the Stagyrite to have accorded his followers the licence to contrive any cometary theory they pleased, as long as it was possible. Sure enough, in the course of his five anti-exhalation arguments Fromondus, though constantly arguing against Aristotelian cometary theory, employed a fundamentally Aristotelian framework to do so. Thus, he showed not a single sign of dissatisfaction with Peripatetic elementary theory: to him, as to Fienus, the cosmos consisted of a heavy, sluggish globe of earth, situated at the centre of the universe, surrounded by several layers of air of increasing rarity, and, just below the lunar orb, by a sphere of extremely rare, pure fire. Exhalations, being rarer than the earth itself, tended to fly away from the centre of the world in straight lines; however, the more earthy material they contained, the heavier they were, and the less high they would be able to ascend. By contrast, celestial motions were invariably circular, and displayed a constancy not to be found in the changeable sublunary regions. The motions of the heavenly bodies were due to the rotation of the *primum mobile,* revolving about the centre of the world every twenty-four hours.

Yet, Fromondus' Aristotelian world view did display certain signs of decomposition. Especially in the course of his discussion of astral scintillation, he implicitly made it clear that he no longer held the heavens and the celestial bodies to be completely incorruptible. His explanation of the various aspects of cometary tails by means of according the hairy stars' heads certain differences of density and opacity points in the same direction. This did not, however, constitute as radical a break with tradition as might be assumed. Throughout the later Middle Ages, scholastics had admitted the fact that some parts of the celestial aether (namely, the stars and planets) were denser than others.[415] Yet, they had not generally admitted variations in density within one and the same celestial body; in doing so, Fromondus went one step further than his predecessors, relying partly on the recent observations of sunspots, but mostly on optical arguments in connection with cometary tails, that had already been put forward by the sixteenth-century adherents of the optical theory of comets.

In his final cosmological chapter, entitled "On the distance and magnitude of the comet", Fromondus abandoned his polemical stance vis-à-vis Aristotle; considering the exhalation theory sufficiently refuted, he presently applied his mind to the investigation of what kind of bodies comets might be, given the fact that they were not burning vapours. Since he had already demonstrated that comets did not consist of any sublunary substance, Fromondus first of all made the rather obvious statement that, therefore, the hairy stars were necessarily formed out of heavenly matter. They were consequently located in the heavens; the present comet, however, had in Fromondus' view stood beyond the moon, but below the sun. Relying on his telescopic investigation of the phenomenon, he stated that the image of the comet's beard, when observed through the telescope, had grown distorted and seemed to incline towards the instrument's orifice, as the moon was wont to do, while, conversely, the head had tenaciously stuck to the heavens, like the sun or Venus. This had convinced Fromondus of the fact that the beard was closer to the earth than the head; but if this were so, then the comet as a whole was necessarily located below the sun, since the beard was caused by solar rays passing through the head. By the same argument, the head could be demonstrated to have been located beyond the moon, because it did not suffer telescopic distortion whereas the moon did.[416]

Thus, to Fromondus the comet was definitely superlunary. How-

[415] See Grant, *Planets, Stars, & Orbs,* 198-202.

[416] Fromondus, *Dissertatio,* 121: "...materiam caelestem facio, & pono tamen infra Solem. ... Istud etiam affirmare possum vel de rudi inspectione per Batavicam Arundinem. Barba enim valde ad orificium tubi se inclinabat, ut Luna fere solet, capite interim satis pertinaciter caelis adhaerente, uti ferme Sol aut Venus. Hinc autem elicio barbam terris quam caput fuisse viciniorem; & hoc tamen infra Solem. Ut ergo barba promitteretur infra caput Sol excelsior fuisse debuit, ut a vertice ferme, per mentum ei radium trajiceret: quod non sic fiat, si aequali a terra distantia

ever, due to lack of instruments, he had not been able to accurately determine its parallax, so that he could not make any definite statements concerning the phenomenon's precise distance. Our author declared himself to be eagerly awaiting the results of German parallax calculations, especially those of Kepler. However, even naked-eye observations sufficed to show that the diversity of aspect had been very small; if the comet had stood in the air below the moon, a certain measure of parallax would definitely have been discerned for it, even without the aid of any instruments.[417]

The most proficient astronomers had, in 1577, determined an altitude of 180,600 German miles, or three times the distance of the moon, for the comet of that year, which distance placed the phenomenon at a location in the heavens beyond Venus. However, Fromondus added addressing Fienus, "Your very learned teacher and predecessor Cornelius Gemma confined it to the lower sphere of Mercury."[418] Yet, this difference of opinion, Fromondus hastily went on, was not as irreconcilable as it might seem; in fact, the comet might well have been seen above Venus and Mercury at some times, and below one or both of these planets at others. After all, these observed facts could be explained by assuming that all the planets revolved around the sun; almost all of the most learned scholars affirmed this to be the case, and only by way of this arrangement could the appearances of the inferior planets be satisfactorily rendered.[419]

Obviously, the assumption that all the planets circled the sun was common to both the Copernican and the Tychonic system alike; therefore, the passage of Fromondus' tract just described does not allow us to decide which of these two systems our author preferred.

---

fuisset, aut inferiori. Eodem argumento caput Lunam superavit: quia, dico, in tubum contumacius se demittebat."

[417] Ibidem, 122: "De parallaxi, unde altitudo eius tenuius inquiri potuit, inopia instrumentorum, non possum certi aliquid. non videtur fuisse valde magna, nec quanta in Luna apparet; & si in aere longe etiam ultra nubes fluctuasset, sine ullo instrumento parallaxim aliquot indubie grad. apertissime oculis ingessisset."

[418] Ibidem, 125: "Itaque ultra Venerem orbem suum duxit; licet eruditissimus magister & praedecessor tuus Cornelius Gemma in caelum Mercurii depresserit."

[419] Ibidem, 125-126: "Non difficulter tamen aliis conciliari poterit, si omnes Planetas circa Solem convertere suos motus, & respicere per modum centri asseramus; quod affirmant doctissimi plerique, & experientia vehementer consentit. sic enim fiet, ut quia Mercurius proxime Solem, deinde Venus, & denique Cometa ille tertius ambiverit, Cometa nunc Venere & Mercurio altior, nunc depressior utroque aliquando fuerit. Sed horum intellectus a nova illa caelestium orbium dispositione pendet; de qua alias."

Yet, his tract, when read in its entirety, leaves no room for doubt that he must have inclined towards the geoheliocentric rather than the completely heliocentric arrangement. First of all, we have already seen that the concepts of Aristotelian physics, especially the notion that earth was the heaviest of the elements, situated at the centre of the universe, were fundamental to his refutation of Peripatetic cometary theory. Also, his belief that the comet of 1618 was located "beyond the moon, but below the sun" was irreconcilable with the Copernican outlook, since, for obvious reasons, in a heliocentric universe, the expression "below the sun" did not make any sense. Thus, Fromondus' remark about the planets circling the sun should be read as a profession of adherence to the Tychonic planetary system (without, however, mentioning Tycho's name).

At this point, we may well ask ourselves what had become of Fromondus' favourable opinion of the Copernican system, as expressed just three years earlier in his *Peregrinatio Caelestis*. Apparently, in 1619, the option of realist Copernicanism had, in Fromondus' outlook, been abandoned to be replaced by emphatic adherence to the Tychonic arrangement. However, while he was composing his dissertation on the comet of 1618, the Copernicans and their opinions were clearly still very much on his mind. Thus, when treating the question of parallax, he remarked that he would very much like to hear the outcome of the parallax determinations of "the most learned Johannes Kepler, the Prince of present-day Copernicans in Germany."[420] This cursory mention of Kepler's name and his cosmological conviction inspired Fromondus to include a lengthy discussion of the vicissitudes and merits of the heliocentric system, on which he spent several pages of his tract, temporarily oblivious of the comet he was in the process of discussing.

His digression on heliocentrism began abruptly and anxiously, with the exclamation: "But what did I recently hear from you about the Copernicans, most honorable Fienus? That they have some years ago been condemned by the most holy Paul V? I had not yet heard about it, nor, I imagine, had any of those many very learned and Catholic men in Germany and Italy, who like Copernicus believe the earth to rotate."[421] Fromondus had not heard any rumours about the

[420] Ibidem, 122: "Libentissime tamen omnium doctiss. Ioan. Keplerum audiam, Principem hodie in Germania Copernicanorum. . ."

[421] Ibidem: "Sed de Copernicanis quid ex te nuper intellexi, Vir Clarissime? Ab

decision at all, while several years had already elapsed since it had been promulgated, and he would scarcely believe that Copernicanism had actually been condemned, unless more certain news were to arrive from Italy. Judging by his phrasing, our author seems to have been genuinely shocked; apparently, he had not anticipated the likelihood of any papal censure on the issue of heliocentrism, and was now trying to come to terms with it.

Much concerned about the spiritual welfare of his many Copernican colleagues, Fromondus remarked that, in his opinion, such papal "definitions"[422] should be especially well publicized in the universities, being the dwelling-places of learned men, who were in greater danger of succumbing to the error of heliocentrism than the common people. After all, it would be difficult indeed to convince any uneducated man of the idea that the earth under his feet rotated at astronomical speeds.[423] Even the great Lipsius, not quite unlearned himself, had rejected the idea of a moving earth as "delirious"; however, he had believed it to have been buried along with its inventor, Copernicus. Fromondus hastened to posthumously disillusion the great humanist: "You are mistaken, great man: even now it [the doctrine of Copernicus] lives and thrives among many."[424]

As for the grounds on which the Pope might have issued his condemnation, Fromondus supposed that he had had in mind certain Scriptural passages, especially Joshua 10, where it was written that the sun had been temporarily cut off from the movement of the *primum mobile,* and Ecclesiastes 1, where it said that it rose and went down and hastened towards the place where it had arisen. If interpreted in the strict sense, it was quite evident that these passages regarded the sun as a moving body. The Copernicans might reply that the Bible, having been written in a kind of language accommodated to ordinary people's sense perceptions, was not to be taken in

uno aut altero anno damnatos a Sanctiss. D.N. Paulo V? Hactenus mihi inauditum. inauditum tot hodie per Germaniam & Italiam doctissimis & catholicis, uti puto, viris, qui terram cum Copernico volvunt."

422 The Latin has "definitiones"; Fromondus' usage of this word in the context of a discussion of a papal decree explicitizing certain cosmological tenets, seems to confirm our hypothesis (see supra, p. 249) that, to Fromondus, any opinion could legitimately be held as long as there did not exist any explicit papal pronunciation on the subject.

423 Fromondus, *Dissertatio,* 123.

424 Ibidem: "Falleris, vir magne: vivit & viget etiamnum apud multos."

its literal sense as far as the natural world was concerned. Fromondus, however, did not understand the need for such an interpretation: after all, he said, the book of Ecclesiastes expressly compared the circular motion of the sun to that of river water, and the latter was unmistakably real, and not accommodated to common sense perception.[425] With this implicit endorsement of the papal decree, Fromondus concluded his digression on Copernicanism to return to the comet.

It has to be said that our author's refutation of the Copernican "theory of accommodation" seems somewhat half-hearted. After all, the fact that the Preacher in one and the same passage spoke both of the circuit of the sun and of that of river water, does not necessarily justify the conclusion that he regarded both movements as real. From the point of view of the Copernicans, it could be argued that, in the case of river water, there was no need for any 'accommodating' language, since it could be ascertained by the senses that the water of rivers circulated. Not so, however, in the case of the sun: to human beings there was no way to establish with certainty whether it was the sun or the earth that moved, so that, when speaking about the sun, the Holy Spirit might well have deemed it necessary to adapt his terminology to the inadequate way in which men perceived the cosmos. Fromondus himself must have realized the weakness of his argument; however, convinced of the authority of the papacy to explicitize Scripture, he chose to submit to the pope's interpretation.

Nonetheless, our author could not clear his mind of Copernican ideas without an effort. When, at the very end of his tract, he got down to expounding his own view of the material, nature, and generation of comets, the first hypothesis he came up with was not quite devoid of Copernican overtones. After having expressed his disbelief in the existence of any solid spheres (ridiculing Fracastoro's proposal of a separate sublunary sphere especially for comets), as well as his agreement with Fienus that the entire heaven below the firmament was fluid[426], he proposed to explain how, in such fluid celestial re-

---

[425] Ibidem, 124. See Ecclesiastes 1:5 ("The sun also ariseth, and the sun goeth down, and hasteth to his place where he arose") and 1:7 ("All the rivers run into the sea; yet the sea is not full; unto the place from whence the rivers come, thither they return again").

[426] Fromondus, *Dissertatio*, 128: "Fracastorii autem caelum Cometarum infra Lunam nemo iam retineat, cum hodie & omni aevo Cometae & sidera alia nascantur supra Lunam. Itaque ego totum a terra ad firmamenti fornicem

gions, comets could be generated. He admitted that to ascertain the hairy stars' material cause was the most difficult problem of comet-ology. Nonetheless, he went on, the Copernicans might put forward the following hypothesis. Since, in accordance with the teaching of Pythagoras, they thought of the earth as a planet, they might well hold that, by analogy with planet earth, the other planets also breathed out exhalations that had been extracted from their bowels by the power of the sun. In the same way that, to Copernicans, the earth by its nature revolved with a circular motion, carrying along with it all of the clouds and exhalations in its atmosphere, so also the planets might revolve around the sun along with their own exhalations.

However, in order to subscribe to this theory, our author went on, it was not at all necessary to convert to Copernicanism. For, even to those who wished to retain the earth as the stable and motionless centre of the universe, it must seem "highly probable" that all the planets, apart from the moon, circled the sun with their own, natural motions. In such an arrangement, Mercury and Venus should be thought of as closest to the sun, not including the earth within their orbits, while Mars, Jupiter and Saturn possessed orbits of such amplitude that they embraced the earth within them.[427] Here, then, Fromondus at last included an unambiguous description of the Tychonic system, in order to demonstrate that, even to those unwilling to subscribe to the newly condemned theory of Copernicus, the generation of comets could be explained by assuming them to consist of planetary exhalations.

If, then, we are willing to recognize the Copernican analogy between the earth and the other planets, and to admit the latter to be alterable bodies from whose bowels substances similar to earth and water might be extracted, these planetary exhalations might provide us with the matter of which comets are formed, Fromondus went on.[428] The fact that we never see the planets grow any smaller need

intervallare spacium tecum, Vir Claris. affirmo liquens & nullo solido orbe praeter nudos & simplices Planetarum globos instructum."

427 Ibidem, 131: "Imo hoc asserere potes, licet Copernicanam mundi ordinationem non totam probes. Nam licet terram in centro universi stabilias, est tamen valde verosimile, omnibus Planetis (excipio semper Lunam) circa Solem, motum proprium & naturalem ire. . .".

428 Ibidem, 132: "Si igitur in Planetis corpora patibilia cogitemus, & quibus aliquid, uti terrae & aquae, detrahi & exhalari possit, hinc forte materiam reperiemus pro Cometis."

not deter us; after all, we do not feel the earth under our feet diminish, in spite of the violent activities of many volcanoes, and, moreover, our eyesight is probably too weak to discern any planetary changes at such enormous distances. If we could scrutinize them at closer range, we might well find that their spherical shape is not really as perfect as it would seem from far off; indeed, judging by the spotted and mountainous nature of the moon, and by the various shapes displayed by Venus, such a derogation from planetary perfection seems quite likely.[429] Any celestial exhalations belched out by the planets, being similar in nature to the planets that produced them, would tend to move naturally in circles round the sun, until the sunrays rarified them to such an extent that they would dissolve to pure, liquid aether.[430]

Despite the plausibility of this explanation of cometary generation, however, Fromondus refrained from expressing his explicit support for the theory. "Those who are full of the opinion they once received from their teachers, and are not curious to learn any more, merely laugh when someone dares to propose something new to arrive at the invention of the truth", he complained. Such people would undoubtedly condemn the planetary exhalation theory as "hallucinations of a sick old man" and dismiss it as ludicrous.[431] In spite of the fact that, clearly, Fromondus himself wished to dissociate himself from such blind partisans of authority, he nonetheless went on to admit, in his very next sentence, that he, himself, also anticipated "many difficulties and obstacles" if the theory of comets as planetary exhalations were to be pursued. Without elucidating what kind of difficulties these might be, he therefore proposed to "try a different way".[432] Thus, apparently, our author was not prepared to profess his explicit adherence to a hypothesis that could, in theory, be upheld within the physical framework of a Tychonic as well as a Copernican universe,

---

[429] Ibidem: "Quis e vicino arbitratus est, an nihil de sphaerica perfectione mutilatum? certe exasperatam verrucis, collibus & montibus totis Lunae superficiem deprehendimus. & de Venere etiam forte an probabile, quam heri 29 die Ianuar. tecum Clar. Domine . . . instar Lunae evidenter corniculantem adspeximus."

[430] Ibidem: "Caelestes ergo illae exhalationes secutae naturam Planetae unde sunt extractae, naturaliter currunt in orbem circa Solem, donec radiis eius in purum & liquidum aethera dissolvantur."

[431] Ibidem, 132-133: "Sed haec aegroti veteris somnia dicent, & Anticyras me sine capite mittant, qui una quam a Praeceptoribus olim acceperunt, opinione pleni, nec ultra curiosi, rident tantum, si quis novi aliquid audeat pro veritatis inventione."

[432] Ibidem, 133: "Et profecto tamen ipse plures difficultates & remeligines in hac via provideo, quam ostendam. Aliud ergo tentemus."

but that, nonetheless, was tainted with Copernican overtones, especially with regard to the far-reaching analogy it presupposed between the earth and the planets.

Sure enough, the "different way" he consequently embarked on was much more easily to be reconciled with an Aristotelian world view. Reminding his readers of the fact that the Stagyrite had supposed comets to draw their material from the Milky Way[433], Fromondus suggested that it might well be possible that the hairy stars came into existence as a result of the coagulation of several of the tiny stars of which the Galaxy had recently been shown to be composed. In this view, it was to be supposed that the Milky Way, rather than forming a part of the firmament, consisted of a band of pale starlets situated in the fluid aether between Saturn and the firmament of fixed stars. Thus, the influence of the sun and the planets was able to act on the galactic matter, and of making it condense and attain a certain measure of gravity, so that, on occasion, parts of it would descend through the fluid aether all the way to the region where the planets moved. On its arrival there, such a dense knot of galactic matter would join in the planetary motions, starting to revolve around the sun, since, being composed of celestial substance, circular movement would come naturally to it. This would explain why comets were observed equally often to move from east to west as from west to east; after all, both of these motions were circular, and there was no reason why comets should prefer either one of them; therefore, they simply took the direction bestowed on them by their initial impetus.[434]

Because the parts of such galactic comets would not be very compact or coherent, they would gradually be forced apart and rarified by the action of the solar rays; thus having been rendered lighter, they would subsequently reascend towards their place of origin in the highest aether beyond Saturn. This was in accordance with the observations, which showed that comets invariably rose higher and higher from the first day of their appearance onwards. It also explained why, initially, comets always seemed to move fastest, while later on they invariably tended to slow down: this retardation was due to the fact that, when they embarked on their journeys back towards their birthplaces, they moved away from the earth in a

[433] See Aristotle, *Meteorology*, I.7.345a.
[434] Fromondus, *Dissertatio*, 134-135.

straight line, so that to observers on earth they appeared to move extremely slowly or even to remain stationary.[435]

Fromondus showed himself more favourably inclined towards this 'galactic matter' theory than towards the earlier 'planetary exhalation' theory. He believed the appearances of the comet of 1618 to be in accordance with it; thus, he professed, that body had not come into existence at the location where he first saw it, but rather in the Milky Way, under Scorpio's tail.[436] Nor did the Aristotelian incorruptibility of the heavens pose any serious problem to Fromondus. He stated that he believed the whole heavenly region below the firmament was naturally corruptible, and that occasionally corruptions actually took place there.[437] However, our author at once went on to qualify this categoric statement: the planets, like the hardest and most unchangeable metals and stones on earth, were resistent to any kind of alteration.[438] Here, then, Fromondus explicitly denied the Copernican analogy between earth and planets with which he had been flirting just a few pages earlier. Apparently, he was unwilling to risk being associated with the condemned heliocentric theory in any way. However, by pointing out the analogy between the unchangeability of the planets in the heavens and certain metals and stones on earth, he did, implicitly, invite comparison between the celestial and the terrestrial regions as a whole: both contained both corruptible and incorruptible materials.

Unlike the planets, Fromondus continued, the galactic matter out of which comets were formed was weak and incoherent, and therefore capable of corruption. With characteristic caution, he hastened to add that, by "corruption", he did not necessarily mean to denote any substantial changes; for the generation of comets out of galactic material, it sufficed to admit that the celestial aether was capable of condensation and rarefaction. To enhance the persuasiveness of his

---

[435] Ibidem, 135.

[436] Ibidem, 134: "Nostrum etiam Cometam non existimo natum, ubi primum vidi, sed longius, & extra Tropicum hibernum in ipsa Galaxia iuxta feram & sub cauda Scorpii; inde enim recte ascendebat & inde factum, ut in Meridie citius adspici valuerit, quam in Belgio."

[437] Ibidem, 135: "Favente ergo generatione Cometarum, ego quicquid caelorum infra firmamentum est natura corruptibile censeo, & facto quandoque aliquid inde corrumpi."

[438] Ibidem, 135-136: "Ut tamen in his Elementaribus quaedam metalla & lapides nullis ignibus violari possunt: sic Planetae tanquam durissimi & excoctissimi a natura adamantes, iniuriam omnem respuunt & detrectationem."

theory, Fromondus pointed out that the existence of density and rarity in the heavens had of old been acknowledged; after all, the Peripatetics had always taught that the stars were denser bits of celestial matter.[439]

Thus, finally, Fromondus, after many detours, arrived at what seemed to him the most probable explanation of cometary generation. They were pieces of condensed aether, drawn from the Milky Way, which remained coagulated for longer or shorter periods, depending on the degree of cohesion of the material. As such, they could be regarded as wandering stars temporarily located in the region of the planets, describing circular motions like the planets—either by virtue of their proper natures, or with the aid of some attached intelligence.[440]

We can conclude that, like that of his colleague Fienus, Fromondus' world view remained predominantly Aristotelian in character. He remained attached to the three main tenets of Aristotelian cosmology, placing the earth in the centre of the universe, surrounded by concentric layers of air and fire, and divided from the celestial regions by the sphere of the moon. Furthermore, he wholeheartedly subscribed to the Peripatetic theory of motion, according to which sublunary bodies moved in straight lines while celestial bodies possessed circular motion. Somewhat less Aristotelian was his conviction that the heavens were fluid and to a certain extent corruptible; however, as we have seen many times before, these tenets could be, and were, relatively easily adopted without breaking with the Aristotelian world view in any fundamental way. Thus, although he relied rather less on Seneca and rather more on recent observations and on thought experiments derived from common sense perception, Fromondus reached conclusions almost identical to those of Fienus: comets were celestial bodies located in fluid and corruptible heavens.

However, perhaps due to the fact that he was clearly better versed in astronomy than his colleague from the medical faculty, Fromon-

---

439 Ibidem, 136: "Aut certe si aetherem substantialiter corrumpi displicet, da saltem rare aliquando fieri, & addensari; & nihil ultra pro Cometis postulo. quidni autem alteratio talis in incorruptibilibus tolerari potest? in caelo praesertim, ubi partium etiam eiusdem speciei altera rari & tenuioris fili, altera densata in stellae orbem?"

440 Ibidem: "Breviter ergo. Cometa peregrinum in Regione Planetarum sidus est, cohaesum de substantia aetheris, & circulum in Sphaera, maiorem, ut Planetae a propria forma (vel alligata intelligentia, si vis) describens."

dus was much less sceptical about human ability to grasp the essence and nature of comets; therefore, unlike Fienus, Fromondus did attempt to discover what kind of bodies the hairy stars really were. In the course of his investigation, he discussed many of the novel theories and observations of the recent decades; thus, he mentioned most of Galileo's telescopic discoveries and did not hesitate to make use of them as evidence for or against certain cometary theories. He was also clearly very well acquainted with the Copernican theory, and maintained contacts with several German Copernicans.

Nonetheless, perhaps due to the papal decree of 1616, about the existence of which he had only recently learned, the mood of his tract remained cautious throughout. Whenever the observed facts allowed for a choice between two cosmological alternatives, Fromondus preferred to stick to the orthodox option—although he invariably also expounded the more audacious alternatives, and never explicitly dissociated himself from them. Thus, he explained to his readers the basics of the Copernican system, the way in which it could be theologically defended, and the analogy it presupposed between earth and planets (on the basis of which the, quite plausible, 'planetary exhalation' theory of cometary generation had occurred to him). However, he hastened to add that all the advantages of the Copernican hypothesis were also contained in the more orthodox Tychonic arrangement of the planets. Furthermore, he sympathetically expounded the idea that comets might consist of planetary exhalations; however, due to unspecified "difficulties", he abandoned it in favour of the 'galactic matter' theory, which was more in keeping both with Peripatetic cosmology and with the dicta of Aristotle, who had held that comets were formed out of material located in the Milky Way. And, finally, despite professing his adherence to celestial corruptibility, Fromondus hastened to add that this need not include anything other than the widely accepted notion that heavenly matter could condense and rarify.

Despite his continual caution, however, our author did arrive at some conclusions that might be regarded as first steps on the road towards unification of terrestrial and celestial physics. In the course of his exposition of the 'galactic matter' theory to which, in the end, he subscribed, he came up with two notions which both amounted to a transfer of terrestrial properties into the heavenly region. First of all, he believed that, as comets were formed through a condensation of galactic matter, they would acquire heaviness and thus descend into the lower regions of the aether, while, conversely, when the sunrays once again rarified them, they would tend to ascend back to the higher regions from whence they had come. Thus, he accorded lightness and heaviness, with their characteristic rectilinear motion, to bodies formed out of pure celestial aether, abandoning the notion that heavenly movements could be nothing but circular. And secondly, as we have seen, Fromondus saw the heavens in their entirety as a region made up of both incorruptible entities (the planets) and corruptible ones (the material of the Milky Way and the comets that were formed out of it); as such, they were fundamentally similar to the sublunary regions, where both incorruptibility (metals and stones) and corruptibility (other elementary mixtures) could equally be found. Thus, albeit implicitly, Fromondus went a long way towards abolishing the dichotomy between sublunary and superlunary regions.

d. *Erycius Puteanus: "Caeli civem esse, Solis sobolem, aetheris partem & partum"*

On 4 March, 1619, Erycius Puteanus, bedridden with a "serious disease" and too ill to write, dictated a short preface dedicating his recently composed work on the comet of 1618 to his patrons, the Archdukes Albert and Isabella.[441] In it, he announced to what purpose he had written the booklet: by demonstrating that what the majority of people believed to be a burning fire and a prodigy bringing death to princes, was in reality a luminescent star entailing good fortune and happiness, he wanted to present *his* Princes, Albert and Isabella, with a sidereal token of prosperity. In order to do so, he proposed to imitate the example of the common people, paying attention to every one of their habitual prejudices, and showing that, all things considered, it could not be maintained that the hairy stars presaged disasters; rather, they should be thought of as messengers of happiness.[442] Thus, Puteanus from the outset made it clear, not only that, in his opinion, comets were celestial bodies, but also that the main concern of his tract was the question of cometary significance.

---

[441] Erycius Puteanus, *De Cometa anni 1618 novo Mundi Spectaculo, Libri Duo Paradoxologia* (Cologne 1619) 5: "Lovanii, in arce vestra, a gravi languens morbo dictavi, adeoque brevis."

[442] Ibidem, 4.

Although, in order to defend his optimistic view of the hairy stars' predictive value, he needed first of all to demonstrate their celestial nature, in the end such physical considerations were subservient to the assurance of peace and prosperity to the Archdukes that constituted the culmination of his tract.

Because of this difference in emphasis, Puteanus' tract bore rather a distinctive character from those of Fromondus and Fienus. Whereas the latter had been chiefly interested in elucidating the essence and physical nature of comets, and had devoted relatively small sections of their treatises to the question of cometary significance, Puteanus' work was divided into two books of almost exactly the same number of pages, each consisting of nine "paradoxes". The first book, dealing with the comet's physical characteristics, was clearly intended as not much more than a preliminary to the second, dealing with its significance; after all, the best way to invalidate the astrological consequences traditionally associated with burning, sublunary comets by Aristotelian-Ptolemaic cometology was to demonstrate that the hairy stars did not, in fact, consist of fiery terrestrial exhalations, but were celestial bodies located beyond the moon. Once the heavenly nature of comets was acknowledged, the entire structure of Aristotelian cometary astrology with its dire predictions would necessarily fall.[443]

Although Puteanus had probably read the dissertations on the comet of 1618 recently published by Fromondus and Fienus[444], the structure of his tract, as well as its purpose, was rather different from those of his colleagues' contributions to cometology. While, as we have seen, both the professors of medicine and of philosophy more or less followed the standard Aristotelian method of expounding arguments on either side of a given question before coming up with a final answer, Puteanus, as a true humanist, employed rather a different strategy. Although he occasionally mentioned (and invalidated) the odd Peripatetic objection to the superlunary nature of comets, most of his tract constituted a learned and somewhat moralistic expo-

[443] See below, § VI.4.b.

[444] This seems probable because Puteanus mentioned the fact that "the most honourable Thomas Fienus. . . has collected the various opinions of great men, and has explained the fluidity of the heavens in his dissertation on the subject of this same comet."; Puteanus, *De Cometa*, 41. The fact that the dissertations by Fienus and Fromondus were published in one volume makes it highly likely that, if, as seems to be the case, Puteanus had read Fienus, he had also read Fromondus.

sition interspersed with quotations from a great variety of classical authors. Whereas, like those of Fromondus and Fienus, Puteanus' tract was composed very systematically, first of all describing the various cometary opinions of the ancients, and then presenting several arguments in favour of celestial comets, based on their motions, duration, parallax, form, light, size, and colour, he did not usually mention the Peripatetic opinion on those subjects before expounding his own; rather, he tended to come straight to the point, without meticulously setting forth, and invalidating, every possible Aristotelian objection in the way his colleagues did.

This difference in approach came to the fore most clearly in the three introductory "paradoxes" with which Puteanus thought it necessary to begin his tract. These also made it clear, from the very beginning of the work, how heavily our author relied on the Stoic world view as set forth by Seneca in the cometological chapter of his *Naturales Quaestiones*. All fear, Puteanus insisted, was aroused by ignorance; therefore, natural phenomena which in some way deviated from the habitual order of the universe, being least well known because of their rarity, were also among the most terrifying occurrences to take place in heaven and earth. The common people, never prone to rationally investigate the causes of such events, tended to remain ignorant and to fall victim to the superstitious impostures of all kinds of astrologers, fortune tellers, and madmen, who foretold all kinds of disasters, but who were, in reality, as ignorant as their gullible customers.[445] This universal ignorance was responsible for the bad reputation of comets; Puteanus, however, proposed to put an end to this injustice, by investigating their nature and essence and thus liberating them from the dark clouds of superstition that had obscured their true identity for too long.[446]

Next to ignorance, another fear-inducing characteristic of rare

[445] Puteanus, *De Cometa*, 12-13: "Vulgus tamen omni aevo formidare maluit, quam scrutari; quicquid non caperet, extimescere. Sic quidem pavoris & horroris mater ignorantia est... Omnino quicquid in obscuro est, percellere solet; adeoque ipsa ignorantia obscura est, & in superstitionem trepidas mentes rapit. Astrologos, Hariolos, Augures, & furentes omnes accuso, qui, ut opinionem scientia velent, dogmata sua fatalibus minis onerant."

[446] Ibidem, 13-14: "Cometa est, ille novus, ille nuperus, ille omnium sermone agitatus: quem examinare, quantum licet, constitui: non ut scire plus velle videar, si curiosior ero sed ut minus metuere si non omnia ignorabo... Gravis enim significatio partim ab ignorantiae caussa est, partim a novitate, quam unam admirari solemus."

natural occurrences, especially comets, was their novelty. Relying heavily on Seneca, who as we have seen began his discussion of comets by pointing out that the reason these phenomena gave rise to such a great deal of debate was their infrequent appearance[447], Puteanus pointed out that anything seemingly out of line with the usual order of the cosmos tended suddenly to open man's eyes, which remained blind to daily events of such magnitude as the rising and setting of the sun. Only when a solar or lunar eclipse occurred, or a comet appeared, did the human race show itself interested in the condition of the sky, alarmed by these deviations from the day-to-day course of nature. Nonetheless, comets, despite appearing at irregular and unpredictable intervals, followed the laws and order of nature in the same way as did other, more customary phenomena, Puteanus insisted.[448] Therefore, he thought it well possible to subject the hairy stars to a closer examination, and set himself the task of answering the three basic questions of what comets were, where they were located, and how they came into existence, announcing that, although he intended to propose solutions that ran counter to the majority opinion, his "paradoxes" would nevertheless approach the truth more closely.[449]

First of all addressing the related questions of what comets were and where they were situated, Puteanus admitted that these were issues about which of old there had been much debate yet few results. Most scholars had, along with Aristotle, believed them to be fires burning in the air; yet many investigators, some of them from the remotest past, had preferred to regard them as stars shining in the heavens. The latter opinion involved many uncertainties, our author pointed out; for, whereas Aristotle had come up with a persuasive theory of how comets might be generated, move, and expire in the air, no-one had as yet been able to do the same with regard to celestial comets. If they were stars, either they had been created in the beginning and existed ever since, or they were newly generated through unknown causes. Both these opinions could be defended on

---

[447] See supra, chapter III, p. 53. See also Seneca, *Naturales Quaestiones*, VII.1.

[448] Puteanus, *De Cometa*, 15-16: "Admiranda haec: naturae tamen legem & ordinem, sequuntur, semel a Deo causâ universi, constituta. Sequuntur Cometae, vano licet terrore passim traducti, & prodigioso omine infames; quia non omnibus annis, aut lustris nascuntur."

[449] Ibidem, 17.

theological grounds, and it was very difficult to make a well-founded choice between them.

In favour of the occasional creation of new stars in the heavens, Puteanus employed the argument of the *potentia Dei absoluta*, well known from medieval theology and natural philosophy, according to which God was capable of making anything happen in the cosmos, short of a logical contradiction. Thus, Puteanus pointed out, Christians were obliged to believe in the possibility of new stars and even new celestial spheres being created at God's command. After all, it was unthinkable that the powers of such an infinite and omnipotent being would be exhausted after the creation of one single work; rather, it was to be supposed that he had created the fabric of the world in such a way as to leave open the possibility of enlarging and embellishing it.[450] In fact, the universe could safely be said to be capable of indefinite growth. It could be increased *ad libitum,* as long as the sum of its parts remained smaller than its Creator, who was infinite, eternal, and limitless, so that, however many new bodies were added to the heavens, the entirety of the cosmos would never equal the magnitude of the God who had made it.[451] Moreover, God's presence was especially pronounced in his heavenly domicile; therefore, the creation of new celestial bodies seemed the most likely instrument for him to employ in order to show the human race his anger and to display his divine power.[452]

Nonetheless, God's omnipotence might equally be saved by arguing that, at the time of the Creation, he had bestowed perfection on the newly-framed universe. If this were the case, then, obviously, no subsequent additions to the fabric of the cosmos were thinkable, as these would inevitably derogate from the perfect state in which the world had been created. From this point of view, any supposed 'new arrival' in the heavens, such as a nova or a comet, should be attrib-

---

[450] Ibidem, 19: "Etenim si stella est, vel fuit semper, vel nova certe origine coepit. Et quae haec tam occulta natura est? Occulta, non repudianda. Iubeat Deus, & nova sidera, imo novi caelorum orbes exsurgent. Ita enim totam hanc mundi machinam condidit, ut augere & ornare etiam possit: nisi exhaustas tantae potentiae vires, unius operis molitione, credas: quod sine scelere nemo arbitratur."

[451] Ibidem, 19-20: "Opus quod videmus, finitum, limitatum, & in ipsa sua magnitudine parvum est: Numen quod credimus colimusque, infinitum, aeternum, omnesque terminos ignorans. Unde si comparatio instituimur, tamdiu augeri opus potest, quamdiu Auctor hoc ipsum superat, nunquam tamen opere aequandus. Quicquid accedet, finitum erit: infinitus ille, a quo accedet."

[452] Ibidem, 20.

uted to the erroneous perception of the human intellect; they should rather be thought of as existing bodies which had been hidden in the recesses of the universe since the creation, and had recently approached the earth so as to become visible to human eyes.[453]

Thus, both the view of comets as newly created celestial bodies and the view of the hairy stars as entities hidden in the depths of space for most of their existence, were theologically defensible. Still, Puteanus expressed a marked preference in favour of the former point of view, believing that it was more easy for new stars to be born in the heavens, than to be hidden there. Once more expressing his belief in the *potentia Dei absoluta*, but this time almost equating God with Nature, he stated that, in every corner of the universe, Nature acted in accordance with God's will. Thus, she subjected both the heavens and the earth to her sway, generating, corrupting and transforming things similarly in both regions. Therefore, new arrivals in the heavens should not be regarded as miraculous; after all, things were continually being born everywhere.[454] Thus, Puteanus, undoubtedly under the influence of the Stoic view of the cosmos as an organic whole all of whose parts were subject to one and the same world soul that worked similar effects everywhere, expressed his belief in a far-reaching analogy between the celestial and terrestrial realms.

Puteanus, then, put forward as his personal view that comets were celestial bodies that were newly generated in some occult yet natural way.[455] Yet, he acknowledged that none of his predecessors had espoused this opinion. Among the ancients, Aristotle and all the Peripatetics had believed comets to be hot and dry exhalations extracted from the earth and set on fire by the rays of the stars; while similarly, the Stoics had thought the hairy stars to be formed of dense air, like other heavenly signs. On the other hand, many ancient authors had believed in the superlunary nature of comets; yet, unlike Puteanus,

[453] Ibidem: "Verumenimvero semel natura completa est, ideoque a Deo constituta, ut perfectionem suam haberet. An dicam? nostro errore sit, ut admittere incrementum videatur. Saepe enim quae miramur, non quia non fuerunt, sed quia latuerunt, prodigia sunt; prodigia, quia ex occulto prodeunt, sive producuntur."

[454] Ibidem, 20-21: "In caelo autem nasci novum sidus, quam condi, facilius dixerim. Agnosco Deum, neque ministram Naturam excludo. Haec quocumque se extendit, morem Auctori suo gerit: terrae, caelique potens, cuncta sibi obnoxia trahit, generando, corrumpendo, transformando. Quid mirum igitur, in nuperis illis aethereisque plagis aliquod novi esse? ubique est."

[455] Ibidem, 22: "Subnasci vero, & occulto quodam Naturae modo, mea iam quidem opinio est."

they had all held the hairy stars to be eternal bodies that were usually hidden in the depths of the universe. This had been the view of the Pythagoreans, the Chaldeans, and the great Seneca, "whose mind and condition were divine rather than human", as Puteanus admiringly added.[456] Democritus and Anaxagoras had also thought comets celestial as well as eternal, believing them to be planetary conjunctions; and Hippocrates and Aeschylus had thought the heads heavenly, yet the tails elementary. However, none of the ancient authors had argued for comets that were celestial, yet transitory.

More recent opinions, too, were divided between heaven and earth, Puteanus added, with most philosophers preferring the sublunary option and the solution proposed by Aristotle, while several mathematicians opted for celestial comets.[457] Needless to say, Puteanus expressed his confirmed allegiance to the latter point of view. With boundless admiration, he mentioned the observations that had recently been conducted by the most proficient mathematicians of the age, convincingly demonstrating the celestial nature of comets; especially Tycho Brahe, "the eye of astronomy", was honoured with lavish praise by our author.[458] He even ventured to suggest that the prediction made by Seneca, that "Some day there will be a man who will show in what regions comets have their orbit, why they travel so remote from other celestial bodies, how large they are and what sort they are"[459], had been fulfilled by Tycho's demonstration that the comet of 1577 had been a celestial body instead of a fiery meteor.[460] Thus, Puteanus triumphantly exclaimed, recent diligence had attained what had been too difficult for ancient science, and recent knowledge had achieved a feat that had eluded the diligence of all previous ages.[461]

Proudly and wholeheartedly aligning himself with the mathematicians he so exuberantly admired, Puteanus repeated that, in his opinion, too, comets were celestial phenomena. Yet, deviating from the

---

456 Ibidem, 23-24.

457 Ibidem, 26.

458 Ibidem, 29: "Certe demonstravit [in quibus Cometae partibus errent] non unus; & inter alios tamen quasi unus, TYCHO BRAHAEUS, ille astronomiae oculus, & RUDOLPHO Caesari nuper carus. Demonstravit, inquam, Cometen anni 1577 non in aeria, sed caelesti regione exstitisse."

459 Seneca, *Naturales Quaestiones*, VII.25.7.

460 Puteanus, *De Cometa*, 29.

461 Ibidem: "Horum [mathematicorum] industriam, an scientiam magis mirer, dubium est. Namque industrià consecuti sunt. quod solertissimae Antiquitatis scientià maius fuit; scientià quod omnis aevi industriam fugit."

view of most ancient authors, including the much-admired Seneca, he did not hold them to be eternal bodies; after all, he had already shown that there was nothing unthinkable about new bodies being generated in the heavens. Convinced of man's capability of acquiring certainty about the nature of the hairy stars, he overconfidently and somewhat enigmatically announced that "after having duly weighed up and examined all the facts" he had arrived at the conclusion that "the comet is a citizen of heaven, a scion of the Sun, part of and born from the aether; a star with regard to its body, a torch with regard to its light, a planet with regard to its motion; its body is its own, its light is not, and its motion is both its own and not its own."[462] Thus, they were transient bodies, that were born and perished in the aethereal region; yet this could take place without in any way damaging the heavens, and without any dire connotations.

Having thus defined his position with regard to cometary theories ancient and modern, Puteanus set out to demonstrate the hairy stars' celestial nature by showing how their etymology, motion, duration, parallax, form, light, size, and colour all led to one and the same conclusion, namely, that they were heavenly bodies. Much the most important of these arguments was the one derived from cometary motion, which Puteanus discussed in his sixth "paradox" and which inspired him to an extensive discussion of recent discoveries and opinions concerning the nature of the heavens. He began this interesting section with a sophism more worthy of an admirer of Isidore of Seville than of a self-professed partisan of Tycho Brahe, stating that comets were necessarily celestial, since they were called "ἀστήρ κομήτης" in Greek and "*stella* crinita" or "*sidus* crinitum" in Latin.[463] However, after this initial casuistry, our author embarked on an investigation of the comet's motion in the course of which he revealed himself as highly knowledgeable about the most topical issues of early seventeenth-century cosmological speculation.

First of all, Puteanus described the path followed by the comet of 1618. The phenomenon had been endowed with a double move-

---

[462] Ibidem, 27: "Etenim expensis, & examinatis omnibus, in eam venire sententiam audeo, ut statuam, Cometam caeli civem esse, Solis sobolem, aetheris partem & partum; corpore stellam, lumine facem, motu erronem: corpore suo, lumine non suo, motu suo & non suo."

[463] Ibidem, 30: "Nomen mihi omen est: aetherem, Deique domum indicat, Graeco & Latino loquendi usu. Graeco quidem ἀστήρ κομήτης, Latino, *Stella,* sive *Sidus Crinitum* appellatur. 'Αστήρ autem Stella, Sidus, ubi, nisi in caelo?"

ment, like the planets: it had followed the daily movement of the firmament, or the "circle of day and night" as our humanist poetically dubbed it, as well as pursuing its own proper motion in the contrary direction, moving northward regularly and proportionately —though decelerating somewhat towards the end of its visibility.[464] Without corroborating his statement with any measurements, Puteanus stated that, when the comet's size had been greatest, it had described three degrees a day, while later it slowed down to less than two degrees daily. Although he professed to have observed the entire movement of the comet, our author's rendition of its path was extremely crude. Without mentioning any instruments employed to determine its positions on various nights, he cursorily stated that the comet had first been seen on 26 November 1618 at five a.m., when it had stood in Libra near the equinoctial circle. From there it had moved northward in between Virgo and the Serpent, and on towards Bootes. Around 13 December it had reached that constellation's left arm, and had moved very close to Ursa Maior, moving past the last two stars of that constellation's tail to become circumpolar. On 21 December, it was seen in the hand of Bootes with faint and shortened rays; it continued moving for several more days, growing progressively weaker, and finally expiring.[465] Clearly, Puteanus had not been observing the phenomenon as assiduously as Fromondus and Fienus, who both agreed that the comet only ceased to be visible around the middle of January.

Despite the scantiness of these data, Puteanus nonetheless thought it possible to deduce from them that comets pursued their own particular paths, conferred on them by Nature, and capable of being grasped. Such regular movements, he went on, were unthinkable in the unstable region of the air, where motions could not even be perceived or expressed in words, and where every movement was fortuitous, uncertain, and vagrant.[466] In an argument derived from Seneca, whom he quoted at length, Puteanus pointed out that fire moved wherever by chance its fuel might happen to lead it, or the wind might blow it, instead of obeying a certain order or law of motion. The Peripatetic idea, that cometary movements were de-

[464] Ibidem, 30-31.

[465] Ibidem, 32-33.

[466] Ibidem, 33: "Haec cum ita sint, peculiares a natura gyros constitutos Cometis censeo, perviosque esse: in aere ne concipi quidem, aut designari posse. Omnia haec fortuita, incerta, & vere palantii sunt."

pendent on the powers of one or another of the stars, which attracted them and in doing so "transferred the law of heaven to the air", did not hold true, because it presupposed the impossible: the nature of the stars was fundamentally different from that of the light, fluctuating, volatile air, so that the regularity of cometary motions could not be explained by placing them in the air, not even by postulating a sidereal force of attraction. The only way of accounting for them was to accord them a heavenly location.[467]

Not content with this demonstration of comets' superlunary position, however, Puteanus went on to disclose to his readers how, in his view, the heavenly region was capable of harbouring the kind of transient phenomena that comets demonstrably were. First of all, he once more expressed his belief in the alterability of the heavens. Why should it not be possible for new things to be born and to perish in the celestial region? he rhetorically asked. After all, it was highly likely that many things changed there, invisible to the weak human eye; and that many things existed, in the faraway recesses of the universe, of which human perception could not conceive. Moreover, probably, many things that *were* visible would, on closer inspection, turn out to conform to an arrangement quite different from that discerned by man. In imitation of Fienus, Puteanus pointed out that, if we were suddenly to be transported to the sphere of the moon, or to any other part of the universe, the whole fabric of the world would look completely different: "The images with which the audacity of mythology has furnished the stars, would vanish."[468]

Utterly convinced that the universe in its entirety had been created perfect, beautiful, and well-ordered, Puteanus could not imagine the reality of the heavens to be as chaotic and fortuitous as it seemed to be when seen by our feeble human eyesight. After all, the image perceived by man was one of disequilibrium: some parts of the celestial regions were filled with an abundance of stars, while others were completely empty, as if the heavens were happy in some places, while

---

[467] Ibidem, 34-35: "Peripatetici nihilominus vi alicuius sideris rapi volunt Cometam, & in aere caeli legem esse. Sed profecto volunt, quod maxime dubium, aut etiam difficile est. Etenim alia sideris, alia aeris natura est, & loco seiuncta. Aer fugax, fluxus, levis: sidus certum, perenne, constans. In summa: aut in caelo Cometa est, aut nulla aer levitate fluctuat. Ergo motus ille caelestis est: & quia motus, ipse quoque Cometa."

[468] Ibidem, 37: "Quod si vel ex orbe Lunae, vel ex alia mundi parte fas homini caelum contueri, novus omnino & ignotus adspectus esset: periissent imagines, quibus fabularum audacia stellas occupavit."

sad and deprived in others.[469] Yet, it was unthinkable that this imbalanced situation should correspond to reality: since the fabric of the world was eternal and divine, it was worthy of the highest degree of order, just as, on earth, the flowers and trees in gardens, the towers in fortifications, and the jewels in necklaces were arranged in a beautifully ordered fashion. Therefore, Puteanus suggested that, probably, not all the stars were attached to one and the same surface; instead, they were all located at different distances in the depths of space, some higher, others lower, but all arrayed in perfect order, albeit a kind of order inaccessible to human sense perception.[470]

Because of the weakness of our senses, Puteanus went on, we fail to distinguish the orderly arrangement of the heavens; however, this does not mean that there exists no such pattern. People tended to dub any occurrence that did not immediately fit in with their own, limited sense of order a 'prodigy'; yet, if their perceptions were enlarged sufficiently for them to be able to see how the unusual phenomenon fitted in with the universal order, the need for the concept of 'prodigy' would disappear.[471] This point of view (which, as we have seen, was to a large extent borrowed from Seneca)[472], had recently been rendered even more probable by several new discoveries of the past few decades. After all, in 1572 a new star had appeared in Cassiopeia, and even more recently, the telescopic observations had finally enabled scholars to see for themselves that many aspects of the heavenly condition diverged sensibly from what had been taught about them by Aristotle. Who, for example, could have dreamt that Venus displayed the same variety as the moon? Or that Saturn was surrounded with several small planets? Or that the paths of the planets intersected each other in heavens filled not with solid orbs, but with a fluid substance?[473]

Clearly, Puteanus regarded the fluid nature of the heavens as a

---

469 Ibidem, 35-36: "Nam ut se imago insinuat, alibi frequentia, alibi raritas est; quasi & felicitate sua alibi exuberet caelum, alibi sterilitate deficiat."

470 Ibidem, 36-37: "Suspicor igitur, non in superficie haerere ac lucere tam admirandos ignes, sed alium atque alium in profundissimo spatio locum occupare. quosdam sublimiores, quosdam humiliores esse; omnes summo & admirando ordine distare: sed ordine, qui oculos nostros fugit."

471 Ibidem, 37-38.

472 See supra, chapter III, p. 53-56.

473 Puteanus, *De Cometa*, 38: "Sed quid? novum in caelo sidus fulgere posse, superiori saeculo Cassiopea ostendit: imo vero pleraque aliter se habere, quam ab Aristotele prodita sunt, sensu tandem exploratum est. Hactenus quis opinatus, Lunae

consequence of the telescopic observations, treating it on an equal footing with the phases of Venus and the 'handles' of Saturn. From his next sentence, his line of reasoning can be deduced: just after mentioning the newly discovered celestial fluidity, our author announced that he had "abandoned the common dogma, and begun to believe that only the moon and the sun revolved round the earth, while the other planets circled the sun, with Venus and Mercury below the earth, and Mars, Jupiter, and Saturn in wider orbits beyond it."[474] Our author thus explicitly expressed himself an adherent of the Tychonic planetary system; apparently, he regarded the telescopic discovery of the phases of Venus as conclusive proof of the truth of the geoheliocentric arrangement, and this in turn inspired him to do away with any solid celestial spheres, and to fill the heavens with fluid matter instead.

Comets, too, pursued their varied paths among and between the orbits of the planets, Puteanus continued; with the sole limitation, however, that they never moved beyond the orbit of Mars, the first planet to include both Venus and the earth within its circuit. This served to explain why several of the comets mentioned in ancient sources, such as Posidonius and Arrianus, stayed in the close vicinity of the sun, and were thus hidden by that body's powerful rays,[475] and also why comets were invariably observed to move around the sun, yet always on the nearer side of the earth.

Thus, to Puteanus, as to Fromondus and Fienus, the heavens beneath the firmament constituted one single fluid continuum, in which there were no solid orbs, but just the intersectible paths of the errant planets and comets. This opinion had been confirmed by the dicta of many venerable authorities, as had been demonstrated by Thomas Fienus in his dissertation on the comet of 1618; thus, Saint Basil had called the heavens "thin as smoke", and Origen had taught how the stars moved just like animals on land, fish in the water, and birds in the sky. Moreover, next to ancient authority, reason, too, had recently concurred in favour of the "new, or indeed ancient opinion"

---

varietate in Venere esse? Quis circa Saturnum minutos Planetas ludere? Quis misceri cursus, liquidamque caelorum naturam expandi?"

[474] Ibidem: "Abeo a communi dogmate, & credere iam incipio, solos circa terram Solem Lunamque volvi, circa Solem reliquos Planetas: sed Mercurium Veneremque citra terram & contractioribus circulis: Martem, Iovem, Saturnum, circa, ampliusque extensis."

[475] Ibidem, 38-39.

of celestial fluidity. Thus, mathematicians had not long ago observed how Venus changed her appearances in the same way as the moon, which variety could only be explained by assuming her to be located sometimes beyond the sun, sometimes below it, and sometimes next to it. The only way to make such appearances possible was to assume Venus to move about freely through fluid space, unhindered by the constraints of any solid orb.[476]

The phases of Venus were not the only novelties to have been discovered by way of the "miracle of the glass". Many celestial bodies which had never been seen before, had been rendered visible by the telescope, so that "what had hitherto been accessible only to the intellect and the acuteness of the mind, had now become perceptible to the eyes."[477] Not only had there turned out to exist many more fixed stars than had always been assumed; also, certain bodies had been detected, not all of the same size, which circled round about the sun and were rendered visible only when they moved in front of it, so that their dark masses contrasted with the splendour of the sun's body behind them. This circumstance made it probable that in other parts of the aether, too, such bodies were moving about, invisible to us because of the absence, in their vicinity, of a luminescent body to render them perceptible to the human eyesight. Such dark bodies, pursuing their own proper motions as well as being carried along with the daily revolution, should in Puteanus' view be thought of as stars, yet imperfect stars with broken light. They should be assumed to come into being and perish in the aether, which could be purged, rendered translucent, and illuminated due to some kind of special circumstance. Comets, then, were such 'imperfect stars'; like ordinary stars, they received their light from the sun, but, unlike ordinary stars, they transmitted some of it in the form of a tail or beard.[478]

Such a cometary globe of aetherial matter, Puteanus continued,

---

[476] Ibidem, 42: "Me profecto praeter auctoritatem inconcussa ratio in novam istam, imo antiquam opinionem trahit. Observant enim hodie Mathematici Venerem Lunae ut dixi vultum esse... Unde haec varietas, nisi & supra Solem (cui semper proxima) & infra & iuxta collocetur? nisi per liquidum, etsi breve cęli spatium erret?"

[477] Ibidem, 43: "Ecce quo ingenium tantummodo ibat, aciesque mentis, nunc oculis pervenitur: ut nesciam, caelum ad nos demissum sit, an nos ad caelum sublati, adeoque sublimati."

[478] Ibidem, 44: "Cogita sidera esse sed imperfecta, sed luce cassa, sidera non sidera. Cogita nasci in aethere, & denasci: quaedam autem peculiari sorte purgari, pellucida reddi & illuminari. Breviter: converti in Cometas, ut cum luce deinde inte-

was surrounded with certain condensed parts of aether, which reflected the rays transmitted by the comets' bodies. This denser aether might either be the remains of the already coagulated material, which had not succeeded in being included in the globe of the comet's body, or it might be a certain flow of aether, which was rendered thicker whenever a new star was generated.[479] In any case, the aether would invariably tend to coagulate in the form of a globe; after all, a globular form was more perfect than the form of liquid and diffused aether, just as the juices of the earth tended to be transformed into precious stones, fruits, and trees.[480] However, as long as such a globe of dense aethereal matter stayed well-assembled and coherent, it remained hidden from view, since excessively dense matter tended to terminate and limit the vigour of the luminescent solar rays; only when such a thick ball of aethereal material began once again to dissolve, was it rendered visible because its progressively rarified matter was capable of absorbing sunlight and assuming a recognizable shape. Therefore, Puteanus concluded, a comet was a dense globe of coagulated aether, which was in the process of being dissolved and rarified; the diffused and progressively thinner aether that, in the course of this process, gathered around the core of the comet's head, constituted its tail, hair, or beard.[481]

After this rather idiosyncratic definition of what, to him, was the nature and essence of a comet, Puteanus ended his paradox on the motion of comets by once again affirming that the changeability of the celestial region was not at all something to be wondered at. Once more emphasizing the analogy between the heavens and the earth, he stated that undeniable experience taught that, although it was true that men and beasts, trees and shrubs, towns and houses came

---

reant. Itaque splendorem a Sole accipiunt haec corpora, haec prodigia: & hactenus velut sidera sunt: transmittunt deinde sive Comae sive Barbae iubar; & a sideribus distinguuntur."

[479] Ibidem, 45: "Densior autem ille aether, vel conglobatae iam materiae reliquiae sunt, quae in eandem compingi massam, ac densari non potuerunt, vel certe affluxus, trahente spissiorem natura liquorem, quotiescumque novum aliquod sideris miraculum generatur."

[480] Ibidem, 46.

[481] Ibidem, 47: "Quid si tamen fieri Cometen statuas a globo iam deficiente? quasi nihil ille aliud sit quam dissolutio eiusmodi corporis; quod cum rarescit, materiamque circumfundit recipiat solem; tamque admirandum luminis spectaculum exhibeat? Globus enim quamdiu densus compactusque, vel recusare videtur lumen, vel tenue admodum recipere, & vix a reliquo caelo distinctum: rarefactus illustrari incipit, & mox a diffuso, ut sic dicam, liquore suo crinitum videas."

up and were cut down again in endless succession, many other and greater things on earth, such as countries and seas, mountains and rivers, invariably remained fixed and immovable. If then, such a variety of modes of existence was observed in the terrestrial region, why should we deny the same to the, much nobler, celestial realm? It ought much rather to be believed that infinitely many things, both small and great ones, were generated and corrupted beyond the moon, unseen by human eyes. The occurrence of such changes would not mean that the heavens would cease being heavens, or the stars stars. After all, no-one would venture to suggest that the many terrestrial alterations resulted in any fundamental change on the level of the universe: the cosmos still remained the cosmos, despite countless sublunary vicissitudes. To Puteanus, heaven and earth were of equal rank, both constituting evenly matched parts of the cosmic whole, and formed out of one and the same material.[482] Furthermore, he added, it had always been acknowledged that, in the heavens, the conditions prerequisite for mutability were continually in existence; namely, both the rarity of the orbs and the density of the stars; both the light of the sun and the darkness of the new moon; and both the motion of the cosmic whole and the rest of its poles.[483]

Puteanus' discussion of the comet's motion, which inspired him to the above digression on the substance and generation of the hairy stars, and on his views about the cosmos in general, constituted by far the most important section of his first book. In the three paradoxes that followed it, our author attempted to argue for the celestial nature of comets on the basis of characteristics such as duration, parallax, colour, and size; however, his arguments here rested largely on the quotation of substantial sections from ancient authorities, especially Pliny and Seneca, without many original contributions.

At times, the data on which Puteanus based his assertions were outright erroneous; thus, he asserted that the comet of 1618 had remained visible for about 30 days, a figure that was far off the mark

[482] Ibidem, 49: "Omnino licet mutetur terra, mundus tamen mundus est: mundi enim pars quaedam a Deo constituta. Sic licet mutetur caelum permanere tamen mundus potest: pars enim & mundi caelum est. Semel effero: Machinam hanc totam quam cernimus, & quam non cernimus, unum corpus esse, & in Aetherem, atque Elementa dividi: partes igitur quaedam elementa sunt, nec a toto natura sua diversae: ut nec totum a partibus. Rerum quidem species aliae aliaeque passim: res eaedem tamen, si materiam spectes, e qua formantur."

[483] Ibidem, 50.

if we recall Fromondus' estimate, based on much more accurate observations, of a total duration of about 52 days for the same comet.[484] However, his faulty data did not prevent Puteanus from employing the standard 'argument from duration' against the Peripatetics: after all, Pliny taught that the longest time a comet had ever been observed was 80 days (which, Puteanus added in a short piece of textual criticism, should probably be read as "180 days"); thus, even if it were admitted that a terrestrial exhalation could remain burning for 30 days, surely 180 days was an impossible time span for fleeting sublunary fires.[485]

Puteanus' discussion of cometary parallax smacked of similar inaccuracy. Although he went to the trouble of explaining the concept to his readership, by means of the example of an imaginary cloud obscuring progressively substantial sections of an observed landscape the higher up it was situated, his subsequent discussion of the parallax displayed by the comet of 1618 was not very illuminating. In fact, it was virtually nonexistent: he merely stated that "Because it has been determined, through continual observation, that the comet's parallax was smaller than that of the moon, it is certain and not at all in doubt that the comet was located beyond the moon."[486] To this all too concise statement, Puteanus added that whether the comet had ascended ten or forty degrees above the horizon, it constantly remained at the same distance from the nearest stars; if it had stood in the sublunary air, a difference of position would have been discernible.

The "proof of the celestial nature of comets, based on their form" with which Puteanus continued his tract, owed everything to Seneca. Puteanus indulged in a lengthy literal quotation of Pliny's subdivision of comets into nine categories, on the basis of the nine different forms in which they could allegedly appear, only to be able to disprove it, with the aid of Aristotle and Seneca. Both these authors had held that only two kinds of comets existed: 'hairy' comets with short rays all around their heads, and 'bearded' comets with long tails or beards stretched out in one direction. Yet, whatever their appearance, Puteanus stated, all comets were, in reality, stars. After all,

---

[484] Ibidem, 51.

[485] Ibidem, 51-52.

[486] Ibidem, 54: "Quia igitur iugi observatione exploratum est, minori Cometam parallaxi, quam Lunam affici: altiorem Luna Cometam esse certum, non arbitrarium est."

Seneca had correctly pointed out that the nature of all comets was globular, whatever shape the splendour of their tails might assume: the body proper of every comet was round, like that of the other stars.[487]

Rather more interesting was Puteanus' discussion of the comet's tail. Emphatically endorsing the optical theory of cometary tails, our author stated that the Aristotelian idea, that they were fiery exhalations of somewhat greater density than that of the heads, was untenable because of the constancy of their direction: they invariably pointed away from the sun. This indicated that the hairy stars were illuminated by celestial light (probably that of the sun, despite the fact that Tycho had believed the tail of the comet of 1577 to have been directed away from Venus), which was refracted in the comet's head. Following the lead of Cardano, Puteanus suggested that comets possessed an intermediate nature, with their matter neither rare enough to allow the sunrays to pass through their bodies unhindered, nor dense enough to reflect them. Unlike Fromondus, Puteanus did not attribute the fact that some cometary tails were curved rather than straight to differences in density within the comet's head, but rather chose once again to emphasize the fallibility of human eyesight. The curved form, he maintained, was nothing more than an optical illusion, due to the fact that the extremity of the comet's tail was closer to the earth, and thus to the human eye, than the comet's head, whenever the planet whose rays created the tail was located above the comet.[488]

Finishing the first book of his treatise in the same vein of Senecan agnosticism, Puteanus stated that the optical theory of cometary tails he had just set forth, could not be invalidated by the objection that, if the tails were indeed formed by sunlight, it was inexplicable how they could assume so many different colours. Puteanus retorted with a quotation in which Seneca pointed out that the splendour of the moon and the planets, too, was very varied and heterogeneous in colour, and that it had never occurred to anyone to inquire how this was possible; paraphrasing another Senecan idea, Puteanus added that even the most ordinary and everyday phenomena had not yet

---

487 Ibidem, 59.

488 Ibidem, 65: "Eadem in Cometis impostura est, eaedem praestigiae: quippe quando extremi & desinentes radii viciniores nobis, quam ipse globus sideris obliquari videntur. Viciniores autem, quia superior loco Planeta est, qui transmisso per Cometam lumine radios efficit."

been adequately investigated, so that it would be vain indeed to inquire too curiously after the nature and essence of such extraordinary appearances as comets.[489]

All in all, the attitude Puteanus displayed in his tract on the comet of 1618 stood midway between Fienus' scepticism and Fromondus' cautious sympathy towards the novel cosmological opinions and discoveries. Like Fienus, he relied very heavily on Seneca (to whom on one occasion he even referred as "The Philosopher"[490]), adopting from the Roman Stoic both a number of arguments against the sublunary nature of comets, and, probably more importantly, a Stoic outlook on the cosmos. Convinced of the unity and interconnectedness of the whole universe, both those parts of it that were situated below the moon and those above it, he insisted that in all likelihood anything that was observed to happen on earth probably could, and did, also happen in the heavens. The celestial region, then, possessed a certain degree of change, just as the earth possessed a degree of constancy; only, most of these heavenly alterations remained invisible to man. However, the fact that they did take place, and that many things in the universe were different from the teachings of Aristotle, had recently been demonstrated by the telescopic observations, the nova of 1572, and Tycho's determination of parallax for the comet of 1577.

Thus far, Puteanus' attitude to a large extent concurred with that of Fienus. However, whereas Fienus' agnosticism had been considerable enough to prevent him from trying to determine the real nature of comets, Puteanus joined Fromondus in his search for the true essence of the hairy stars. He was able to theoretically justify this through repeated references to another Stoic idea put forward by Seneca: the belief in the existence of fixed laws of nature which bestowed on the cosmos a pattern of harmonious order. Even seemingly 'extraordinary' phenomena, such as comets, conformed to these laws, Puteanus stressed: thus, it was possible to diligently investigate the orderly pattern of the universe, and thereby, in the end, to eliminate the need for any notion of 'prodigiousness'. In order to reach a true understanding of comets, ignorance, the foremost cause of cometary fear, should and could be attacked.

Despite this optimistic attitude, and despite his great admiration

[489] Ibidem, 67.
[490] Ibidem, 17.

for the newly made discoveries, however, Puteanus' own observations left much to be desired. Nonetheless, he displayed considerable familiarity with recent cosmological debate; thus, he mentioned the phases of Venus (and the reason why these could only be explained by assuming that planet to circle the sun), the handles of Saturn, the sunspots, the fluid nature of the heavens, and the Tychonic system, to which he explicitly subscribed. His view of the nature of comets had been suggested to him by the analogy with the sunspots, which he assumed to be caused by certain dark aethereal bodies wandering through the fluid heavens.

Of the three Louvain tracts of 1618, that of Puteanus was the least Aristotelian in character, at least as far as methodology was concerned. The humanist did not employ the traditional *quaestiones*-structure, unlike Fromondus and Fienus failing to systematically refute all possible counter-arguments before setting forth his own opinion. Nonetheless, the basic elements of his world view were not substantially different from those of his two colleagues. He endorsed the fluidity and corruptibility of the heavens, the absence of solid orbs, and the Tychonic system (thereby implicitly subscribing to geocentricity as well). Also, like Fromondus, in emphasizing the analogous nature of heaven and earth, and the essential equivalence between the two, he went a long way towards abandoning the Aristotelian dichotomy between celestial and terrestrial regions.

e. *New Universities at Leiden and Groningen*

As we have seen, the university of Louvain, from which issued forth the cometary tracts by Fienus, Fromondus and Puteanus, had from its foundation borne a staunchly Roman Catholic character. Despite the manifold insecurities and frequently changing situations caused by the wars of the Dutch Revolt, the university's official policy had always been to maintain loyalty to the papacy, and the institution remained a stronghold of the Counter-Reformation. Not surprisingly, therefore, as soon as the rebels, led by the Prince of Orange, started to gain a foothold in the northern provinces, they were in urgent need of an institution where they might educate their own intelligentsia; especially the training of reformed ministers was regarded as vital. Thus, at the instigation of Orange, in February 1575 a university was rather hastily founded in the Holland town of Leiden, to provide the youth of the northern Netherlands with the education they could no longer pursue at Louvain; in 1585, the Frisian town of Franeker followed the Leiden example, and in 1614, another university was established at Groningen.

It might seem logical to deduce from this account that Leiden and the other northern universities were characterized by intellectual atmospheres quite distinct from that of their unswervingly Catholic southern counterpart, Louvain. Obviously, to a certain extent this was indeed the case: both the Leiden and Groningen acts of foundation required that all members of the university swear an oath of adherence to the reformed religion. At Leiden, however, this requirement was dropped as early as 1578 for all students except those studying theology[491], after which date a number of Roman Catholic students enrolled at the university—among them Thomas Fienus.

Yet, despite the evident religious difference between the new northern universities on the one hand and Louvain on the other, as far as the educational programme was concerned they were all remarkably alike. The university of Groningen from the outset emphasized its adherence to Aristotelian doctrine: in article XL of the *Leges Academicae,* the philosophers were admonished "not to deviate from Aristotle's philosophy" and informed that "champions of absurd paradoxes and inventors of new dogmas at variance with Aristotelian doctrine" would not be tolerated.[492] And, although the Leiden educational programme, contrived in 1575 by the French theologian Feugueray, seems to have been influenced by antischolastic currents such as humanism and Ramism[493], education in the philosophical faculty soon assumed a more or less emphatically Aristotelian character. Initially, for the teaching of logic and physics, Aristotelian textbooks by Augustinus Hunnaeus and Cornelius Valerius from Louvain had been used; however, in 1582, a number of students appealed to the academic Senate, demanding that the use of these books be discontinued in favour of the teaching of the text of Aristotle himself, on the grounds that Hunnaeus and Valerius did not accu-

---

[491] Edward G. Ruestow, *Physics at Seventeenth and Eighteenth Century Leiden: Philosophy and the New Science in the University* (The Hague 1973) 4.

[492] *Leges Academicae Illustrium Ordinum Groningae et Omlandiae* (Groningen 1619) art. XL: "Philosophi ab Aristotelis Philosophia non recedunto: propugnatores absurdorum paradoxorum et inventores dogmatum novorum ab Aristotelica doctrina discrepantium non feruntor." Published in W.J.A. Jonckbloet ed., *Gedenkboek der Hoogeschool te Groningen, ter gelegenheid van haar vijfde halve eeuwfeest* (Groningen 1864) 216.

[493] See Paul Auguste Georges Dibon, *L'enseignement philosophique dans les universités néerlandaises à l'époque pré-cartésienne (1575-1650)* (Leiden 1954) 8-10.

rately render the Stagyrite's opinions. This appeal was granted; after all, the Acts of the university stated that Leiden was to uphold the doctrine of Aristotle.[494]

The first professors to teach physics at Leiden—Trutius, Molinaeus, Vorstius, Bontius, and Jacchaeus—explicitly associated themselves with the medieval scholastic tradition, at least, so it would appear from the disputations over which they presided. These were all based on Aristotle's physical works, such as the *Physica, De anima, De caelo, De mundo, De generatione et corruptione,* and *Meteorologica*, supplemented with quotations from J.C. Scaliger's *Exercitationes in Cardanum,* the Paduan Aristotelian tradition as represented by Zabarella, the Collegio Romano Jesuit Benedictus Pererius, and the *Physiologiae peripateticae* by the Marburg professor Johannes Magirus; from 1604 onwards, when Jacchaeus started to teach physics, there were also repeated references to the Jesuits of Coimbra.[495] Virtually the only non-Aristotelian work to be used in early Leiden physics teaching was Lipsius' *Physiologia Stoicorum*, which appeared at Louvain in 1604.[496]

Thus, the philosophy taught at Leiden was overwhelmingly Aristotelian in nature; moreover, Leiden philosophers did not hesitate to make extensive use of the various new developments that were taking place in Roman Catholic Aristotelianism, such as the Paduan method of *regressus,* the Jesuit tendency to incorporate the exact sciences within the framework of Aristotelian metaphysics, and the monumental commentaries on the *Organon* prepared by the Coimbran Jesuits.[497] In all probability, therefore, the physical theories taught at Leiden during the early decades of the university's existence did not deviate in any substantial way from those current at the arts faculty of Louvain. If anything, the Leiden physics course may have been rather more rudimentary than its southern counterpart; after all, the arts faculty at Leiden initially bore a very propaedeutic and utilitarian character, with logic, physics and ethics being regarded first and foremost as tools by means of which could be attained the more impor-

---

[494] Ibidem, 12-14.

[495] Ibidem, 57-59; M.J. Petry, "Burgersdijk's Physics" in E.P. Bos & H.A. Krop eds., *Franco Burgersdijk (1590-1635). Neo-Aristotelianism in Leiden* (Amsterdam/Atlanta 1993) 83-118: 92-93.

[496] Dibon, *Enseignement,* 63; Petry, "Burgersdijk's Physics", 92.

[497] See Petry, "Burgersdijk's Physics", 96.

tant education of the higher faculties, theology, medicine, and law.[498]

We find this general tendency of meticulous yet somewhat superficial adherence to the Stagyrite's writings confirmed when we subject the few surviving disputations which treated cometology to a close examination. Two early Leiden philosophical disputations either exclusively or partly dealt with the nature and properties of comets: Petrus Pilius' *Disputatio de Cometis,* defended (unsuccessfully it would seem[499]) on 24 February 1598 under the presidence of Trutius, and Theodorus Orthius' *Theses Physicae de Meteoris,* presided over in 1600 by Petrus Bertius. Pilius' disputation was unmistakably Peripatetic in character: he stuck very closely to the Philosopher's cometological doctrine, asserting comets to be elementary, burning bodies. Consisting of hot and dry exhalations extracted from the earth by the heat of the sun, they belonged to the genus of long-lasting fiery meteors. They were brought into being by a twofold efficient cause: their remote efficient cause was the intrinsic heat of the earth, preparing a suitably combustible material, along with the power exercised by the sun and the stars, by means of which this material was attracted and carried up on high, while their more proximate efficient cause was the fiery principle that set the exhalation alight once it had arrived in the upper region of the air, or in the sphere of fire. If such a burning exhalation extended equally in all directions the phenomenon was called κομήτης, whereas if it stretched out on one side only it was named πωγωνίας.

Thus, Pilius' entire account of the hairy stars' generation and essence was no more than a considerably simplified rendition of the seventh chapter of Book One of Aristotle's *Meteorology.* The only other author whose influences can be detected in the disputation was Pliny, whose popular classification of comets into several different categories according to their shapes Pilius more or less faithfully rendered; not, however, without adding the remark that "Either all these kinds of comet can be reduced to Aristotle's two species, [namely, κομήτης and πωγωνίας], or else Aristotle did not think it necessary to expound all varieties of comet."[500] As for the portentous value of comets, Pilius professed his adherence to Aristotle's physical approach,

---

498 Dibon, *Enseignement,* 248-249. See also Ruestow, *Physics,* 11.

499 According to Dibon, *Enseignement,* 37.

500 Petrus Pilius, *Disputatio Physica de Cometis* (Leiden 1598) thesis XI: "Sed hasce ad duas Aristotelis species referri posse dicemus, vel Aristoteli non fuisse curae, ut omnem varietatem cometarum exponeret."

according to which they were signs of winds and great drought; however, he emphatically denied that they were either causes or effects of any of the consequences usually ascribed to them by Arabo-Ptolemaic astrology, such as wars, deaths of princes, and the like; this can undoubtedly be regarded as an attempt to purge the Stagyrite's dicta of any medieval accretions, inspired by the humanist environment at Leiden.[501]

Orthius' theses on meteors, defended in 1600, displayed an equally punctilious loyalty to the Aristotelian corpus on which they were based. Meteors, Orthius stated, belonged to the genus of imperfectly mixed bodies; they could be subdivided into fiery, watery, and airy meteors. Not surprisingly, he counted comets among the fiery variety; more specifically, they belonged to the category of real fiery meteors located in the upper region of the air, which possessed a certain measure of density. They were generated out of copious, dry, fat and compact exhalations which, having risen up to the upper region of air, caught fire and stretched out their rays either in one direction or in many; they were under the dominion of one particular star. Some were round and therefore called hairy, while some were bearded. Briefly, Orthius quoted Seneca and Pliny, in connection with the maximum duration comets could attain, which was a question on which Aristotle had not pronounced his verdict.[502] Unlike Pilius, however, Orthius did give evidence of being well-informed about more or less contemporary astronomical developments: in a "paradoxon" appended to his theses, he stated that "The star which appeared in the year 1572 near the constellation of Cassiopeia was not a comet, and was not located in the region of air, but rather

---

[501] Ibidem, thesis XVI and "Corollarium": "Cometam nec caussam, nec signum certum bellorum, excidii urbium ac gentium, interitus regum & magnatum, mutationis religionum affirmamus: multo minus earundem calamitatum effectum." On humanistic tendencies at Leiden, see J.A. van Dorsten, "Èen humanistische universiteit" in *Leidse universiteit 400. Stichting en eerste bloei 1575-ca. 1650* (Amsterdam 1975) 10-11, and C.L. Heesakkers, "Leids humanisme in de vroege Gouden Eeuw", *Hermeneus* 66 (1994) 186-196.

[502] Theodorus Orthius, *Theses physicae de meteoris* (Leiden 1600) thesis VIII: "COMETA est meteoron igneum ortum ex halitu copioso, sicco atque viscoso, arcteque compacto, atque ad supremam aeris regionem evecto, ibidemque inflammato, ex se omnes vel aliquas in partes radios emittens, certae alicuius stellae dominio subjectum. Estque duplex, Crinitus seu rotundus, & πωγωνίας, qui Barbatus vocatur. Tempus durationis varium est: alii ad summum sex menses durare tradiderunt, ex quorum numero Seneca: alii 80 dies, quod placet Plinio, sunt tamen interdum breviores etiam durationis."

stood in the firmament, or the heaven of the stars."[503] However, Orthius did not underpin this statement with any written arguments.

The first few decades of physics teaching at Leiden passed rather chaotically; the chair of physics at times remained vacant, and none of the early professors who did occupy it stayed for very long. Moreover, none of them produced any substantial publications, apart from the theses defended by their students in public disputations, which, as we have seen, were characterized by strict adherence to Aristotle's text. Only with the advent of the Scot Gilbertus Jacchaeus, who succeeded Bontius to the chair in 1617, did the teaching of physics lose its exclusively textual character: Jacchaeus' text-book of physics, the *Institutiones physicae,* which appeared in 1614, constituted the first Leiden attempt at systematizing natural philosophy for educational purposes.[504]

As far as cometology was concerned, however, this did not make much of a difference. To Jacchaeus, as to Pilius and Orthius, comets were "copious, fat, coherent and dense exhalations, that kept the flames which set them alight burning for a long time".[505] Because they participated in the daily movement of the firmament, comets could not possibly be located in the lower or middle regions of air; instead, Jacchaeus maintained, they were situated in the upper airy region. Their duration, as well as their colour and their shape, depended on the degree of greasiness of the material out of which they were formed. Their motion could be in any direction, east to west or north to south. The reason that they portended winds, tempests, earthquakes, and illnesses was that, for a comet to be formed, a great amount of exhalations was required, which, before having ascended into the upper reaches of the air to create a comet, got a chance to work their pernicious effects in the lower atmosphere.[506]

This Jacchaean cometology was based on Aristotle's text, as well as on the scholastic tradition of commentaries on the *Meteorology;* thus, the Stagyrite himself had believed comets to be portents of wind and drought only, while the other inconveniences adduced by Jacchaeus, earthquakes and illnesses, were medieval additions. Also, the idea that their colour depended on the nature of the material

---

503 Ibidem, "Paradoxon".
504 Dibon, *Enseignement,* 59.
505 Gilbertus Jacchaeus, *Institutiones physicae* (2nd ed., Leiden 1624) 177.
506 Ibidem, 178.

from which they were formed was not mentioned by Aristotle. Yet, although Jacchaeus' text-book relied more heavily on the scholastic tradition than the disputations defended around the turn of the century, the basic cometology he expounded remained the same: with only very slight deviations, he adhered to Aristotle's exhalation theory.

Nonetheless, apparently Jacchaeus could not bring himself to treat comets in a completely traditional way, without at least mentioning the recent developments that had shed a rather different light on the hairy stars. At the very beginning of his chapter on comets, he bade a scathing farewell to those who believed comets to possess a celestial nature, maintaining that, if they did, they would not be seen to be generated and to perish; moreover, he stated, the mathematicians had discovered that they were generated in the elementary region.[507] Thus, implicitly, Jacchaeus upheld the unchangeability of the heavens; fleeting bodies such as comets could not possibly be located there. Furthermore, he also unquestioningly accepted the medieval tradition of faulty parallax measurements, by means of which such mathematicians as Regiomontanus and Vögelin had found comets to be located below the moon. However, in the final sentence of his cometological chapter, our author put his entire account into perspective by adding that "Although such is the common opinion on comets among the Peripatetics, everyone admits that the [comets] that have recently appeared were not located in the elementary region, but in the heavens."[508] Yet, like Orthius, Jacchaeus did not attach any consequences to this acknowledgment of the discoveries made by Tycho; his cometology remained as Peripatetic as ever, probably because, in accordance with the Aristotelian division of disciplines, he did not think it fit to introduce any mathematical reasoning into a text-book of natural philosophy.

For the same reason, it need not surprise us that neither of the two treatises devoted exclusively to the comet of 1618 to have been produced in the universities of the northern Netherlands were written by natural philosophers. Although, in the tradition of the *Corpus*

[507] Ibidem, 177: "Valeant ergo qui cometam naturae coelestis esse arbitrantur: nam tum nec generarentur, nec corrumperentur. Quid? quod Mathematici deprehenderunt in regione elementari gigni."

[508] Ibidem, 178: "Et haec quidem communis est Peripateticorum, de cometis sententia; quanquam qui nuper apparuerunt, omnium confessione, non in regione elementari, sed coelesti apparuerunt."

Aristotelicum, comets belonged to the class of meteors and were therefore part of the subject matter to be treated by natural philosophy, monographs occasioned by one particular cometary appearance were not usually produced by natural philosophers, who were often primarily interested in expounding the general theory of comets in accordance with the Stagyrite's dicta, instead of investigating one specimen in particular. The latter activity was traditionally reserved for physicians, who needed to observe comets in order to ascertain what possible astrological effects they might have on human disease, or, from the fifteenth century onwards, to mathematicians carrying out measurements for each individual comet.[509] As we have seen, both Dutch tracts on the comet of 1577 were written by professors of medicine, while only one of the three Louvain authors on the comet of 1618, Fromondus, was a philosopher, and even he was atypical in having, from an early age, been very interested in observational astronomy.

While, therefore, the cometology taught at Leiden by natural philosophers such as Trutius and Jacchaeus was completely traditional and did not address recent astronomical developments in any meaningful way, we might reasonably expect to discover rather more state-of-the-art discussions of the essence and nature of comets in the tracts written by Nicolaus Mulerius, professor of medicine at Groningen, and Willebrord Snellius, mathematician at Leiden, on the occasion of the comet of 1618. After all, both men, but especially Snellius, the son of one of the foremost Ramists of his day, are usually credited with a practical attitude to the investigation of nature, according great importance to mathematics and to autopsy, as opposed to the qualitative reasoning and uncritical reliance on authority usually associated with Peripatetic science.

Much has been made of this alleged practical orientation of scientific activity in the northern Netherlands during the decades around 1600. The general picture emerging from studies of the natural sciences as practised in the early Dutch Republic is one of backwardness and Aristotelian conservatism at the universities, starkly contrasting with the bustling scientific activity displayed outside or on the margins of the academic environment. The idiosyncrasies of the northern Netherlandish mentality, deriving partly from the humanist

[509] See supra, p. 157 and § III.9.

tradition with its aversion to scholastic subtleties and partly from the practical needs of a seafaring, trading, swampy country engaged in near-continual warfare with Spain, are thought to have fostered the eminently practical and empirical sciences of surveying, navigation, cartography, fortification, and hydraulic engineering, with which were associated the names of Adriaan Anthoniszoon, his son Adriaan Metius, Willem Blaeu, and above all, Simon Stevin.[510]

At the instigation of Prince Maurice of Orange, in 1600 the university of Leiden was enriched with an educational programme named "Duytsche Mathematicque", intended for the training of the engineers of whom the young country was in great need. Simon Stevin, mathematician and inventor as well as tutor and counsellor to Prince Maurice, was charged with the elaboration of the subject-matter to be taught there: arithmetic and geometry, use of surveying instruments, design of fortifications, as well as practical application of the knowledge thus acquired, all taught in the Dutch language.[511] The first lecturers at the Duytsche Mathematicque were Ludolf van Ceulen, mathematician and fencing master, and Simon van der Merwe; among its first students was Adriaan Metius, the son of the Alkmaar fortification master Adriaan Anthonis, who went on to become professor of mathematics at Franeker and to enrich the cur-

---

[510] See for example K. van Berkel, "Vernuftelingen rondom Stevin (1580-1620)" in his *In het voetspoor van Stevin. Geschiedenis van de natuurwetenschap in Nederland 1580-1940* (Meppel/Amsterdam 1985); Frans Westra, *Nederlandse ingenieurs en de fortificatiewerken in het eerste tijdperk van de Tachtigjarige Oorlog, 1573-1604* (Diss. Groningen 1992); P.J. van Winter, *Hoger beroepsonderwijs avant-la-lettre. Bemoeiingen met de vorming van landmeters en ingenieurs bij de Nederlandse universiteiten van de 17e en 18e eeuw* (Verhandelingen der Koninklijke Nederlandse Akademie van Wetenschappen, Afd. Letterkunde, Nieuwe Reeks, deel 137, Amsterdam 1988); K. van Berkel, "Universiteit en natuurwetenschap in de 17e eeuw, in het bijzonder in de Republiek" in H.A.M. Snelders & K. van Berkel eds., *Natuurwetenschappen van de Renaissance tot Darwin. Thema's uit de wetenschapsgeschiedenis* (The Hague 1981) 107-130; E.J. Dijksterhuis, *Simon Stevin. Science in the Netherlands around 1600* (The Hague 1970); H.A.M. Snelders, "Science in the Low Countries during the sixteenth century. A survey", *Janus* 70 (1983) 213-227; H.A.M. Snelders, "Alkmaarse natuurwetenschappers uit de zestiende en zeventiende eeuw" in *Van Spaans beleg tot Bataafse tijd. Alkmaars stedelijk leven in de zeventiende en achttiende eeuw* (Alkmaarse historische reeks 4, Zutphen 1980) 101-122. Sadly, hardly any of these studies makes an effort to put the northern Dutch situation into its proper context by comparing it to the state of the sciences in other European countries; at most, the impression is advanced (by van Berkel and Snelders) that the flourishing of practical science in the Dutch Republic was a continuation of the sixteenth century tradition in the southern Netherlands, especially Antwerp.

[511] See van Winter, *Hoger beroepsonderwijs*, 16-17.

riculum there with the teaching of navigation, surveying, fortification, and astronomy.[512]

The emphasis on the use of mathematical knowledge for practical purposes, which in Leiden was so much in evidence, on the margins of the academic environment, in the Duytsche Mathematicque, was not entirely absent from the university proper, though. The first professor of mathematics to be appointed at Leiden, in 1580, was Rudolf Snellius, one of the most important Ramists of his day. The doctrines of Peter Ramus, the infamous despiser of Aristotelianism who became a Protestant martyr through falling victim to the Saint Bartholomew Massacre, dealt primarily with logic and educational reform.[513] They were characterized, among many other things, by a marked emphasis on practical utility, which Ramus considered by far the most valid motive for the acquisition of knowledge. Very important were observation and experience, which, in Ramus' view, ought to form the basis of any investigation of nature; crucial to Ramist method was the adduction of effects, or instances of natural phenomena, from which the scholar could arrive at knowledge of the underlying principles by means of induction.[514] In the same vein, Ramus strove to create an astronomy without hypotheses, based solely on observation and experience, believing that such had been the astronomy of the ancient Egyptians and Babylonians, who had been closer to natural reality than the Greeks after Eudoxus, who had started contriving all kinds of astronomical theories. Also, Ramus preferred practical mathematics to the theoretical variety, and gave examples of the practical utility the mathematical sciences could yield.[515]

Rudolf Snellius had become acquainted with these Ramist ideas while studying and teaching at Marburg university during the decade between 1565 and 1575, and had enthusiastically embraced them. When he returned to Holland and assumed the chair of mathematics

---

512 Ibidem, 49.

513 On Ramism, see W.J. Ong, *Ramus. Method and the Decay of Dialogue* (Cambridge, Mass./London 1958). On the distinction between Ramism in logic and Ramism in philosophy, see P. Dibon, "L'influence de Ramus aux universités néerlandaises du 17e siècle", *Actes du XIème Congrès International de Philosophie, Bruxelles, 20-26 Août 1953* XIV (Amsterdam/Louvain 1953) 307-311.

514 R. Hooykaas, *Humanisme, science et réforme. Pierre de la Ramée (1515-1572)* (Leiden 1958) 57.

515 Ibidem, 66-84; J.A. Vollgraff, "Pierre de la Ramée (1515-1572) et Willebrord Snel van Royen (1580-1626)", *Janus* 18 (1913) 595-625: 602-605.

(as well as that of Hebrew) at Leiden, his teaching reflected his Ramist ideals: emphasis on practical applications, predominance of applied over theoretical mathematics (thus, in his lectures on optics, he explained to his students the principle of the telescope as early as 1609), and well-ordered arrangement of subject matter in the interests of educational clarity.[516] The importance, to Snellius, of the latter ideal came to the fore clearly in his very numerous commentaries on Ramus's text-books of dialectics, physics, arithmetic, and geometry.[517] Thus, Rudolf Snellius can be said to have advocated, within the walls of the university, ideas similar to those being taught, from 1600 onwards, at the Duytsche Mathematicque.

Nonetheless, Ramism never obtained a firm footing at Leiden; at least, not Ramism of the unadulterated variety vociferously defended by Snellius. His relations with his colleagues always remained somewhat tense, and when, in 1598, the chair of physics became vacant due to the departure of Molinaeus, the university Curators appointed Everard Vorstius to the post, in spite of the fact that the Senate had recommended Rudolf Snellius. Sassen ascribes this turn of events to the eagerness of the Curators to maintain orthodox Aristotelianism, and not to let any Ramist ideas influence the teaching of physics[518]; however, the official reason given by the Curators for passing over the candidacy of Snellius was the avoidance of uniting the teaching of the two very distinct disciplines of mathematics and physics in one person.[519] Although Snellius did teach physics to a selected number of pupils in private[520], it seems fair to say that, on the whole, natural philosophy at Leiden around 1600 was based on the traditional Aristotelianism of Jacchaeus rather than on the practically oriented Ramism of Snellius.

Still, there was a different, more humanistic side to Ramism. We have already seen how Ramus had advocated a return to the pristine astronomical wisdom of the ancient Egyptians and Babylonians; such

---

[516] Van Berkel, "Vernuftelingen", 31; Petry, "Burgersdijk's Physics", 87.

[517] A complete list of these editions is to be found in K. van Berkel, "De geschriften van Rudolf Snellius. Een bijdrage tot de geschiedenis van het ramisme in Nederland", *Tijdschrift voor de geschiedenis van de geneeskunde, natuurwetenschappen, wiskunde en techniek* 6 (1983) (4) 185-194.

[518] Ferdinand Sassen, *Het oudste wijsgeerig onderwijs te Leiden* (Mededeelingen van de Koninklijke Nederlandse Akademie van Wetenschappen, Afd. Letterkunde, Nieuwe Reeks, 4 (1941)) 22.

[519] See Dibon, *Enseignement*, 28.

[520] Petry, "Burgersdijk's Physics", 87.

an idea was in perfect accordance with the general attitude of the humanistic *ad fontes* movement. Ramus's own contributions to physics, too, were a far cry from the applied mathematics and technical skills with which Snellius' Ramism is usually associated: he wrote elaborate, humanistic commentaries on Virgil's *Georgics,* Ovid's *Metamorphoses,* Pliny's *Natural History,* and Seneca's *Natural Questions*, which have led one historian to brand Ramus' approach to natural philosophy as "armchair physics".[521] Also, although he was fiercely critical of indiscriminately quoting ancient authorities, his own ideas on the proper way to attain natural knowledge were themselves of an unmistakably Stoic inspiration. Central to Ramus's thought was the idea that truth revealed itself through natural reason, human intelligence, and the cosmic order of things, so that man, to reach a true conception of things, needed only to follow his own human nature[522]; this belief was in marked accordance with the Stoic world view that was going through a vigorous revival during the sixteenth century.

Thus, there were two related yet distinct sides to Ramist natural science. At the root of both lay the belief that, in order to reach a true understanding of natural phenomena, empirical observation, rather than intricate intellectual subtleties, was required. However, this idea could lead scholars of Ramist inspiration to two different, but not necessarily mutually incompatible, outcomes: on the one hand, they might apply themselves to practical science of the empirical, utilitarian variety, as practised at the "Duytsche Mathematicque"; on the other hand, however, at a more academic level, they might instead, or in addition, concentrate on the application of natural reason leading to the discovery of pristine, unpolluted wisdom, as well as true knowledge of nature. The latter tendency, however, was not at all particular to Ramism, but borrowed in large part from the Stoic conception of cosmic order. Thus, the practical current that can be detected in the science of the northern Netherlands as practised by engineers such as Stevin was probably not rigidly isolated from the more bookish concerns current at the universities during the same period; such men as Rudolf Snellius and Adriaan Metius, who were university professors well acquainted with the ins and outs of contemporary scholarship, while at the same time sympathetic to

---

[521] Hooykaas, *Humanisme, science et réforme,* 41.
[522] Ibidem, 51.

the applied sciences of engineering, fortification, and astronomy, may well to a certain extent have bridged the apparent divide.

The same can be said for our two northern Netherlandish authors on the comet of 1618, Willebrord Snellius and Nicolaus Mulerius. As mentioned before, Willebrord Snellius was Rudolf's son. He was born in Leiden in 1580, and from an early age showed a keen interest in mathematics. He enrolled at the university of his home town at the age of ten, to study law; yet his most prodigious achievements lay in the field of mathematics, and in 1600, through the influence of his father, Van Ceulen, and Stevin, he received permission to teach this subject at the university.[523] During his four-year academic peregrination, the young Snellius met the famous Flemish mathematician, Adriaan van Roomen, in Würzburg; he conducted observations at Prague with Tycho Brahe and Kepler, and also met Michael Mästlin at Tübingen. After a stay in Paris to study law, and a visit, with his father, to the court of Prince Maurice of Hesse-Cassel, one of the foremost aristocratic patrons of alchemists and Paracelsians, as well as a convinced Ramist[524], Snellius returned home to Leiden in 1604.

In 1613, after the death of his father, Willebrord succeeded to the chair of mathematics at Leiden. Having adopted Rudolf's enthusiasm for Ramist ideas, he applied himself to practical science; his two most notable feats in this field were his determination of the length of the meridian by means of a developed version of Gemma Frisius' triangulation techniques, published in his *Eratosthenes Batavus* in 1617, and his discovery of the law of refraction of light rays which now bears his name, formulated around 1621 as the result of many years of experimentation and study.[525] He had many instruments made for his geodetic work by Willem Blaeu, among which was a huge, 210-centimetre quadrant; also, he possessed a telescope of his own as early as 1610.[526] His teaching, too, seems to have been practically oriented; he translated Stevin's textbook of practical science, the *Wisconstighe Ghedachtenissen* ("Thoughts on Mathematics") into Latin,

[523] Dirk J. Struik, "Snel, Willebrord", in Charles Gillespie ed., *Dictionary of Scientific Biography*, 499.

[524] On the scientific interests of Maurice of Hesse-Cassel, see Bruce T. Moran, *The Alchemical World of the German Court. Occult Philosophy and Chemical Medicine in the Circle of Moritz of Hessen (1572-1632)* (Stuttgart 1991).

[525] Struik, "Snel", 501.

[526] Vollgraff, "Pierre de la Ramée et Willebrord Snellius", 596.

published a commentary on Ramus' *Arithmetica*, and had an unspecified "instrument" made for the benefit of his students.[527]

Nicolaus Mulerius, too, accorded great importance to applied science. He was born in 1564 in Bruges, in the southern Netherlands; however, due to the Calvinist convictions of his parents, the Mulerius family were forced to flee their home town when Nicolaus was two years old. After several years of persecution, during which Mulerius' mother and baby brother were captured and executed by the Inquisition, the family left Flanders for the northern Netherlands. The experience of being persecuted on account of Calvinist opinions must have made a lasting impression on the young Mulerius; in any case, he was to maintain an unwavering loyalty to orthodox Calvinism throughout his life, and to strike up a warm friendship with Franciscus Gomarus.

At Leiden, Nicolaus pursued his education at the newly founded university, initially studying philology, philosophy, and theology, but soon preferring the more exact disciplines of mathematics and medicine.[528] At Leiden, Johannes Heurnius, Justus Lipsius, and Rudolf Snellius counted him among their pupils. In 1589, Mulerius received his doctorate in medicine, and left Leiden to become town physician at the Frisian town of Harlingen, a profession which he does not seem to have particularly enjoyed: in 1594, he wrote several letters to Heurnius in an unsuccessful attempt to obtain a position at Leiden university.[529] In 1603, despite attempts by the citizens of Harlingen to retain him, Mulerius left Frisia for Amsterdam, but was shortly afterwards appointed provincial physician in Groningen. In 1608, he returned for a few years to Frisia to assume the post of headmaster of the Leeuwarden gymnasium; during his stay there he declined an offer to succeed to Johannes Drusius' chair of Hebrew at Franeker

[527] P.C. Molhuysen & P.J. Blok eds., *Nieuw Nederlandsch Biografisch Woordenboek* (Leiden 1911-1937) VII 1160.

[528] *Effigies & vitae Professorum Academiae Groningae & Omlandiae* (Groningen 1654) 66: "Hic igitur *Nicolaus* inter Academicos nomen professus, Philologiae primum ac Philosophiae, dein etiam Theologiae mentem manumque applicuit. Quo minus tamen hanc sibi Spartam unice ornandam sumeret, cum aliis de causis, tum secreto quodam genii ingeniique ductu factum est, quibus ad Medicinam ac Mathematicas disciplinas excolendas fuit advocatus."

On Mulerius, see also Eric Jorink, "Tussen Aristoteles en Copernicus. De natuurfilosofische opvattingen van Nicolaus Mulerius (1564-1630)" in H.A. Krop, J.A. van Ruler & A.J. Vanderjagt eds., *Zeer kundige professoren. Beoefening van de filosofie in Groningen van 1614 tot 1996* (Hilversum 1997) 69-83.

[529] Huisman, "Boeken en brieven van Nicolaus Mulerius", 287.

university. However, at the foundation of the university of Groningen in 1614 Mulerius' friend Ubbo Emmius did manage to entice him to accept a professorship there; when Emmius allowed him the choice between the chairs of Greek and mathematics, Mulerius opted for the latter, in addition to the chair of medicine.[530]

Mulerius thus became the first professor of mathematics, as well as the first professor of medicine at Groningen. However, he seems to have displayed a marked preference for the former of these two subjects, paying little attention to medicine either in his teaching (there were only very few medical students at Groningen in Mulerius' day) or in his publications, which all dealt with mathematical and astronomical issues. We have already seen that, from about 1604 onwards, Mulerius annually produced almanacs and prognostications in the vernacular, which were in great demand. These booklets had a high utility value, and were eminently practical in nature. The same can be said about most of his more serious publications: many of these were (sometimes explicitly) intended for the benefit of sailors, engineers, and other craftsmen. Thus, he wrote a Dutch-language manual on the use of the astrolabe, as well as a series of astronomical tables based on the data of Ptolemy, Alfonso, Copernicus, and Tycho, a text-book of astronomy, and a treatise on fortification techniques.[531] This practical orientation may well be partly due to the fact that Mulerius had studied mathematics under Rudolf Snellius. On the other hand, however, he also published a chronological work named *Iudaeorum Annus Lunae-Solaris, & Turc-Arabum mere Lunaris* (1630), at the instigation of J.J. Scaliger, in which he indulged in the humanistic activity of dating the most important events of universal history as accurately as possible. Like Snellius, Mulerius regularly made telescopic observations; in a letter to Roelof Wicheringe, written on 13 March 1613, he wrote that he scrutinized the sun, with its dark spots, every day with the aid of a telescope.[532]

Both Snellius and Mulerius were very well-informed about the recent revolutionary developments in astronomy. As we have seen, as a young man, Snellius had met Kepler, Mästlin, and Tycho, and had even assisted the latter with his observations. Also, he was well-ac-

---

[530] *Effigies & vitae*, 67; L. Elaut, "Nicolaus Mulerius uit Brugge, de eerste medische hoogleraar te Groningen (1564-1630)", *Scientiarum Historia* 1 (1959) 3-13.

[531] See the list of Mulerius' works in *Effigies & vitae*, 69.

[532] Huisman, "Boeken en brieven", 286.

quainted with Stevin's *Wisconstighe Ghedachtenissen,* which he translated into Latin, and which included a defense of the realistically interpreted heliocentric system. Furthermore, Snellius corresponded with Philip van Lansbergen, the Middelburg Copernican who was to become embroiled in the above-mentioned violent polemic with Fromondus on the issue of heliocentrism. As for Mulerius, he was even more closely involved with the question of Copernicanism. In 1617, he saw through the press the third, and for a long time the most authoritative, edition of *De Revolutionibus,* to which he added his own commentary and notes, as well as many corrections.[533] The main purpose of this enterprise was to rid the great work of the many computational and typographical errors which had marred the previous two editions. Thus, Mulerius placed himself squarely within the tradition of the Wittenberg Interpretation: his admiration for Copernicus' achievement confined itself first and foremost to the merits of heliocentrism as a calculational tool, without committing himself to the reality of the system. Unlike his friend Emmius, however, who in a 1608 letter cried out in indignation at the news that Stevin had dared, in a book dedicated to Prince Maurice, to defend his "more than fanatical and absurd opinions" about the motion of the earth[534], Mulerius did not take any polemical stance in the debate on the true world system; while himself remaining loyal to geocentrism, he did, in his works, acknowledge the alternatives proposed by Copernicus and Tycho.

When the comet of 1618 appeared, both Snellius and Mulerius were at the peak of their scholarly activity: the year 1617 had seen the publication both of Mulerius' monumental Copernicus-edition, and of Snellius' masterpiece *Eratosthenes Batavus.* Both men had displayed a life-long interest in astronomical developments, were well acquainted with the ideas of the foremost astronomers of their day, and had been making telescopic observations for several years. Not surprisingly, therefore, each of them decided to subject the new hairy arrival to close scrutiny, and to impart the results of their investiga-

533 *Nicolai Copernici Torinensis Astronomia Instaurata, Libris Sex Comprehensa, qui* de Revolutionibus orbium caelestium *inscribuntur. Nunc demum post 75 ab obitu authoris annum integritati suae restituta, Notisque illustrata, opera & studio D. Nicolai Mulerii Medicinae ac Matheseos Professoris Ordinarij in nova Academia quae est Groningae* (Amsterdam 1617).

534 Ubbo Emmius to Sibrandus Lubbertus, 29 September 1608, in H. Brugmans ed., *Briefwechsel des Ubbo Emmius* II (The Hague 1923) 51-52.

tions to the general public. Snellius chose to compose a lengthy Latin tract, while Mulerius used a Dutch-language pamphlet as a vehicle for his opinions; yet both writings could not but reveal their authors' attitude to contemporary cosmology and cometology. Let us now, therefore, proceed to investigate to what extent the Aristotelian world view could retain its plausibility, to Snellius and Mulerius, in the light of the characteristics of the comet they so diligently observed.[535]

f. *Willebrord Snellius: "Solis ἀποσπασμάτιον"*

Unquestionably the most self-assured work to have been occasioned by the comet of 1618 in the Netherlands, was Willebrord Snellius' *Descriptio Cometae qui anno 1618 mense Novembri primum effulsit.* The work was dedicated to Maurice, Landgrave of Hesse-Cassel, whose father Wilhelm had been a very keen and able amateur astronomer and one of only four men to have located the comet of 1577 above the moon. Although Maurice himself was more interested in alchemy[536], he still had at his disposal the excellent collection of astronomical instruments started at Cassel by his father, and, as we shall see, Snellius' relationship with the Landgrave was so good that he did not hesitate to request the use of these instruments whenever he needed them.

It was Maurice, too, who had asked Snellius to commit to paper his thoughts on the essence of the comet of 1618. As our author mentioned in the dedicatory preface to his tract, he himself had until 1618 always considered it sufficient to describe cometary paths as accurately as possible, without making any more definite pronouncements on their natures; after all, he said, he knew only too well that the heavens were subjected to the same amount of variety and changeability as the sublunary region, and had therefore always believed it unviable to attempt to discover the essence and causes of heavenly occurrences. Thus, initially, the extension of human eyesight provided by the telescope had had the same, largely negative, effect on Snellius as it had on Fienus, leading him to scepticism with regard to the investigation of the universe. However, our author continued, as soon as Maurice's questions on the nature of the 1618

535 On astrological and religious ideas in Snellius' and Mulerius' writings on the comet of 1618, see also Eric Jorink, "Hemelse tekenen. Nederlandse opvattingen over de komeet van 1618", *Gewina* 17 (1994) 68-81.

536 See Moran, *Alchemical World*, passim.

comet had reached him, he had started to apply himself in earnest to the aetiology of comets, temporarily oblivious of all else.[537]

The above passage, taken from the very first paragraph of Snellius' dedication, makes it clear that, from the very beginning of his tract, our author left no room for doubt as far as his opinions on the nature of the heavens were concerned: their changeability, to Snellius, was a fact so firmly established that it needed no further corroboration. Elsewhere in his tract, too, he time and again repeated his aversion to the natural philosophy of the Peripatetics, chiding them for "thoughtlessly allowing the truth to be violated by the concoctions of one single philosopher", and for suffering the "inventor of so many useless orbs" to "press down comets in our common air, and thus confine them to dungeons."[538] Even recently, Snellius sneered, many learned men had not dared to speak out against commonly accepted Peripatetic cometology, "preferring to render themselves less trustworthy rather than to diminish faith in the trivial conclusions of the philosophers, and rather doubting their own experiences than other people's conjectures."[539]

Thus, from the outset, Snellius left no room for doubt that, to him, Aristotelian cometary theory and Peripatetic, solid-orb cosmology were completely untenable; they violated truth, as brought to light by observations and experience, two concepts our author considered of paramount importance. Throughout his tract, he displayed an almost obsessive preoccupation with calculation and precise measurement, emphasizing the importance of approaching natural phenomena with the aid of mathematics, and of observing them as accurately as possible. Underlying this methodology was a zeal for discovery of

---

[537] Willebrordus Snellius, *Descriptio cometae qui anno 1618 mense Novembri primum effulsit* (Leiden 1619) "Dedicatio": "Ego quoque hactenus sufficere existimabam, si illorum vestigiis insisterem, neque ultra tenderem; ut scirem etiam coelum illi vicissitudini obnoxium, quam indies in his terris experimur in tanta rerum omnium varietate. Simul ac vero a T. C. tam nobile zetema mihi propositum vidi, omnia ex eo tempore mihi minora visa sunt, atque ad ipsarum causarum investigationem summa alacritate contendendum existimavi."

[538] Ibidem, 1: "Sed nescio quam infelici astro tot doctorum animi siderati fuerint, & obsessi; dum unius philosophi figmentis veritati vim fieri tam secure patiuntur. Ut omnes iam inde antiquitus ab admirando tot inutilissimorum orbium fabro, quique iis adeo firma lacunaria imposuerit, cometas in hunc nobis cognatum aërem propemodum deturbari, & istis tanquam carceribus concludi aequissimo animo sint passi."

[539] Ibidem, 3: "Multi magna eruditionis fama & celebres viri, nihil contra receptam opinionem pronuntiare ausi sunt; ut sibi potius, quam levibus philosophorum conclusiunculis fidem derogare; & suae potius experientiae, quam alienis conjecturis diffidere maluerint."

the truth, to be attained by means of diligent investigation and to be rendered in plain language unadorned by excessively ornamented speech.[540] Despite the fact that, like his Louvain colleagues, he did make use of Aristotelian concepts such as matter, form, and final cause (thus, summarizing his own cometology in the dedication to the Landgrave, he stated that, as matter, the comet was a burning solar exhalation, as form, a conflagration, and that the purpose to which it had been created was the glory of God and the chastisement of mankind), the rhetoric pervading Snellius' tract was the most consistently anti-Aristotelian of all the Dutch cometary treatises.

The tract commenced with a chapter on "Various ancient and more recent opinions on the distance of comets to the earth". If, Snellius stated, the uncommon lustre of comets had, in the past, induced scholars to examine their nature and causes as diligently as it had through the ages made everybody stare at the sky, their essence would have been investigated and understood long ago, and conjectures and false arguments would have been eliminated. Learned men should have made it a habit to regularly lift up their eyes to the heavens, supported by "mathematical wings"; if they had, they would long ago have discovered that things were not as Aristotle would have them.[541] There had, nonetheless, from the outset been a number of men who had remained unconvinced by the Stagyrite's "petty little arguments"; quoting Seneca, Snellius mentioned Charmander, Epigenes, Apollonius Myndius, and Posidonius among their number. Even more praise, however, was due to Regiomontanus, the "restorer of long-neglected mathematics": despite the fact that he, too, had been affected by the "Sacred Disease" of believing comets to be sublunary, at least he had made an effort to reduce their hitherto inexplicable phenomena to numbers, and contrived a method of measuring their distances.[542] Finally, the nova of 1572 and the comets of 1577, 1580 and 1585 had, thanks to the

---

540 Ibidem, "Dedicatio": "Nihilominus, Illustrissime Princeps, etsi maxime T. C. iudicium pertimescam, tanti apud me veritatis studium & investigatio est, ut sententiam meam nudam, nullo verborum lenocinio adornatam T. C. proponere non erubescam, quo ipsa se ipsam ista simplicitatis specie tanto magis commendet."

541 Ibidem, 1.

542 Ibidem, 2: "Magnus ille Regiomontanus, Mathematum, quae negligentius habebantur, instaurator (etsi opinionis sacro hoc morbo obsessus teneretur, cometas infra lunaris ambitus cortem nasci & denasci) rationem excogitavit, qua incomperta haec & inextricabilia ad numeros veniret, & distantiam adeo (ut creditum est) propinqua demonstratione geometrica numquam fallaci definiret."

diligent observations of several men (most notably Tycho Brahe, Mästlin, and Wilhelm of Hesse-Cassel) finally been proved to be superlunary, and had thus demonstrated the corruptibility of the heavenly region. However, Snellius refused to rest content with the achievements of these men: the present comet provided him with an opportunity of ascertaining for himself its heavenly nature, and delivered him from having to rely on other people's observations only: "For, as the old saying goes, Ἡ γὰρ ἐμπειρία τέχνην ἐπόιησε".[543]

Sure enough, in the next three chapters Snellius went on to describe in great detail the observations to which he himself had diligently subjected the comet. He started out by giving a concise description of the comet's course, "based on the phenomena" ("διὰ τῶν Φαινόμενων"), after which he went on, in the next chapter, to determine with the maximum possible accuracy the path of the comet in its great circle, its intersection with the ecliptic, and its angle of declination, "based on the observations" ("διὰ τῶν τηρήσεων"). Lastly, he devoted a separate chapter to the calculation of the comet's parallax, and its distance from the earth.

The description of the comet's course, based on the phenomena (by which Snellius meant its movement as it could be observed with the naked eye, as opposed to the calculation of the precise position it occupied at specified moments in time) was clearly of minor importance to Snellius; he devoted only two and a half pages to it, as opposed to thirteen and a half in which he treated the motion as based on the observations. Cursorily, he described the account given by those who had witnessed the comet's first appearance on 27 November: that day before sunrise, the sky had been alight with scattered fires, as if somewhere far off in the countryside a hut was burning; this later turned out to have been the comet's tail. Afterwards, the comet itself had appeared, more or less hidden in the glow of the rays it emitted on all sides. Snellius himself first observed the phenomenon on 30 November, when its tail, stretched out westward over a length of about 25°, was at first sufficiently rare for stars to shine through it. From time to time, however, it seemed to increase in density. At the time, the tail was outstretched along the fifteenth and twenty-sixth stars in Virgo, although it did not quite reach the fifteenth; the comet's distance to Spica was 25° 8'. From these data, Snellius deduced that at that moment in time, the tail was not di-

[543] Ibidem, 2-4.

rected away from the sun, but from Venus, although several days later it had become antisolar. The movement of the comet could be summarized as leading from Libra to Bootes, passing between the star in his right leg and the one in his elbow, afterwards crossing the stars in Bootes' back, and moving to the right of his left shoulder towards the tail of Ursa Maior. Finally, on 17 December it had passed the polar circle to become circumpolar, which it remained until its disappearance around the middle of January.[544]

However, Snellius, with his great confidence in the power of mathematics to elucidate physical problems such as the constitution of comets, did not rest content with such a scanty rendition of the phenomenon's course; following Regiomontanus and Tycho, he, too, wanted to subject 'his' comet to a thorough mathematical examination. Unlike his Louvain colleagues, who had none of them been able to dispose of as much as a Jacob's staff, and were therefore obliged to confine themselves to descriptions "based on the phenomena" instead of adding any considerations "based on observations", Snellius as quickly as possible equipped himself with state-of-the-art instruments. When the comet first appeared, he informed his readers, he had had at his disposal no more than a mere Jacob's staff with which to conduct his observations of the phenomenon; however, he realized very well that the data this instrument yielded were not at all up to his high standards of accuracy. However, some of his other instruments were not very well suited for the observation of distances between stars, either because of their weight or, in the case of his great copper quadrant, because it was too heavy to be handled by one person only. Therefore, Snellius had sent for a sextant made by the famous instrument-maker Jobst Bürgi, from Maurice of Hesse-Cassel's collection, which was duly installed on 11 December. Although even this high-quality instrument was not ideal because of certain unspecified "impediments in its motion", several of the observations made from 11 December onwards found favour in Snellius' highly critical eyes, and could serve to answer the questions under investigation.[545]

Just three of these observations, those made on 13, 14, and 24 December, were so precise as to be entirely above suspicion, and

[544] Ibidem, 5-6.
[545] Ibidem, 7.

capable of "illustrating Truth".[546] Snellius recounted how, in the early morning of 13 December, he measured the comet's distance from two nearby fixed stars, whose positions were accurately known from Tycho's star catalogue; next, he provided his readers with an accurately rendered calculation, based on the rules of triangulation, of the comet's longitude and latitude at that moment, meticulously describing every step of his procedure, and illustrating his reasoning by means of a diagram. In a similar way, he computed the comet's position for the other two 'accurate' days, 14 and 24 December, as well as for 11 December; although the latter day's observation was not really up to Snellius' standards of precision, he nonetheless included it in his reasoning because it could serve to deduce the ratio of the comet's velocity.[547] He even ventured to calculate the comet's longitude and latitude on 2 December, at which moment he had not yet received the Cassel sextant and had had to rely on the inaccuracy of the Jacob's staff; yet, Snellius assured his readers, although the observation in itself was therefore worthless, it was needed to determine the comet's variation in speed, and, when compared to the other, more accurate data, it would turn out to be not that far off the mark. Furthermore, once the comet's entire path had been calculated, nothing would be easier than to derive from it the true position the comet had occupied on 2 December.[548]

Sure enough, to demonstrate the suitability of the observation of 2 December, Snellius used it as the point of departure for his next calculation: the determination of the great circle through the heavens

---

[546] Ibidem, 8: "E quibus ternas solum selegimus de quarum fide nulla est dubitatio, ad 13, 14 & 24 Decembris, quas antea instituimus, si opus erit, ad illustrandam veritatem huc quoque advocabimus."

[547] Ibidem, 12: "Huc eam, quam 11 decembris institui observationem, quoque aggregabo, quae etsi non tam diligens sit atque istae superiores, tamen ad motus rationem & velocitatem non fuerit contemnenda."

[548] Ibidem, 13: "Etsi ante hunc diem [11 Decembris] omnes observationes a me factae, non tantum fidei apud me habeant, ut illas ad accurationem epilogismum quidquam momenti afferre posse existimem, tamen si unicam quam sub initium institui huc advocem, ut inde diurni motus velocitatis varietas nobis constet, non existimo me ideo in cuiusdam reprehensionem incursurum, cum ipse praemoneam observationem hanc non ad tenue elimatam, sed rudiore minerva, quod aiunt, institutam. neque enim si cometae locum hunc cum plurimum dierum profectu contendam, admodum longe a vero me collineaturum opinor. Et si veritatem, aut observationis huius lucubritatem explorare libeat, ex cometae tramite, quem accurate satis definiemus, id non esset admodum operosum."

which the comet had followed.[549] Once again, explaining his reasoning by means of accurately rendered calculations, Snellius used triangulation to determine the distance covered by the comet between 2 and 11 December, between 2 and 13 December, between 11 and 13 December, and between 13 and 24 December. After having placed the results in a neatly arranged table for the sake of clarity, Snellius pointed out to his readers that the results of his computations revealed a flaw in the observation made on 11 December: whereas the data for the periods 2-11 December, 2-13 December, 13-14 December, and 13-24 December showed a gradual decrease in the daily distance covered by the comet, varying from about 2 3/4 degrees during the interval between 2 and 11 December to about 2.10 degrees between 13 and 24 December, the data based on the forty-eight hours between 11 and 13 December yielded a covered distance of 3 1/4 degrees a day, which was obviously incompatible with the results for the previous and succeeding periods (unless, of course, it were to be believed that the comet's velocity had suddenly increased between 11 and 13 December, after which it had equally unexpectedly resumed its former gradually decreasing speed).[550] Thus, Snellius was only too well aware that a reduction of the phenomena to mathematical calculation could cast a very different light on the reliability of the observations: the observation of 11 December, made with Bürgi's sextant, turned out less accurate than the one made on 2 December with an ordinary Jacob's staff.

In conclusion to his chapter on the comet's course in its great circle, Snellius calculated the intersection of that course with the ecliptic and its angle of inclination. Using the observations of 13 and 24 December, because these were sufficiently remote in time (for, Snellius pointed out, it was a well-known fact that the use of observations made within a short period of time was risky, because a small

---

549 Ibidem, 16: "Sed ut de Cometae motu & velocitate nobis constet, primum ex data longitudine per calculum antecedentem quoque spatium maximi circuli inter has epochas interjecti ex suis principiis derivabimus, ut quantum inter proximas quasque nobis constet. Equidem a prima observatione nostra quam 2 dec. instituimus, ut vel hinc collato calculo constet nos in tanto intervallo haud multum (propemodum dixeram nihil) a vero motu abivisse diversos."

550 Ibidem, 17-18: "Hinc lapsus diei 11 manifestus est. . . . Ego quae ad hanc rem conferre arbitrabar in medium protuli, etsi diei undecimae observatio hic nullius admodum debeat esse momenti, quae adeo enormiter contra antecedentium & consequentium dierum leges peccet."

inaccuracy could lead to great errors of calculation[551]) he, again by means of triangulation, established the point of intersection at ♏ 13° 20' 30". As far as he could remember from the rather crude observations he had made for the comet of 1607, Snellius added, that comet had cut the ecliptic at almost exactly the same point; however, he did not go on to attach any general conclusions to this observation.[552]

Snellius' second mathematical chapter was devoted to the important question of how far the comet had been removed from the earth. Its distance from the surface of the earth was very great, our author assured his readers; this could be demonstrated in several different ways. First of all, he resorted to a number of qualitative considerations based on the usual order in the cosmos. Thus, the phenomenon had been observed to move on a great circle, from which it had never deflected, and on which its motion had been invariably constant and regular; such a movement was characteristic of heavenly bodies. Also, the comet's relatively low velocity was proof that it had stood far above the moon; after all, the further heavenly bodies were removed from the earth, the slower they moved.[553] However, Snellius would not be Snellius if he had relied exclusively on such rather unmathematical considerations; sure enough, in the very next sentence, he stated that, in order to eliminate every possible doubt, geometrical arguments and accurate observations ought to be invoked.[554]

By means of triangulation, Snellius demonstrated that, on the assumption that the comet had stood a mere 10 earth radii from the earth's centre, it should have risen in altitude by a distance of 1° 19' 25" between the first observation made by Snellius at 4.30 a.m. in the morning on 13 December, and the second observation, made two hours later. Thus, at the time of the second observation, it should have been spotted far beyond the upper of two chosen reference stars, since, at the time of the first observation, the comet had

---

[551] Ibidem, 18: "Verum cum in observationibus nimium propinquis e re minima in istis maximus error possit obrepere. . .".

[552] Ibidem, 20: "Ex quibus liquido constat istum cometam [1607] qui secundum signorum successionem ferebatur sui tramitis ductu continuato eclipticam incidisse pauculis gradibus ultra novissimi cometae nodum."

[553] Ibidem, 20-21.

[554] Ibidem, 21: "Sed ista ut probari possint, tanquam vero assideant, ita invitum & reluctantem cogere non possunt, aut scrupulum omnem eximere: Quamobrem a Geometria nobis ad hanc rem argumenta erunt mutuanda, & observationes huc nobis avocandae, ut quid illae nobis addicant radio & pulvere explicemus."

been removed from that star by a mere 52' 23". However, this had not at all been the case; at 6.30 a.m., the comet had not covered more than a quarter of the distance between the two reference stars, so that the difference in parallax between the two moments of observation could not have been more than 12, 13 or 14 minutes. These data demonstrated conclusively the fact that the comet had not changed its position with regard to the two reference stars, especially considering the fact that the entire distance of "12, 13 or 14 minutes" could not entirely be ascribed to parallax: after all, the comet's own motion in the direction of the higher of the two reference stars, which according to Snellius' calculations amounted to 12 1/5 minutes between the first and the second observation, should also be taken into consideration.[555]

Moreover, even assuming Snellius' second observation to have been inaccurate, and the comet to have moved 17' instead of 12 1/5' during the interval between the observations (so that this movement could not in its entirety be ascribed to the phenomenon's proper motion) the difference in parallax between the two altitudes would still be very small.[556] In order to demonstrate this, Snellius went on to calculate the comet's distance from the earth's centre, departing from the assumed difference in parallax of 4'. Making use of the theorem that "The complements of the observed heights bear the same ratio to one another as the parallaxes themselves, which can be deduced from their sines", our author computed that the comet's parallax at the time of the first observation had been 13' 33" and at the time of the second observation 9' 33". From these data could be derived, by means of triangulation, a distance of the comet to the earth's centre of 211 68/100 earth radii.[557] Thus, even if some of the comet's apparent difference in altitude was to be attributed to parallax instead of to its proper motion, it would still be located very much further from the earth than the moon.

After these two very mathematical chapters, Snellius went on to set forth his opinions on the comet's physical characteristics. First of

---

555 Ibidem, 22-24.

556 Ibidem, 26: "...fingamus omnino nos hallucinatos in loco cometae consignando, & eum plus quarta parte intervalli confecisse; sit sane pars tertia, essent ea 17 scrupula, motus cometae interea propemodum 13 scr. differentia fuerit scrupulorum 4 scr. ut nunc demus hanc in circulo verticali assumenda, tantilla igitur esset eo loco parallaxeos in maxima & minima observatae altitudinis differentia."

557 Ibidem, 26-28.

all, he devoted an extensive section to the vicissitudes of the phenomenon's tail, and to the refutation of the optical theory of comets as proposed by Cardano. By way of an introduction, Snellius rendered some of his predecessors' findings on cometary tails: Apian and Gemma Frisius had discovered their antisolarity, yet a few decades later, Tycho had on the basis of his own observations of the comet of 1577 denied that this was an iron-clad rule, since his comet's tail had pointed away from Venus. Tycho had attributed the tail's lustre to refraction of solar rays inside the head, while Regiomontanus had held comets' trains to be flames invariably tending to fly upward because of their lightness.[558] Rendered curious by so many conflicting opinions, and inspired above all by Tycho's diligence, Snellius had himself devoted a great deal of attention to the tail of the comet of 1618, "so that I would not need to rely on other people's testimonies", as he put it.[559]

Initially, when Snellius first observed the comet on 30 November and saw that at that time the tail was extended in the direction opposite to the sun, while Venus then found herself far removed from the sun, our author believed that Tycho had been quite wrong.[560] However, on closer scrutiny, things turned out rather different; when he reduced the phenomena to numbers, determining the intersection of the line that could be drawn through the comet's head and the ecliptic, as well as its angle of inclination with the ecliptic, and comparing these data to the precise positions occupied by the sun and Venus at the time, Snellius found that, in actual fact, on 30 November the tail had indeed been turned away from Venus instead of the sun, "which did not seem probable to me before I had reduced everything to numbers."[561] However, after a few days, the comet's train did become antisolar.

After this account, serving to once more persuade his readership

---

558 Ibidem, 29.

559 Ibidem, 30: "Ego quoque istorum studio incitatus, haud oscitanter caudae positum observare institui, ut non haberem necesse alienis testibus credere, quippe qui ad αὐτοψίαν tanti miraculi quoque evocarer, movebat me super caeteros Tychonis diligentia, & hac in parte fides."

560 Ibidem: "Itaque cum primum cernerem 30 novembris caudam a plaga Soli opposita declinare, & scirem venerem a Sole longius abesse, non tamen e vestigio illius [Tychonis] sententiae prima fronte mihi subscribendum existimabam."

561 Ibidem, 30-31: ". . . Quamobrem eo ipso momento hanc comam solam venerem respectavisse planum est: quod initio antequam ista ad numeros revocassem mihi verisimile non videbatur."

of the necessity of the use of mathematics in matters astronomical, Snellius addressed the question of where the tails of comets were located, and what was their essence. In order to ascertain whether they were in the heavens or in the atmosphere below the moon, our author introduced his readers to the distinction made in the science of optics between on the one hand real phenomena located in the sky and visible to the human eye, such as the sun, the moon, the stars, and the clouds, and, on the other, apparent phenomena caused by refraction or reflexion of light rays, such as rainbows, haloes around the sun or moon, or double and triple suns.[562] Despite the fact that most of these phenomena tended to display a circular shape, Snellius stated that he did not doubt that, often enough, they might appear rectilinear.[563] According to the science of optics, as well as experience, the second category of apparent phenomena such as rainbows and mock suns were all located in the atmosphere close to the earth. They were seen only in cloudy skies, and never remained once the clouds disappeared, thus showing that their causes were merely optical: if it were possible to travel towards the place were they seemed to be situated, one would find nothing but air there, and the phenomena themselves would turn out to have disappeared. Conversely, the tails attached to comets were not seen in the vicinity of clouds; on the contrary, they were best visible on clear and cloudless nights. Furthermore, when observed by people placed at various locations on the earth's surface, the cometary tails were invariably seen to point at the same fixed stars, as the observations made by Tycho, Mästlin, and Rothmann had conclusively demonstrated. All this was proof enough to show that the trains displayed by comets were not apparent optical illusions, but real entities located in the heavens.[564]

Having thus satisfactorily proved that cometary tails were

---

[562] Ibidem, 32: "...unius generis sunt illa quae per se subsistunt, & a nobis cernuntur ut Sol, Luna, sydera, nubes. Illorum autem κατ'ἔμφασιν apparentium sunt ea, quae opticam refractionis causam habent, parelia, paraselena, gemini & tergemini soles, ut historici loquuntur; vel ea quoque quae per refractionem & reflexionem simul suae essentiae causam trahunt, halones & irides, quarum discrimen non satis commode opticis cognitum existimo."

[563] Ibidem, 33: "Huiusmodi autem Φάσεις admodum varias posse existere, ut etiam in directum exporrigantur, mihi non sit dubium & saepius est observatum. quamvis fere semper in orbem diffuse appareant...".

[564] Ibidem, 33-34: "Atqui caudam hanc cometarum appendicem non in nubilo, sed aere vidimus serenissimo; & quidem tanto ampliorem, quanto aer ipse erit faecatior. Et, quod rei caput est, omnes etiam locis diversissimis eodem in caeli tractu,

superlunary phenomena, Snellius went on to refute the theory proposed by Cardano to explain their causes. Paraphrasing the Italian's theory, that the effluxions from the comet's head were due to its intermediary density, not dense enough to absorb all light rays nor rare enough to allow them to pass unrefracted, our author stated that although this theory seemed very probable, it nonetheless deviated from the truth. Himself a renowned opticist, Snellius realized only too well that, in order to become visible, light rays needed to pass from a less dense medium to a denser one. Otherwise, he pointed out, we should see sunrays everywhere in the sky even during the night, when the sun stood far below the horizon. In reality, however, at night we see the rays of the sun only insofar as they reach denser bodies, such as the moon and the planets. In one and the same medium, only rays travelling in straight lines are visible, and unless they encounter a denser medium, they cannot generate vision.[565]

But, if they did not consist of refracted solar rays, why then did cometary tails almost invariably point in an antisolar direction? Snellius devoted a separate chapter to this question; however, he realized only too well that it was very difficult to come up with a satisfactory answer, stating that he would be very satisfied if only he could say something probable on the subject, without directly uncovering the truth of the matter.[566] Indeed, the chapter following this uncharacteristically modest remark was the most rhetorical of the entire tract, culminating, after many eloquent detours, in the implicit confession that Snellius did not know the reason for the tails' antisolarity any more than his refuted colleagues. This section is, however, extremely interesting in that it shows us how Snellius held the authority of the ancients in very high esteem. His radical and fierce rejection of Aristotle and Peripatetic science was replaced partly with an emphasis on meticulous observation and mathematical calculation, but in addition Snellius also yearned for the recovery of knowledge in its pristine state, as it had resided among scholars of ancient

quemadmodum in earum descriptionibus apud Tychonem, Mestlinum & Rhotmannum videre est. Unde efficitur utique hanc appendicem comam non in aere κατὰ Φαντασίαν apparere, sed in ipso caelo existere."

565 Ibidem, 35: "manet enim illud immutabile effatum, visibile in eodem medio radiis tantum rectis ferri. Et, nisi radii incidant in medium densius, hoc est a superficie terminentur, visionem nullam facere."

566 Ibidem, 45: "Ego si in re adeo ardua verum non assequar, saltem verisimile aliquid dixisse doctioribus me visum iri non despero. . .".

times far before the advent of Aristotle. His preoccupation with mathematics and experience can even be said to have been inspired to a certain extent by the desire to recover, through calculation, the lost wisdom that had been almost intuitively known to the pristine sages.

In a Stoic vein, our author stated that in the present chapter he addressed only those scholars who were prepared to observe the world with mathematical and physical eyes, not just the small globe on which they lived, but the entire cosmos in its interconnectedness.[567] Only they could judge Snellius' opinions, in aid of which he invoked the authority of the most ancient sages to serve as intercessors in the imaginary lawsuit to which he willingly subjected himself.[568] After this programmatic introduction, our author went on to bestow great praise on the achievements of the ancients. Berosus, the Babylonian priest who came to Greece around 280 BC and composed an account of Babylonian wisdom, the *Babyloniaca*, for the benefit of the Greeks, was held in high esteem by Snellius. Although his opinion about the moon (a globe catching fire whenever it passed below the sun, not through direct contact, but through natural affinity) was clearly faulty, as was apparent from the spots on the moon which always retained the same shape, Snellius declared that nonetheless, it showed how diligently the ancient Babylonians had investigated nature. After all, in stressing the mutual affinity between the body of the sun and that of the moon, Berosus had shown that he knew about the existence of a magnetic force permeating the universe.[569]

Indeed, Snellius believed the ancient Babylonians to have been better informed about the order of nature than his own contemporaries; how else could they have provided each of the planets with its own proper qualities and effects, so aptly that no-one could have done better? Moreover, they had also realized that the motions of the planets were dependent on their approaching and receding from the sun, and possibly they had known even greater things which had in

[567] Ibidem: "Verum ad illorum tribunal provoco, qui universi huius mundi complexum Mathematicis & vere Physicis oculis contemplantur, neque huic puncto solum quod incolunt, tanquam terrestria animalia adhaerescunt."

[568] Ibidem: "Sed vetustissimorum auctoritate me muniam, si forte istis intercessionibus impetrare possim, ut tanto aequiora iudicia experiar."

[569] Ibidem, 46.

the meantime fallen into oblivion.[570] However, Snellius could not believe that the Babylonians had obtained all this knowledge simply and solely by observing the phenomena; rather, he thought they had possessed a doctrine inherited from their forefathers, which like a mystery they handed down to the most intelligent among them, who were deemed most worthy of receiving such a great secret. If Josephus' story, that Seth had foretold to his offspring that the world would once perish by a flood and a second time by a conflagration, was true, Snellius believed that in doing so he had not so much prophesied, as interpreted the secrets of nature, to which he had been an initiate.[571]

Sadly, though, most of the ancient wisdom had been lost in the course of the ages; when it was handed down from the Babylonians and Egyptians to the Greeks, much of it had perished, and in the transmission from the Greeks to the Romans and to us even more was forgotten, because the initiates were reluctant to disclose too many of their secrets for fear of them falling into the hands of the common people, who would through ignorance misemploy them and use them to bad purposes instead of the benefit of mankind for which they had been intended.[572] Consequently, Snellius added, we now possess mere shadows of the great things that were known to the ancients.[573] The idea of lost pristine wisdom was very common during the Renaissance, especially with authors sympathetic to Neoplatonism[574]; more specifically, the belief that in the course of time

---

[570] Ibidem, 46-47: "Existimo utique illis Naturae rationes melius quam nobis fuisse perspectas. Neque enim temere est, quod singulis planetis suas quasdam qualitates atque effectus primi attribuerint: & quidem ita, ut vix potuerit aptius. Quamvis enim planetas ipsos motus suos pro solis accessu, aut ab iisdem discessu componere animadverterent & forsan his maiora, quae nunc obliterata iacent. . .".

[571] Ibidem, 47: "Habuerunt omnino doctrinam πατροπαράδοτον, quam non omnibus sui ordinis hominibus: verum primariis, ut opinor, & ingenio praestantibus, tanquam μυστήριον per manus tradiderint, quosque ipsi tanti arcani capaces existimarent. Nam si verum est. . . Zetum posteris praedixisse terram hanc semel diluvio, iterum incendio perituram, non tam prophetico spiritu actum, quam naturae secreta interpretatum existimo."

[572] Ibidem, 47-48.

[573] Ibidem, 48: ". . .ut maximarum rerum solas umbras possideamus."

[574] See D.P. Walker, *The Ancient Theology* (London 1972). See also Charles B. Schmitt, "Perennial philosophy: from Agostino Steuco to Leibniz", *Journal of the History of Ideas* 27 (1966) 505-532; and idem, "Prisca theologia e philosophia perennis: due temi del Rinascimento italiano e la loro fortuna" in *Il pensiero italiano del Rinascimento e il tempo nostro. Atti del V Convegno internazionale del Centro di studi umanistici, Montepulciano, 1968* (Florence 1970) 211-236 .

much of this pristine philosophy had been lost can be found in the writings of William Gilbert, Peter Ramus, and Simon Stevin, all of them well known to Snellius.[575]

Yet, Snellius saw no reason for despair. Although most ancient knowledge had fallen into oblivion, and, especially during the recently elapsed centuries of the Middle Ages, God had punished mankind's neglect of the investigation of nature by making the human race recede into the most profound mental darkness, the many new stars and comets that had appeared in the recent past to Snellius were tokens of hope, sent by God to encourage scholars to recover the knowledge of nature that had been lost for so long. After all, such massive bodies would never be born in the heavens unless they had a purpose, our author teleologically added.[576] Thus, Snellius believed that the intuitive knowledge of nature, which had been possessed by the ancient sages, but obscured by the slavish imitation of Aristotle, could be recovered through the diligent observation of nature and the application of mathematics to physical processes. By advocating the importance of mathematics and autopsy, he hoped to regain an insight into the secrets of the cosmos, seen as a single entity pervaded by natural affinities between its parts.

One such affinity, and a very important one at that, was the power exercised by the sun on the motions of the planets. Clearly influenced by Kepler, Snellius declared that the initiated ancients had called the sun "Heart of the World", not only because of its biblical dominion over light and darkness, day and night, and seasons, but also because it ruled over the orbits followed by the planets. For, whenever the sun moved closer to the three superior planets, they found themselves at apogee, the farthest removed from the earth as possible, while whenever the sun receded from them, Mars, Jupiter

---

[575] See William Gilbert, *De magnete*, Book VI, chapter 3; on Ramus, who applied the idea of lost ancient wisdom specifically to the sciences of dialectic and mathematics, see Ong, *Ramus*, 47 and Hooykaas, *Humanisme, Science, et Réforme*, 78; on Stevin, who applied it to heliocentric astronomy, see R. Hooykaas, "The Reception of Copernicanism in England and the Netherlands" in *The Anglo-Dutch Contribution to the Civilization of Early Modern Society. An Anglo-Netherlands Symposium* (Oxford 1976) 33-44: 38. See also Léontine Zanta, *La Renaissance du Stoicisme au XVIe siècle* (Paris 1914) 190 for similar ideas to be found in the work of Justus Lipsius.

[576] Snellius, *Descriptio Cometae*, 48: "Haud dubie enim sublimia ingenia splendidissimo illorum siderum iubare ad veritatis investigationem & naturae arcanae eruendum excitabuntur. Neque enim unquam temere nullove effectu tantae moles in summo aethere exoriuntur, & a Deo Opt. Max. tanquam signa ad bene sperandum in ipso caelo omnium oculis exposita eriguntur."

and Saturn moved closer to the earth, towards their perigee.[577] The inferior planets, Venus and Mercury, moved round about the sun in very close circles, like bodyguards, Snellius added; although Tycho's suggestion, that the other planets also orbited round the sun, but in wider circles encompassing the earth, was not inept either.[578] Thus, Snellius professed his adherence to a world system in which at least the two inferior planets circled the sun. Such a system had been contrived by the fifth-century scholar Martianus Capella, and was therefore known as the Capellan system; it gained a certain measure of support during the seventeenth century.[579] Snellius' phrasing did not make it clear whether he was also prepared to endorse the complete Tychonic system; at the very least, he thought it well possible. He did not, however, mention the Copernican alternative of a completely heliocentric system, despite the fact that he most certainly knew about it, having met Kepler at Prague while the latter was working on his *Astronomia Nova*.

Whatever his opinion on the true world system, Snellius, like Kepler, thought it beyond doubt that the motions of the planets were ruled by an innate power of the sun. Moreover, he was convinced that this same solar power was responsible for directing the tails of comets in an antisolar direction.[580] Sadly, though, hardly anything was known about the sun's mysterious potency; literally quoting Seneca (though without mentioning the author's name) he sighed that "One lifetime does not suffice to investigate such great things." Although, unlike Kepler, Snellius did not go on to explicitly identify the power exercised by the sun on the planets (and the cometary tails) with the terrestrial phenomenon of magnetism, he did suggest an analogy between the two innate forces by inserting a lengthy ac-

---

[577] Ibidem, 48-49: "Sol ille, quem Veterrimi Cor caeli vocitabant, non tantum quod dierum noctiumque inaequalitatem dispenset & totius anni gratam vicissitudinem temperet, planetisque omnibus suam lucem foeneret: verum etiam quod ipsorum cursus moderetur ac regat ita dictus. Ipso enim accedente tres superiores a terris altissimi in apogaeo, recedente in perigaeo versantur."

[578] Ibidem, 49: "Venus & Mercurius tanquam laterones eum proximi ambiunt. Etsi reliquos superiores idem quoque facere, sed amplissimis conversionibus quibus terram quoque includant, non inepte Tychoni sit visum."

[579] See Christine Schofield, "Tychonic and semi-Tychonic world systems" in René Taton & Curtis Wilson eds., *Planetary astronomy from the Renaissance to the rise of astrophysics. Part A: From Tycho Brahe to Newton* (Cambridge 1989) 33-44: 38.

[580] Snellius, *Descriptio*, 49: "A Sole inquam horum motus regi nemini intelligenti dubium esse potest. Hac eadem insita Solis vi istas cometarum caudas ita dirigi, ut a Sole aversae jaceant, omnino mihi persuadeo."

count of the history of the discovery and investigation of magnetism.

The discovery of the northward inclination of the loadstone was made a mere four hundred years ago, he said; before, it had only been known to attract iron, a property said to have been discovered by Hercules. By lending his name to the magnet, Hercules was greatly honoured for his rather useless discovery; yet whoever found out about the constant inclination of compass stones, a quality so important to navigation, had been praiselessly forgotten, although he deserved much greater honour, Snellius indignantly recounted, once again displaying his utilitarian bent of mind. Much later again, it was discovered that the loadstone did not always point straight north, but at times deviated, albeit constantly, about 20 or 30°. People with too much leisure and too much imagination had come up with fanciful explanations for this deviant behaviour, such as the existence of a great magnetic mountain situated near, but not quite at, the North Pole. However, in the end the learned William Gilbert had, after long and daily observation and investigation, discovered the true cause, which was to be sought in the fact that sometimes the properties of compasses were hampered by the great magnet Earth, whose magnetic force was not equally distributed over her surface.[581]

This lengthy account of the history of magnetism served a dual purpose. First of all, it was intended to demonstrate how painstaking the investigation of natural phenomena could be, and how very time-consuming: it could take centuries to discover the true cause of things. Emphasizing these difficulties gave Snellius an excuse for not being able to come up with a conclusive reason why cometary tails were not always quite antisolar, which was, after all, the subject of the chapter. Secondly, Gilbert's explanation of the deviation of compasses provided Snellius with an analogy which he could use to suggest a possible reason for the inconstancy of the tails: the solar power, too, might not be equally forceful in every region of the heavens. Thus, according to Tycho, the tail of the comet of 1577 had pointed away from Venus instead of the sun, while at the same time, that comet had never moved beyond the boundaries of the zodiac. Similarly, the tail of the comet of 1618 had not been precisely antisolar as long as the comet was located in the zodiac. Thus, Snellius seems to suggest that, possibly, within the zodiac the solar power was less strong, allowing the tails to deviate. However, he dared not make

---

581 Ibidem, 49-51.

any definite statements on the subject, since he considered his observations of the tail too inaccurate and therefore unreliable.[582]

Although, then, in Snellius' account the behaviour of the tail remained largely unexplained, he displayed much more self-confidence as to the properties of the matter out of which comets' heads were formed. One thing was absolutely certain: they could not possibly consist of terrestrial exhalations. Snellius devoted a separate chapter to a conclusive demonstration, underpinned with calculations, as to why it was unthinkable for a comet with the characteristics of the 1618 one to be formed out of sublunary matter. For, even if one were to imagine the entire airy region to be compressed into a single globe, this globe's mass would still not equal the present comet in size. Obviously, even common sense arguments served to show that such an immense volume of air could not possibly leave the atmosphere to form a comet, without us noticing its departure, and with it the departure of the air we breathe and live on, Snellius argued; but, in order once and for all to cure gullible people of their foolish misconceptions, he in addition proposed to reduce his thought experiment to numbers.[583]

Employing his own *Eratosthenes Batavus* value for the earth's radius, as well as Posidonius' estimate, that the region of air rose up to a height of 13 miles above the earth's surface, Snellius calculated that the earth's diameter was three times that of a globe consisting of all the air contained in the atmosphere. Thus, such a globe would be about 1 1/2 times the size of the moon. If this body were located at a distance of 300 earth radii, it would occupy a space of about 0° 3' 50" in the heavens; therefore, the comet's body, if located at the same distance and seen to occupy a space of 4', would already be much larger than the airy globe. Tycho's comet, for one, occupied a space of 7', so that its diameter was about twice that of the imaginary airy globe, and therefore its body was about eight times as large as the whole atmosphere compressed into one.[584] Not to mention, Snel-

[582] Ibidem, 51: "Non ausim de caeteris diebus quicquam certi pronuntiare, nam caudae positum non ea diligentia observavi, quam quidem nunc ipse desidero. . .".

[583] Ibidem, 37: "Et quis tantam congeriem, adeo vastas & copiosas exhalationes a nobis abire, & illam ipsam vitalem auram a toto terrarum orbe subduci non sentiat? . . . Id vero quoque calculo experiri libet, ut hunc alterum veternum hominibus nimium credulis excutiamus."

[584] Ibidem, 38: "Tycho 7 scrupula suo cometae attribuit, ut iste diametro propemodum fuerit dupla. unde efficitur corpus istud totius atmosphaerae fuisse octoplum."

lius added, the immense mass of the tails, which in the case of the comet of 1618 was diffused over 50° in latitude; if this volume were to be compressed into a globe, its size would be several thousand times larger than the earth.[585]

Obviously, if comets did not consist of any sublunary matter, they must somehow be formed out of the celestial aether above the moon. To Snellius, this region was not filled with the perfect quintessence postulated for it by Aristotle; rather, it was a stuff possessing qualities like those to be found below the moon. At the very least, it was capable of catching fire; thus, Snellius argued, the sun should be thought of as one great fiery ball, burning and generating an immense heat. In ancient Athens, the philosopher Anaxagoras had almost been convicted to lifelong exile for advocating this opinion, our author indignantly recounted, and even nowadays, adherents of "the philosophical sect that has filled the study of nature with its conjectures" would still not admit the sun's fiery nature.[586] Yet, the idea was clearly consistent with nature and truth, Snellius insisted; after all, the natural state of affairs was for light to be invariably accompanied by heat. The sun, undeniably the primal, indeed the only source of light in the universe, by whose lustre all the other planets were illuminated, would never be able thus to shine with its own light if it were not burning and fed by fuel.[587]

Moreover, the sun's fiery nature had recently been proved by the novel discoveries made with the aid of the telescope. These had shown how varied errant spots moved round the sun's body in the same way as clouds moved through the earth's atmosphere, passing in front of the solar disc from east to west in a few days, and not possessing any parallax, so that they had to be located very close to the sun itself. Nor were they attached to the solar surface; if they were, the same spots would return from time to time, which was not the case. Rather, they displayed a very varied spectacle, assuming many different shapes and densities, and not always the same in

---

585 Ibidem, 38.

586 Ibidem, 39: "... Anaxagoram virum summum vix eripuit Pericles ... quod diceret, Solem μύδρον διάπυρον globum massamve candentem, & ex duro concretum ignem suum pascere, quod ne nunc quidem sibi persuaderi patietur ea philosophorum secta, quae totam rerum naturam suis conjecturis implevit."

587 Ibidem, 40: "At Sol ille qui fons lucis & sedes sit propria, a quo reliqui omnes planetae collustrantur... cum luce luceat sibi insita & omnino sua, non potest eam habere sine fomite imo flamma."

number. Judging by these appearances, Snellius thought it highly likely that the sunspots ought to be explained by assuming them to be exhalations of the burning sun, which it belched out from the extremely dense interior of its body, just like earthly volcanoes such as Etna and Vesuvius belched out lava. This also explained why, through the telescope, some parts of the solar disc were seen to be more resplendent than others.[588]

Comets, then, were nothing else than a particular kind of exhalation of the sun, formed whenever the ordinary solar effluxions carried with them lumps of more solid matter torn loose from the sun's interior, and dense enough to catch fire.[589] For this reason, Posidonius had been right in saying that most comets were never seen, because in the vicinity of the sun they were rendered invisible by its rays, and Ptolemy had had good reason to point out, in his *Tetrabiblos,* the necessity of observing comets especially during eclipses.[590] Furthermore, Snellius was in no doubt either about the matter of which cometary tails consisted: they were also solar exhalations, but of a rarer and more volatile kind; after all, the so-called sunspots, as observed through the telescope, were seen to be in a continuing state of flux, with denser and rarer parts interacting with one another. This volatile state was in perfect accordance with the behaviour of the tail of the comet of 1618, which Snellius himself had observed to change its density and its colour, at times allowing the human eye to perceive the stars lying behind it, and at other times obscuring them because of its temporarily increased density. The exhalations forming the tail were illuminated by the sun and thus rendered visible to observers on earth, since the sunrays, when meeting the cometary tails, passed from a less dense medium to a denser one.[591]

Due to the tails' variations in brightness and intensity, Snellius supposed them to be continually supplemented with new matter

---

588 Ibidem, 40-42.

589 Ibidem, 42: "Illa igitur Solarium exhaltationum [sic] volumina, si quid materiae solidioris per ipsos crateras secum ferant, aut ad flammas perferendas aptioris; tale inquam Solis ἀποσπασμάτιον cometa est."

590 See Ptolemy, *Tetrabiblos,* II.9.

591 Snellius, *Descriptio cometae,* 42-43: "... ut mihi caudae fulgor non aliud sit, quam rarior exhalatio, etiam ab ipso capite flagrante, Solis radiis illustrata; quae non secus atque tenuiorum nubium in hoc nostro aere strigmenta collucent, ut non sint radii Solares per corpus cometae pellucidum refracti: sed Solis radii in medium densius allapsi, propter quos illud collustratum a nobis cerneatur."

from the head. Like a burning piece of charcoal, that emitted rays all around its body like a lamp in moist air, so also the comet's head sent out a burning exhalation in the form of a tail or beard; in the end, when all the matter contained in the head had been thus emitted, the head would be entirely consumed by its flame.[592] A very similar view of cometary tails, which is in fact basically correct, was to be proposed by Kepler in the Appendix to his *Hyperaspistes*, which appeared in 1625, six years after the publication of Snellius' tract.[593]

Snellius was very confident about the correctness of his 'solar exhalation' theory of comets. However, he himself pointed out one major anomaly, to the resolution of which he devoted a separate chapter. For what was to be made of comets that did not originate in the vicinity of the sun, but were born in an entirely different region of the heavens?[594] The resourceful Snellius had thought of a solution even to this thorny problem, although he admitted that he was proposing a mere hypothesis, which would need to be corroborated by many and frequent future observations. He was prepared to give up his theory if, but only if, such observations were to render it untenable, "for to persist in one's opinion even if experience goes against it, comes down to trusting no-one but one's own witnesses, and wrongfully relying on them", our author added in one of his many legal metaphors.[595]

In any case, until his theory was falsified Snellius believed that comets born far away from the sun did not really differ very much from their counterparts in the solar vicinity. They, too, were solar

[592] Ibidem, 44: "Atque istud profecto satis indicii arbitror, comam semper nova materia sustentari, idque ex ipso Cometae capite, quod Solis ἀποσπασμάτιον flagrans esse veritati consentaneum censeo. Ipsum enim Cometam ab igne suo depasci obtusum, tanquam in carbone accenso lumen ostendit, quod circum circa corpus medium radios de se in orbem spargit, ut lucerna in aere roscido. Caudae autem tractus eius conflagrantis exhalatio, ut circa Solem maculae. . .".

[593] See C. Doris Hellman, "Kepler and Comets" in A. Beer & P. Beer eds., *Kepler: Four Hundred Years. Vistas in Astronomy* 18 (1975) 789-796: 792.

[594] Snellius, *Descriptio*, 53: "Sed omnem difficultatem, utut haec de cometarum quorundam exortu verissima sint, nondum videbor expedivisse, qui duntaxat ostenderim, quosdam cometas sub sole nasci. Atqui ut istos illic natos his legibus teneri concedi possit, quid illis fiet, qui sub alto septentrione aut zodiaci plaga Soli opposita primulum exarserunt."

[595] Ibidem, 54: "Ego utique magni existimo referre in re nova & hactenus incerta aliquid vero affine excogitavisse, donec ista crebra & diutina observationum historia aut approbet, aut refellat: pertinaciter enim suam sententiam invita & reluctante experientia tueri, est suis testibus tantum fidem habere, & iisdem importune insistere."

exhalations, yet of a variety that did not immediately catch fire on being ejected from the sun's body. Instead, they moved on and on through the exceedingly cold celestial aether; for Snellius was convinced that the entire celestial region was pervaded with the most intense cold, except, of course, in the vicinity of the sun or the stars. In those parts of the aether where there were no solid bodies to catch and reflect the solar or stellar rays, there could not be found a single source of heat. Thus, the above-mentioned, not yet enflamed solar exhalations could travel immense distances through the icy cold without catching fire, in the direction in which they were propelled by the force of their expulsion from the sun, or perhaps by some innate quality of motion. Whenever, in the inhospitable outer regions of the aether, several such exhalations encountered one another, they might merge and catch fire, either by means of an innate force of their own or stimulated by the power of the stars.[596] Such a fusion of exhalations served to explain the enflaming of comets in regions far removed from the sun.

All in all, Snellius' tract was the most consistently anti-Aristotelian of the Dutch cometary treatises of 1618. It was pervaded with biting anti-Peripatetic rhetoric, and the author on several occasions scathingly criticized the conjectural and ill-founded nature of scholastic scholarship. Instead of such reliance on other people's questionable authority, Snellius advocated autopsy and accurate observations, as well as the reduction of physical phenomena to mathematical calculations, as the only sure way of discovering the true nature of things. In accordance with this methodology, he made use of state-of-the-art instruments, triangulation techniques, diagrams, and tables in order to investigate the characteristics of the comet of 1618. Accordingly, his tract contained more, and more accurate, data for the comet than those of his Louvain colleagues; most notably, he determined the phenomenon's great circle, its parallax, the point of intersection with the ecliptic, and the angle of inclination.

Nonetheless, his conception of the universe was not very different from that of Fienus, Fromondus, or Puteanus. Like them, he endorsed the changeability of the heavens and the absence of solid

[596] Ibidem, 55: "Ut mihi mirum non videatur, si vagae illae & per universum aetherem palantes rariores ἐκκρίσεις aliquando coëant singulae, aut etiam plures in unum concurrentes condensentur: atque inde insita vi sua, quam ab ipso exortu secum trahunt, vel etiam siderum vi exstimulatae accendantur, & in manifestum ignem prorumpant."

orbs, though unlike the Southern Netherlanders, he did not think it necessary to corroborate these tenets with any kind of argument, either from common sense or from authority. To Snellius, the fact that generation and corruption could take place in the heavens had been conclusively demonstrated by the comet observations of Tycho, Mästlin, and the Landgrave, and any further effort to prove it would be wasted on those who were too stubborn to believe it. Curiously enough, this unwavering belief in celestial alterability at one point led Snellius astray: towards the end of his tract, he stated that he knew about the opinion of Apollonius of Myndus, according to which comets were periodically returning objects describing closed orbits, but added that he could not see any necessity for thus postulating the hairy stars to be eternal bodies, since the possibility of heavenly generation and corruption had been proved conclusively enough.[597] Therefore, Snellius preferred to accord comets a very ephemeral nature.

His own theory of comets relied heavily on the recent telescopic observations of sunspots, which he believed to be solar exhalations. Snellius' idea, that comets came into being whenever such an exhalation carried with it a lump of the sun's material solid enough to catch fire, however, closely resembled the cometary theory of the much-despised Aristotle himself; in fact, all Snellius did was to transfer the Stagyrite's theory from the earth to the sun. He did at one point suggest that, possibly, the fixed stars might also produce exhalations suitable to the formation of comets, failing, however, to mention the possibility, entertained but in the end discarded by Fromondus[598], that the planets, too, might exhale cometary matter. Thus, he did not seem too keen to endorse the close analogy between the earth and the planets on which the 'planetary exhalation theory' was based, and which smacked of Copernicanism.

In fact, the Copernican theory was not at all mentioned by Snellius. He did acknowledge the Tychonic arrangement of the planets, and thought it "not inept"; nonetheless, he explicitly went no further than the Capellan system, in which only Venus and Mercury circled

---

[597] Ibidem, 57: "Non sum nescius Apollonium Myndium cometas in altissimos caeli recessus quibusdam circulorum conversionibus tanquam ad apogaeum subducere ex oculis nostris, & rursum eosdem circa perigaeum oculis nostris admovere, atque ita perennitatem illis machinari. Sed multa in caelo ipso nasci & denasci iam diu experientia ipsa comprobavit."

[598] See supra, p. 297-298.

the sun. He did, however, following Kepler, recognize the important role the sun played in controlling the motions of the planets; implicitly, he seemed inclined to identify this solar power with the terrestrial phenomenon of magnetism in the same way Kepler had done. To Snellius, the cosmos was to be regarded as a single whole with interconnected parts, full of occult forces and secrets of nature that had been known to the pristine sages and might well be rediscovered if only the correct method of continuing observation and mathematical calculation were used.

g. *Nicolaus Mulerius: "Rather an Eternal Creature of God"*

If Snellius' tract was the most openly anti-Aristotelian of all the Dutch cometary treatises, the one written by his Groningen colleague Nicolaus Mulerius was the most Aristotelian; and if Snellius' tone was the most self-assured, Mulerius' was the most hesitant. This marked difference may partly be due to the fact that Snellius addressed his fellow scholars in Latin, while Mulerius wrote in Dutch for the general public; thus, the latter may have thought it more suitable to render the most common opinion on comets, without expounding the many novel explanations in too much detail. Also, the contrast between the two writings may be ascribed to the fact that Snellius was an outsider at Leiden university, while, conversely, Mulerius was a very prominent member of the academic establishment at Groningen, with its self-professed loyalty to the Stagyrite's teachings.[599]

In any case, Mulerius, addressing a non-academic audience, adopted a simplifying and didactic tone throughout his tract. Like the writings of Gemma and Heurnius on the comet of 1577, the main purpose of Mulerius' pamphlet seems to have been to urge his readers to better their ways and to repent, this being the only means of averting God's anger and chastisement; hence the tract's title: "Heavenly Trumpet, Matutinal Arouser, or Comet with a Long Beard". In a short preface "To the Christian Reader", Mulerius emphasized that the purpose to which comets, in their quality of "extraordinary messengers of God", appeared, was to warn mankind to change their sinful ways. The pamphlet's last (and longest) chapter, dealing with comets' significance, was also pervaded with calls for repentance, as well as with dire predictions and historical examples

[599] See above, section V.4.e.

of the pernicious effects wrought by comets in the past. Nonetheless, in three earlier chapters respectively treating "What comets are, and some varieties of comets"; "The opinions of learned men on the generation of comets"; and "On two comets seen in this year 1618", Mulerius made an earnest effort to set forth, in simple terms, the most common tenets of contemporary cometology. Throughout these sections, the author displayed his didactic qualities, lucidly explaining difficult concepts such as "parallax", and translating any Latin or Greek phrases he thought it necessary to quote. Despite the simplifying character of these three chapters, they do provide us with a fair amount of information on Mulerius' world view: the recent astronomical developments, though not as extensively treated as they might have been in a Latin tract meant for scholarly peers, were not quite absent from our author's account.

In the pamphlet's first, untitled, chapter, Mulerius recounted what he believed to be the justification for wanting to investigate the nature and essence of comets. In a long paraphrase of the opening section of Book VII of Seneca's *Natural Questions*, our author pointed out that small, yet novel natural events invariably raised more curiosity than such enormous, yet daily recurring phenomena as the incredible speed with which the sun pursued its daily path round the earth. Comparing the comet to a comedian and the sun to a king, on the grounds that the common man was more interested in the former than in the latter, Mulerius somewhat inconsistently argued that, although great things, such as kings and the sun, deserved more attention than less reverent things such as comedians and comets, nonetheless the *raison d'être* of the hairy stars ought also to be diligently examined, since God had created everything to a purpose.[600] Therefore, in the present pamphlet, the author intended to investigate the learned opinions on comets, their location and course, and their significance.

Mulerius commenced his chapter about the diverse scholarly opinions on comets with a remarkable profession of loyalty to Aristotle.

600 Nicolaus Mulerius, *Hemelsche Trompet Morghenwecker, ofte Comeet Met een Langebaert Erschenen anno 1618. in Novembri ende Decembri* (Groningen 1618) chapter I: "So bevinden wy oock huydens'daegs aen dese Comeet tegen de Son te rekenen noch minder te achten is, als een Comedien-Speler by een Coninck, nochtans hier is onghelijck meer sprakes ende vragens nae die Comeet als nae de Sonne. Maer wy behoren onse sinnen meer te setten op 'tgene dat meer te achten is. Ende so wanneer onsen Heere Godt ons wat nieuws laet sien, dat moet mede niet veracht zijn."

Admiringly, he stated that "Among all the ancient philosophers whose writings have come down to us, there is none who has treated natural phenomena more capably and more naturally than Aristotle. Therefore he is rightly called *Naturae interpres, Naturae antistes,* which means an interpreter or translater, a high-priest in the Temple of Nature."[601] Clearly, an author thus singing the Stagyrite's praises can hardly be expected to deviate from the ancient master's theories in any substantial way, so that, at this early point in the tract, the reader might reasonably expect Mulerius to set forth, and endorse, the Aristotelian exhalation theory of comets.

Indeed, our author went on to describe in considerable detail what he considered to be the correct interpretation of Aristotle's cometary theory, maintaining that many scholars had failed to grasp the true meaning of his words.[602] According to Mulerius, the Stagyrite had made a distinction between two kinds of comet. The first and rarest kind appeared whenever one of the fixed stars itself turned into a comet, displaying 'hair' all around its body; such a comet always remained in the same place. This was, indeed, rather an idiosyncratic exegesis of Aristotle's text: most commentators had held the Stagyrite to have meant, by the phrase "But when the exhalation is formed by the movement of one of the stars. . . then one of them becomes a comet"[603], not that the star itself started emitting hairy rays, but that by its power it attracted some sublunary vapours situated far below it, which then appeared to follow its course. Thus, Mulerius here seemed to be suggesting that Aristotle had himself acknowledged the possibility of superlunary comets, albeit in a minority of cases.[604]

Most comets, however, our author went on, (once again fostering the impression that, by contrast, the first kind of comets *were* indeed celestial bodies) were not stars, nor were they situated in the

---

[601] Ibidem, chapter III: "Onder allen ouden Philosophen die wy noch hebben behouden, isser niemant die van natuyrlijcke dinghen bequamer ende natuyrlijcker geschreven heeft als *Aristoteles*. Daerom wordt hy te rechte ghenoemt *Naturae interpres, Naturae antistes,* dat is een Tolck ofte Taelsman, een Opperste Priester inde Tempel der Natuyr."

[602] Ibidem: "Dit is *Aristotelis* eygen ghevoelen vande Cometen, van veele geleerde niet recht verstaen, ende so verclaert als wy nu ghedaen hebben."

[603] Aristotle, *Meteorology*, I.7.344a.

[604] Mulerius, *Hemelsche Trompet*, chapter 3: "Sijn gevoelen is, dat t'ghene wy Comeet noemen is somtijts een sterre somtijts gheen. Hy verhaelt self gesien te hebben dat een van die vaste sterren in de grote hondt is een Comeet geworden, hebbende rontom uythangende hayr."

heavens.[605] They moved about in the air below the sphere of the moon, hatched in that region out of fat, viscous vapours which had been extracted from the earth by the power of the sun or the other planets. Every day, Mulerius added, such exhalations were separated from the upper air by the circular motion of the heavens, and attracted by the stars in the Milky Way, which fed on them; this was the reason why the galaxy was white in colour. Whenever the earth exhaled so many vapours that the stars in the Milky Way, saturated, could not cope with them, they remained in the air, coagulated into a single mass, and caught fire through an innate tendency towards combustion; such a mass, then, was called a comet, because it resembled a star with long hair.[606]

Of the opinions deviating from this Aristotelian account, Mulerius chose to mention only that of the Pythagoreans. According to them, our author said, there existed many more planets than the five commonly seen, which, however, usually remained invisible because they were hidden in the recesses of the universe. Occasionally, such a hairy planet appeared, only to vanish again shortly afterwards to return to its far-off habitation, rising to such altitudes that it could no longer be seen. Mulerius added that he thought it unnecessary to recount any of the other opinions the ancients had entertained on the subject of hairy stars[607]; significantly, he omitted the views of Anaxagoras, Democritus, and Hippocrates of Chios, that comets were due to optical effects produced either in the heavens or in the atmosphere. As we shall see shortly, to Mulerius, the Pythagorean view of comets as a special kind of planets had assumed particular significance in the light of recent developments.

Despite his self-professed admiration for Aristotle, these newly made astronomical discoveries had made a considerable impression on Mulerius, as can be gathered from the appreciative way in which he described them. Four diligent observers of the comet of 1577, Wilhelm of Hesse-Cassel, Tycho Brahe, Cornelius Gemma, and Michael Mästlin, had with strong evidence demonstrated that that comet had not been located in the air, but in the aether, Mulerius

605 Ibidem: "Maer de meeste deel vande Cometen sijn wandelaers, ende sodanighe nae *Aristotelis* gevoelen en sijn gheen sterren, ende hebben haer wandelplaets niet inden Hemel. . .".

606 Ibidem.

607 Ibidem: "Daer sijn noch verscheyden opinien by die ouden gheweest die wy onnodigh achten te verhalen. . .".

informed his readers.[608] Tycho had computed that the phenomenon had been five times further away than the moon, at 250,000 miles from the earth's surface, and had been the size of one hundredth part of the terrestrial globe. It moved from south-west to north-east, with its tail nearly constantly pointing away from the sun.

Clearly, Mulerius accepted all of these revolutionary findings, believing them to have been "proved with strong evidence". Consequently, he had to admit that "After such an account, the opinion of Aristotle, both with regard to comets' location and with regard to their material and generation, must necessarily fall."[609] After all, it seemed inconceivable that elementary vapours, being intended to act as a purification of the air, should rise up to an altitude of 250,000 miles, and coagulate into such an immense mass, one hundredth of the size of the earth (which, Mulerius calculated, amounted to a globe of twenty-six million miles in size). Furthermore, most terrestrial exhalations were consumed by the stars in the Milky Way by way of fuel; if this were not the case, Aristotle had taught, we ought to see comets in the sky every day. Thus, there was no way in which it could be maintained that the earth produced enough vapours to keep the galactic stars alight and, in addition, to constitute the enormous mass of a comet in the superlunary aether.

If, then, the comet's material was not elementary, Mulerius continued, it had to be some kind of celestial matter, whose precise nature was as yet unknown to mankind. Unless, that is, one were to uphold, along with the Pythagoreans and Seneca, that comets were not generated beings, but eternal bodies coeval with the other stars, which remained unseen for long periods of time but were occasionally rendered visible, either by their ascending and descending movements or by other unknown causes. Such an opinion had recently been corroborated by the phenomena which could be discerned, by means of the "newly invented spectacles", against the solar disc. Through the telescope, dark bodies could be seen wandering across the luminescent background provided by the sun. By extrapolation, Mulerius concluded that it was not at all implausible that similar

---

[608] Ibidem: "Dese vier mannen [Wilhelm, Tycho, Gemma, and Mästlin] ghetuyghen ende bewysen met vaste redenen dat die selve Comeet niet inde lucht (als *Aristoteles* meent) maer in *Aethere,* dat is inde Hemel boven de Mane soude geweest syn...".

[609] Ibidem: "Nae sulcke verhael moet *Aristotelis* ghevoele vallen, niet alleenlijck so veel aengaet de plaetse vande Cometen, maer oock die stoffe ende oorspronck."

bodies also roamed about elsewhere in the universe, remaining invisible to earthly observers until they attained a position from which the sunlight could render them discernible.[610]

Thus, in the light of recent discoveries, especially the superlunary comet of 1577 and the sunspots, Mulerius had grown favourably inclined to the Senecan option of comets as eternal creatures, special planets describing highly elongated courses. Yet, as we have seen, he was also very impressed with the achievements of Aristotle, whom he believed to have been the best interpreter of nature ever. Indeed, immediately following his sympathetic account of Senecan/Pythagorean cometology, our author hastily added that he had expounded the 'eternal bodies' theory only so that the readers might ponder it, and not in order to affirm its truth: "This theory is here proposed *deliberativè tantum non affirmativè*, so as to be able to retain the postulate of Aristotle and other scholars, that *in aethere nihil generari nec corrumpi*. That is, that above the moon nothing new can be generated, nor can any thing perish there."[611] Thus, Mulerius resorted to the scholastic habit of carefully setting forth an opinion incompatible with Aristotle's, only to conclude that this was no more than a possible way of looking at things, and that the true state of affairs could only be rendered by resorting to the Stagyrite's dicta.[612] However, the reader of Mulerius' pamphlet cannot but feel that, if he were free from the constraints of authority, the author would have preferred the Senecan option.

Indeed, as became apparent from Mulerius' next chapter, the 'eternal bodies' theory was rendered even more plausible by the author's own observations of the comet of 1618. Mulerius professed to have himself observed the phenomenon very diligently by means of an unspecified instrument. He first saw it on 19 November at six o'clock in the morning, in the southeastern sky near Libra, with a long antisolar tail. The comet shared in the daily revolution of the *primum mobile*, moving from east to west; however, the phenomenon

[610] Ibidem: "Dit gevoelen conde gesterckt worden door t'gene datmen nevens de Son sien kan door middel van de nieu gevonden brillen. Het sijn duystere lichamen om de Son swervende diemen nergens siet, als tegen den Son om dat het daer so licht is. Ende men soude moghen dencken dat sodanighe lichamen niet alleenlijck om die Son maer oock op meer plaetsen haer omganck hebben, ende blyven onsichtbaer voor onsen oghen, tot datse in sodanighe positie gheraken van hoochte ende leechte datse van de Son sichtbaer ghemaeckt worden."

[611] Ibidem.

[612] See above, chapter IV, p. 142.

also had a motion of its own, moving in a north-northwesterly direction by about three of four degrees a day. The fact that the phenomenon thus moved west instead of east, to Mulerius disproved Aristotle's idea that comets always moved in an easterly direction.[613]

Its tail had at first covered 54° or more in length; however, after the first week it began to diminish. Mulerius had made several observations, until 8 December, and had found that the comet could not possibly have had a large parallax; he was certain that its parallax had never exceeded that of the moon, although he could not give any precise value for it.[614] Also, he had established, by stretching a string on a celestial globe from the sun through the position of the comet, that the end of the string touched the same star on the globe as the comet's tail pointed at in the sky; therefore, he concluded, the tail had been consistently antisolar. This fact had convinced Mulerius of the truth of the optical theory of comets: "The tail was formed by solar rays shining through the comet's body."[615]

Thus, Mulerius' own findings concerning the comet of 1618 corroborated the conclusions attached to the comet of 1577 by its foremost observers. Both bodies were located far above the moon; they did not consist of elementary matter; their tails were formed by sunrays passing through the head; and they could move in many different directions, not just from east to west. Therefore, Mulerius concluded, "I believe that, if Aristotle were alive today, he would willingly change his opinion about the entire subject of comets. And perhaps he would adopt the above-mentioned view of the foremost astronomers, that comets are located in the heavens higher than the moon, and are not fires kindled from elementary matter. And he might join Seneca in professing not to believe that a comet is a newly kindled fire, but rather an eternal creature of God."[616] Once again,

---

613 Mulerius, *Hemelsche Trompet*, chapter 4: "Daer na is die Comeet uyt de rye ghelopen nae N. Noortwest toe ofte wel so Noordlijck, blyvende altijt by een rechte baen: Teghen t'gevoele van *Aristoteles*, welcke meent dat geen Comeet Westwaerts wandelen maer altijts Oostwaert gaen. . .".

614 Ibidem: "Wy hebben verscheyden observatien gedaen, tot op de dach toe synde den 8. Decemb. oude stijl, ende daer mede bevonden dat sy geen groot *Parallaxin* conde hebben, altoos minder als de Mane, doch hoe groot connen wy niet seggen."

615 Ibidem: ". . .ende dat die voorschijnsel bestaet uyt die Sonnen stralen, door schynende door het corpus van den Comeet."

616 Ibidem: "Also dat ick wel gelove, waert saecke dat *Aristoteles* wederom leefde, hy soude syn gevoelen vande gantschen handel der Cometen geerne veranderen.

then, our author showed himself reluctant to discard Aristotle, yet keen nonetheless to introduce the recent observational discoveries into the Aristotelian framework. His continued loyalty to Aristotle undoubtedly explains why, in his third chapter, he preferred the Senecan 'eternal bodies' theory to the alternative explanation, according to which the hairy stars were newly generated out of a "heavenly material as yet unknown to us". After all, if comets were eternal bodies coeval with the rest of the universe, at least the Stagyrite's celestial inalterability could be saved, while discarding only the Aristotelian cometary theory, which, in itself, was of minor importance to his world view.

The world view of Mulerius, then, remained predominantly Aristotelian. Although he was well aware of the Copernican system, having just seen through the press the third edition of *De Revolutionibus*, he nowhere mentioned the possibility of a heliocentric universe. He remained committed to geocentrism, on several occasions mentioning the movement of the *primum mobile*, and numbering the sun among the planets.[617] Nor did he breathe a word about the geoheliocentric system of Tycho, though nothing in his pamphlet excludes the possibility that Mulerius adhered to the Tychonic scheme. He continued to endorse the Aristotelian incorruptibility of the heavens, and thereby, implicitly, the dichotomy of celestial and terrestrial realms; however, the viability of this belief was strained to its limits by the recent superlunary comets. Still, it could for the time being be saved by assuming comets to be eternal bodies not in need of generation. Finally, by implication, postulating the existence of many celestial bodies roaming about freely in the aether entailed the absence of any solid orbs, as well as the fluidity of the heavenly aether.

Mulerius concluded his chapter on the comet's astronomical char-

---

Ende soude misschien annemen t'gevoele vanden treflijcksten Astronomi, hier boven verhaelt, als dat de Cometen in den Hemel staen hogher als de Mane, dat het geen vuyr en is uyt Elementarischen stof anghesteken. Ende met *Seneca* spreken: *Ego non existimo Cometam subitaneum ignem, sed inter aeterna opera Naturae:* Dat is te seggen: ic en kan niet geloven dat een Comeet soude sijn een nieu angesteken vuyr, maer veel eer een eeuwich schepsel Godes." This remark was rather a commonplace during the Renaissance; it was also used by Tycho, Galileo, and Ramus. See Jones Schofield, *Tychonic and semi-Tychonic World Systems*, 90; A. Rupert Hall, *From Galileo to Newton* (Dover edition, New York 1981) 41; Hooykaas, *Humanisme, Science, et Réforme*, 14.

617 For instance in a passage in chapter 4 in which he spoke of a "procession of all the planets", mentioning the "planet" Phoebus among them.

acteristics with yet another Senecan thought, extensively quoting the great Stoic to the effect that in the investigation of natural phenomena, great reticence was required, in order to avoid making any impious assertions which were not at all in conformity with the true nature of things. Like all the other Dutch authors of 1618, Mulerius, too, stressed human ignorance in front of the great majesty of Nature, modestly, or perhaps slightly despondently, affirming that "*In rebus divinis et Coelestibus facilius est dicere quid non fit, quam docere quid fit.*"

5. *Conclusion: Seneca Saves Aristotle*

At the beginning of the sixteenth century, the compromise between Aristotle and God was still generally valid. The vast majority of scholars accepted the rule that, although, in theory, God was capable of bringing about any natural effect whatsoever, nonetheless he normally acted in conformity with the inexorable laws of the Christianized Aristotelian world view. However, as we have seen in our previous chapter, in the course of the century serious inroads were made on the stronghold which the partnership of God and Aristotle had tenaciously kept for such a long time. Among the most successful aggressors of the Aristotelian fortress were the newly recovered Stoics, whose ideas of the cosmos as a single continuum with mutually corresponding parts, the identity of sublunary and superlunary matter, the corruptibility of the heavens, and the absence of any fundamental barrier between heavens and earth, went through a vigorous revival with a number of sixteenth-century thinkers. By the time the comet of 1577 arrived, the Aristotelian world view was already in the process of trying to adapt to this intrusion of Stoic tenets.

The purpose of our present chapter has been to investigate how successfully this adaptation took place, among Dutch scholars, in the light of the appearances of the comets of 1577 and 1618, many of whose characteristics were found to be incompatible with traditional Aristotelianism. After our examination of the ideas which the comets inspired the Netherlandish authors to commit to paper, it can safely be concluded that, at least as far as their cosmological aspects are concerned, the Dutch cometary tracts can be read as a series of, conscious or unconscious, attempts at reconciling the novel discoveries to the time-honoured Aristotelian world view. This could most efficiently be achieved through summoning, by way of auxiliaries, none other than the self-same Stoics who had been menacing Aristotelianism for over a century.

Of the three tenets basic to the Aristotelian world view—geocentricity, circularity of celestial motions, and fundamental dichotomy between sublunary and superlunary regions—the latter had always been by far the most seriously threatened by the unifying cosmology of the Stoics; and it was precisely this dichotomy whose plausibility was considerably weakened by the appearances of superlunary comets. Not surprisingly, then, the Dutch authors' world views deviated most from traditional Aristotelianism on precisely this issue. The fundamental difference between incorruptible and unchangeable heavens formed out of their own special kind of matter, and alterable, continually changing sublunary regions filled with the four known elements of earth, water, air, and fire, became increasingly untenable as the years progressed and new discoveries continued to be made; while, in 1577, Gemma and Heurnius were not yet prepared to completely do away with the Aristotelian dichotomy, forty years later the authors who discussed the comet of 1618 were all agreed that the heavens were as corruptible as the earth.

In 1577, both Gemma and Heurnius acknowledged the superlunary position occupied by the comet of that year. However, Gemma stressed the exceptionality of this location; most comets, he believed, were sublunary and elementary, in accordance with Aristotle's dicta. Yet, even though they were placed below the moon in the changeable atmosphere, their nature was not quite as distinct from that of their superlunary counterparts as might be expected: because, to him, comets were entities whose purpose it was to maintain contact between heaven and earth, Gemma conceived of a special, non-Aristotelian sphere, just below the lunar orb, filled with a kind of matter that partook of both celestial and terrestrial qualities. For similar reasons, Heurnius, too, posited the existence of a region not mentioned by Aristotle, filled with a substance intermediate between heavenly unchangeability and terrestrial inconstancy. Although, unlike Gemma's, Heurnius' special cometary zone was located above rather than below the moon, the underlying idea was the same in both cases: unwilling to radically part with the Aristotelian dichotomy constituted by the lunar orb, both authors conceived of a kind of transitional region, in which celestial and terrestrial qualities might mingle without upsetting the incorruptibility of the heavens.

Gemma, however, was prepared to go one step further: occasionally, even the celestial region itself might suffer a certain degree of change. Thus, he suggested, invoking the medieval *potentia Dei absoluta*, it was well possible for God to make use of natural causes in order to create new superlunary beings. In this manner, each celestial orb might be capable of generating novel beings; or God might call forth previously hidden stars from the far recesses of the universe. Clearly, although he did not categorically endorse celestial corruptibility, he was prepared to allow the heavens a certain measure of change. The same can be said for Heurnius, who, despite explicitly affirming the incorruptibility of the celestial aether, did think it capable of coagulating, or at least of parts of it to separate themselves from the mass of the sun to assume courses of their own in the form of comets. Thus, as early as 1577, the Aristotelian incorruptibility of the heavens had lost some of its credibility. Gemma, however, did still believe in the reality of solid planetary orbs.

In 1618, after the telescopic discoveries of Galileo, the Stoic idea of celestial alterability and the fundamental unity and similarity of heavenly and earthly regions, had won a decisive victory over the time-honoured Aristotelian dichotomy. Apart from Mulerius, all the early seventeenth-century authors believed the heavens to be corruptible and fluid. The superlunary comets of 1577 and 1618, as well as the sights seen through the telescope, had convinced them of the truth of this condition; however, they also appealed to ancient non-Aristotelian authorities, especially the Church Fathers. Puteanus and Fromondus in addition pointed out the fundamental similarity of sublunary and superlunary regions: after all, both authors claimed, the earth possessed many unalterable features, just as the heavens were capable of changing to a certain degree. Clearly, the Stoic idea of the cosmos as a single continuum, governed by a single set of laws of nature rather than by two fundamentally diverse series of rules, proved particularly appealing in the light of the recently observed novelties in the heavens. Snellius, for one, in addition subscribed to the Stoic tenet that the terrestrial element of fire was to be found above the moon as well as below it, witness the fact that he believed the sun to be a ball of fire.

However, not all 1618 authors were prepared to accord the corruptible fluid aether precisely the same, far-reaching, degree of change that was to be found in sublunary bodies. Mulerius showed himself most reluctant, paying lip service to the Aristotelian celestial incorruptibility; however, he made it sufficiently clear that he was very favourably inclined towards the idea of Seneca, that comets might be eternal bodies moving in the outermost recesses of the universe, thereby eliminating the need for breaking with Aristotle's unchanging aether while at the same time accounting for the appearance of superlunary comets. Fienus adopted a similar stance, favourably rendering the theory of Apollonius of Myndus, that comets were eternal bodies describing extremely eccentric orbits, and therefore rarely visible. Fromondus and Puteanus went one step further, in coming up with cometary theories that did affirm comets to be newly created entities, yet postulated that no *substantial* changes were needed in the heavens for their generation: they could be accounted for by resorting solely to differences in density and rarity, qualities of old associated with the heavens. Snellius was the only one to actually ascribe 'sublunary' alterations to the heavens in order to explain the generation of comets: endorsing the fundamental similarity between the earth and the planets, he believed the hairy stars to be formed out of burning solar exhalations.

Slowly but surely, then, the unification of terrestrial and celestial physics, championed by their contemporary Johannes Kepler, began to creep its way into the world views of our Dutch authors. Gemma and Heurnius had already proposed special regions filled with intermediary matter, in which heavenly and earthly qualities were allowed to meet; by 1618, the whole of the planetary heavens under the solid firmament had turned into such an intermediate zone, a region filled with fluid matter in which all kinds of changes could take place. Fromondus, in positing the idea that variations in density might be found within one and the same celestial body, and in asserting comets to be bodies which in the course of their existence gained heaviness and lightness, as well as the concomitant rectilinear motions appropriate to these qualities, went some way towards assigning to the heavens physical processes as experienced on earth. Snellius, moreover, suggested a Kepler-like analogy between the terrestrial phenomenon of magnetism and the innate power of the sun directing the movements of the planets; however, believing both to be occult forces about which as yet virtually nothing was known, he made no attempt at further elucidating, or quantifying, them.

Insofar as the world views of our Dutch authors deviated from Aristotle, then, they did so in a Stoic sense. The non-Aristotelian tenets they adopted (fluidity and corruptibility of the heavens, cosmic unity, absence of a fundamental difference between the celestial and the terrestrial region) were precisely the points on which the Aristotelian world view had been under attack ever since the early sixteenth century. Their plausibility had been greatly enhanced by the behaviour of the comets of 1577 and 1618, as well as the telescopic discoveries; also, they were in conformity with the dicta of the Church Fathers, whose authority, at least in Louvain, was very great. The two other fundamental constituents of the Aristotelian world view, however (geocentricity and circularity of celestial motions), as well as other elements such as the several concentric layers of the atmosphere and the rectilinear nature of sublunary motions, survived the era of revolutionary novelties virtually unscathed.

All of the Dutch authors, both those discussing the comet of 1577 and those writing forty years later, were geocentrists. However, due to Galileo's discoveries, especially that of the phases of Venus, the authors of 1618 were favourably inclined towards the geoheliocentric system of Tycho. Fromondus and Puteanus explicitly professed their adherence to the Tychonic system, while Snellius, too, thought the completely geoheliocentric scheme not implausible; however, he did not explicitly subscribe to it, but confined himself to the Capellan system. Neither Fienus nor Mulerius spoke out on the issue of the rival world systems; at most, we can gather from their tracts that they conceived of the earth as the centre of the universe. The only author to mention Copernicanism was Fromondus; although, in his youth, he had been favourably disposed towards heliocentrism, and although even in 1618, after having heard of the papal condemnation of 1616, he flirted with the Copernican conception of the earth as one of the planets, in the end he cautiously stuck to the Tychonic system.

None of our authors seems to have broken with the Aristotelian description of the sublunary region in any fundamental way. Gemma, Heurnius, Fienus, and Fromondus all subscribed to the division of the atmosphere into layers of different densities, with the heavy earth at the centre and the spheres of water, lower air, middle air, and upper air situated concentrically around it. Only Gemma explicitly called into question the existence of an outermost sphere of fire just below the lunar orb, stating that in his opinion this region should be conceived of as filled, not with fire, but with the above-mentioned intermediary substance of which sublunary comets were formed. The authors of 1618 failed to mention the Aristotelian sphere of fire; to Snellius, at least, such a sphere had become quite superfluous since he believed fire to exist above the moon as well as below it, so that there was no need to confine it to a single sublunary sphere.

The Aristotelian world view, then, remained intact only insofar as it concurred with the Stoic outlook on the cosmos. The elements common to both world views — geocentricity, sublunary heaviness and lightness, circularity of heavenly motions, explanation of atmospheric phenomena — were not called into question either in 1577 or in 1618. The appearance of superlunary comets, and the telescopic discoveries of Galileo, induced the Dutch authors merely to adopt the Tychonic world system, which allowed them to continue to subscribe to Aristotelian physics, and to endorse the corruptibility and fluidity of the celestial region, in support of which they could call on the formidable authority of the Church Fathers and the Stoics, and which had a pre-Aristotelian, venerably ancient ring to it which proved extremely appealing to men of an age in search of the most pristine wisdom. Especially the satisfactory explanation of superlunary comets, without breaking with Aristotelian physics in any fundamental way, was facilitated by the availability of a formidable example, set by the revered Stoic Seneca in the extensive cometological book of his *Natural Questions*.

The world views emerging from the Dutch cometary tracts of 1577 and 1618 are all remarkably similar. Despite small deviations, especially in the authors' own explanations of the way in which comets were generated, the basic constituents of their cosmological outlooks (Tychonic system, corruptibility and fluidity of the heavenly regions, continued adherence to Aristotelian physics) were very similar, with no fundamental differences between the northern and the southern Netherlands. The only issue on which there was some dissension between individual authors, was the epistemological question to what extent the human intellect was capable of grasping the true state of affairs in the universe and the laws that governed them.

To a certain extent, all the authors were influenced by the epistemological remarks made by Seneca in book seven of the *Natural Questions,* where he stressed that many things remained invisible to the human senses and ungrasped by the human intellect. The cosmos was a highly harmonious continuum governed by steadfast laws of nature; yet, it was very difficult for mere mortals to uncover them, and the investigation of nature, though not impossible, was a highly complicated and time-consuming affair; nature would yield her secrets only after many generations of continual examination and observation. Thus, the Senecan example could lead the authors relying on it in two different directions: on the one hand, it might foster an agnostical stance, and a sense of insufficiency of human capabilities, while, on the other, it might encourage a more optimistic quest after the laws of nature. Moreover, the telescopic discoveries made by Galileo displayed the same kind of duality as Seneca's text; on the one hand, they might overawe the observers and strike them with a sense of utter inadequacy and despondency, while, conversely, they might also inspire them and imbue them with a sense of slow yet unstoppable progress.

Both positions are apparent from the Dutch cometary tracts. On a scale from predominant agnosticism to confidence in laboriously unravelling the secrets of nature, we can place Fienus at the one end and Snellius at the other, with Puteanus, Gemma, and Fromondus somewhere in between. To Fienus, the universe remained a mystery which could be marvelled at and contemplated, but not grasped. Accordingly, he made no attempt at framing any cometary theory of his own, but rested content with repeating the suggestions made by some of his predecessors, without explicitly committing himself to a single one of them. Puteanus and Gemma showed themselves somewhat less pessimistic; both authors emphasized the *potentia Dei absoluta*, stating that God might at any given moment go against the rules which governed nature under normal circumstances; yet they added that He might well use natural instead of supernatural causes to create such 'special effects'. Therefore, these causes, being in conformity with the harmonious order of nature (though contrary to the limited order preceived by the human senses) could be tentatively investigated by the human intellect, and suggestions could be made concerning them. Puteanus even pointed out that, once everything had been subjected to diligent and continuing observation and investigation, in the end the need for any notion of 'prodigiousness' would vanish. Fromondus, though not as outspoken on the issue, seems to have agreed that even uncommon appearances such as comets might well be rationally explicable. At least, he himself made an earnest effort in this direction, stating that Aristotle himself had granted his followers the licence of coming up with any possible explanation on the especially complicated subject of the hairy stars. Finally, Snellius, the convinced Ramist, occupied by far the most optimistic epistemological stance: taking up the gauntlet thrown by Seneca, he expressed his belief in the capability of human reason in the end to grasp the laws of the universe, using continued and highly accurate observations, as well as mathematical calculation, as its tools.

Thus, although, for the present, their world views were very much alike, our authors displayed a certain measure of dissension as far as their plans for the future were concerned. There was one further issue, however, on which they were all agreed: the explanation of a natural phenomenon, such as a comet, was never complete without an attempt at elucidating that phenomenon's final cause: to what purpose had it been created, and what could we expect of it? In this respect, the authors were still very much Aristotelians: they all subscribed to the tenet that nothing was ever created in vain, and that therefore the reasons for every phenomenon's generation were valid, indeed essential, topics of investigation. To some of the authors, such as Gemma, Heurnius, Puteanus, and Mulerius, the examination of the comet's final cause constituted their principal incentive for honouring it with a tract. Others, such as Fienus, Fromondus, and Snellius, were primarily interested in its physical characteristics and its location; yet, they, too, devoted separate chapters to the phenomenon's significance.

CHAPTER SIX

COMETARY CONCOMITANTS: ASTROLOGY AND TERATOLOGY IN THE DUTCH TRACTS

1. *Introduction*

a. *The Mysterious Demise of the Final Cause*

One of the central problems of the history of ideas is the mysterious demise of the final cause. Although many objections can be raised against the thesis, often encountered in introductory textbooks of early modern history, that during the seventeenth century natural causes came to replace supernatural influences[1], the fact remains that sixteenth- and early seventeenth-century works on natural phenomena, such as Cardano's *De Subtilitate* (1554), or J.C. Scaliger's commentaries on this book (1612)[2], somehow "read" differently from late seventeenth-century works such as Bayle's *Pensées diverses sur la Comète* (1683) or Fontenelle's *Histoire des Oracles* (1686). Although both Scaliger and Bayle were emphatically opposed to endowing comets with any significance whatsoever, and to a certain extent even adduced identical arguments against the hairy stars' predictive value, the impression gained from reading both books is fundamentally different, and can perhaps best be expressed by saying that Bayle somehow seems to be "one of us", a Modern, while Scaliger, in spite of the reasonableness of his arguments, incontrovertibly remains one of the Ancients.

Obviously, to point out a vague difference in spirit between two texts is one thing; but to elucidate how, within a single century, this difference could come about, never to have disappeared since, is quite another. Many historians of ideas have tried their hands at explaining this enigmatic watershed, a task which in the end comes down to clarifying one of the central problems of the birth of modern

---

[1] See for example D. H. Pennington, *Europe in the Seventeenth Century* (2nd ed., London 1989) 123, 182.

[2] *Julii Caesaris Scaligeri Exotericarum Exercitationum Liber XV De Subtilitate, ad Hieronymum Cardanum* (Frankfurt 1612).

science: how did metaphysics come to be separated from physics, and how did God come to be eliminated from scientific discourse?

A rather unsatisfactory solution to the problem was given by Sara Schechner Genuth in her dissertation on the "assimilation of comet lore into natural philosophy".[3] Although Genuth recognizes that "successful prediction of a comet's return by itself was no more effective in combatting the old views than [earlier] developments in astronomical theory"[4], her attempt at tracing the causes of the disappearance of cometary divination remains somewhat unconvincing. She points out that, since "People generally had recourse to divinatory practices to counter their own helplessness in circumstances on which they were dependent"[5] it should come as no surprise that "a decline in divinatory rituals should coincide with improvements in the control of the natural and social environment."[6] Explicitly acknowledging her debt to Keith Thomas's *Religion and the Decline of Magic*[7], Genuth indicates the increase in standard of living, the growth of trade and rise of industries, the disappearance of the plague, the improvements in communication, the rise of the sciences of economics and statistics, and so-called "self-help schemes" such as deposit banking and insurance as factors that, during the seventeenth century, "stepped in to fill the gap" left by the replacement of animistic philosophy with a mechanical world-view.[8]

However, clearly this argument does not explain anything. First of all, all Genuth does is to replace the question of why cometary divination declined during the seventeenth century, with the equally pressing and indeed almost identical question of why the animistic philosophy Genuth associates with Kepler was replaced with a mechanistic one. Moreover, Keith Thomas, to whose views Genuth professes to owe much, in a later section dissociates himself from the idea, embraced by Genuth, that technological improvements were responsible for the decline of magic.[9] Pointing out that, at least in

---

[3] Sara Schechner Genuth, *From Monstrous Signs to Natural Causes. The Assimilation of Comet Lore into Natural Philosophy* (unpublished Ph.D. dissertation, Harvard 1988).

[4] Ibidem, 335.

[5] Ibidem, 339.

[6] Ibidem, 340.

[7] Ibidem, 331 n. 164.

[8] Ibidem, 340-341.

[9] See Keith Thomas, *Religion and the Decline of Magic. Studies in Popular Belief in Sixteenth and Seventeenth Century England* (London 1971) chapter 22.

England, magic lost its appeal before technological advances were made that were capable of satisfactorily replacing it, Thomas concludes that such technological advances could be made precisely because magic had ceased to be credible.[10] Therefore, the change that took place in the seventeenth century was mental rather than technological.[11] Thus, on Thomas's view as much as, implicitly, on Genuth's, we are left with the difficult task of explaining an intellectual development by means of intellectual, rather than social or economic, factors.

Thomas himself recognizes the difficulty of such an enterprise. He points out that, throughout the ages, there had been scholars who advanced a rationalistic approach to nature, such as Hippocrates, who denied that epilepsy was due to supernatural causes, Aristotle, who believed so-called prophetic dreams to be mere coincidences, and Cicero, who denounced divination. The efforts of certain sixteenth-century scholars, such as Pomponazzi and other members of the Paduan school of natural philosophy, at explaining even the most unusual phenomena from natural causes, had been inspired by such rationalistic works of Antiquity, rather than by a new approach to nature anything like that of the mechanistic philosophy of the seventeenth century.[12] Thus, it still remains to be explained why the 'new science' succeeded in disenchanting the world, whereas all previous rationalistic philosophies had failed. The closest Thomas gets to an answer to this question, and to pinpointing what exactly it was that changed so fundamentally during the seventeenth century, is his remark that, before the watershed, scholars "tended to accept every story, no matter how bizarre, and then to devote their energies to finding an explanation for it"[13], whereas towards the end of the seventeenth century, this attitude disappeared to be replaced with the postulate that phenomena be directly experienced before attempts should be made to explain them. Gradually, scientists lost their reverence for the pristine wisdom of the past and started to rely on nothing but "unaccompanied human aid"[14].

This is precisely the development we have examined in chapter II,

---

[10] Ibidem, 786.
[11] Ibidem, 791.
[12] Ibidem, 773.
[13] Ibidem, 771.
[14] Ibidem, 771 and 793.

where it has been referred to as the 'cessation of dialogue'.[15] Somewhere during the late seventeenth century, scholars lost their age-old faith in the trustworthiness of their predecessors, and for the first time refused to take them seriously. This is one of the main factors responsible for the difference in tone between, for instance, Bayle's *Pensées diverses sur la Comète* and Scaliger's *Exotericae Exercitationes*. It is not the purpose of the present chapter to offer an account of the reasons for the cessation of dialogue; as will have become clear from the above, such attempts are doomed either to an unjustifiably high level of generalization, like skeletons without flesh, or to inaccuracy. Rather, this chapter, like the previous one, is an attempt at reconstructing the lines of thought of the authors of the Dutch cometary tracts of 1577 and 1618. Where, in our preceding chapter, we have tried to look at the comets with contemporary eyes, unhindered by the knowledge with which later astronomical science has provided us, our present concern is to venture on a mental retreat to a time when purpose and end where not yet excluded from scientific discourse; a time that was well-acquainted with the rationalism of Antiquity and the naturalism of the later Middle Ages, but yet refused to abandon teleology; a time when the elucidation of the final cause of natural phenomena was still considered a vital topic of investigation; a time on the verge of the cessation of dialogue.

However, before embarking on this enterprise, it seems advisable first of all to tarry for a short while in the period during which dialogue first began to cease. To elucidate the demise of final causality associated with cometary appearances, we are lucky enough to have at our disposal an eminently competent guide. In 1683, Pierre Bayle published a collection of thoughts occasioned by the comet of 1680, in which he eloquently refuted the common opinion that comets were portents of evil. Some historians have credited Bayle with dispelling the clouds of superstition and inaugurating a new era in which comets predicted nothing.[16] Although, as we shall see, such a view cannot be maintained, since earlier examples of scepticism regarding comets' meaning can be found with relative ease, the fact remains that Bayle was one of the first to set forth the traditional

---

[15] See supra, p. 20-25.

[16] See for example Paul Hazard, *La crise de la conscience européenne 1650-1715* (Paris 1935), Part II, chapter 2; and James Howard Robinson, *The Great Comet of 1680. A Study in the History of Rationalism* (Northfield, Minn. 1916).

sceptical arguments in a 'modern' way immediately recognizable to the present-day reader. However, despite this modernity, he was still very much in touch with the old ideas, and was well aware of the whole range of arguments which had, in the past, been proposed in defence of cometary significance. Consequently, his work on the comet of 1680 offers us a unique trait d'union between our own indifferent attitude and the ancient body of ideas to which the Dutch authors of 1577 and 1618 were still so manifestly heirs.

b. *Pierre Bayle as a Guide*

The title of Bayle's *Pensées diverses sur la Comète* is somewhat misleading. Although the work was occasioned by the comet of 1680, it in fact contains a host of arguments which have no bearing on the hairy star at all, dealing instead with idolatry and atheism, virtues and vices, and the contemporary political scene in Europe. Obviously, in the present section we shall concern ourselves exclusively with those of his arguments which are relevant to the topic under investigation; more particularly, we shall concentrate on Bayle's views on the two main problems that had always been of central importance to those investigating cometary consequences: first, whether they were capable of physically causing damage on earth; and second, whether they could plausibly be regarded as signs, sent by divine agency, whose purpose it was to forewarn mankind of disasters to come.

Bayle himself made a clear distinction between the two problems: in the summary of his arguments with which he concluded his work, he stated that "if comets are portents of any evils, this would be either because they are efficient causes of these evils, or because they are signs of these evils."[17] Therefore, his whole book was devoted to demonstrating that they were neither physical causes, nor miraculous signs.

First of all, Bayle believed it impossible to regard comets as efficient causes, physically wreaking havoc on earth. It could not be satisfactorily proven that the hairy stars possessed any virtue which rendered them capable of physically affecting the earth. The idea

[17] Pierre Bayle, *Pensées diverses sur la Comète* (1683, edited with an introduction by A. Prat, re-edited by Pierre Rétat, Paris 1984) II 311-312: "Pour réduire en abregé toute ma Dispute, je dis que si les Cométes [sic] étoient le presage de quelques malheurs, ce seroit ou parce qu'elles sont la cause efficiente de ces malheurs, ou parce qu'elles sont un signe de ces malheurs. On ne doit point nier cela."

that they could send earthward any of the matter out of which they were formed, was highly implausible, since according to Cartesian physics, particles did not tend to move towards the centre of the world, but rather to move away from the centre around which they revolved.[18] Moreover, even if cometary exhalations were able to move towards the earth, these would not have a very notable effect, since on traversing the immensity of space, they would break up into infinitely many tiny particles, of which only a very small number would actually reach the terrestrial atmosphere.[19] And if, like the Aristotelian opponents of Cartesianism, one were to maintain that comets exert physical influence, not by emitting streams of atoms, but by means of non-material qualities, it would still be impossible to uphold that their influence would last after the comet itself had disappeared.[20]

Moreover, even if cometary particles were able to reach the earth and to make their effects felt there, it would be impossible to ascertain what these effects would be. More particularly, to Bayle it seemed no more reasonable to believe them capable of producing horrors traditionally associated with comets, such as pestilence, war, and famine, than to uphold that they could give rise to health, peace, and abundance, since nobody knew the natures, figures, movements, or other qualities such particles might possess.[21] How, for instance, could it seriously be believed that the present comet, incapable of preventing a spell of extreme cold during its visibility, would go on to cause a war three years after its disappearance, by heating people's blood and thus rendering them more irascible? Might it not just as well have the opposite effect, cooling men's blood and causing them to become more sedate, so that the comet's appearance could be said to have heralded peace rather than war?[22]

Nor could it be maintained that it was possible to prove the fact that comets caused physical changes on earth by any argument *a*

---

[18] Ibidem, I 45-48.

[19] Ibidem, I 48-51.

[20] Ibidem, I 52-53.

[21] Ibidem, I 53-54: "On peut dire en second lieu, que supposé que les Cometes repandent jusques sur la terre des corpuscules capables d'une grande action, il n'y a pas plus de raison à soûtenir qu'ils doivent produire la peste, la guerre, la famine, qu'à soûtenir qu'ils doivent produire la santé, la paix, et l'abondance, parce que personne ne connoit la nature de ces corpuscules, la figure, le mouvement, ou les autres qualitez de leurs parties."

[22] Ibidem, I 54.

posteriori; by saying, for instance, that the experience of many ages showed that the appearance of a comet was invariably followed by a host of calamities. After all, to conclude, from the simple fact that one occurrence invariably followed another, that the first occurrence was therefore the cause of the second, was an example of "pitiful reasoning", remarked Bayle.[23] Who, for example, would wish to deduce from the fact that whenever a certain inhabitant of Paris would look through his window at the St Michel bridge, he would see people passing in the street below, that therefore, the man's looking out of the window was the cause of the people passing?[24] Such a *post hoc ergo propter hoc* argument could find no favour in Bayle's eyes, who further invalidated this line of reasoning by pointing out that it could not be deduced from the histories that more or worse calamities happened after the appearance of a comet than in other years, and that, moreover, many instances could be given of good fortune following cometary appearances (partly, of course, because what was good in the eyes of one party in, for instance, a war, was invariably bad for the opposing party, so that one and the same event could be both good and bad at the same time).[25]

Rather than being caused by comets heating the blood of certain irascible princes, wars, according to Bayle, were due to an immense number of fortuitous causes arising from the free will of men. For it could not possibly be maintained that comets caused the myriad of coincidences that affected the course of events, like intercepted letters, orders carried out too late, the death of a single man reversing the fortunes of the warring parties: such completely fortuitous events could never be caused by cometary particles flying through the air.[26] The more heavily occurrences depended on chance whims and passions, the more unlikely it was that they should be subject to the influence of a necessary and blind cause, such as the stars.[27] If earthly disorders can be attributed to any stars at all, Bayle concluded, these ought surely to be terrestrial 'stars', the princes with their multitude of intentions, interests, and forces; so that an intelligent man, learned

[23] Ibidem, II 312.
[24] Ibidem, I 83.
[25] Ibidem, I 84-85; 91.
[26] Ibidem, II 243.
[27] Ibidem, II 249.

in history and the ways of the world, is capable of making more trustworthy predictions of the future than any astrologer.[28]

Thus, the 'science' of astrology, generally regarded as the foundation on which cometary predictions were based, found no favour whatsoever in Bayle's eyes. He believed it to be "the most ridiculous thing in the world", "impertinent, chimerical, and ignominious to human nature"[29], founded on completely implausible rules and suppositions. Why should a comet passing through a group of stars which the Ancients, led by poetic fictions, had quite arbitrarily named 'Virgo', predict infertility, miscarriages, or the inability of young women to find husbands?[30] And why should the pestilences and famines traditionally associated with cometary appearances affect one particular region or people only (as was often asserted by the astrologers)? After all, if, as was usually assumed, comets were the physical causes of such calamities, they ought to inflict famine and death on all the regions over which they passed, not on just one in particular.[31]

As will have become clear from the above, Bayle regarded the theory that comets were physical causes of ensuing calamities as entirely unfounded: there was no philosophical reason why they should be able to affect the earth, while there were many arguments that rendered such physical cometary efficacy quite improbable; the empirical argument *a posteriori*, that the experience of many ages showed that more calamities occurred after cometary appearances than at other times, was false; and the rules of astrology were arbitrary and implausible. Terrestrial events, Bayle believed, were due to terrestrial causes to do with mere coincidences and the free will of man.

However, this abundance of arguments against comets as physical causes did not suffice to silence all proponents of cometary significance, as Bayle well realized. There existed another, equally ancient tradition of cometary divination, much more difficult to refute, ac-

[28] Ibidem, II 252-253: "Un homme d'esprit et qui se souvient de loin, qui connoit par l'histoire et par l'usage du monde, les Principes sur lesquels roulent les affaires generales, et qui de plus est instruit à fond de l'air dont on se gouverne dans chaque Pays, fait souvent des conjectures si justes sans l'aide de l'Astrologie, que quand tous les Astrologues du monde uniroient leurs forces pour decouvrir ce que les etoiles predisent du changement des Etats, ils ne diroient rien qui vaille, en comparaison de l'autre."

[29] Ibidem, I 57.

[30] Ibidem, I 64.

[31] Ibidem, II 313.

cording to which the hairy stars were not so much physical causes as divinely sent signs of future events. After having invalidated the opinion of comets as causes, Bayle went on to address the question of whether, as was often maintained, they could be regarded as signs of impending evils, sent by God in order to incite the human race to do penitence and to mend their ways. Those who wish to uphold this opinion, Bayle went on, will necessarily have to admit that comets are created by God in a miraculous way, and that they are thus unconnected to the chain of secondary causes. After all, comets produced by secondary causes could signify only those things with which they were necessarily linked, not wars, pestilences, and famine, which were calamities dependent on the human free will or on other causes that had nothing to do with the appearance of hairy stars.[32]

The idea that comets might be natural signs of future ills was dismissed by Bayle as utterly implausible. For how could it be maintained that, whenever natural causes created a comet (for example, by igniting a conglomeration of planetary exhalations, which was one of the ways scholars believed comets might come into existence) physical conditions on earth at once arranged themselves in such a way as to be capable of originating pestilences, wars, and the other conditions comets were supposed to signal? After all, changes on earth arose from terrestrial causes completely independent of those heavenly alterations that gave rise to cometary appearances, just as there was no plausible causal connection between, for example, the temperature of the planet Saturn and the blowing of certain winds on earth.[33] "We believe", concluded Bayle, "that the planets produce a comet without the aid of the earth, and that, in turn, the earth produces a number of ills without the aid of the planets, or of the comet. . . . And consequently, if comets are a sign of coming evil, it has to be maintained that God produces them specially, whenever he realizes that the earth is ready to give rise to that evil; for according to the laws of nature, it might never occur that at the precise moment that the earth found itself in such a condition, a suitable inflammable exhalation found itself in heaven to create a comet there."[34]

---

[32] Ibidem, I 157-158 and II 184.

[33] Ibidem, II 188-189.

[34] Ibidem, II 192-193: "Nous supposons que les Planétes produisent une Cométe sans le concours de la terre, et que la terre produit à son tour quantité de maux sans le concours des Planétes, ou de la Cométe. . . . Et par consequent, si les Cométes sont un signe de quelque mal à venir, il faut que Dieu les produise tout exprez, lors

Thus, if comets were signs of future events, they were necessarily to be regarded as divinely produced miracles, since they had no natural connection whatsoever, either causal or signal, to any events on earth. But the idea of miraculous comets was repugnant to Bayle, who devoted an extensive portion of his tract to invalidating it. For obvious reasons, the arguments he employed in doing so were theological rather than physical in nature. Thus, he stated that, by supernaturally producing comets, God would only encourage idolatry, since whenever anything unusual occurred in nature, the pagans were invariably quick to attribute it to the anger of their false gods, and therefore would resort to their superstitious ceremonies for averting it.[35] Indeed, demons had always been glad to use monsters and other unusual natural occurrences to frighten the pagans, and to incite them to continue their superstitious ceremonies in order to avert the anger of their false deities; for ignorance of the laws of nature rendered the majority of people prone to fear anything slightly deviant from the ordinary course of nature, even though, in Bayle's opinion, it was quite clear that the birth of a monstrous animal was due to natural causes just as much as was the birth of a normal specimen.[36]

Another circumstance pleading against the view of comets as miraculous creations of God, was the fact that, in any case, all the comets that had appeared before the advent of Christ had necessarily been due to natural causes. Therefore, it was very unlikely that, suddenly, after Jesus's birth, comets had changed their natures and turned into divine prodigies, whereas before they had always been mere natural occurrences.[37] Also, if God had intended comets as signs warning mankind, he would undoubtedly have endowed them with certain specific qualities enabling earthlings to discover for whom exactly they were intended, and what exactly they were supposed to indicate; for it would be impious to uphold that God showed all mankind signs that actually menaced only a few people, thereby causing the others to fear without reason.[38] Indeed, whenever God

qu'il voit que la terre est prête à faire éclorre ce mal là; car selon les loix de la Nature, il n'arriveroit peut-être jamais, que la terre étant en cet état, une matiere inflammable se trouvast tout à propos dans le Ciel pour y former une Cométe."

[35] Ibidem, I 162-163.

[36] Ibidem, I 163, 170.

[37] Ibidem, I 186.

[38] Ibidem, I 191.

did intend certain occurrences to be read as signs addressed to humankind, he always took care to make sure that people would understand that the sign in question was not just an ordinary natural effect; thus, for the solar eclipse that occurred during Christ's crucifixion, he chose a time when no natural eclipse could possibly have taken place.[39]

Thus, Bayle deprived comets of every kind of meaning they had ever carried. Taking on a tradition as old as the collective memory of mankind, he systematically ruled out all the ways that had ever been proposed of accounting for their final cause, the goal with which they came into existence. It was not true that they caused wars by heating princes' blood and rendering them irascible; it was not true that they caused pestilences by infecting the atmosphere; it was not true that they caused famine by desiccating the air. Nor did historical experience show that more calamities occurred after cometary appearances than at other times; nor could it be said that, somehow, the condition of the stars naturally indicated the conditions shortly to arise on earth; in fact, the whole art of astrology, based on arbitrary and unproven rules, was ridiculous. And, finally, even the age-old *refugium miserorum*, the absolute power of God to go against the ordinary laws of nature, was invoked to no avail by proponents of cometary significance: to Bayle, assuming comets to be miraculous creations of God was unfounded and even impious nonsense. If anything, the hairy stars signified neither more nor less than each and every other part of the Creation: like the whole of nature, they testified to God's power and magnificence, whose greatest feat was the well-regulated course of nature according to simple and unwavering rules, to which all parts of the creation, be they common or rare, invariably obeyed.[40]

Apart perhaps from the remarks about demons, the whole of Bayle's reasoning, as summarized above, makes a very sensible impression. Taking into account the reasonableness and common sense with which the author proposed his remarks on the nature and meaning of comets, it is not all that surprising that authors such as Paul Hazard and James Howard Robinson have praised him as one of the first combatants of cometary superstition. Below, we shall be able to evaluate whether such a view is entirely justified; at present, however, we are not interested in Bayle's *Pensées diverses* for the au-

[39] Ibidem, I 172-173.
[40] Ibidem, II 230-233.

thor's own opinion's sake so much as for what his argument can teach us about the standard view on cometary significance which he took such great pains to refute. Apparently, at the end of the seventeenth century, the ideas to be attacked were the following:

1. Comets can be regarded as physical causes of certain events on earth. More particularly, they produce hot and dry conditions in the terrestrial atmosphere and in the human body, which in turn give rise to wars, famine, and pestilence. The truth of this proposition can be deduced from the historical fact that cometary appearances are always followed by more wars, famines, and pestilences than ususal.

2. Comets are natural signs of a range of calamities. The heavens and the earth correspond to each other, if not in a directly physical way, then at least by means of a certain sympathy between the two, which makes it likely that whenever conditions in the heavens are suitable to produce a comet, conditions on earth are prone to cause wars, famine, and pestilence.

3. The meaning of comets can be deduced by means of astrological rules. For instance, a comet moving through the sign of Virgo denotes infertility and miscarriages.

4. Comets are miracles, supernaturally created by God in order to warn mankind of impending punishment. They are not part of the ordinary course of nature, and are not subject to natural laws.

Thus, there traditionally existed two fundamentally different ways of classifying comets and their consequences. According to one outlook, they were natural entities, just like the stars, the clouds, or thunderbolts. Whether they themselves caused the calamities usually attributed to them, or whether they were natural signs merely indicating that other calamitous causes were about to come into action, is not important; in either case, they were parts of a natural chain of events, subject to the laws of nature and proper subject-matter for the sciences of astronomy, astrology, and meteorology. According to the second outlook, conversely, comets were supernatural entities on a par with other divinely sent signs such as speaking statues, bloody rains, or resurrections from the dead. On this view, they were not subject to the ordinary laws of nature, and the only sciences that could make sense of them were theology and teratology.

The fact that comets are such outstanding examples of entities in the twilight zone between the domain of astrology and the natural sciences on the one hand and that of teratology and theology on the other, is responsible for the complexity of most early modern attempts at elucidating the character of their predictive value. Therefore, in order more easily to understand the various traditions and possibilities with which our Dutch authors saw themselves faced, it is advisable first of all to give a short outline of the two divinatory strands to which comets could belong: that of physical astrology and that of teratology.

2. *Theoretical Background: The Issues Facing Cometologists*

a. *The Two Faces of Divination: Astrological and Teratological Strands*

i. *Their Origins in Antiquity*

In his well-known study of Renaissance astrology, *The Zodiac of Life*[41], Eugenio Garin has pointed out the existence of two fundamentally different ways of approaching astrology, between which there prevailed a continuous tension. Adherents of the first approach, according to Garin, tried to reduce heavenly intelligences and souls to necessary principles of rationality, while those in favour of the second outlook placed more emphasis on the personality of the divine and on the free individuality of souls.[42] Aby Warburg, too, alluded to the dual nature of the astrological art, when he defined it as "the meeting and confrontation point between the demands of a rational order, as in Greek science, and the myths and superstitions inherited from the east: between logic and magic, between mathematics and mythology, between Athens and Alexandria."[43]

Indeed, most accounts of the history of astrology, often implicitly, acknowledge the existence of this tension, so eloquently articulated by Warburg, between two rival traditions whose outlooks on the world and the ways in which man could attain knowledge of future events were fundamentally different.[44] Both strands originated during Antiquity; during the Middle Ages, they vied with each other for dominance over scholarly thought, with now the one, now the other

[41] Eugenio Garin, *Astrology in the Renaissance. The Zodiac of Life* (Bury St Edmunds 1984).

[42] Ibidem, 61.

[43] Quoted by Garin, ibidem, xi.

[44] See for example S.J. Tester, *A History of Western Astrology* (Bury St Edmunds 1987) 68-69; Lynn Thorndike, *A History of Magic and Experimental Science* I (New York 1923) 100-114; Tamsyn Barton, *Ancient Astrology* (London 1994) 7, 35, 109-111.

coming out on top; and together, they reached the era of Renaissance and Reformation, where they both became topics for heated debate among the learned. The relationship between the two strands remained problematic throughout, and at times, they intertwined to such an extent that it became difficult to disentangle them. Thus, during the sixteenth century, Philip Melanchthon and Caspar Peucer, while trying to found astrology on a demonstrative basis as one of the physical sciences, did not make this task any easier for themselves by attempting to include in it all kinds of prodigies, portents, and divine signs, as well as the ordinary courses of the heavenly bodies. In so doing, they saw themselves faced with the dilemma of how, by means of a natural science, with its rational methods and demonstrations, to account for portents directly inspired by divine intention.[45]

In fact, the two strands, which may tentatively be dubbed 'astrological' and 'teratological', were often confused and conflated throughout history; yet, since they were based on two fundamentally different outlooks on the world and the relationships between its parts, for the sake of clarity the distinctions between them should be pointed out before embarking on an account of cometary theory, which provides the most outstanding example of a science at the crossroads between the physical and the teratological.

The teratological strand was undoubtedly the more ancient of the two. The earliest evidence of divinatory activity that has come down to us, laid down in cuneiform texts from Mesopotamia, teaches us that, to the old Babylonian fortune-tellers, divination from celestial phenomena was of secondary importance in relation to extispicy, the inspection of the entrails of sacrificial animals. Although the famous *Enuma Anu Enlil*, a huge collection of Babylonian predictions found in the royal archives in Nineveh, consists solely of omens based on celestial divination, it should not be concluded that sky-omens occupied any special position among the various Mesopotamian divinatory techniques. Rather, celestial omens were part of the same category as monstrous births, animal behaviour, or burning incense: these were all techniques of interpreting the messages sent by the gods.[46]

Thus, to the Babylonians, not just the heavens, but the whole of

[45] Pierre Freyburger, "Le problème du fatalisme astral dans la pensée protestante en pays germaniques" in *Divination et controverse religieuse en France au XVIe siècle* (Cahiers V.L. Saulnier, 4, Paris 1987) 35-55: 48.

[46] Barton, *Ancient Astrology*, 12-13.

nature was replete with divine intentions, and each and every natural occurrence could be regarded as a message from the gods. Contemplation of the stars, on this view, was only one way of discovering the gods' designs, and the study of the heavens did not occupy any privileged position fundamentally different from the investigation of other natural phenomena. Later, in Greece and Rome, a basically similar attitude to the art of divination was adopted by the members of the Stoic school of philosophy. They, too, regarded the stars and the rest of nature as one single entity filled with omens; to them, too, astrology was just one of many different branches of divinatory science.

Before the founding of the Stoic school, the Pythagoreans and Plato had already stressed the association between gods and men by means of a world soul pervading the whole universe; the Stoics later physicalized this doctrine, emphasizing the connection between the material of which the soul was made, the *pneuma,* and the different parts of the universe.[47] According to Cicero, the Stoics believed that "in the beginning, the universe was so created that certain results would be preceded by certain signs, which are given sometimes by entrails and by birds, sometimes by lightnings, by portents, and by stars, sometimes by dreams, and sometimes by utterances of persons in a frenzy."[48] Thus, clearly, the Stoic outlook saw the whole universe as filled with signs, be they in the heavens or on earth. Cicero in his first book *On Divination* provides us with ample examples of instances regarded by the Stoics as omens. A sacrificial ox without a heart, the cries or flight of birds, the eclipses of sun and moon, a girl born with two heads, mysterious voices issuing from temples: all these were portents and could be interpreted by skilful diviners: "These signs do not often deceive the persons who observe them properly."[49]

The Stoics underpinned their belief in divination by means of several arguments. According to Cicero's account of Stoic divinatory theory, the first and most important reason for believing man capable of reading the signs of the future was a religious one: the gods

---

[47] Ibidem, 109-110; S. Sambursky, *Physics of the Stoics* (Princeton 1959, paperback 1987) 66.

[48] Marcus Tullius Cicero, *De divinatione* I.lii.118 (English translation by William Armistead Falconer, Loeb Classical Library edition, Cambridge, Mass./London 1923) 351.

[49] Ibidem, 353.

(whose existence, to the Stoics, was beyond doubt[50]) truly cared for man, being "the friends and benefactors of the human race"; therefore, since it was advantageous to man to know about future happenings, the gods sent mankind signs of these happenings so that they might prepare themselves for the future as best they could; and from this it followed that, along with the signs themselves, the gods also provided mankind with means of understanding those signs; if not, they would be useless.[51]

The other Stoic postulate that made divination and the foretelling of future events possible, physical rather than religious in nature, was the existence of "an immanent and pre-established order in the world by which the succession of single events and the interconnection of phenomena, including those related through divination, [was] determined once and for all."[52] Since, according to the Stoics, all events were predetermined, and happened in accordance with Fate (defined as "an orderly succession of causes wherein cause is linked to cause and each cause of itself produces an effect"[53]) everything that was bound to happen found within nature every efficient cause of its happening. This meant that diligent observation of natural occurrences could lay bare a certain pattern of particular effects always or mostly following particular causes: "Hence it is that it may be known by observation what effect will in most instances follow any cause, even if it is not known in all. . .".[54]

Sambursky is right in pointing out that, to the Stoics, there was no essential distinction between (what we would call) scientific inference, and inductive divination.[55] The only difference between the two kinds of 'prediction' was that in the case of scientific inference, every chain in the causal nexus could be understood and elucidated by human reason, whereas in the case of divination, the precise nature of all the causes leading to a certain effect was beyond the intellective capacity of the human mind. Thus, Cicero pointed out that signs of bad weather conditions[56], such as the flight of the herons towards the

---

[50] For the Stoic arguments in favour of the existence of the gods, see Marcus Tullius Cicero, *De Natura Deorum* ("On the Nature of the Gods"), Book II.

[51] Cicero, *De divinatione*, I.xxxviii.82-83.

[52] Sambursky, *Physics of the Stoics*, 68.

[53] Cicero, *De divinatione*, I.lv.125.

[54] Ibidem, I.lv.126.

[55] Sambursky, *Physics of the Stoics*, 67.

[56] Such weather signs were not, to Cicero, properly divinatory signs, yet the ana-

shore whenever a storm was impending, were due to causes quite ungrasped by the human intellect: "Hardly ever do we see such signs deceive us and yet we do not see why it is so."[57] All men could do was to diligently observe such natural signs in order to establish an empirical basis on which the art of divination could be founded.[58] However, Sambursky, in stressing only the Stoics' alleged "positivistic approach to the phenomena", and in attributing to them the idea that "the main task of a scientist is to correlate results and to stick to 'observables' rather than to inquire into the nature of causes"[59], fails to appreciate the overriding importance of the role played by the gods in the Stoic divinatory beliefs, and their essentially religious nature.

We have already seen that the first foundation of divination, according to the Stoics, was religious: the care of the gods for mankind. Although this divine component played the largest role in what Cicero called 'natural divination' (i.e., the kind of divination practised by "those who unaided by reason or deduction by signs which have been observed and recorded, forecast the future while under the influence of mental excitement"[60]) the agency of the gods was not absent either from "artificial divination", practised by "those... who, having learned the known by observation, seek the unknown by deduction."[61] The divine origin of divination was most obvious in the case of the 'natural' variety, because when predictions were made by persons in a state of frenzy or by means of prophetic dreams, the innate power of the human soul to foretell the future was abnormally highly developed: "Therefore the human soul has an inherent power of presaging or of foreknowing infused into it from without, and made a part of it by the will of God. If that power is abnormally developed, it is called 'frenzy' or 'inspiration', which occurs when the

---

logy was valid because they were "instances which, although outside the category of divination, yet resemble it very closely"; see *De divinatione*, I.vii.13.

[57] Cicero, *De divinatione*, I.ix.15.

[58] Ibidem, I.vii.12: "Such signs as these have been observed for an unlimited time, and the results have been checked and recorded. Moreover, there is nothing which length of time cannot accomplish and attain when aided by memory to receive and records to preserve."; and ibidem, I.lvii.131: "Now, if it is known by observation and experience that these means of divination have their source in nature, it must be that the observations made and records kept for a long period of time have added much to our knowledge of this subject."

[59] Sambursky, *Physics of the Stoics*, 67.

[60] Cicero, *De divinatione*, I.xviii.34.

[61] Ibidem.

soul withdraws itself from the body and is violently stimulated by a divine impulse. . .".[62]

However, the practice of 'artificial' divination, too, was regulated and made possible by the same "inherent power of presaging" which had been "made a part of the human soul by the will of God". Thus, Sambursky is unjustified in placing meteorology, or the prediction of the weather, in one and the same category as artificial divination such as augury, extispicy, or astrology.[63] In fact, the Stoics made a clear distinction between predictions made entirely through the use of human reason, on the one hand, and those made as a result of direct heavenly inspiration, on the other. According to Cicero, those men who could foretell certain events, be they natural disasters, such as floods, or political events such as the impending rise of tyranny, could at most be called "foresighted" or "able to foresee the future"; however, they were no "diviners". Likewise, many things were often foreseen by physicians, politicians, and farmers; yet their predictions could not properly be called "divination".[64] Real divination, the reading of the signs sent by the gods out of solicitude for the human race, was based not on human reason alone, but first and foremost on the divine power present in all the parts of the universe:

> Assuming the proposition to be conceded that there is a divine power which pervades the lives of men, it is not hard to understand the principle directing those premonitory signs which we see come to pass. For it may be that the choice of a sacrificial victim is guided by an intelligent force, which is diffused throughout the universe; or, it may be that at the moment when the sacrifice is offered, a change in the vitals occurs and something is added or taken away. . .[65]

---

[62] Ibidem, I.xxxi.66.

[63] See Sambursky, *Physics of the Stoics*, 67: "Basing themselves on their strictly deterministic doctrine, the Stoics denied any essential difference in method between scientific inference and inductive divination: meteorology, for instance, also attempts to forecast storms by certain signs of the weather."

[64] Cicero, *De divinatione*, I.xlix.111-I.l.112: "Some of these men make predictions, not as the result of direct heavenly inspiration, but by the use of their own reason. For example, by means of a natural law, they foretell certain events, such as a flood, or the future destruction of heaven and earth by fire. . . . Such men we may call 'foresighted'—that is, 'able to foresee the future'; but we can no more apply the term 'divine' to them than we can apply it to Thales of Miletus. . . There are many things foreseen by physicians, pilots, and also by farmers, but I do not call the predictions of any of them divination."

[65] Ibidem, I.lii.118.

After this clarifying passage, Cicero went on to provide his readers with numerous examples of such direct interventions of the Divine Will, which caused the particular premonitory movements or cries of birds, the portentous eclipses of the sun and moon, the births of children with two heads, and the advent of prophetic dreams. In each of these instances, he explicitly added that the sign in question had been sent by the gods with a specific purpose. Thus, the heartless ox that was sacrificed by Julius Caesar just before his death was "sent by the immortal gods to Caesar that he might foresee his death"[66]; and the movements of the birds observed by the augurs were influenced "by a god, whose divine will all things obey"[67].

To the Stoics, then, divine agency was an essential condition for the viability of divination. The solicitude of the gods for men was the *ultima ratio* of the human capability of reading the signs of the future. Yet, at the same time, the Stoic belief in an inexorable and unwavering Fate, which preordained all things in an unbreakable chain of necessary cause and effect, meant that even the interventions of the gods were part and parcel of the roll of fate. Thus, every single natural occurrence, be it the regular movement of the heavenly bodies, the volatile blowing of the winds, or an event as unusual as the birth of a monster-child, represented a link in the never-ending causal nexus, and could as such, on the basis of careful observation and interpretation, act as a sign of the future. Seneca expounded the Stoic views on the relationship between divine agency and necessary fate as follows:

> "But how do things indicate future events unless they are sent to do so?" In the same way as birds provide favourable or unfavourable auspices even though they are not, in this respect, moved in order to appear to us. "But god moved them," he says. You make god too unoccupied and the administrator of trivia if he arranges dreams for some people, entrails for others. None the less, such things are carried out by divine agency, even if the wings of birds are not actually guided by god nor the viscera of cattle shaped under the very axe. The roll of fate is unfolded on a different principle, sending ahead everywhere indications of what is to come, some familiar to us, others unknown. Whatever happens, it is a sign of something that will happen. Chance

[66] Ibidem, I.lii.119.
[67] Ibidem, I.liii.120.

and random occurrences, and without a principle, do not permit divination. Whatever has a series of occurrences is also predictable.[68]

It will have become clear from the above that the Stoic outlook on divination can be described as 'teratological'. The whole universe was filled with signs from which, after careful scrutiny, impending events might be deduced. The Greek word τέρας could mean either "portent" or "marvel"; from the Stoic point of view, any natural event could be regarded as such a portent (although in practice, by far the most attention tended to be accorded to those τέρατα which were truly unusual or seemed to run counter to the ordinary course of nature).[69] In principle, the whole universe was accorded predictive value.

On the teratological view, then, astrology, or prediction from the positions of the heavenly bodies, was just one of many possible techniques of practising divination.[70] Cicero mentions astrology as one of the forms of artificial divination, along with the interpretation of lightnings, prophecy, augury, and the interpretation of prodigies.[71] Yet, as even Cicero and Seneca themselves in their expositions of Stoic divination implicitly admitted, astrology did represent a special case, insofar as its divinatory possibilities were potentially much more promising than those of techniques such as augury or the interpretation of lightning. Thus, Cicero, when explaining how future events could be deduced from the chain of necessary cause and effect, stated that "Persons familiar with the rising, setting, and revolutions of the sun, moon, and other celestial bodies, can tell long in advance where any one of these bodies will be at a given time."[72] Although he added that "the same thing may be said of men who, for a long period of time, have studied and noted the course of facts and the connexion of events, for they always know what the future will be", he imme-

---

[68] Lucius Annaeus Seneca, *Naturales Quaestiones* II.32.3-4 (translated by Thomas H. Corcoran, Loeb Classical Library edition, Cambridge Mass. 1971) 151-153.

[69] See Thorndike, *History of Magic* I, 103: "[According to Seneca] all unusual celestial phenomena are to be looked upon as prodigies and portents. But no less truly do the planets in their unvarying courses signify the future." See also Gerard Verbeke, "Ethics and Logic in Stoicism" in M. Osler ed., *Atoms, Pneuma and Tranquillity: Epicurean and Stoic Themes in European Thought* (Cambridge 1991) 11-24: 19.

[70] See also Thorndike, *History of Magic* I, 103, who points out that in Seneca's view, astrology was one of many divinatory arts, along with augury, sacrificial divination, and divination from thunder.

[71] Cicero, *De divinatione,* I.vi.12.

[72] Ibidem, I.lvi.128.

diately qualified this latter statement by the remark: "or, if that is putting it too strongly, they know in a majority of cases; or, if that will not be conceded either, then, surely, they sometimes know what the future will be."[73] Clearly, the movements of the heavenly bodies offered a much more secure basis for the prediction of future events than any of the other divinatory techniques.

Seneca, too, pointed out the superiority of astrology, at least from the human point of view of comprehensibility, in comparison with other signs. Although every part of nature could be regarded as a sign, many of these signs went unnoticed because of the incapability of the human intellect to grasp them or properly to observe them. Thus, the observations of the Chaldaean astrologers were restricted to the powers of the five planets; although Seneca was convinced that the countless other stars in the firmament were, in essence, equally meaningful, their powers were too difficult for human observers to discern, since they were too manifold to discover.[74]

To Seneca, this insufficieny of human reason to chart each and every stellar influence threatened the trustworthiness of the astrologers' predictions, seeing that they took into account the influences of only a very small number of stars. However, not all ancient philosophers were so negatively inclined towards the astrological art. After all, in essence, astrology as a predictive technique had far more potential than any of the other divinatory arts, since the movements of the stars and planets through the firmament obeyed certain fixed and knowable laws, and, moreover, the effects of some of them (most notably the sun and moon) on the earth were immediately obvious to the senses, giving rise to an argument from analogy according to which the other planets, too, sent earthward their own specific physical influences.

One of the most authoritative ancient theories of celestial influence on the terrestrial region was that of Aristotle. In his work *On the Heavens,* the Stagyrite advanced a theory on the relationship between heaven and earth that ran along the following lines. First, he demonstrated, on the basis of the circular motion of the heavenly bodies as opposed to the straight motions natural to the terrestrial elements, that the substance of which the heavens were made was a fifth element distinct from the four that made up the sublunary world. Sec-

[73] Ibidem.

[74] Seneca, *Naturales Quaestiones,* II.32.6-8.

ondly, he argued, on the basis of the perfection of circularity of motion, that this celestial quintessence was prior in nature to, and therefore more divine than, the sublunary elements. It was ungenerated and incorruptible, since it possessed no opposite into which it could change, unlike the sublunary elements, which suffered continual alteration and decay.[75] On the basis of this fundamental difference in nature and nobility between the celestial and the terrestrial regions, Aristotle concluded that "It therefore seemed fitting that the incorruptible, divine, celestial substance should exercise an influence over the behaviour of the corruptible and ever-changing sublunary bodies."[76]

Similarly, in the first book of his *Meteorology,* Aristotle stated that the capacity for movement and change of the terrestrial region was regulated by the motions of the heavens, and that, whereas "fire, earth and the kindred elements must be regarded as the material cause of all sublunar events... the driving power of the eternally moving bodies must be their cause in the sense of the ultimate source of their motion."[77] Also, in several passages, the Stagyrite explicitly named the action of the sun as one of the principal causes of changes on earth. Thus, in the *Meteorology,* he declared that the sun's motion was in itself sufficient to produce all the heat in the terrestrial region: "To produce heat [by rarefying and inflaming the air], a motion must be rapid and not far off. The motion of the stars is rapid but far off: that of the moon close but slow: but the sun's motion has both required characteristics to a sufficient degree."[78] And in a rather cryptical passage in *On Generation and Corruption,* Aristotle pinpointed the role played by the sun in causing terrestrial bodies to come into being *as well as* to perish again. Since generation and corruption should be regarded as two different processes, these must be caused by two different motions (because, according to Aristotelian physics, motion was the primary cause of change). Therefore, it was not the "primary locomotion" of the firmament as a whole that was responsible for terrestrial change and decay, as that movement was in one direction only. Rather, the Sun's motion along the inclined circle

[75] See Aristotle, *On the Heavens,* I.2 and I.3.

[76] Edward Grant, *Planets, Stars, and Orbs. The Medieval Cosmos, 1200-1687* (Cambridge 1994) 571-572.

[77] Aristotle, *Meteorology,* I.2.339a (translated by H.D.P. Lee, Loeb Classical Library edition, Cambridge Mass./London 1952) 9.

[78] Ibidem, I.3.341a (translation p. 23).

(i.e., the ecliptic), which to Aristotle seemed to imply a variation in distance from the earth, caused the opposing processes of generation and corruption below the moon.[79]

The heating faculty of the sun was the most obvious to the senses; however, it was not the only celestial influence recognized by Greek natural philosophers. The moon's influence on the tides, for example, was also acknowledged from at least the time of Seleucus in the second century BC, while the moon was also thought to be responsible for the dew which allowed plants to ripen, as well as for a variety of effects within the human body. Thus, it was believed that the moon affected generative capacities in humans (Aristotle, for instance, thought that it controlled menstruation and ensured that conception took place most often in the middle of the lunar month), while it was also associated with recurrent fevers, epilepsy, and periodically occurring madness.[80] There existed several theories about the physical processes involved in these lunar influences: on one view, recorded by Plutarch, "the Moon sent down heat to Earth and sent upwards exhalations from Earth, refining and purifying them within itself", while Ptolemy thought that "The Moon, too, as the heavenly body nearest the earth, bestows her effluence most abundantly on mundane things, for most of them, animate or inanimate, are sympathetic to her and change in company with her; the rivers increase and diminish their streams with her light, the seas turn their own tides with her rising and setting, and plants and animals in whole or in some part wax and wane with her."[81]

Thus, in ancient Greece, next to the teratological tradition that saw the whole universe as replete with signs, there also existed a much more physically oriented astrological tradition, according to which at least the two great luminaries, sun and moon, physically affected life on earth with their influences. The scope of this tradition was much enlarged by the theories of Claudius Ptolemy, who, during the second century AD, ventured to include the influences, not just of sun and moon, but of the other planets as well, in an elaborate scheme of astrological causation. In one of the introductory chapters

---

[79] Aristotle, *On Generation and Corruption*, II.10.336a; John D. North, "Celestial influence-the major premiss of astrology" in Paola Zambelli ed., *Astrologi Hallucinati. Stars and the End of the World in Luther's Time* (Berlin/ New York 1986) 45-100: 45.

[80] Barton, *Ancient Astrology*, 102-103.

[81] Quoted by Barton, *Ancient Astrology*, 103.

to his *Tetrabiblos,* aptly entitled "That Knowledge by Astronomical Means is Attainable, and How Far", Ptolemy set forth his reasons for believing in the influence exercised by the heavenly bodies on the terrestrial region.

Following Aristotle, he asserted that there existed a certain power emanating from the eternal ethereal substance, which permeated the whole region about the earth, and affected the terrestrial elements, since "fire and air are encompassed and changed by the motions in the ether, and in turn encompass and change all else, earth and water and the plants and animals therein."[82] The most obvious example of the workings of this power was provided by the sun, which continually affected earthly conditions, both by the "changes that accompany the seasons of the year to bring about the generation of animals, the productiveness of plants, the flowing of waters, and the changes of bodies", and by "its daily revolutions, furnishing heat, moisture, dryness, and cold in regular order and correspondence with its positions relative to the zenith."[83] The moon, too, influenced sublunary nature, parts of which tended to wax and wane with her.

However, Ptolemy did not rest content with having demonstrated the universally recognized influences of sun and moon on the terrestrial region. Rather, he believed that the passages of the stars and planets, too, signified hot, windy, and snowy conditions of the air, and thus affected sublunary beings. To make things even more complex, these planetary influences could mingle and mix with each other, bringing about many complicated changes in the process. The most important of these changes, signified by the more obvious configurations of sun, moon, and stars, were usually known beforehand by diligent observers, just as farmers could derive the quality of their future crops from the winds prevailing at the time of the sowing of the seed.[84] Obviously, the greatest and least complex of the heavenly influences, such as the annual variations of the seasons, could be understood by even the most ignorant of men; somewhat more difficult to ascertain were "things that are not of so general a nature", such as "the special signs of storms and winds that arise periodically by reason of the aspects of the moon and fixed stars to the

---

[82] Claudius Ptolemy, *Tetrabiblos*, I.2 (translated by F.E. Robbins, Loeb Classical Library edition, Cambridge Mass./ London 1940) 7.

[83] Ibidem.

[84] Ibidem.

sun"[85]. But, Ptolemy wondered, why should not a man accurately acquainted with the movements of all the stars, the sun, and the moon, as well as with their natures and potentially effective qualities, be able to tell on each given occasion the characteristics of the air from the relations of the phenomena at the time? Why should not such a man be able to forecast the weather, to perceive the general quality of a man's temperament from the heavenly conditions at the time of his birth, and even to predict occasional events?[86]

Ptolemy went much further than Aristotle in according the heavenly bodies influences over terrestrial events. Whereas the Stagyrite had confined their efficacy to such general natural processes as the change of the seasons or the turning of the tides, his Alexandrian colleague thought it possible to predict, from careful observation of the heavens at the time of people's births, each subject's own idiosyncratic temperament, as well as his fortunes insofar as these depended on the position of the stars at the time of the nativity. Although Ptolemy admitted that, next to the ambient, other factors (such as differences of seed, places of birth, rearing and customs) all contributed to influence the particular way in which a life was lived[87], and that, therefore, the astrologers' predictions of necessity had to remain incomplete, since they could take into consideration only those circumstances that had "a directive and natural connection with the original gifts of nature"[88], he nonetheless firmly believed in the usefulness of the art of casting nativities. After all, foreknowledge of future events could prepare the soul to "greet with calm and steadiness whatever [came]"[89], and, moreover, one who knew in advance what misfortunes were to befall him might provide remedies in order to prevent or mitigate these misfortunes.[90] Thus, to Ptolemy, the

---

[85] Ibidem, 11.

[86] Ibidem, 11-13.

[87] Ibidem, 17-19.

[88] Ibidem, I.3, p.21: "In somewhat summary fashion it has been shown how prognostication by astronomical means is possible, and that it can go no further than what happens in the ambient and the consequences to man from such causes—that is, it concerns the original endowments of faculties and activities of soul and body, their occasional diseases, their endurance for a long or short time, and besides, all external circumstances that have a directive and natural connection with the original gifts of nature, such as property and marriage in the case of the body and honour and dignities in the case of the soul, and finally what befalls them from time to time."

[89] Ibidem, 23.

[90] Ibidem, 27-29.

stellar influences were not inescapable: precautions might be taken to undo them. This part of Ptolemy's doctrine was to become the most well-known and most frequently called upon by future defenders of the astrological art, and the phrase "Vir sapiens dominabitur astris" ("A wise man may overcome the influences of the stars") was to become the most often quoted of all Ptolemy's astrological dicta.

After these introductory remarks on the nature of astrological prediction in general, Ptolemy went on to provide his readers with an extremely detailed description of the manifold physical characteristics of the different stars and planets: of whether they possessed heating or cooling, drying or humidifying qualities; of whether they were beneficent or maleficent, masculine or feminine, diurnal or nocturnal, and what consequences could be assigned to these qualities; and of the way in which the positions of the planets within the constellations, and their whereabouts with regard to each other, could alter, moderate, or enhance their powers. Next, he went on to apply these principles to the two branches of astrological prediction described above: the 'general' variety dealing with whole races, countries, and cities, with wars, famines, pestilences, and earthquakes, as well as with weather conditions such as changes in temperature and variations in the quality of crops[91], and the individual or 'genethlialogical' variety dealing with the natural temperaments of individual men.

Ptolemy's entire astrological edifice, as laid down in the highly systematic *Tetrabiblos*, was based on natural causation. The author made a great effort to express all planetary powers in physical terms, and stressed that he had striven to omit all things which did not have a probable natural cause.[92] Thus, for example, he gave a physical account of the exaltations of the planets[93], explaining that the sun had its exaltation in Aries, because when in that sign, its heating power and the length of the day began to increase. Similarly, because Saturn, a pre-eminently cold planet, was the physical opposite of the hot sun, it was assigned Libra as its exaltation, since Libra was the sign diametrically opposed to Aries, the sun's exaltation.[94]

Thus, despite the fact that Ptolemy's astrology was more far-reaching than Aristotle's, since he took into account the fortunes of indi-

---

[91] Ibidem, II.1.

[92] Thorndike, *History of Magic* I, 112-114; Tester, *History of Western Astrology*, 67-68; Barton, *Ancient Astrology*, 107-109.

[93] I.e., the signs in which their influences were enhanced.

[94] Ptolemy, *Tetrabiblos*, I.19.

viduals as well as general atmospheric conditions, the Alexandrian's approach fitted in quite well with the ideas of the Stagyrite and his followers of the Peripatetic school. They all regarded physical properties (most notably the four qualities of hot, cold, wet and dry) as basic to the astrological art, with the aether acting on the elements of fire and air, and these in turn altering the other elements. No mention at all was made of any God or gods who used the stars to convey their messages or to issue warnings to man on earth.[95] Though both the Peripatetics and Ptolemy recognized the divine character of the stars, being formed out of the godly and perfect celestial aether, this divinity did not subsequently play any major role in their naturalistic accounts of astrological influence—apart, perhaps, from being the ultimate cause of the heavenly bodies' motions.[96]

Not surprisingly, then, the Peripatetics were not favourably disposed towards the manifold forms of divination advocated by Stoics such as Cicero and Seneca. Cicero in his *On Divination* alluded to this Peripatetic disapproval several times, stating that such artificial means of divination as the inspection of entrails, augury, and the interpretation of dreams, were "disapproved of by the Peripatetics and defended by the Stoics"[97], and that "Dicaearchus, the Peripatetic, though he accepted divination by dreams and frenzy, cast away all other kinds", while "The Stoics, on the other hand, . . . defended nearly every sort of divination".[98]

However, the distinction between Stoics and Peripatetics was not as clear-cut as the above account may suggest. For one thing, most Stoics, while in favour of all kinds of divination, and of the teratological stance of regarding the whole universe as filled with divine signs, also recognized the special status of astrology as a divinatory technique capable of being rationalized and subjected to rules in the way Ptolemy had done in his *Tetrabiblos*.[99] Indeed, the Stoic idea of the oneness of nature and of the universe, including both gods and men, and their belief that the same physical laws applied throughout the cosmos, was certainly conducive to astrological ideas as articulated

---

[95] See also Jean Céard, *La nature et les prodiges. L'insolite au XVIe siècle, en France* (Geneva 1977) 3-12.

[96] Barton, *Ancient Astrology*, 111.

[97] Cicero, *De divinatione*, I.xxxiii.72.

[98] Ibidem, I.iii.5-6.

[99] Tester, *History of Western Astrology*, 68-69; Barton, *Ancient Astrology*, 34-35; F.H. Sandbach, *The Stoics* (2nd ed., London/Indianapolis 1989) 80-81.

by Ptolemy and other authors in the same tradition.[100] Ptolemy himself probably derived many of his theoretical arguments in favour of astrology from a lost work on divination by the famous Stoic author Posidonius.[101] And, on the other hand, the idea of a universe filled everywhere with divine signs was not the exclusive property of the Stoics: Neoplatonists such as Plotinus were also convinced that the unity of nature under the control of a sovereign and divine entity rendered divination possible, be it from stars, birds, or other parts of nature.[102]

It would, therefore, be wrong to equate the 'teratological' strand of divination exclusively with Stoic philosophy, or, conversely, to associate the physicalizing, 'astrological' strand only with the Peripatetics. Bearing this in mind, however, the purpose of the present section has been to demonstrate how, during Antiquity, two very different approaches to the art of divination came into existence. On the one hand, to those who regarded the universe as an integrated whole, ruled by an omnipresent god (or gods) who was solicitous for the well-being of mankind, divine signs might be discerned in every part of nature. On the other hand, the prediction of future events might be approached in a much more physical way, by attempting to systematically chart the physical influences emitted by stars and planets, without allowing for any possible divine intelligence associated with these influences. At different points in time, both strands made their entry onto the intellectual scene of medieval Europe, where both underwent alterations or adaptations due to the all-pervading presence of Christianity. It is to these alterations that we must now turn our attention.

ii. *Their Fortunes under Christianity*

The advent of Christianity brought about an important change in outlook with regard to divination and the foretelling of future events. Of the two strands described in the previous section, the astrological was the first and most severely affected by the rise to prominence of

---

100 Tester, *History of Western Astrology*, 68-70; North, "Celestial influence", 48-50.

101 Franz Boll, *Studien über Claudius Ptolemäus* (Leipzig 1894) 131 ff.

102 North, "Celestial influence", 51; Plotinus, *Enneads*, II.3: "There must be differentiation—eyes and so forth—but all the members will be in sympathy with the entire animal frame [the Universe] to which they belong. Only so can there be a unity and a total harmony. And in such a total, analogy will make every part a sign." (Translation by Stephen MacKenna, Penguin Classics edition, London 1991) 79.

the omnipotent Christian God. The Christian emperors of late Antiquity saw the repression of astrology as part of the struggle against paganism, and issued a number of laws forbidding it.[103] Although these laws were not always very harshly enforced, astrology's respectability as a legitimate means of discovering the future did suffer serious decline during the first centuries of Christianity in the Roman Empire. It was seen as a threat to the authority of the Church; after all, if the stars determined human behaviour, what role was there left for God to play? The astrological art was often associated with magical practices, as well as with heretical sects such as Gnosticism and Manichaeism, which the early Church saw it as its task to eradicate.[104]

The early Christian authors usually included their attacks on astrology in the context of discussions of fatalism and free will, stressing that all kinds of determinism, the astral determinism of astrology among them, should be condemned as infringing on God's omnipotence and human free will. Many of them, such as Tatian and Tertullian of Carthage, ascribed the invention of astrology to fallen angels or demons, although Origen conceded that the stars were to be regarded as a kind of divine handwriting which was intended for angels and good demons to read God's intentions. Scathing attacks on astrology were mounted by many of the most influential Church Fathers: Diodorus of Tarsus, Gregory of Nyssa, John Chrysostom, Gregory of Nazianzus, Augustine, Basil and Ambrose all regarded prediction of future events by means of observation of the stars as intolerable forms of determinism representing a dire disregard for God's omnipotence.[105]

As time wore on, the Church became less and less tolerant of rivals. The first Church Council forbidding clergy to be astrologers or magicians was the Council of Laodicea, held in 365; and in 572, the Second Council of Braga pronounced the anathema against those using astrology before building houses, planting trees, or marrying.[106] The Church's attempts at uprooting astrology seem to have been quite successful, as can be learned from the fact that, until the twelfth century, astrological works dropped out of the library lists in

[103] Barton, *Ancient Astrology*, 64-65.
[104] Ibidem, 72-75.
[105] Ibidem, 75-76.
[106] Ibidem, 78-79.

the western half of the Empire.[107] The two most popular liberal arts books of the early Middle Ages, Martianus Capella's *Marriage of Philology and Mercury* and Macrobius' *Commentary on the Dream of Scipio,* featured only a small amount of astrological theory. Thus, under the influence of the Church's hostility, as well as the general decline of learning in the West, scientific astrology all but disappeared; there is no evidence for practitioners of the astrological art before the mid-twelfth century.[108]

The advent of Christianity also had its consequences for the teratological tradition. According to Christianity, God, having created nature, surpassed it in greatness. Since a creature ought not to be equated with its creator, the Hellenistic principle of immanence, according to which God and Nature were to a large extent synonymous, and the natural order was in a sense deified, was anathema to the Christian faith, which held God to be transcendent and fundamentally above the natural world.[109] Thus, Christianity broke the bond between God (or: the gods) and nature on which the Stoic theory of divination had been based: no longer could it be maintained that the will of God could be known simply by observing the regular, divine order of the universe. Instead, the natural theology of the Stoics was replaced with the belief that, ultimately, the universe was not intelligible to man by means of human reason alone. In order to discover the real purposes of God, observation of natural processes was in itself insufficient: divine revelation was the only means of attaining knowledge of the future.

However, the transition was gradual rather than abrupt, and it is possible to detect the influence of the ancient ideas on divination, as expounded by Cicero and Seneca, in the views of that most authoritative of Church Fathers, Saint Augustine. Although, obviously, Augustine regarded God and nature as two fundamentally distinct entities, his account of portents and prodigies owes a great deal to the Stoic ideas of harmonious correspondence between all parts of the universe, regulated by a divine power solicitous for man's well-being.

[107] Ibidem, 79.

[108] Tester, *History of Western Astrology*, 113-114, 142.

[109] See William J. Bouwsma, "The Two Faces of Humanism: Stoicism and Augustinianism in Renaissance Thought" in Heiko A. Oberman & Thomas A. Brady Jr. eds., *Itinerarium Italicum. The Profile of the Italian Renaissance in the Mirror of its European Transformations. Dedicated to Paul Oskar Kristeller on the Occasion of his 70th Birthday* (Leiden 1975) 3-60: 9-10.

According to Augustine, the whole of creation was both natural and, at the same time, miraculous. Even the most everyday events, such as rainfall or the growth of plants, demonstrated the mysterious creative power of God at work in the universe.[110] Yet, because men had grown so accustomed to these so-called 'daily miracles', they were no longer moved to awe by them and needed to be provoked by unusual manifestations of God's power. Therefore, Augustine believed, God had created the 'seminal reasons', entities which could be regarded as 'hidden causes'—hidden, that is, within the nature and appearance of things, only to become manifest, and work so-called miracles, under special circumstances, such as the prayers of the saints.[111]

Thus, to Augustine, events that struck man as miraculous were in fact part of nature just as much as 'ordinary' events such as, for instance, the daily rising and setting of the sun: "They [miracles] were wonderful acts of God shown as events in this world, not in opposition to nature but as a drawing out of the hidden workings of God within a nature that was all potentially miraculous."[112] Therefore, although usually, the general public regarded as portents only those natural occurrences that seemed contrary to nature, they in fact were no less natural than everyday events, since both were equally due to the will of God. For that reason, Augustine believed that "A portent . . . does not occur contrary to nature, but contrary to what is known of nature."[113] Although the seminal reasons were usually hidden and only acted under special circumstances, yet, like all other natural causes, they had been implanted in the world at the creation, and therefore the miracles they worked were unusual but not strictly supernatural.[114] Like Cicero and Seneca, then, Augustine ascribed the extraordinarity of miraculous events, not to any counternatural quality on their part, but to the incapacity of the human intellect to grasp their causes.[115]

The period between Augustine and the revival of the Aristotelian

---

[110] Benedicta Ward, *Miracles and the Medieval Mind. Theory, Record and Event, 1000-1215* (Philadelphia 1982) 3.

[111] Ibidem; Robert M. Grant, *Miracle and Natural Law in Graeco-Roman and Early Christian Thought* (Amsterdam 1952) 218.

[112] Ward, *Miracles*, 3.

[113] Saint Augustine, *City of God*, XXI.8 (translated by Henry Bettenson, Penguin Classics edition, London 1984) 980.

[114] Grant, *Miracle*, 218-219.

[115] Compare the following passages: "Those unbelievers therefore must not throw dust in their own eyes in this matter of the knowledge of nature, and assume that it is

corpus in the twelfth century, while, as we have seen above, devoid of any substantial interest in astrology, witnessed a great rise to prominence of the teratological strand. Interest in miracles became very widespread, while in the process, the teratological tradition as expounded by Augustine underwent a number of alterations. Although some early medieval authors tried to provide their accounts of miracles and prodigies with a kind of rational underpinning, such as the Stoic and Neoplatonic idea of universal sympathy between all parts of nature[116], on the whole they were more interested in the role such portentous occurrences had to play in the history of salvation. The birth of monsters, and other unusual natural events, were seen as signs of upheaval in the natural order, caused by the sins of mankind; in her capacity as servant of God, Nature dispensed rewards and punishments for the acts of man.[117] Also, monsters and prodigies were seen as signs of God's continual involvement in nature, and served to remind man of the existence of the Creator. Because of their importance with regard to human salvation, medieval chroniclers saw it as their task to dutifully record all prodigious occurrences.[118]

Gradually, the Augustinian idea of the whole of nature as miraculous, and of miracles as ultimately natural occurrences, began to lose

---

not possible for something to occur in any substance through the exercise of divine power, which conflicts with the property of that substance as known to them in their own human experience. And yet the natural phenomena known to all men are no less wonderful, and would be a source of astonishment to all who observe them, if it were not man's habit to restrict his wonder at miracles to the rarities." (Saint Augustine, *City of God*, XXI.8); "For god has not made all things for man. How much a part of god's immense work is entrusted to us? The very one who handles this universe, who established it, who laid the foundations of all that is and placed it around himself, and who is the greater and better part of his work, has escaped our sight; he has to be perceived in thought. Moreover, many things related to the highest divinity or allotted a neighbouring power are obscure. . . . And yet we are surprised if we imperfectly understand some little bits of fire, even though the greatest part of the universe, god, remains hidden!" (Seneca, *Naturales Quaestiones*, VII.30.3-4); "But what? You ask, Carneades, do you, why these things so happen, or by what rules they may be understood? I confess that I do not know, but that they do so fall out I assert that you yourself see." (Cicero, *De divinatione*, I.xiii.23). On Stoic influences in Augustine's theory of miracles, see Grant, *Miracle*, 218-219.

116 Thus, Isidore of Seville in the sixth century still expounded Augustine's theory of miracles as "not . . . contrary to nature, but contrary to nature as known"; see Charles M. Radding, "Superstition to Science: Nature, Fortune, and the Passing of the Medieval Ordeal", *American Historical Review* 84 (1979) 945-969: 958.

117 Céard, *Nature et prodiges*, 35.

118 Ibidem, 37-38.

ground in favour of the view that miracles were truly supernatural. Thus, Saint Anselm defined miracles as "those things which are done neither by created nature nor by the will of the creature but by God alone", and distinguished between "three ways in which things happen, that is, the miraculous, the natural and the voluntary".[119] To Anselm, then, there existed a category that had been unknown to Augustine: that of *mirabilia* or miracles caused by the direct intervention of God, worked without recourse to any natural causes and not subject to the laws of nature.[120] Other eleventh and twelfth-century authors, such as Peter Abelard and Gilbert de la Porrée, made the same sharp distinction between miracles, seen as events directly belonging to divine omnipotence, and other matters that belonged to the natural and created order. This rift between the divine and the natural entailed a new freedom to examine natural events, which were in a sense 'desacralized'.[121]

The recovery of the works of Aristotle during the twelfth century further reinforced this severing of faith and reason. As we have seen in Chapter III, the revival of the Aristotelian corpus gave rise to a tendency to rely on human reason alone, unaided by revelation, in the investigation of natural processes. The view that gained prominence during the twelfth century, that the universe was run by the operation of natural forces that did not require the regular intervention of God, was strongly reinforced by the attitude of scholastic philosophers, who, relying on the logical coherence of the Aristotelian system, ventured to explain all observed phenomena by means of natural rather than supernatural causes.

This attitude had important consequences for both the astrological and the teratological strands of divination. First and foremost, the rationalism of the schoolmen tended to restrict the scope of the miraculous. No longer was the whole of nature regarded as one great miracle; instead, the Augustinian view of miracles as essentially natural occurrences was replaced with a sharp distinction between miracles, proceeding directly from God and occurring only rarely, and all other natural events, which had purely physical causes that could be grasped by the human intellect. Scholastic philosophy claimed that

---

[119] Saint Anselm of Canterbury, *De Conceptu Virginali*, 2.11 (quoted by Ward, *Miracles*, 4).

[120] Ward, *Miracles*, 4-5.

[121] Ibidem, 6; Radding, "Superstition to Science", 959-60.

the universe was rational, and that events could be predicted by human reason, thereby implicitly limiting God's absolute power over nature—in the sense that, in the vast majority of cases, nature operated according to laws that were fixed once and for all, and that God only very rarely employed his omnipotence to deviate from these laws.[122]

Thus, the teratological strand suffered a decline in prominence during the twelfth and thirteenth centuries. Although, for an age still unanimously committed to Christianity, it was of course unthinkable to do away with the supernatural in all its manifestations, since the absolute power of the omnipotent deity had to be provided for under all circumstances, the category of supernatural occurrences was nonetheless drastically reduced due to the all-pervasive tendency to explain phenomena first and foremost by means of natural causes, with divine agency being invoked only as a last resort.[123] This desacralizing of nature tended to severely limit the category of supernatural phenomena: even if certain natural qualities were insensible or idiosyncratic, the scholastics still believed it possible to find rational, physical explanations for them. As the author of the thirteenth-century treatise *De mirabilibus mundi,* wrongly ascribed to Albert the Great, phrased the general attitude of the day: "The philosopher's work is to make marvels cease."[124]

The selfsame attitude that so categorically reduced the scope of the teratological strand, however, was responsible for an unprecedented rise to prominence of its astrological counterpart. The desire to explain all natural occurrences by means of physical forces found itself an outstanding means of expression in the newly recovered system of physical astrology. As was described at length in chapter III, the secular masters at the Paris arts faculty were easily won over to the astrological theories of Ptolemy and the additions of the Arabs, most notably Albumasar, who had to a large extent disregarded human free will and advocated a far-reaching stellar determinism.[125]

According to Albumasar, the alterations of the elements which lay at the root of all sublunary events were caused by the motions of the heavenly bodies. He thought that the heat produced by the circular

---

122 William Eamon, *Science and the Secrets of Nature. Books of Secrets in Medieval and Early Modern Culture* (Princeton 1994) 63. See supra, chapter IV, p. 142.

123 Ibidem.

124 Quoted by Eamon, *Science and Secrets,* 73.

125 See supra, p. 64.

movement of the *primum mobile* made the world more subtle, so that changes in bodies could take place. The natural action of the heavens, on this view, was transmitted to the terrestrial region by means of some invisible intermediary, somewhat like the action of the magnet being transmitted through the air.[126] The diversity and variety of the sublunar world were accounted for by postulating that each of the planets exercised its own specific influences.[127] The more rapid their movements, Albumasar believed, the more powerful were the planets' effects, so that the Moon, especially, produced notable changes in the sublunary sphere.[128]

During the twelfth and thirteenth centuries, a number of authors in some way or another adopted a similar, physicalized approach to celestial influence. Hermann of Carinthia stated that the celestial sphere provided what was stable in the world, while the complex of spheres to do with the Sun's oblique motion provided variety; Robert Grosseteste in his light metaphysics explained the earth's centrality in the cosmos by pointing out that this position had been allotted it so that it might be a focus for the light from the celestial spheres; and John Buridan believed that the Sun's sphere caused a quality in the other spheres other than heat, which, on reaching heatable things, generated heat in them.[129] Peter of Abano, too, argued that God acted on the sublunary world through the heavenly bodies, whose spheres, with their different natures and qualities, caused the phenomena that depended on them according to precise rules and natural causes. Although he admitted the possiblity of direct interventions of God, who might produce miraculous results by circumventing the laws of nature, Peter insisted that the science of astrology, dealing with the *natural* influences of the stars, was rational and mathematical in character.[130]

After the recovery of the Aristotelian corpus in the twelfth and thirteenth centuries, it was not just the Stagyrite's own theories of celestial causation that went through a revival. Along with the re-

---

126 North, "Celestial influence", 54.

127 John North, "Medieval concepts of celestial influence: A survey" in Patrick Curry ed., *Astrology, Science and Society. Historical Essays* (Bury St Edmunds 1987) 5-16: 7-8.

128 North, "Celestial influence", 57.

129 North, "Medieval Concepts", 9-14.

130 Graziella Federici Vescovini, "Peter of Abano and Astrology" in Patrick Curry ed., *Astrology, Science and Society* (Bury St Edmunds 1987) 19-39: 20-22, 38.

newed interest in the idea of the heavenly bodies affecting the terrestrial region by means of physical forces that were susceptible to investigation by the human intellect, the physicalized astrological theories, as laid down by Ptolemy and adapted by the Arabs, also made a significant comeback. After an absence of centuries, the art of astrology made its reappearance on the Latin medieval scene, supported by the weight of the Aristotelian world system with its insistence on the fact that all sublunary change was the result of and dependent upon motions in the heavens. Before long, the theory of astrological influence came to be recognized as one of the sciences studied by scholastic philosophers, and the astrological art became an integral part of the Aristotelian world view.[131]

The rise to prominence of the astrological strand during the high Middle Ages, however, was not accepted indiscriminately by all scholars. From several quarters, objections were raised to the art, with some philosophers protesting that the aspirations of the astrologers, though commendable in principle, were aimed too high, while theologians took exception to the curtailment of God's absolute power, as well as of the free will of man, which was implied by the stellar influences on which the astrological art was based. Yet, despite such demurrals, astrology of the Arabo-Ptolemaic variety formed an integral part of the Aristotelian world view that conquered the Latin West in the Middle Ages and that still had such a firm hold on our late sixteenth and early seventeenth-century authors.

For the time being, then, physical astrology was there to stay. But at the same time, as a result of the same tendency to explain natural phenomena first and foremost by way of physical causes, the teratological strand underwent a number of important changes. A number of authors of the late Middle Ages took pains to restrict the scope of the category of the supernatural as much as possible, insisting that even the rarest and strangest occurrences could be explained by natural causes. According to Aristotelian science, phenomena regarded as 'natural' were only those which occurred most of the time, following the usual course of nature.[132] As a consequence, scholastic philosophers often had considerable trouble in accounting for unusual events, such as the birth of Siamese twins or of deformed children,

---

[131] Tester, *History of Western Astrology*, 176 ff.

[132] Bert Hansen, *Nicole Oresme and the Marvels of Nature: The "De causis mirabilium"* (Toronto 1985) 63.

which deviated from the ordinary course of nature and as such could not be explained by the Aristotelian scheme. At most, these phenomena could be established by empirical observation and ascribed to unspecified 'occult causes'.[133] However, during the fourteenth century, attempts were made to elucidate these causes, to strip the occult causes of their mysteriousness, and thus to accentuate the fundamental gap between God's supernatural actions, which were acknowledged in only a very small number of cases, and all other phenomena, which were due to natural causes, be they rare or common.

One of the most elaborate efforts at naturalizing marvels was made by the Paris nominalist Nicole Oresme. In his treatise *De causis mirabilium,* dating from the second half of the fourteenth century, Oresme strove to make his readers aware of the fact that almost all of the events considered 'marvellous' by the general public were in fact just as natural as all other, regular occurrences. They had acquired their status as marvels only because they were due to causes that were generally overlooked, for instance because these were very small compared to the effects they worked, or because they were wrongly perceived.[134] In Oresme's view, for example, the birth of a monster was due to "the failing of something needed for proper generation": after all, for a person to be generated, "there is needed such material as is not too hot . . . not too watery etc., not to little etc. Second, it is required that the plastic spirit which shapes the material be strong and well proportioned, not too hot nor etc., not too subtle, . . . etc. Third, there is needed a temperate place having such and such conditions, not slippery nor etc. And if any one of these or of the other requisites is lacking, there is not generated a person . . .".[135] Thus, monstrosities were entirely due to natural deficiencies and, accordingly, lost much of the predictive value as divine warnings which they had possessed before the recovery of Aristotle.

By the end of the Middle Ages, then, the general tendency was to ascribe as many events as possible to natural causes. The Aristotelian world view, with its insistence on physical celestial influences as causes of changes on earth, could and did accommodate the

---

[133] Lorraine Daston, "Wunder, Naturgesetze und die wissenschaftliche Revolution des 17. Jahrhunderts", *Jahrbuch der Akademie der Wissenschaften in Göttingen* (1991) 99-122: 105.

[134] Hansen, *Oresme and Marvels,* 70-73.

[135] Nicole Oresme, *De causis mirabilium* III, translated by Hansen, *Oresme and Marvels,* 229.

Ptolemaic-Arabic system of physical astrology, according to which the stars predicted future events by means of a complex of wholly natural influences. At the same time, the teratological tradition, according to which unusual natural events could be regarded as divine signs of the future, suffered serious decline. Although Christian philosophers could never quite abolish the category of supernatural events called into being by God for some special purpose, the scope of this category was drastically reduced by the tendency to search for the causes of all unusual events within the natural order. In practice, the direct agency of God, which had been so prominently present in the teratological tradition of antiquity and the early Middle Ages, became a topic of little interest, all attention being focused instead on the effects worked by Nature on her own, be they the physical celestial influences of the astrologers or the naturalized marvels of Oresme. The connection between the natural and the supernatural, that was such an essential trait of the outlooks of the Stoics, Saint Augustine, and early medieval thinkers such as Isidore of Seville, had been savagely severed.

By the end of the Middle Ages, the astrological tradition, though immensely popular, saw itself faced with a number of grave criticisms. The most serious of these was the theologically-inspired objection to the prediction of future events in the lives of individual human beings, as practised by genethlialogical astrology. Thinkers such as Thomas Aquinas and Nicole Oresme warned of the dangers such practices could pose for the Christian faith: if the stars ruled the actions of every human being, how could man, by choosing to do good and avoid evil, gain salvation? And how could the omnipotence of God be upheld, if all that happened happened in accordance with the decrees of the stars? However, these theological arguments were not the only reasons for condemning the prediction of the fortunes of particular individuals. There was also a more down-to-earth problem, to do with the limitations of the human intellect and sense perception. Already in Antiquity, authors such as Cicero, Sextus Empiricus, and Saint Augustine had pointed out the impracticability of the astrological method. Their arguments were repeated and elaborated during the Middle Ages by Oresme and Henry of Hesse.

Both Oresme and Hesse regarded the investigation of sublunary causes as more worthwhile than the astrologers' search for celestial influences, which, they believed, was doomed to failure. During the fifteenth century, a similar stance was adopted by Giovanni Pico della Mirandola, who in his *Disputationes adversus astrologiam divinatricem* (1496) did not reject the existence of a general celestial influence on terrestrial phenomena, but did deny that this influence could be resolved into discernible relations between particular heavenly causes and corresponding earthly effects. Pico stated that the astrological influence worked only on material objects, including man's body but not his will; and like Oresme and Hesse, he urged to reform astrology by naturalizing it, for example by restricting the forms of celestial power to heat, light and motion.[136]

One of the fiercest responses to Pico's *Disputationes* was written in 1520 by the Italian Aristotelian Pietro Pomponazzi. In his *De naturalium effectuum causis sive de incantationibus* Pomponazzi asserted that, far from being unattainable or impious, astrological knowledge was both possible and commendable. Like Pico, he advocated the search for natural causes of all occurrences; however, this aspiration led him to very different results from his Florentine predecessor. Whereas Pico had concentrated on observable forces such as those of light and heat, which could be perceived by the senses, Pomponazzi believed that *all* the powers exercised by the motions and interrelationships of the heavenly bodies could be identified. According to him, providence, or the stars, not only inclined, but also compelled the human will with the same necessity as an agent acts upon its material instrument. Endorsing the Stoic theory that all terrestrial events were due to the unique, universal and divine efficient cause expressed in natural and cosmic laws, Pomponazzi thought it possible for astrology, by discovering these laws, to predict the destinies of individuals and of entire civilisations.[137] Thus, though both Pico and Pomponazzi were moved by the desire to naturalize astrology, the results of their efforts were very different: whereas Pico's outlook drastically limited the astrologers' pretensions, Pomponazzi's greatly enlarged the scope of astrological causation, since, according to him, all terrestrial events had celestial causes.[138]

Throughout the sixteenth century, both adherents and opponents of astrology continued the late medieval tendency of naturalizing

---

[136] Brian P. Copenhaver, "Astrology and Magic" in Charles B. Schmitt & Quentin Skinner eds., *The Cambridge History of Renaissance Philosophy* (Cambridge 1988) 264-300: 269-270.

[137] Antonino Poppi, "Fate, Fortune, Providence and Human Freedom" in Schmitt & Skinner eds., *Cambridge History of Renaissance Philosophy*, 641-667: 653-657.

[138] Copenhaver, "Astrology and magic", 273.

celestial influences as much as possible. Basically, they only differed as to the extent to which such an enterprise was feasible. Whereas astrological enthusiasts such as Pomponazzi adopted the optimistic stance that all stellar forces could be charted, their more sceptical counterparts held that the aims of astrology should be adjusted, to the effect that only those celestial influences open to observation by the human senses and subjectable to rational rules tested by experience constituted suitable subjects of astrological investigation.

Earlier, we saw that the subject matter of teratology, too, had to a large extent been naturalized during the period of late scholasticism. Oresme had tried to prove that most events usually perceived as marvels or prodigies by the ignorant masses, were in fact reducible to entirely natural causes, thereby to a large extent eliminating their teratological function as divine signs. However, during the sixteenth century, the status of portentous occurrences changed rather dramatically, and the ancient tradition of teratology made a significant comeback. Whereas the approach of authors such as Oresme had served to reduce the scope of the supernatural (those events caused by God's direct interventions going against the ordinary course of nature) and to create a fundamental rift between truly divine miracles on the one hand and occurrences reducible to natural causes on the other, with most attention being focused on the latter, in the course of the sixteenth century interest in God's immediate actions, and in the status of unusual events, began to resurface.

Several developments contributed to this change of emphasis. First of all, the advent of the Reformation brought in its wake a renewed attention paid to the status of the stars as signs, rather than causes, warning of subsequent earthly events. Many biblical texts affirmed the signalling function of unusual natural occurrences, almost invariably in connection with the end of the world or of a great empire. Thus, the gospel of St Matthew described Christ's second coming as being preceded by ominous portents: "Immediately after the tribulations of those days shall the sun be darkened, and the moon shall not give her light, and the stars shall fall from heaven, and the powers of the heavens shall be shaken: and then shall appear the sign of the Son of man in heaven. . .".[139] And the Old Testament, too, was full

139 St Matthew 24:29-30; for corresponding passages in the other gospels, see St Mark 13:24-25 and St Luke 21:25-26. See also St Luke 21:11: "And great earthquakes shall be in divers places, and famines, and pestilences; and fearful sights and great signs shall there be from heaven."

of prophecies and dire omens, such as the passage in 2 Esdras predicting the downfall of Babylon: "The sun shall suddenly shine again in the night, and the moon three times a day. Blood shall drop out of the wood, and the stone shall give his voice... There shall be a confusion in many places, and the fire shall oft break forth, and the wild beasts shall change their places, and menstrous women shall bear monsters...".[140]

The emphasis placed by the reformers on the text of the Bible, and the need for all believers to become acquainted with it, greatly amplified the impact made by such scriptural predictions on the general public. Moreover, the newly invented art of printing contributed greatly to the diffusion of knowledge about ominous occurrences, and brought in its wake an enormous passion for broadsides about prodigies and monsters among large sections of the population.[141] Since such portents invariably carried associations with the apocalypse, general world reformation, and the vindication of the elect, the vogue for prodigious occurrences was grist to the mills of the reformers. Luther and Melanchthon, and a great number of minor Protestant pamphleteers following in their footsteps, made good use of the new medium of prodigy literature. Together, the Wittenberg reformers published a polemical broadside featuring two horrible monsters: the so-called 'Monk-Calf' and the 'Pope-Ass', hideous creatures the first of which was based on a real calf born with what looked like a cowl around its neck, while the other was a fictitious monster reputedly fished out of the Tiber in 1496.[142]

The pamphlet in which the two reformers described and interpreted this pair of monsters, *Deuttung der czwo grewlichen Figuren, Bapstesels czu Rom und Munchkalbs zu Freijberg ijnn Meijsszen funden* (1523), became tremendously popular and influential. Although it dealt with at least partly fictitious monstrosities, and, like many other pamphlets, served an almost exclusively propagandist purpose, it did enhance the interest in genuine marvels and monsters, both among the

---

140 2 Esdras 5:4-8.

141 See Hans-Joachim Köhler, "The *Flugschriften* and their Importance in Religious Debate: A Quantitative Approach" in Zambelli ed., *Astrologi Hallucinati*, 153-175; and Helga Robinson Hammerstein, "The Battle of the Booklets: Prognostic Tradition and Proclamation of the Word in early sixteenth-century Germany", ibidem, 129-151.

142 Katharine Park & Lorraine J. Daston, "Unnatural Conceptions: The Study of Monsters in Sixteenth- and Seventeenth-Century France and England", *Past & Present* 92 (1981) 20-54: 26.

learned and the barely literate.[143] Whenever a deformed child or animal was born, and whenever 'battles' were seen being fought in the clouds, blood dripping from the sky, or a comet appeared, the market would be flooded with large quantities of broadsides describing the prodigy in question and interpreting the message it was supposed to convey. Moreover, the political and intellectual turmoil created by the aftermath of the Reformation, as well as the humanist revival of interest in ancient divination as an element of Roman culture, also inspired a number of more learned authors to commit to paper their thoughts about contemporary prodigious appearances. These writers produced voluminous Latin treatises on the subject, as well as 'prodigy books' in the various vernaculars, a new genre that rose to enormous popularity in the course of the sixteenth century.[144]

Common to all these works, learned and popular alike, was a fundamental similarity of interpretation. Basically, they were all interested first and foremost in discovering God's intentions with the world and with mankind. They all but abandoned the search for the proximate causes of unusual natural occurrences, with which Oresme had been so preoccupied two centuries earlier; rather, they regarded nature as a mirror of God's will, which could be deciphered by man since it was intended by God as a message or warning to the human race.[145] According to a widespread belief fostered by the analogy with the story of the Creation, the history of the world was usually conceived of as comprising six periods of a thousand years each, after which the seventh, or eschatological, era would commence. Since the science of chronology usually placed the reformation after the fifth millennium in the sequence, during the sixteenth

[143] Ibidem, 26, 28.

[144] Ibidem, 30. Examples of Latin treatises on prodigies are Caspar Peucer's *Commentarius de praecipuis divinationum generibus* (1553), Conrad Lycosthenes' *Prodigiorum ac ostentorum chronicon* (1557), Cornelius Gemma's *De naturae divinis characterismis* (1575), Levinus Lemnius' *De miraculis occultis naturae* (1559), and Hieronymus Cardanus, *De rerum varietate* (1557). Examples of vernacular prodigy books are Pierre Boaistuau's *Histoires prodigieuses* (1560-1598) and Ambroise Paré's *Des monstres et prodiges* (1573). Recent literature on monstrous births and prodigies includes Ottavia Niccoli, *Prophecy and People in Renaissance Italy* (Princeton 1990); D. Wilson, *Signs and Portents. Monstrous Births from the Middle Ages to the Enlightenment* (London/ New York 1993); I. Ewinkel, *De monstris. Deutung und Funktion von Wundergeburten auf Flugblättern im Deutschland des 16. Jahrhunderts* (Tübingen 1995); G.-H. Schumacher, *Monster und Dämonen. Unfälle der Natur. Eine Kulturgeschichte* (Berlin 1993); and M. T. Jones-Davies ed., *Monstres et Prodiges au temps de la Renaissance* (Paris 1980).

[145] Park & Daston, "Unnatural Conceptions", 34.

century it was widely believed that Christianity was being prepared for the impending Day of Judgment. This belief went hand in hand with an "acute sense of imminent reward and punishment"[146], and Charles Webster, writing in 1982, has even professed the belief that "God's retribution seemed more imminent to the contemporaries of Paracelsus and Newton than the possibility of a nuclear holocaust seems to us."[147]

Because of such widespread apocalyptic expectations, the significance of unusual natural occurrences was considerably enhanced. Scholars were expected to come up with a solution to questions relating to the immediate future of Christian Europe, and to the "precise place of their age in the preordained pattern of events forming the climax of universal history".[148] For this reason (along, perhaps, with the emphasis laid by the Reformation on the direct contact between God and man, as opposed to the mediations considered necessary by Catholic theology), the power of God to go against the ordinary course of nature, which had been effectively bridled by the tradition of Aristotelian scholasticism[149], returned as a central topic of interest. In this context, Webster has pointed out the importance, during the sixteenth and seventeenth centuries, of "the idea of God as a monarch, not content with *ordinary* providence, but exercising *extraordinary* powers over the destiny of His creation, wielding this monarchy to determine eventualities proportionately to the deserts of every section of the human race"[150], while Katharine Park and Lorraine Daston have called attention to the fact that, to many sixteenth-century authors, "all enquiry into the proximate physical causes of monstrous births [was] wasted time", because, to them, "God shape[d] and altere[d] the natural order in accordance with his pleasure, so that nature [became] a cipher, a mirror of his will."[151]

Thus, under the influence of the Reformation and the humanist rediscovery of the Ancients, the teratological tradition described by

[146] Charles Webster, *From Paracelsus to Newton. Magic and the Making of Modern Science* (Cambridge 1982) 16.
[147] Ibidem.
[148] Ibidem.
[149] See supra, p. 142.
[150] Webster, *Paracelsus to Newton*, 16.
[151] Park & Daston, "Unnatural Conceptions", 34. See also North, "Celestial influence", 96, who points out that "Both Luther and Calvin... played down the importance of natural powers, and put more emphasis [than their medieval predecessors] on immanent divine causation."

Cicero, Seneca, and Augustine, made a spectacular comeback onto the European scene. Once again, the natural order, and the deviations it occasionally displayed, became the centre of attention, and was regarded as the stage on which God revealed to mankind his intentions in order to warn them of impending disasters.

By the end of the sixteenth century, then, those striving to foretell future events from natural occurrences saw themselves faced with a complex range of possibilities. They might turn to the astrological tradition, based above all on the works of Aristotle and Ptolemy as transmitted to the Latin West by the Arabs, and fitting in nicely with the still almost universally accepted Aristotelian world view, with its emphasis on physical causation and its explanation of sublunary generation and corruption as results of celestial influences. However, the credibility of the astrological art was being undermined by a number of grave criticisms. Although no-one would deny the existence of a general celestial influence on the sublunary world, many thinkers pointed out that the precise nature of this influence could never be perceived by human reason. Some believed that, at most, the effects of light and heat could be traced; others would in addition allow the practicability of 'natural' astrology, dealing with the prediction of the weather and the effects of the stars on the human body. But any attempt to chart every one of the celestial influences, in the way the astrologers professed to do, was regarded by many as unattainable. Especially the prediction of the fortunes of individual persons was regarded with suspicion, not least because of the theological dangers such genethlialogy entailed: to Catholics, it impaired human free will, to Protestants, it made a mockery of original sin, and to both, it unjustifiably curtailed God's omnipotence.

The latter defect could be avoided by taking recourse to the teratological tradition, which went through a vigorous revival during the sixteenth century. Once again, nature came to be seen as a mirror of God's intentions, in which every deviation from the ususal was charged with an urgent significance. However, here, too, the practitioner saw himself faced with a pressing problem: namely, how to distinguish false marvels produced by demons or bad spirits, from genuine divine signs called into being by God, either naturally or supernaturally.

The very fact that so many problems were raised, and debates conducted, with regard to the prediction of the future testifies to the amplitude of the support both the astrological and the teratological traditions could still boast. Whereas astrology, at least in its basic assumptions, could quite easily be fitted into the Aristotelian world view, and became one of its basic constituents, the ideas underlying teratology were more in keeping with the outlooks of Stoicism and Neoplatonism that went through a revival during the sixteenth century. As we have seen before in connection with cosmology, the period under consideration was marked above all by the confrontation between the Aristotelianism that had become firmly entrenched during the Middle Ages, and the rival ancient philosophies which had recently been revived by the humanists. And, as in the case of cosmology, comets once again present us with an outstanding opportunity to examine at close range the gradual process that was eventually to lead to the demise of the Aristotelian world view.

c. *Comets at the Crossroads*

Justus Lipsius, inspired by the appearance of the comet of 1577, articulated his opinion on the meaning of comets in the following way:

"These days we, too, have seen a comet, enormous, ominous, and with an unusual appearance. You ask what it presages? I am not an astrologer, and I have not studied the celestial signs; but I do know this old saying: 'Not a comet exists that is not accompanied by evil'. And I believe this verse more readily than I do Aristotle, who refers these matters to fortuitous and inferior causes."[152]

In these few sentences, the eminent humanist sums up in a nutshell the two basic stances which, at the end of the sixteenth century, it was possible to adopt vis-à-vis cometary signification: the more

[152] "Cometam & nos per hos dies vidimus, immanem, dirum, & facie haud saepe visitatâ. Quid portendat, quaeris? Chaldaeus non sum, nec caeli signa cognovi: tamen hoc vetus novi, Οὐδεις κομήτης ὅστις οὐ κακὸν φέρει. Cui versui magis quam Aristoteli credo, qui haec talia ad caussas fortuitas refert & infernas." Letter by Lipsius to Hieronymus Berchemius; the modern editors of Lipsius' correspondence (*Iusti Lipsi Epistolae*, A. Gerlo/M.A. Nauwelaerts/H.D.L. Vervliet eds., Brussels 1978, I, 75. 76 10 13) have dated this letter 13 October 1576, adding, however, that this date is doubtful, among other things because the comet Lipsius refers to only appeared in November 1577. If, indeed, the allusion is to the comet of 1577 (which was also the belief of Libertus Fromondus, who quoted the passage in his *Meteorologicorum Libri Sex* (2nd ed., Louvain 1646) 192) the letter ought to be dated at least two years later. The Greek aphorism quoted by Lipsius is a proverb; see Leutsch-Schneidewin eds., *Paroemiographi Graeci* (Göttingen 1851) II, 819.

astrological and physical stance of Aristotle, referring cometary significance to "fortuitous and inferior causes", and the more Stoically-oriented, teratological stance, with its firm conviction that, whatever their causes, any unusual natural event was bound to carry an ominous meaning. At the end of the sixteenth century, Lipsius clearly much preferred the latter standpoint.

The history of cometology up to the first half of the sixteenth century has been treated at length in chapter III, and it would therefore be superfluous to enter into it in detail again. The purpose of the present section is rather, to remind the reader of the most important divinatory aspects of the comet lores of the different historical periods prior to the sixteenth century, and to relate these to the alternations of the astrological and teratological strands as described in the two previous sections.

The cometology of Antiquity was for the most part teratological in character. All the influential authors of the Roman empire in whose works the hairy stars featured in one way or another, adhered to the Stoic tenet, according to which the cosmos was one single living entity whose parts stood in harmonious correspondence with one another, so that any change in one part invariably indicated impending changes elsewhere. Seneca saw comets as signs of the future, linked to the laws of the universe; Manilius believed them to be heavenly tokens of impending doom, appearing whenever great upheavals were at hand; and Pliny conceived of them as very rare natural occurrences, whose causes could not be grasped by the human intellect, but which beyond doubt were signs of imminent misfortunes. The great historians of the Roman empire, Suetonius, Lucan, Dio, Josephus, and Tacitus, provided these theoretical presuppositions with a wealth of historical examples: at the most significant moments in history, especially the deaths of rulers and the fall of empires, large quantities of portents had appeared, among which comets featured very prominently. Thus, the teratological tradition firmly associated comets with governmental upheavals, thereby affirming their potential as tools of political propaganda.[153]

On the whole, in Antiquity there were only two prominent dissenters who preferred a physical explanation of cometary concomitants to its teratological counterpart. Aristotle came up with a very level-headed account of cometary significance: according to him, comets

153 See supra, p. 47-49.

could be regarded as natural signs of coming wind and drought. Since, in his view, comets consisted of many hot, dry exhalations, their appearance testified to the presence of an abundance of such exhalations in the air. These would, by reason of their hotness and dryness, desiccate the air, so that droughts would ensue; and, moreover, since, to Aristotle, winds were caused by dry terrestrial exhalations[154], the appearance of a comet could also be seen as an indication of impending windy weather conditions.

During Antiquity, this purely physical Aristotelian reasoning did not find much adherence, at least not in an undiluted form. Although Ptolemy certainly owed a great deal to the Stagyrite's cometary and other meteorological theories, his treatment of the hairy stars' significance, though still overwhelmingly physical in character, paved the way for a more astrological approach. Like Aristotle, and for similar reasons, Ptolemy held that the appearance of a comet portended wind and drought; however, he in addition endowed them with productive powers closely akin to those of the celestial bodies, stating that they "naturally produce[d] the effects peculiar to Mars and to Mercury—wars, hot weather, disturbed conditions, and the accompaniments of these."[155]

This rather terse Ptolemaic statement was to prove immensely influential, and, combined with the Aristotelian theory of cometary generation it was to give rise, from the twelfth century onwards, to a general tendency to treat the effects of comets in an astrological, rather than a teratological way, as resulting from a chain of natural causes carrying earthward stellar and atmospheric influences. During the early Middle Ages, before the recovery of the Aristotelian corpus and the astrological works by Ptolemy and the Arabs, comets were typically regarded as divine signs, and the authors from this period were more interested in their prophetic function than in their physical characteristics. However, from the twelfth century onwards, the cometologies of Aristotle and Ptolemy reemerged, and their theories were elaborated to yield an intricate edifice of physical cometary causation.

Scholastic authors made an effort to confine themselves to physical causes in the explanation of cometary consequences. Although the majority of them did not regard comets as the direct causes of the

---

[154] See Aristotle, *Meteorology*, II.iv.360a.
[155] Ptolemy, *Tetrabiblos*, II.13.

phenomena that were supposed to follow their appearances, they did incorporate the hairy stars into a logical sequence of purely physical effects. Thus, Albert the Great, though admitting that comets were not the immediate causes of anything, believed that they could be regarded as natural signs of Martial and Mercurial effects, since the very fact that they appeared denoted the activity of those two planets. On this view, comets were generated by Mars and Mercury (a tenet taken from Ptolemy), and, therefore, their generation was a sign of great Martial and Mercurial activity, making it likely that other effects traditionally associated with these two planets would soon follow.

Other authors in addition mentioned a more Aristotelian-based theory, according to which comets not only denoted particular planetary effects, but also certain atmospheric conditions. The most complete rendition of this line of reasoning that has come down to us, is that of John of Legnano. To him, a comet was a burning terrestrial exhalation, which heated the air through which it passed. It would also leave behind there some small particles of its matter, which would cause winds to arise. Furthermore, the heat caused by the comet led to desiccation of the air, crop failure, scarcity, and famine. In addition, on occasion some of the hot and dry exhalations rising from the bowels of the earth to form a comet were prevented to emerge due to the cold of the exterior air; whenever they tried nonetheless to escape from the earth's interior, this struggle caused an earthquake; and this might in turn lead to a displacement of the waters of the sea, and thus to inundations. Finally, an abundance of hot exhalations caused people to become choleric, to wage war, and to strive for changes of government; because of their delicate condition, princes were especially prone to fall victim to this increase in choler.[156]

In this way, the scholastics rationalized and naturalized the consequences that had of old been associated with comets, such as the fall of empires, pestilences and famine, and the death of kings. However, not all of them could agree with this reasoning. Henry of Hesse, though adhering both to the sublunary nature of comets and to the tendency to explain phenomena by physical causes, was very sceptical about cometary significance. He did not believe in any affinity

---

[156] See supra, p. 81-82.

between comets and planets, of the kind that Albert the Great had postulated, and therefore assigned no astrological consequences to them. Moreover, he also denied the validity of such meteorological explanations as that offered by Legnano: if comets were kindled only on reaching the sphere of fire, then, to Henry, there was no reason why they should heat the underlying layers of the air. Thus, much of Legnano's and similar reasonings became implausible. At most, comets, as well as winds, earthquakes and similar phenomena, were symptoms of one single underlying cause: an abundance of unhealthy exhalations in the air.

Thus, like general theories with regard to the prediction of future events, cometology, too, was, during the late Middle Ages, characterized by a marked preference for the Aristotelian-Ptolemaic mode of explanation, according to which physical causes could be adduced to account for the predictive value of certain natural phenomena. However, again like divination in general, cometary prognostication, too, was affected by the advent of the Reformation and the revival of non-Aristotelian philosophies.

We have already seen how Marsilio Ficino thought comets to be entities produced by the ethereal spirits to welcome great men home after their deaths, and how Paracelsus believed the hairy stars to be fabricated at God's behest by the superior Penates as signs of future events.[157] Tycho Brahe, too, had recourse to supernatural powers in order to account for the generation of the comet of 1577; in his German treatise on the subject, he stated that "comets do not have their origin or signification from any natural course of the stars nor from any eclipse of sun or moon, but they are rather a new and supernatural creation of God the Almighty, placed in the heavens in His good time, the significance and influence of which not only cannot have anything in common with the influence of the planets, but actually works against them and violently upsets their normal workings, for they overwhelm the natural signs with much greater powers and bring about their own effects instead."[158] A few decades later, Johannes Kepler, too, ascribed the significations of novae and comets

---

[157] Supra, p. 87.

[158] Tycho Brahe, *Von der Cometten Uhrsprung was die alten vnnd neuen Philosophi inn denselben vermaint vnnd dauon zuhalten sei* (first published by J.L.E. Dreyer in *Tychonis Brahe Dani Scripta Astronomica* IV (Copenhagen 1922) 381-396) 390; English translation by J.R. Christianson, "Tycho Brahe's German Treatise on the Comet of 1577: A Study in Science and Politics", *Isis* 70 (1979) 110-140: 137.

directly to God, who called them into being in order to attract the attention of the astrologers, and to promote human salvation. Although Kepler did not go so far as to deny the physical powers ascribed to the planets by the astrologers, he did believe these constituted only part of the picture: as important a role was played by the harmonious relations that existed between the different parts of the cosmos.[159]

This renewed emphasis on the direct action of God in nature, and on the status of comets as divine signs or warnings, rather than parts of a physical chain of cause and effect, points to a resurgence of the teratological kind of cometary divination that had been so prominent in Roman antiquity. Moreover, like their Roman counterparts, the sixteenth-century adherents of this form of foretelling soon discovered the propagandistic value that interpretations of cometary appearances might carry. Thus, J.R. Christianson has called attention to the political overtones of Tycho's German treatise on the comet of 1577, written at the behest of king Frederick II of Denmark. Tycho's predictions from the hairy star's appearance can be read as an advocacy of an irenic and pacific striving for universal peace, as well as a rational exercise of the free will, leading to action which could moderate or control the predictable effects of the comet and other cosmic events. This plea stood in marked contrast to the predictions attached to the comet by Tycho's colleague Dybvad, who had different political convictions which led him to propose "anguished prayers for deliverance from divine wrath".[160]

Paracelsus' commentary on the comet of 1531 was probably also intended to play a political role. He sent it to the Zürich reformers, Leo Jud and Ulrich Zwingli, at an especially portentous time: the height of the conflicts between Swiss Anabaptists, Zwinglians and Catholics. Immediately arrangements were made to have Paracelsus' tract printed and distributed, so that its dire predictions of impending conflict, and disruption of monarchies, might contribute to Zwingli's campaign for a better state of war preparedness.[161]

We can conclude that, while the physico-astrological strand of cometary divination, based on Aristotle and Ptolemy, still enjoyed widespread support by the end of the sixteenth century, the terato-

---

159 Thorndike, *History of Magic*, VII, 21-27.
160 Christianson, "Tycho's German Treatise", 130.
161 Webster, *Paracelsus to Newton*, 26.

logical tradition based on Stoico-Platonic theories of cosmic correspondence and on an awareness of God's direct interventions in nature also came to be an essential ingredient of comet lore around 1600. In the opening sentences of the present section, we have seen Justus Lipsius express his preference for the latter approach. Presently, the time has come for us to inquire into the opinions of his fellow countrymen, who devoted separate treatises to the comets of 1577 and 1618. In so doing, we shall be led by the following questions: Which of the two strands of cometary divination did our authors prefer? Can a difference be discerned between the attitudes in the 1577 tracts from those of 1618? Was there a difference between learned and popular writings, and, if so, how can this be explained? Did the opinions of the authors from the northern Netherlands deviate from those of their southern colleagues? Did the political situation in the years 1577 and 1618 have an impact on the predictions attached to comets? And, since our previous chapter has led to the conclusion that, in cosmology, those tenets of the Aristotelian world view that our authors abandoned were usually replaced with Stoic alternatives: can we discern a similar development (abandonment of the physical astrology of the Aristotelian world view in favour of a more teratological, Stoically-inspired form of divination) with regard to cometary prognostication?

3. *The Netherlands, 1577*

a. *Current Affairs: Confusion and Deception*

Pestilence, famine, invasions by foreign powers, civil strife, the deaths of great rulers, changes in government, storms and inundations, and earthquakes: such were the calamities that had of old been associated with the appearance of comets, both by adherents of the astrological and by those of the teratological tradition. They constituted the worst disasters with which any society could be afflicted, and throughout the ages, these were the catastrophes people feared most, whether they thought them to be caused by natural powers or called into being by God to punish his people. Apart from the earthquakes (at least, in the literal sense of the word), the Netherlands, during the decade prior to 1577, were experiencing the full range of 'cometary concomitants' to an unprecedented extent. The appearance of the great comet of 1577 must have confirmed people's darkest fears, and have given rise to the feeling that worse was still to come.

Pestilence, in the guise of the infamous Black Death, ravaged the Low Countries on many occasions during the late Middle Ages and the sixteenth century. During the earlier part of the sixteenth century, the dreadful disease visited the province of Holland almost once every two years; during the 1570s, every year except 1576 and 1578 were plague years in one or more of the Holland towns.[162] And during the years 1578-1579, a major outbreak hit the university town of Louvain in Brabant, costing Cornelius Gemma his life. Leo Noordegraaf and Gerrit Valk have recently demonstrated how disruptive these recurrent bouts of plague must have been of the entire society. Because of the generally recognized infectiousness of the disease, the authorities usually issued bans on imports and exports of goods, as well as quarantines, so that trade was seriously hampered.[163] Moreover, the outbreak of plague in one town would frequently induce its inhabitants to flee to safer places; these refugees would often be ordered to return by the town magistrates, who feared the social and economic consequences of an exodus of rich and highly qualified citizens.[164] And finally, the advent of the plague invariably gave rise to a widespread and terrible fear among all strata of the population. After all, in a town afflicted by the Black Death, one was constantly reminded of the disease's presence: many houses would be vacant, the streets would be filled with funeral processions, the death bells would be continually tolling, and the whole town would be filled with the smell of corpses.[165]

One of the factors contributing to the outbreak of the plague may well have been scarcity and high prices of corn and other necessities, leading to famine and susceptibility to disease.[166] In the Netherlands of the second half of the sixteenth century, the years 1545-46, 1551-52, 1556-7, 1565-66, 1570-75, 1587-88 and 1590-99 were marked by scarcity and high cost of living.[167] From about 1555 onwards, the lower strata of society had been confronted with ongoing inflation

[162] See Leo Noordegraaf & Gerrit Valk, *De Gave Gods. De pest in Holland vanaf de late middeleeuwen* (Amsterdam 1988) 225-230.
[163] Ibidem, 136-158.
[164] Ibidem, 159.
[165] Ibidem, 102-103.
[166] Ibidem, 81.
[167] Ibidem, 80.

and a decreasing standard of living. Despite a brief recovery after 1557, the years 1564-65 were marked by a serious economic crisis. The extremely severe winter led to soaring corn prices, accompanied by unemployment and famine among the lowest strata of society, as well as a decrease in living standards among the middle classes. The so-called 'Miraculous Year' 1566 was marked by widespread famine.[168]

These hardships may well have contributed towards the general feeling of discontent and unrest that pervaded the Low Countries during the 1560s. Ever since Philip II had succeeded his father Charles V as king of Spain and lord of the Netherlands, in October 1555, tension had been mounting among his subjects in the Low Countries. For a number of reasons, the relationship between Philip and the different social groups in the Netherlands steadily deteriorated during the second half of the sixteenth century. The estates of the different provinces objected to the increasing demand made on their financial resources by the king in order to finance his war effort against France[169]; the high nobility felt offended by the curtailment of their power due to the growing influence of the non-noble bureaucrats in the Council of State; the lower nobility saw Philip's policy of centralization as a threat to ancient local privileges; and all the above-mentioned groups, as well as the incumbent clergy, were much opposed to the reorganization of the Netherlands Church initiated by the king and his closest adviser, Granvelle.[170]

But perhaps the most important source of discontent, and one common to very large groups from all layers of the Netherlands population, was the harsh and intransigent way in which Philip II ventured to repress heresy. We have already seen how different reformatory currents, such as Lutheranism, Calvinism, Anabaptism, and a number of spiritualist sects, made headway in the Netherlands during the earlier decades of the sixteenth century.[171] Philip, a champion of Catholic orthodoxy and one of history's staunchest comba-

[168] M. Baelde, "De Nederlanden van Spaanse erfopvolging tot beeldenstorm, 1506-1566" in I. Schöffer, H. van der Wee & J.A. Bornewasser eds., *De Lage Landen van 1500 tot 1780* (Amsterdam/Brussels 1978) 38-101: 84-85; Geoffrey Parker, *The Dutch Revolt* (revised ed., London 1985) 56-57.

[169] Jonathan I. Israel, *The Dutch Republic. Its Rise, Greatness, and Fall 1477-1806* (Oxford 1995) 133-136; Parker, *Dutch Revolt*, 38-39.

[170] Israel, *Dutch Republic*, 141-144; Baelde, "Nederlanden", 78-79; Parker, *Dutch Revolt*, 47-50.

[171] See supra, p. 145-147.

tants of heresy, realized only too well that the advance of Protestantism was one of the most serious problems with which his government was faced.[172] Not only was he opposed in principle to any compromise with what he regarded as heretics, he furthermore had to confront the complication that several members of the nobility, already discontented with Philip's attempt at centralization of the government, in addition began to reveal Protestant sympathies, or at least to tolerate Protestant activism in their domains.[173]

The Netherlandish nobility made several attempts at persuading the king to mitigate his harsh anti-heresy measures; however, Philip, in his 1565 letters sent from the woods of Segovia, flatly refused to call a halt to the Inquisition's activities. This refusal cost him dearly. Already in the early 1560s, magistrates in many parts of the Netherlands were no longer prepared to proceed against heretics who did not disturb the peace. Public opinion, too, was much opposed to the inhumane way in which Protestants were treated by the authorities: there were riots at the execution of heretics in Valenciennes in 1562 and in Antwerp in 1564, and Philip's laws were effectively sabotaged by the widespread refusal to denounce heretics to the Inquisition.[174]

Thus, the king's letters from the Segovia Woods ran counter to the sentiments of large parts of the population. The lesser nobles reacted by drawing up a document known as the Compromise of the Nobility, calling for the dismantlement of the Inquisition and the abolition of the placards against heresy, in the hope that in this way a religious peace could be established.[175] The governess-general, Margaret of Parma, under the pressure of public opinion, which was overwhelmingly in favour of the nobles' demands, reacted by way of a 'Moderation' according to which the persecution of heretics for their private beliefs was to cease—although public meetings were still prohibited.[176]

These moderate concessions led to a serious decline in royal authority, as well as to widespread support for the nobles of the Compromise, henceforth adorned with the sobriquet of 'Beggars'. Many of those outlawed for heresy returned from exile, and before long, in defiance of the Moderation, they began to organize large-scale open-

[172] Israel, *Dutch Republic,* 137.
[173] Ibidem, 141.
[174] Parker, *Dutch Revolt,* 63.
[175] Israel, *Dutch Republic,* 145.
[176] Parker, *Dutch Revolt,* 71.

air Calvinist services in the countryside. This so-called hedge-preaching was tremendously successful. The long-standing dissatisfaction with the harsh inquisitorial measures against Protestants combined with the economic malaise to unleash a wave of violent image-breaking which contemporaries dubbed the 'iconoclastic fury'. Starting on 10 August, 1566, in the small town of Steenvoorde in western Flanders, the fury spread over the Flemish countryside; on 20 August, the image-breakers reached Antwerp, and during the subsequent weeks, a number of towns in the northern Netherlands were also affected, although there, the intensity of the iconoclasm was less than below the great rivers.[177]

Often, the burgher guards of the afflicted towns refused to act in order to prevent the image-breaking, and small numbers of Protestants, usually led by a pastor or merchant or by the local lord, were able to destroy the Catholic paraphernalia unhindered—sometimes even in front of large crowds "who watched in silence and lifted not a finger".[178] Nonetheless, there was also a more violent side to the iconoclasm: after the actual fury itself had petered out, small groups of former 'stormers' continued to roam the countryside, terrorized the local population, and even managed to gain control of the town governments of 's Hertogenbosch, Tournai, and Valenciennes.[179]

Dismayed by these demonstrations of her loss of control, Margaret of Parma recruited troops to suppress such insubordinate movements. In March 1567, she defeated a rebel force near Antwerp, and after a terrible siege, she recaptured the short-lived Calvinist stronghold of Valenciennes.[180] In the light of these developments, the policy of Prince William of Orange, who tried to mediate between Catholics and Calvinists and to negotiate local religious settlements, proved untenable. His ideal of a country united by religious tolerance and freedom of conscience, governed by the local nobility, was unworkable as long as the governess-general refused to cooperate; after her successful repression of open adherence to Calvinism, Orange, along with many other prominent nobles and committed Protestants, went into exile.[181]

The defeat of the moderates who, for humanitarian and political

[177] I. Schöffer, "De opstand in de Nederlanden, 1566-1609" in Schöffer, van der Wee & Bornewasser eds., *Lage Landen 1500-1780,* 103-165: 103.

[178] Parker, *Dutch Revolt,* 78-79; Schöffer, "Opstand", 105.

[179] Schöffer, "Opstand", 107.

[180] Israel, *Dutch Republic,* 153.

[181] Ibidem, 152-153.

reasons, favoured a milder stance vis-à-vis the Protestants, became even more evident when, in the summer of 1567, Margaret of Parma was succeeded as governor-general of the Netherlands by the Duke of Alva, who was convinced that the only proper way to govern the unruly provinces was by violent constraint. However, Alva's harsh measures (the most loathed of which were the institution of the Council of Troubles, whose task it was to punish, without regard to local priviliges, all those who had been involved in the troubles, and the introduction of new taxes, notably the highly unpopular Tenth Penny) served only to aggravate the situation. Thus, William of Orange, as well as several other exiled noblemen, were judged guilty of treason by the Council of Troubles, and condemned to forfeit all their property in the Netherlands. This sentence, pronounced *in absentia*, left Orange no choice but to resort to military action, if he was ever to regain his possessions.[182]

Thus, Alva's draconian measures were in the end completely counterproductive: they provoked the levying of troops and the organization of armed resistance on the part of the Dutch exiles, and they served to alienate all those who, like Orange, favoured a moderate stance of religious tolerance, from the Brussels regime, and to drive them into the arms of the Protestants, who constituted the majority of the exiles.[183] Although Orange's four 1568 attempts at invading the Netherlands with forces from Germany, England, and France failed, four years later the Prince and his allies enjoyed an unexpected stroke of luck that was to have momentous consequences. While, in the spring of 1572, Orange was contriving a well-planned campaign against Alva's forces, according to which the Duke would be attacked simultaneously from France, by the Huguenots, from Germany, by Orange and his troops, and from the sea, by the Sea Beggars, a Beggar fleet under the command of Count Lumey de la Marck, recently expelled from England, unexpectedly captured the small town of Brill, which had recently been abandoned by its Spanish garrison.

The capture of Brill came at a portentous moment. Alva's Tenth Penny, along with an exceptionally severe winter which had killed off most of the fruit trees, had caused widespread famine and unemploy-

---

[182] Ibidem, 161; Parker, *Dutch Revolt*, 108.

[183] J.J. Woltjer, "De Vrede-makers", *Tijdschrift voor geschiedenis* 89 (1976) 299-321: 308-309.

ment, a condition aggravated by the trade war that had prevailed between England and the Hispanic world since 1568, as well as by the disruption of commerce in Holland and Antwerp due to the Sea Beggars' privateering.[184] The bleak economic situation gave rise to widespread despondency; thus, a merchant observed that at the 1571 edition of the festival held annually at Brussels, nobody smiled and nobody sang, while in Ghent, the magistrates forbade dancing in the streets and the public performance of dance music from July 1571 onwards "on account of these oppressive times".[185] And, to make matters worse, on 1 November 1570 great storms and high tides had destroyed several sea-dikes around the coasts of the Netherlands and flooded a good deal of farmland, while plague struck the Low Countries during the summer of 1571 with greater severity than at any other time during the sixteenth century.[186]

Although, politically, the capture of Brill and the subsequent defection of other Zealand and Holland towns, such as Flushing, Enkhuizen, Leiden, Haarlem, and Dordrecht, were tremendously important and constituted the first occasion on which the exiles gained a firm foothold on Netherlandish soil, on a strictly human level the events of the spring of 1572 did not make that much of a difference. Undoubtedly, there existed a good deal of pro-Beggar sentiment among the local population; yet this may well have been due to a not insignificant extent to the hated Tenth Penny and the food shortages, rather than to feelings of genuine sympathy vis-à-vis the *Gueux*.[187] Indeed, the opening of the gates to the Beggars was often considered to be the lesser of two evils: people would rather put up with the raucous seafarers than with the Spanish garrisons, who had in the recent past caused a great deal of harrassment.[188] On the whole, the attitude of the population was apathetic, rather than actively hostile towards the Brussels regime.[189]

Fear of the Spanish soldiers, underpaid and prone to mutiny, played a decisive role in determining the attitudes of town govern-

---

184 Parker, *Dutch Revolt*, 127-128.

185 Ibidem, 128.

186 Ibidem, 128-129.

187 See J.J. Woltjer, "Opstand en onafhankelijkheid" in idem, *Tussen vrijheidsstrijd en burgeroorlog. Over de Nederlandse Opstand, 1555-1580* (Amersfoort 1994) 9-63: 49 on the attitude of the population in Flushing.

188 Schöffer, "Opstand", 116.

189 Parker, *Dutch Revolt*, 134.

ments in Holland and Zealand. The magistrates, most of whom had been appointed in 1566 on the basis of their unwavering loyalty to the king, were usually not at all keen to open their gates to the Beggars, who had in recent years undertaken numerous raids in the coastal areas of the Netherlands, pillaging churches and monasteries and kidnapping prominent citizens.[190] For this reason, after the capture of Brill, the governors of other Holland and Zealand towns, confronted with a choice between the Spanish troops under the command of the Holland stadholder Bossu, and the Sea Beggars, were faced with a difficult dilemma. The problem was usually solved by the lower strata of the towns' populations, whose recent experiences with billeted Spanish troops were such that they preferred anything to a renewed Spanish presence in their towns. However, not all towns that refused to allow entry to the troops of Bossu would subsequently join the revolt against the Brussels regime: thus, the citizens of Amsterdam, while not prepared to admit any Spanish garrison, still remained loyal to the king and defended their city against the Beggars.[191] And many of the towns that did allow entry to the troops of Lumey, did so on the express condition that all secular and clerical magistrates remain in office, and that churches and monasteries be left undamaged.[192]

In view of the current situation, such reluctance to wholeheartedly join the revolt is hardly surprising. Although, obviously, in the Netherlands there existed a group of convinced Calvinists, intent on forcing a total rupture with Catholic Spain, these militants never constituted a majority of the population. There still remained a segment of the population (admittedly, rather a small one) favouring strict Catholic orthodoxy as laid down in the Tridentine reforms; from this group were recruited the intransigent Inquisitors such as Ruard Tapper and Pieter Titelmans. And, finally, by far the most numerous were the so-called 'middle groups', made up of people who were not themselves Protestants, but who to a certain extent sympathized with the ideas of the reformers and who in any case were much opposed to the harsh repression of heresy as practised by the Brussels regime. Although these moderates were just as appalled by

---

190 Woltjer, "Opstand", 47; Schöffer, "Opstand", 116.

191 Woltjer, "Opstand", 50.

192 Ibidem, 53.

Alva's measures as were the Calvinists, they were also suspicious of the rebels' radicalism and their intolerance of dissident opinions.[193]

These were the people who, in 1572, were faced with a thorny dilemma: on the one hand, it was becoming increasingly clear that the revolt in Holland and Zealand was controlled by Lumey and the Beggars, rather than by the moderate William of Orange, and that, consequently, it was assuming a rigidly Calvinist character; on the other hand, however, the government of Alva had brought nothing but hardship. Although, in most cases, the scales tipped to the side of the rebels, this choice was often a wholly negative one: within the rebel provinces of Holland and Zealand, there remained large groups of Catholics whose rights were being flagrantly violated, while some important towns, notably Middelburg and Amsterdam, refused to join the revolt at all. The conflict was fast assuming the characteristics of a civil war.[194]

Meanwhile, the southern and eastern Netherlands, too, were in turmoil. In May 1572, Count Louis of Nassau, Orange's brother, had been able to enter the capital of Hainault, Mons, without meeting any resistance: the numerous Protestants in the town had opened the gates to him. Another force of Huguenots and Netherlands exiles had been able to capture Valenciennes; Orange's German forces had taken the important eastern town of Zutphen with scarcely a struggle; and within six weeks, almost all of Overijssel and Drenthe and much of Friesland was in the hands of Orange's supporters.[195] Orange himself then invaded Brabant with the intention of relieving Mons, which was being besieged by Alva. His campaign started out extremely successfully: Weert, Diest, Tienen, Louvain, Nivelle and Dendermonde opened their gates to him, and Malines and Oudenaarde even revolted spontaneously and asked the Prince for a garrison.[196] However, the St Bartholomew Massacre ruined Orange's plans, preventing the Huguenot reinforcements, on whose assistance the Prince's campaign had hinged, to arrive, and forcing him to retreat to Holland.

Alva's revenge was terrible. He reconquered Mons and Louvain, and allowed his troops to sack and plunder Malines for three days on

---

[193] Ibidem, 21; Schöffer, "Opstand", 118.
[194] Woltjer, "Opstand", 53-57; Woltjer, "Vrede-makers", 309.
[195] Parker, *Dutch Revolt*, 135-136.
[196] Ibidem, 140.

end, after which atrocity the other Brabant towns hastened to make peace with him; furthermore, he sent his son, Don Fadrique, to recapture the defected towns of Holland and Zealand. Although this campaign was not much of a success (only Haarlem was forced to its knees) Fadrique did sack the unfortunate towns of Zutphen and Naarden in a horrible way.

Hostilities continued over the next few years, with Alva's successor Requesens recapturing several Holland and Zealand towns for the king, and severely chastising them. Nor was this 'authorized' pillaging the only way in which the local population suffered the cruelty of the Spanish soldiers. The years 1573, 1574, 1575 and 1576 were all marked by mutinies, induced above all by overdue wages. The mutineers would terrorize the local inhabitants in a terrible way, and because they were usually well-organized, the authorities could do nothing to stop them.[197] After Philip II declared his bankruptcy in 1575, it became obvious to everyone that the recent wave of mutinies was not yet at an end; the threat was so imminent that some southern towns even began to arm their citizens in order to be able to ward off the mutineers.[198] These precautions could not prevent that a large band of Spanish troops, after capturing the Zealand town of Zierikzee, resolved to mutiny and streamed into Brabant. After sacking Aalst, only a few miles from Brussels, and routing an army raised by the States of Brabant near Tienen, they moved on to Antwerp, which town they reached on 4 November and subjected to the most appalling atrocities.

The 'Spanish Fury' in Antwerp only served to confirm the conviction, which had already taken root in Brussels (where, in the power vacuum following the death of governor-general Requesens in March, the States of Brabant had taken matters into their own hands), that it was essential to call an assembly of the States General in order to negotiate with Orange and the rebel provinces. Although the way in which this meeting of the States General had been convened was revolutionary, since it had happened behind the king's back, the purposes of most of its members were not. They hoped

[197] On mutinies in the Spanish army of Flanders, see Geoffrey Parker, "Mutiny and Discontent in the Spanish Army of Flanders, 1572-1607" in idem, *Spain and the Netherlands 1559-1659. Ten Studies* (revised ed., Glasgow 1990) 106-121.

[198] Schöffer, "Opstand", 128.

that, in reaching an agreement with Holland and Zealand, it would prove possible to establish reconciliation and religious peace throughout the Netherlands territory, with Catholicism remaining the predominant religion and the Protestants acquiring certain rights.[199]

Only four days after the Spanish Fury, on 8 November 1576, the States General reached an agreement which was to enter history as the 'Pacification of Ghent'. In it, it was agreed that, henceforth, the States General would fix the times and frequency of their own meetings; that government should consist only of Netherlandish magistrates; that all Spanish troops should leave the Netherlands; and that, for the time being, the religious status quo would be maintained (so that the public practice of Protestantism would be allowed in Holland and Zealand, while elsewhere, only private Protestant practice and the possession of Protestant books would be tolerated). Henceforth, the States General became the central organ of government, negotiating with foreign powers, legislating, concluding treaties, and raising their own army.[200]

The Pacification of Ghent constituted the moderates' final attempt at reaching an agreement with the rebel provinces of Holland and Zealand. Effectively, all the provinces of the Netherlands were now united in their opposition to their sovereign ruler's policies, and in their striving for moderation in religious matters and for the re-establishment of their ancient privileges of self-government. For a short while after the signing of the Pacification, it seemed as if the struggle against Philip II would turn into a general revolt in which all the provinces, be they Catholic or Protestant, would cooperate concertedly.[201] The new governor-general sent by Philip II, his half-brother Don Juan of Austria, had no choice but to accept the Pacification as it stood. On 12 February 1577, he signed the Perpetual Edict to this effect; on 28 April, the Spanish troops left the Netherlands for Italy; and although Orange had from the outset been distrustful of the new governor, and had tried his best to discredit him with the States General[202], the aimiable way in which Don Juan behaved vis-à-vis

[199] Woltjer, "Vrede-makers", 312.
[200] Parker, *Dutch Revolt*, 179.
[201] Woltjer, "Opstand", 60.
[202] See K.W. Swart, *Willem van Oranje en de Nederlandse Opstand 1572-1584* (edited by R.P. Fagel, M.E.H.N. Mout & H.F.K. van Nierop, The Hague 1994) 120 ff.

the States, offering to plead with the king for financial support, served to remove much of the suspicion with which he had at first been regarded.[203]

However, Don Juan soon betrayed himself unworthy of the trust which the States General had put in him. Convinced that the "affairs of these provinces cannot be remedied in any other way than by means of fire and blood"[204], he occupied the citadel of Namur, with the intention of organizing a new campaign against the States General, and summoned back the Spanish troops. Orange's worst suspicions had now been confirmed; the States General, shocked at Don Juan's action, asked the Prince to come to Brussels, where he was given a hero's welcome. A resumption of hostilities with Spain was now inevitable, especially in the light of the radicalization which was now becoming manifest in many areas: in the north, Utrecht and Amsterdam were forced to defect to the side of the rebels, and Calvinists took the opportunity of taking control of town governments, while in the south, very radical Calvinist regimes seized power in the towns of Ghent, Bruges, Ypres, and Arras.

The Pacification of Ghent had failed. The idea on which it had been based (that it ought to be possible to solve the religious and political problems of the Netherlands by means of a peaceful and tolerant settlement), though supported by a large group of moderates, had proved untenable due to the intransigence of small but militant groups of radicals. Don Juan, Philip II, and a royalist minority among the Netherlanders refused to accept the religious settlement of the Pacification, as well as the curtailment of royal power it entailed; and radical Calvinists, though relatively few in number, demanded much more than the half-hearted 'religious status quo' arrangement the Pacification provided them with.[205] The stance of the 'middle groups' had demonstrated its utopian character: whereas, in November 1576, a serious effort had been undertaken to pacify the Netherlands without defecting from the king's authority, just fifteen

---

[203] Pieter Bor, *Oorsprongk, begin, ende vervolgh der Nederlandsche Oorlogen, beroerten, en burgerlyke oneenigheden, beginnende met d'opdracht der selve Landen, gedaen by keyser KAREL den Vijfden, aen sijnen Soon Konink PHILIPPUS van Spanjen, en eindigende met het einde van 't jaer MDC, Waerachtelijk en onpartijdelijk beschreven* (1595, reprint Amsterdam 1679) Book X, 824-825.

[204] Ibidem, 827: "... dat de saken van dese Landen niet en waren te remedieren dan door vier en bloed ...".

[205] Woltjer, "Vrede-makers", 320-321.

months later, war had resumed, and the troops of Don Juan inflicted a crushing defeat on those of the States General at Gembloux, on 31 January 1578.

When the comet of 1577 appeared, the Netherlands were going through an extremely difficult phase. They were being faced with the worst imaginable hardships: recurrent bouts of plague, scarcity and famine, inundations and an almost continual state of war were ravaging the country. The political situation was highly unstable throughout the decades prior to 1577, and it remained unclear what exactly was going to be the outcome of the tensions which opposed nobles to bureaucrats, the king and his governors to the States General, Catholics to Protestants, and loyalists to rebels. To a large extent, the troubles assumed the character of a civil war: thus, most of the Brussels bureaucrats were themselves Netherlanders; most of the magistrates of the Holland and Zealand towns that opened their gates to the Beggars in 1572 were loyal to the Spanish regime and were more or less forced to defection by the lower ranks of the population; and some towns, such as Amsterdam and Middelburg, refused to defect at all. Although there did exist (relatively small) groups of convinced loyalists and radical Calvinists, the majority of the population of the Low Countries remained undecided as to which was the best course to steer: they were as averse to the intolerant stance of the radical Calvinists as they were to the inhumane regime of Alva, and gladly welcomed the Pacification of Ghent as a peaceful and moderate solution to the troubles.

When, after Don Juan's coup, it became clear that the Pacification had failed, this must have come as an enormous deception to the majority of Netherlanders, who longed above all for a return to peace and quiet. The contemporary historian, Pieter Bor, writes that, when Juan's capture of the Namur citadel became publicly known, "this brought about a great clamour, despondency, and reversal in all people's hearts; everybody utterly abandoned the hopes which they had placed but a short while before in his [Don Juan's] person, fearing that a much fiercer, more bitter, and crueler war would break out than ever before."[206] No wonder that, when at this ominous

---

206 Bor, *Oorsprongk*, Book X, 832: "Als dese subijte inneminge van het voorsz Casteel van Namen . . . in 't openbaar en onder 't volck quamen, so maekte 't selve een seer groten roep, verslagenheid en alteratie in de herten van eenen yegelijken, en

moment in history a comet (with all its connotations as a harbinger of war, deaths of rulers, changes of government, and civil strife) appeared, several Netherlandish authors ventured to interpret its significance.

b. *The Revival of the Teratological Tradition: Gemma and Heurnius*

When discussing the views of Cornelius Gemma on the cosmological characteristics of the comet of 1577, we have already seen that, read in its entirety, the tract which Gemma devoted to the hairy star conveys the impression of a moral and ethical exhortation, rather than an astronomical or astrological treatise.[207] In the light of the political developments described in the previous section, this is quite understandable. We know that Gemma frequented the circle of humanists centred around the printing press of Christopher Plantin at Antwerp.[208] Although, as we have seen[209], it cannot be established whether, like Plantin and Ortelius, Gemma was a member of the Family of Love, his contact with men who were may well have familiarized him with the tenets of Familism, prominent among which was the idea that the end of the world was at hand. This expectancy gave rise to a heightened vigilance as to the possible signs which might precede the final catastrophe, chief among which was the archetypal forerunner of the Last Days, the comet.

Beyond doubt, most of the Antwerp humanists, whether Familists or not, fostered a *politique* attitude towards the contemporary upheavals.[210] The importance they attached to charity and religious peace was at odds with the harsh policy of intolerance of dissidents which

---

lieten geheel vallen de hope die sy so onlangs te voren op syn persoon gehad hadden, vresende datter een veel felder, bitterder en wreder oorlog soude ontstaen, dan daer noch ooit geweest was...".

[207] See supra, p. 170.

[208] See supra, p. 165.

[209] See supra, p. 165.

[210] See A. Hamilton, *The Family of Love* (Cambridge 1981) ch. IV, 65-74, for the *politique* attitudes of Plantin and Ortelius; J. Kluyskens, "Justus Lipsius' levenskeuze: het irenisme", *Bijdragen en mededelingen betreffende de geschiedenis der Nederlanden* 88 (1) (1973) 19-37, for Lipsius's lifelong attempts at reuniting divided Christendom; see also M.E.H.N. Mout, "Humanists and the Horrors of War: Neostoicism and the Revolt of the Netherlands" in H. de Schepper & P.J.A.N. Rietbergen eds., *España y Holanda. Ponencias de los coloquios Hispano-Holandeses de historiadores 1984-1988* (Madrid/Nijmegen 1993) 109-118; and J.A. van Dorsten, *The Radical Arts. First Decade of an Elizabethan Renaissance* (Leiden 1970) 50-61.

had in recent years been practised by the Brussels regime. The members of Plantin's circle belonged to the group of moderates described above, who strove to overcome religious differences and to oppose the growing polarization.[211] They hoped that it would prove possible to reconcile the divergent views in matters of religion, and to restore peace, by means of studying God's creation and uncovering the divine plan that underlay it.[212] According to J.A. van Dorsten, the activities in the field of cosmography displayed by such members of the Plantin circle as Abraham Ortelius and his cooperators, were fostered by a sense of the inability of theology to unite mankind, and were aimed at replacing the futile efforts of the different religious factions with the study of nature, an activity not related to any church. The contemplation of natural phenomena was meant to bring to an end the ceaseless arguments of the day, and to restore peace.[213]

For this reason, the treason of Don Juan and the failure of the Pacification must have come as an especially bitter disappointment to the circle of humanists around Christopher Plantin. They abhorred fanaticism, religious intolerance, and war in all its manifestations, stressing the importance of attaining both inner and outer peace.[214] In the early 1570s, Benito Arias Montano had pleaded in favour of the abolition of the Council of Troubles and the appointment of moderates to important offices, and an Antwerp Familist had lent Requesens a large sum of money to finance his conciliatory policy.[215] However, with the failure of the Pacification, the ideal of peace and universal harmony, which for a fleeting moment in late 1576 had seemed to be round the corner, must have appeared more unattainable than ever. Many disillusioned Familists fled the country and

---

[211] N. Mout, "The Family of Love (Huis der Liefde) and the Dutch Revolt" in A. Duke & C.A. Tamse eds., *Britain and the Netherlands* VII: *Church and State since the Reformation* (The Hague 1981) 76-93: 92.

[212] J.A. van Dorsten, "*Temporis filia veritas*: wetenschap en religievrede", *Tijdschrift voor geschiedenis* 89 (1976) 413-419: 415. An English version of this article can be found in J.A. van Dorsten, *The Anglo-Dutch Renaissance. Seven Essays* (J. van den Berg & Alastair Hamilton eds., Leiden/ New York/ Copenhagen/ Cologne 1988) 38-45. See esp. pp. 42-43.

[213] Van Dorsten, *Radical Arts*, 50-61, esp. 56-57.

[214] Mout, "Humanists and the Horrors of War", 113-114; Hamilton, *Family of Love*, chapter IV ("The Antwerp Humanists"), passim.

[215] Mout, "Humanists and the Horrors of War", 115-116.

sought refuge in Germany, England, or the northern Netherlands.[216]

The chaotic and troubled situation in their fatherland greatly vexed the members of the Plantin circle. Justus Lipsius was horrified at the never-ending violence with which the Netherlands were afflicted. He was constantly forced to flee in order to avoid becoming entangled in the hostilities, and, although he did not have much sympathy for the cause of the rebels, in the early months of 1578, after the battle of Gembloux, he nonetheless accepted an appointment at the university of Leiden: to him, what mattered most was peace and quiet.[217] The appalling war ravaging the Low Countries also inspired the great humanist to compose his most popular work: *De Constantia in publicis malis*, first published in 1584.

Starting from the almost unbearable conditions in his home country, Lipsius advocated a Stoic attitude of firmness and patience vis-à-vis the manifold hardships of the age. Chief among the 'public ills' referred to in the title of his work, Lipsius mentioned war, pestilence, famine, tyranny, and massacres, precisely those calamities of old associated with comets. However, he maintained that all these horrors, though at first sight unbearable, nonetheless served a useful purpose. Since all events proceeded from God, they were necessarily directed towards the ultimate good. Even terrible events such as the recent afflictions of the Netherlands served a good purpose: they could strengthen the brave and punish evil-doers. And even the treacherous governor, Don Juan, served as the agent of a divine purpose.[218] Nor were the contemporary troubles particularly grave, or particularly new, Lipsius pointed out: ancient times, too, were full of wars, and especially the tyranny of the Roman emperors provided a striking parallel with the widespread intolerance rife in the Netherlands of Lipsius' day.[219]

This was the atmosphere in which Gemma's tract on the comet of 1577 came into existence. During the years prior to 1577, many

[216] Ibidem, 116.

[217] M.E.H.N. Mout, "In het schip: Justus Lipsius en de Nederlandse Opstand tot 1591" in S. Groenveld, M.E.H.N. Mout & I. Schöffer eds., *Bestuurders en geleerden. Opstellen over onderwerpen uit de Nederlandse geschiedenis van de zestiende, zeventiende en achttiende eeuw, aangeboden aan Prof. Dr. J.J. Woltjer bij zijn afscheid als hoogleraar van de Rijksuniversiteit te Leiden* (Amsterdam 1985) 55-64: 57.

[218] See Gerhard Oestreich, *Neostoicism and the early modern state* (Cambridge 1982) 24-25.

[219] Ibidem, 26-27; as early as 1572, Lipsius had called Alva a present-day Tiberius: see Mout, "In het schip", 57.

prominent members of the Plantin circle which he frequented had belonged to the large group of moderates who had striven to appease the warring factions and to overcome religious differences without bloodshed. They combined Neostoic ideas such as the importance of patience and virtue, and the sense that even the worst calamities served a useful end foreseen by God, with the eschatological expectations central to the Familist creed, and with the determination to uncover the divine plan that underlay the cosmos in order to appease the warring Netherlands, and to bring to an end the unbearable conditions of suspicion and treason that accompanied the civil strife in the Low Countries.

When the comet appeared, Gemma's home town of Louvain had just lived through an especially turbulent phase. In the summer and early autumn of 1572, it had for a short while been occupied by the troops of William of Orange; in the same year, Alva had recaptured the town; but after the Pacification and the treason of Don Juan, it had, along with many other Brabant towns, joined the revolt. However, after the battle of Gembloux, in February 1578, the old university town was conquered once again by Parma and Don Juan. It is impossible to ascertain exactly when Gemma's tract was written; its *terminus a quo* is Gemma's last sighting of the comet, on 18 January 1578. Thus, our author must have finished work on the treatise just before or just after the battle of Gembloux and the subsequent recapturing of Louvain.

The chaotic events of the recent past manifestly had their repercussions on the way in which Gemma regarded the appearance of the comet. He commenced his tract by an allegorical poem ("Idyll on Diseased Europe") in which he laid down the most important themes he meant to address. In it, Jupiter, watching the earth from on high, observed that the whole world was suffering from a new and very grave illness. The seeds of the disease, which he had hoped to have tempered by administering some medicine during the 'patient's' previous crisis, had on the contrary spread throughout the body. The heart, source of vitality, which under normal conditions brought health and salubriousness to the entire body, had now become the cause of ruin; not knowing its own affliction, it dragged down with it all the other parts, arms, legs, feet and all. Therefore, Jupiter decided, the earth must be swept clean from the highest mountain-top; a messenger should be sent from the gates of Capricorn towards the south, carrying a ghastly, fiery mop of hair, in order to serve as a physician for sickly Europe. The best cure for her would be for the kings to mitigate their way of governing, and for the people to submit themselves to their own proper rulers. Then, the fury of war would cease, and steadfast charity would get a chance to patch together everything that had been torn apart by hatred.[220]

From this brief allegory can be gathered all the major themes of Gemma's tract. Underlying the whole treatise was the metaphor of the body politic, which, wielded by Gemma, is certainly to be taken in its literal sense. In the sixteenth century, it was not uncommon for authors, especially those with medical backgrounds, to hold that "illnesses affect states as they do human bodies, but they can be cured by wise statesmen in the same way that skilled physicians cure bodily ills."[221] As we shall see, Gemma extended this way of reasoning to include the situation of the entire cosmos, so that, to him, there existed a close correspondence between 'illness' or 'health' in the heavens and 'illness' or 'health' on earth, be it on the level of state government or on the level of individual human bodies. Hence Jupiter's decision to send a 'messenger with fiery hair', a comet, in order to exterminate the evils rife in the states of Europe, and to renew the reign of charity, which in its turn would bring back health to the body politic. Clearly, Gemma believed that there existed a godhead who looked down with benevolent intentions on humankind; that this godhead made use of comets as signs in order to persuade people to change their ways; and that the proper cures for the current afflictions were charity, benign royal government, and popular subordination. The rest of his tract can be read as a detailed elaboration of these central ideas.

The first chapter of the treatise contained a description of "The characteristics of the present age, and the common investigation of the whole coming alteration by means of those things which now appear as critical signs of the suffering world".[222] Gemma com-

[220] Cornelius Gemma, *De prodigiosa specie naturaque cometae, qui nobis effulsit altior lunae sedibus, insolita prorsus figura, ac magnitudine, anno 1577. plus septimanis 10. Apodeixis tum Physica tum Mathematica* (Antwerp 1578) 2.

[221] Peter Burke, "Tacitism, scepticism, and reason of state" in J.H. Burns & Mark Goldie eds., *The Cambridge History of Political Thought 1450-1700* (Cambridge 1991) 479-498: 486; the quotation is from Filippo Cavriana, *Discorsi sopra i primi cinque libri di Cornelio Tacito* (Florence 1597) 12.

[222] Gemma, *De prodigiosa specie*, 5: "De praesentis aetatis Characterismo: deque totius venturae vicissitudinis communissima notione; per ea quae nunc apparent ceu mundi laborantis critica signa."

menced by explaining what he believed to be the theoretical underpinnings of his notion, that impending events could be deduced from natural signs. Endorsing Plato's idea, that the world in its entirety was a living animal with a soul, a mind, and a body, Gemma stated that, since the different components of an animal were in close correspondence with each other, so too were the different parts of the cosmic animal. Because of this intimate connection between all parts of the world, the upheavals, actions, and passions which took place in the constitutions of individual animals, happened also in the vast reaches of the exterior world, in the heavens and on the earth. The Maker of the universe had arranged everything in such a way, that imminent events were so to speak prefigured in all the parts of the world by way of an example. As a logical consequence of this theory of cosmic correspondence, Gemma added that the causes and signs of disease and health, and the methods for curing these illnesses, were also identical for every single part of the universe.[223]

Therefore, Gemma went on, a state could be judged to be healthy whenever it was constituted in a harmonic way; in the same way, that is, as the bodily structure of man, and the design of the universe as a whole. Similarly, a state's illness could be deduced from an imbalance between its different elements; in other words, whenever the relationships between the different groups within a given society deviated from the harmonious coherence exemplified in nature, that society should be diagnosed as suffering from some disease.[224] Clearly, our author stated, at present the Christian commonwealth was afflicted by a most serious and dangerous illness: an insatiable

---

[223] Ibidem, 5-6: "Mundi fabricam velut animal quoddam ex anima sua, suoque spiritu & membris corporeis apte compositum, ex divina Platonis sententia nemo ignorat, qui gentilitiae Theologiae fontes vel a limine salutarit. . . Ex quo demum & illud effici ratione conclusimus [in libro de Naturae divinis characterismis] quaecunque in animalibus vel circa partium distributiones atque consensus, vel actionis passionisque varietatem solent accidere, eadem prorsus antiquius paulo in orbis exterioris ambitu vasto, in caelo, terraque; & partibus singulis universi ab ipso opifice primo per exemplaris modum esse depicta: Morbi & sanitatis vicissitudines, causas & signa, curandique methodum, & scopos endeicticos in utroque genere passim vel usque ad scrupula consentire."

[224] Ibidem, 6: "Quod si igitur regni alicuius, vel civitatis & Reipubl. harmonica constitutio, tum ad humanae compagis normam, tum ipsius universi oeconomiam aptissime compatetur: eademque dum sibi consonat, sanitatis ideam optimam repraesentent: Morbi vero, id est vel ametriae vel dyscrasiae alicuius, ubi elementorum divulsa societas, figurarumve contextus in manifestum divortium permutatur . . .".

lust for power had struck society and had brought in its wake numerous other pestilences, which had found their origin in the heart and the other principal organs of the body politic, but had quickly spread all over, "through veins, nerves, and arteries, all the way down to every single muscle and bone".[225]

The present disease of society could even be compared to a hectic fever, Gemma maintained; a fever, that is, that ruined and corrupted the very roots of the body, with the patient all the while failing to realize he was being devoured by fever at all.[226] This simile had been suggested to Gemma by the hypocrisy and dissimulation he saw all around: through all layers of society, people put on just and honest faces, and manifested piety and loyalty to Christ, king, and fatherland in their habits, on their faces, and in their gestures, yet if it were possible to look inside the deepest dungeons of their hearts, nothing but a cauldron filled to the brim with fury, hatred, cruelty, ambition, avarice, cowardice, gluttony, and lust would be found.[227]

Despite this state of gruesome imbalance and disease within society, however, Gemma believed that hope of a better future should not be abandoned. After asking the rhetorical question of whether there was no hope left, and whether there were no divinities that looked down on the pious with benevolence, our author affirmed that "Certainly, the eye of the Lord, who does not spurn the prayers of the just, is alive, and looks down from the heavens on the children of man, to see if there is anyone who understands or who seeks God."[228] God, then, did not stand idly by; he would not allow those of his people who earnestly strove for the good to be dragged down

[225] Ibidem: "...patere cuivis arbitror, nunc illud esse vel maxime quod acutissimo pariter morbo, ac summe periculoso laboret passim respub. Christiana; posteaquam dominandi cupiditas insatiabilis pestes innumeras trahens, instar generalis Epidimiae, non aliter a corde caeterisque principibus locis per venas eiusdem, nervos, arterias, usque ad lacertos & ossa singula pervagata est...".

[226] Ibidem, 7.

[227] Ibidem: "In eo conveniunt horum pars maxima, quod aequi atque honesti titulum ore tenus, ac pertinacissime tueantur: imo pietatem ipsam in habitu, fronte, gestuque circumferant, nec minus quam sese fideles Christo, regi, patriae hac ratione contendant, cum si quis reclusis pectorum cancellis introspiciat forte, nil nisi inversam quandam Sileni faciem deprehendat foris magnificam & quantivis pretii, larvatae pietatis, & praetextatae religionis insignia: intus ollam furoris, odii, crudelitatis, ambitionis & avaritiae, omnis ignaviae, gulae, & sceleratae libidinis ita succensam ...".

[228] Ibidem, 8: "Vivit profecto, vivit oculus Domini, qui non spernit iustorum preces: verum (ut Psalmographus inquit) de caelo prospicit super filios hominum, ut videat si sit intelligens aut requirens Deum."

by the majority of hypocrites who had brought the state to the edge of ruin. Gemma firmly believed that God would shortly act in order to cure the world of its disease; more specifically, he expected some sort of a cosmic cataclysm, by means of which the corrupt and incurable parts of society would be separated from the healthier organs, to prevent the pestilential influence from contaminating the parts that were as yet unaffected.[229]

For such a decontamination to occur, society would necessarily first have to go through a 'crisis' (a medical term from the Hippocratic corpus by which was denoted a phase in the illness when marked changes took place in the patient's symptoms), and according to the theory of cosmic correspondence to which Gemma subscribed, such a crisis in society was invariably preceded by 'crises', or significant changes, in other parts of the cosmos. Sure enough, our author went on to point out that the expected "alteration and supreme vicissitude of all things" was to be preceded by "critical signs", which matched the impending cataclysm in dignity: "voices of men and gods, omens, monsters, prodigies, of a kind which have certainly not been seen since the birth of Christ, or even since the creation of the world."[230] This last remark demonstrates that the alteration Gemma so eagerly expected could be nothing other than the second coming of Christ; surely, no other event would merit to be announced by signs equal in dignity to the star of Bethlehem.

By far the most important of these recent critical signs, Gemma went on, had been the new star which had shone forth in 1572 in the constellation of Cassiopeia. When discussing Gemma's Dutch-language pamphlet dealing with the nova, we have already seen how he believed it to be a divine prodigy, supernaturally created by God in order to announce to mankind a great change, such as the Lord's

---

[229] Ibidem, 8-9: "Nam quia iuxta Medicinae communissimum theorema, Quicquid corruptum penitus & immedicabile est, sola ablatione curandum, ne serpens latius pestiferum malum ad membra saniora permanet, tandemque in habitum universi corporis defigatur... Proinde & in illo furoris calice vinum merum diluitur mixto, insanis regibus, atque imprudenti populo largius hauriendum: donec humores per inflammationem noxii concoquantur, & ad expulsionem criticam stimulante natura, tandem & partes penitus desperatae, a sanis & integris aliquando discedant, vel etiam ab illis, ne sint molestae in posterum, amputentur."

[230] Ibidem, 9: "Precedunt hanc sane mutationem rerumque vicissitudinem summam, quae & pro dignitate respondent critica signa, voces hominum atque divorum, omina, monstra, prodigia, cuiusmodi sane non vel a Christo nato, non etiam ab orbe condito, conspecta a quopiam celebrantur."

second coming, or the accession of peace, grace, and forgiveness.[231] In his tract on the 1577 comet, our author once again repeated that, in his opinion, those who believed the new star to have arisen from natural causes were quite unskilled at the arts of mathematics and geometry.[232] Rather, the nova was to be accorded the most majestic status; all the other signs which had appeared of late, including the present comet, were to be interpreted in the light of that most wonderful prodigy, as if they were subordinated to the new star in the same way that military leaders were subordinated to their commander-in-chief.[233]

However, the nova was by no means the only divinely sent portent to have appeared during the past few years of hardship in the Netherlands, Gemma hastened to add. Prior to the comet of 1577, the most significance-laden of these had been two 'chasms'[234], both of which had been seen at Louvain in the year 1575: one in February, the other on 28 September. The first of these had portrayed the recent events and changes of fortune in the Netherlands like a mirror; Gemma, who had himself watched the spectacle, assured his readers that "the parts of the hours [during which the different sights of the chasm followed each other] corresponded exactly to the time spans of years and months [during which the real-life events in the Netherlands had taken place], following an astronomical ratio."[235] The vision had consisted of two curved arches, one of which, with its curvature towards the north, soon split into a number of different

231 See supra, p. 153-154.

232 Gemma, *De prodigiosa specie*, 9.

233 Ibidem, 9-10 & 18: "... placuit nobis & huius [cometae] venturo saeculo memorabilem theoriam illi subiungere, quam edidimus nuper de sidere novo, cuius ea fere praestantia est, ut quicquid portentorum est uspiam his locis atque temporibus efflorescens, quasi ad unius ambitum & capitale iudicium referantur, sub illo militet, non aliter quam videmus diversos duces atque praefectos in unam expeditionem sub uno imperatore, aut belli totius capite recenseri."

234 The word 'chasma' ('chasm' or 'abyss') was used by Latin-writing authors to denote scenes like battles, lances, crucifixes, or other sights seen in the clouds, often by a crowd of people, and were thought to have a portentous meaning. These phenomena were probably caused by the aurora borealis. For a discussion of several 'chasmata' seen in the Netherlands during the late Middle Ages and the sixteenth century, see G.A. Evers, "Voorteekenen in de lucht", *Historia* 12 (1947) 249-254.

235 Gemma, *De prodigiosa specie*, 10: "...quorum prius eodem ordine figurarum quo sese exhibuit, tum oculis nostris omnem quoque consecutionis faciem deinceps, fortunarumque vicissitudinem in status Belgicos, tanquam e specula demonstravit: adeo ut & horarum partes annorum mensiumque spatiis exactissime convenirent, iuxta astronomicam rationem."

arches which shone forth from a dark pool with very bright light, while the other, multicoloured like the rainbow, was more inclined towards the South and moved from east to west past the belt of Orion. The extremities of both arches were joined together like a pair of horns above the spot where the moon, which at the time was new, stood. Then, the fiery colour of the moon started to glow ever more fiercely and turned into genuine purple, so that it seemed to be coloured, not by the nearby ocean, but by the bloodshed of an enormous massacre.[236]

These visions, Gemma was convinced, constituted the first 'monograms' of two different treaties, the results of which so far were already known to everybody.[237] By this remark, our author probably alluded to the Pacification of Ghent and the Perpetual Edict with Don Juan, both of which had been concluded within the two years following the chasm. After all, Gemma's remark that the results of both treaties were known to everybody sounds rather cynical, which is quite understandable if we take into consideration the fact that, by the time Gemma's treatise was composed in early 1578, the Edict had already been broken by Don Juan, the Pacification had failed, and war was about to break out once again. Hence, in all probability, the prominent role accorded to the purple, bloody colour of the moon in Gemma's description of the chasm.

Moreover, the sights which had followed the two arches were no less martial: Gemma and the other spectators had seen the figures of military expeditions, armies of cavalry and infantry rushing up and down, as well as a vivid image of a great conflict and a battle fought with lances and spears. When the lances of the army situated in the northwestern region of the sky, in the direction of Antwerp, suddenly broke and turned bloody, thus clearly showing on which side victory was to be expected, the hairs of all the onlookers were made to stand on end. Gemma self-professedly shuddered at having to describe how, from the thick of the above-mentioned massacre, an ominous vapour resembling spilled blood rose up and flew off in a westerly

---

[236] Ibidem, 10-11: "Amborum extrema in occasus cardine, ac supra Lunam tum temporis novam ad instar cornuum apposite iungebantur, tum Lunae flammeus color magis magisque in puram purpuram excandescens, ut non tam vicini Oceani fluctu, quam stragis ingentis tingi effluvio videretur."

[237] Ibidem, 11: "Erant haec nobis diversissimorum foederum monogrammata prima, quorum successus quorsum evaserit, hactenus & lippis & tonsoribus in confesso est."

direction, where, near the western pole, it spread out across the breadth of the sky and turned from red to purple, blinding the spectators' eyesight, and agitating their minds at the thought of this ominous prediction of future mourning.[238]

All the stars in the sky, except the Pleiades, were eclipsed by this enormous bloody cloud; however, quite suddenly, a number of very bright and multicoloured clouds from the northern chasm appeared and tore apart the previous vision of spilled blood. Sadly, this more felicitous sign was rather short-lived, since shortly afterwards, new spectacles of lances and flames made their appearance, and the whole northern sky seemed to be on fire. Finally, the aspect of the entire sky changed into a confused vision like a boiling cauldron or a dice-box, turning from darkness to light and vice versa, and wildly spinning around.[239] Rather than explicitly elucidating what he believed to be the significance of these images, Gemma added simply that "How precisely these [visions] accord so far, becomes obvious from the fortunes which are reigning at the moment."[240]

The second chasm, seen on the eve of Michaelmas, 1575, was somewhat less horrendous but much more confused. It consisted of a number of bright arches, from which protruded lances, citadels, and battle arrays. Several battles ensued, after which a number of rays joined together in the image of a tent, below which could be seen a chaotic rushing about and skirmishing. Next, this image was torn apart, and sulphurous rays were hurled in a northwesterly direction, towards the coastal region. Then, as in the previous chasm, the sky was filled with armed combat and with a fiery colour, and a burning mountain was seen, moving very slowly from east to west, where it came to a halt. It had even been reported that, after midnight, a horrible massacre had been seen, the clash of arms had been heard, and drops of blood had fallen from the sky. However, Gemma added that he had not been awake to verify the latter assertions, though they were affirmed by many eye-witnesses.[241]

About the meaning of this second chasm, Gemma was even more

---

238 Ibidem, 11-12.

239 Ibidem, 12-13.

240 Ibidem, 13: "Et haec quidem hactenus quam probe respondeant, etiam quae nunc tenet fortuna eloquitur."

241 Ibidem, 14: "Infanda sunt, ac nimis prodigiosa, quae post noctis medium spectata in aethere memorantur versus Eurontum. quae quoniam ipse non viderim, tum fatigatior paulo, tum deceptus ex vacuo temporis intervallo, nolim hic latius recensere. quanquam tunc editam apertissime stragis ideam, conflictus atrocissimos etiam,

vague than he had been about the first, confessing that he "did not at all know to what it appertained."[242] He did admit that, in all probability, most of it could be ascribed to physical causes; however, he could not believe that such precise images appeared to him and his fellow countrymen without good reason. On the other hand, it could not be maintained either that any part of the apparition corresponded to present-day situations, despite the fact that certain incompetent soothsayers had misused it, saying it represented the expedition currently going on in Zealand.[243] After all, Gemma stated, God's warnings invariably came well ahead of his punishments, so that there was always a considerable time span between the sign sent and the events signified, the amplitude of which corresponded more or less to the magnitude of the occurrences foretold.[244]

According to Gemma, the surest way of predicting future events was to study the divine signs in nature, especially when these were clearly ordered, and when physical occurrences manifestly displayed their symbolic character.[245] Nevertheless, he strongly believed that a Christian philosopher ought not to attempt to predict certain precisely specified events, mentioning the places and times at which they were to occur. The foreknowledge of such particularities should rather be left to divine providence; human beings should content themselves with the prognostication of generalities, and in all other things submit their intellects to the will of the godhead.[246] Therefore, Gemma much appreciated the stance of the holy men who had writ-

---

armorum sonitus, & caelo labentis sanguinis guttas, uno ore multorum consentiens observatio confirmarit."

[242] Ibidem, 15: "De toto hoc Chasmate quorsum pertineat, prorsus ignoro: & horum plurima referri ad physicas causas non infitior.".

[243] Ibidem: "Nec tamen temere tantam ordinis oeconomiam figurarumque varietatem & speciem certam in eiusmodi nobis ostendi, quisquam homini peritiori facile persuadebit, neque rursum illius pars aliqua in hunc diem usque pro dignitate quicquam quod responderet obtinuit; etsi ineptissimi plures eodem quo apparuerat die, in expeditionem Zelandicam detorquerent." The "Zealand expedition"probably refers to the 1575 siege of the Zealand town of Zierikzee by Mondragon.

[244] Ibidem: "...cum Deus ante minetur quam feriat, ac fere pro magnitudine rerum inter & signum & res signatas amplum constituat temporis intervallum...".

[245] Ibidem: "Nam etsi meo iudicio iudicandi de futuris nulla sit certior via vel ratio, quam ex divinis characterismis, praesertim toties feruntur ordinibus certis, & ad figurae symbola physicarum rerum consensus tanquam in sententiae partem advocatur...".

[246] Ibidem, 15-16: "Munus tamen reor Philosopho dignum (maxime Christiano) praedictionem quae circa singula est, quae pariter modis locorum, ac temporum differentiis definita est, ipsi relinquat dispensatrici providentiae rerum, ac generali notione contentus ... intellectum de caetero submittat numinis voluntati...".

ten about the portents seen before the fall of Jerusalem: they had not invented any fixed and precise interpretation of the omens, but, terrified by these certain indications of divine wrath, they had preferred to pray to God to mitigate his anger.[247]

Conversely, those who ventured to predict, as soon as an unusual occurrence had appeared, what regions and nations would be affected by it, and exactly at what times its evil consequences would make themselves felt, did not find any favour in Gemma's eyes. Such soothsayers, who often hardly observed the sky at all, and failed to investigate the essence of the phenomenon under consideration, were ridiculous and methodless: instead of first ascertaining what were the general characteristics of the comet, they at once jumped to conclusions about the particularities of its significance.[248]

Rather than vainly attempting to penetrate into the details of the divine plan, Gemma contended, we should give heed to the urgent warnings of the divinity, cease to resist him with our own wicked opinions, abandon our former sins, and join together in a bond of one love and one religion.[249] This was also the main reason for which he had decided to write a tract on the newly appeared comet, our author added; convinced that it was a highly majestical and admirable portent of the eternal godhead, he wanted to prevent the hairy star from being appropriated by those false prophets, who abused the art of divination and attached to the comet all kinds of nonsensical predictions.[250]

In our previous chapter, we have already seen why Gemma thought the present comet to possess such a prodigious status. After all, to him, the vast majority of comets were ordinary, sublunary meteors; this was the reason why the one observed in 1577, located

247 Ibidem, 16.

248 Ibidem, 16-17: "Sed nae tales ridiculi prorsus atque ἀμέθοδοι, postquam illi constare debuerit e genere toto priusquam ad singula transferantur, videre debuerint prius cui regioni maxime cometa decernat, quam eius definiant aliquam civitatem."

249 Ibidem, 17: ". . . relictis pristinae vitae erroribus, in unius amoris ac religionis vinculum coeamus, si spreta ventris illuvie, fastu, & aemulatione & diffidentia mutua, & vindictae cupiditate, caeterisque opinionibus pravis nosmetipsos vocanti numini contumaces minime praebuerimus."

250 Ibidem, 17-18: "Neque vero tam facile de hac re scribendi animus foret, nisi & prodigii magnitudo & formae raritas instigaret, magisque multo scriptorum plurium ea inconsulta temeritas, qua & arti convicium faciunt... Itaque ne tanta maiestas rei, portentumque aeterni numinis admirandum pervulgatis naeniis & fucorum ignavia nimis depressum tandem ex posterorum memoria penitus exolescat, placuit nobis & huius venturo saeculo memorabilem theoriam illi subiungere, quam edidimus nuper de sidere novo. . .".

above the moon, was so exceptional, and was to be considered a highly significative prodigy called into being by God by means of an intertwinement of natural and metaphysical causes.[251] To Gemma, the comet of 1577 and the nova of 1572 had been the first new heavenly bodies situated in the superlunary region since the new star allegedly seen by Hipparchus around 125 BC. This Hipparchian nova had been the harbinger of great alterations on the political and religious scene, bringing in its wake the Jewish diaspora, the weakening of the power of the Greeks and the advance of the Roman empire, as well as the incarnation of Christ and the rise of Christianity.[252]

Therefore, Gemma contended, at least as great an alteration in secular and religious affairs was to be expected now that a second nova had shone forth, especially since it had been followed by many other portents, among them the present comet.[253] Since, in Gemma's view, all parts of the universe corresponded with each other, the ominous change in the heavens represented by the appearance of the comet was to be followed by a similar change in the earthly realm: if any more such heavenly portents were shortly to appear, our author added, it should not be doubted that the end of the present age was at hand.[254] Furthermore, he was convinced that the nova and the comet were the signs alluded to by the Sibylline Oracles, which were to announce the Last Days, and the coming of false prophets and the Antichrist, accompanied by pestilence, war, and famine.[255]

Thus, Gemma accorded the comet great significance because God had played an important part in its generation. We have already seen that, to him, it was not unthinkable that novel celestial bodies might be called into being at God's instigation, and that in order to accomplish this, God might make use of natural causes, such as a compressing of celestial matter facilitated by a suitable planetary con-

---

[251] See supra, p. 182-183.

[252] Gemma, *De prodigiosa specie,* 42-43.

[253] Ibidem, 43: "Quid porro nunc de ventura in Christi Rempub. mutatione sperandum est? cui non portentum unicum tale, sed confertim plurima contigere, quorum tamen princeps est ille phosphorus novus affixus nuper stellarum inerrantium caelo in annum integrum & menses 4 perseverans...".

[254] Ibidem, 41-42: "Huic autem Cometae concedunt alteras partes: ut si pleraque propediem ex eodem genere caelitus adiungantur, profecto iam nunc advesperascentis instare saeculi finem, iuxta vaticinium Domini nostri Iesu Christi, non dubitemus...".

[255] Ibidem, 44.

junction.[256] However, in the case of the present comet, he confessed that he had been unable to trace any sufficiently dignified and significant celestial causes which might have generated it.[257] Nonetheless, he does seem to have accorded some influence to the heavenly configuration during the months previous to the comet's appearance. After a short description of two lunar eclipses and a number of planetary aspects seen in recent months, he added that "In what way these things are to be investigated, I shall describe elsewhere in my Meteorography."[258]

Thus, next to the indisputably divine component in the comet's generation and significance, Gemma also recognized a role played by natural causes such as planetary aspects, eclipses, and other celestial configurations. Although he once again stressed that any attempts at explaining in detail the specific effects the comet was to have, were doomed to failure, since the knowledge of such particularities could only be attained by means of divine inspiration, and chided those idle soothsayers who, in accordance with Ptolemaic rules, dared to predict exactly which region or town was to be afflicted by the comet's consequences[259], his own stance vis-à-vis the complex rules of Ptolemaic astrology was somewhat ambiguous. On the one hand, he emphasized that "a mind suited to knowledge by divine inspiration will understand more of the truth, than a man who has overzealously studied science"[260]; yet on the other hand, he admitted that, next to the method of divination from oracles and prodigies, there also existed another way of discovering the alterations imminent in the Christian Republic: the way, that is, of the astrologers, who read the "book of the universe."[261]

According to astrological lore, within seven years a memorable

---

256 See supra, p. 183.

257 Gemma, *De prodigiosa specie,* 47: "Habes finem & generationis causam προηργουμένην, quanquam ad huius exortum clarissimae facis utcumque quis quaerat altius, non videantur caelestes causae pro dignitate ac magnitudine suffecturae."

258 Ibidem: "Illa vero qua ratione debeant vestigari, alias in meteorographiae methodo superis faventibus demonstrabo."

259 Ibidem, 48.

260 Ibidem: "... animaque ad cognitionem divinitus apta, veri plus assequatur, quam qui supra modum in scientia se exercuerit."

261 Ibidem, 57: "Quod autem summae mutationes impendant toti Reipub. Christianae, si quis forte non tantum oraculis atque prodigiis fidem adhibens ex universitatis libro postulet erudiri, nunc illud insuper animadvertat, quod omnium astrologorum hodie consentientium ore profertur: cuncta haec incidere ferme in exactam periodum omnium trigonorum totius orbis signiferi."

and very significant conjunction of all the superior planets was to take place in the extremities of the zodiacal sign of Pisces, the end of the watery trigon. The next such conjunction would then take place in the sign of Aries, the first of the fiery triplicity, so that within seven years' time, the sceptre of dominance would, as it were, be transferred to the "power of fire".[262] The significance of this transition was heightened by the fact that the nova had appeared in the vicinity of the sign of Aries, and by the course of the comet of 1577, which had moved through all the signs between Capricorn and Aries. After all, according to the rules of Ptolemy, tropical signs (Capricorn and Cancer) denoted a change of rulers, so that the fact that the comet had started in Capricorn could be interpreted as the sign of an impending alteration on the political level. This reading was reinforced in its turn by the fact that the three planets which dominated the comet were Jupiter, Saturn, and Mercury, the former two of which ruled over realms, while the latter, Mercury, ruled over laws.[263]

Thus, the astrological rules based on Ptolemy led to the same outcome as Gemma's own theory of cosmic correspondence: a great change in the Christian Republic was imminent. Nonetheless, he did believe the latter method to be the more reliable: after all, the astrological method could easily lead to excesses, abuses, and thus to totally unfounded prognostications. We have already seen how Gemma cautioned his readers against attempting to forecast coming events in too great a detail; not only was this impossible, since foreknowledge of particularities was the prerogative of God and of those he chose to inspire; furthermore, Gemma added, the methods of astrology in themselves were not to be relied on too heavily. After all, many of the rules and theories allegedly drawn up by Ptolemy, had in fact been invented by the Arabs and become contaminated with all kinds of superstition and nonsense, not based on certain experi-

---

[262] Ibidem: "Nam ante annos abhinc septem planetarum superiorum coitus admirandus & longe maximus circa extrema piscium aqueae triplicitati supremum terminum figit, novoque principio trigonorum ab Ariete facto, velut igneae potestati sceptra transscribit." On the significance accorded to the transition from the watery to the fiery trigon, see Germana Ernst, "From the watery Trigon to the fiery Trigon: Celestial Signs, Prophecy and History" in Zambelli ed., *Astrologi Hallucinati*, 265-280.

[263] Gemma, *De prodigiosa specie*, 57: "[Confer] . . . nunc & cometae cursum a Sagittarii extremis partibus fere usque ad piscium fines. Quod si enim e tropico signo Imperiorum mutationes definiat Ptolemaeus, quid vel de puncto illius primo vel propter idipsum excurrente linea statuemus? praesertim quando & Saturnus cum Iove imperiis, ipse Mercurius inprimis legibus praesit, fax autem modo conspicua velut illorum soboles sit etiam in caelo Mercurii deprehensa."

ence or on any kind of demonstration.[264] Furthermore, Gemma repeated some of the well-worn arguments employed time and again by critics of astrology: there was no rational ground on which the attribution of certain earthly regions to particular sections of the heavens could be based, and more generally, the art of the astrologers was not based on any solid experience, and they were not even agreed among themselves as to the parallel and meridian lines, or the trigons of the different signs.[265] Much more sensible, to Gemma, was the attitude of Charlemagne, who did not fear the celestial signs themselves, but only God, the creator of those signs, who in his ineffable goodness called them into being to warn sinful man of his anger.[266]

Gemma's view, then, can be summed up in the following way: new stars and superlunary comets, being created in a normally unchangeable region out of celestial matter, were such rare and majestic occurrences that they could not be ascribed to purely natural causes. They were called into being by God, who used both natural and supernatural causes to create them, with the purpose of warning mankind to repent of their sins and to change their ways; if they failed to do so, he would strike the wicked with his punishment. However, to Gemma it was possible to attain somewhat more precise knowledge of what in particular the celestial sign in question had come to announce. Although the prediction of particularities, as practised by those astrologers who relied on rules of the Arabo-Ptolemaic tradition, was a highly untrustworthy business, our author did believe human reason capable of discovering at least the general thrust of the portents' messages. This could be attained by relying on the same rules used by physicians to diagnose the degree of health or illness of the human body, that is, by closely studying the "essences and intimate causes of external indications."[267]

However, despite his abnegation of Ptolemaic astrology, the con-

---

264 Ibidem, 48-49: "Quid igitur est quod male feriati vates audent non solum modos, & calamitatum tempora praefinire, verum & Machlinum, Gandavum, Artesiam, suis innectunt praedictionibus sigillatim? Quasi vero illa signorum ad certas terrarum regiones a Ptolemaeo tradita non tam ex Arabum & Chaldaeorum referta nugis innumeris officina, omnique superstitionis genere delibuta, quam usu certo vel demonstratione profluxerit."

265 Ibidem, 50.

266 Ibidem.

267 Ibidem, 50-51: "Porro si quid hic Physicis ac Mathematicis tribuendum est, quaeratur in primis decreti genus & species, deinde tempus, & modus, & locus, ea

clusions which Gemma reached as to the consequences of unusual celestial occurrences did not deviate very much from those generally accorded them by regular astrologers. He believed that all the elements of the world, being interconnected, experienced a "chain of evils" which had been initiated by the influence of the nova. The new star had given birth to a rare intemperacy, which, according to the natural law of reciprocal influence, had spread from one extreme to the other. This had brought about a varied range of calamities: pestilence, famine, terrible thunderstorms and other unusual weather conditions, storms, lightning, earthquakes, whirlpools, and inundations. Moreover, the same influence had also penetrated into the bodies and souls of human beings, giving rise to violent passions. And, since society was composed of many souls together, this was where the evil influences would remain felt for a long time, and where it would prove the most difficult to remove the contagion.[268] Moreover, the comet, too, having passed through constellations made up of stars of the nature of Mars and Mercury, denoted tempests, drought, and earthquakes.[269]

Thus, it would seem as if Gemma's theory of cosmic correspondence, and study of the divine signs in nature, led to precisely the same outcome as the Ptolemaic, astrological lore of cometary consequences. On both views, albeit for somewhat different reasons, the appearance of a comet brought about pestilence, famine, death, inundations, earthquakes etcetera. However, Gemma did not leave it at that. To him, the characteristics of the comet portended, not only such limited and local inconveniences as inundations or pestilences, but in addition announced "highly ineffable, varied, and daily alterations throughout the most part of Europe, greater than could be wondered at or expected."[270] To Gemma, the coming of the comet

methodo, qua per externa indicia crises ad salutem interitumque humani corporis hauriuntur: id est, ab ipsa essentia rei vel intimis causis, atque a tota peristasi, quantitate, qualitate, situ & motu."

[268] Ibidem, 52: "Afficiuntur protinus elementa mundi afflatu Sideris novi. Nascitur intemperies rara ab uno extremorum naturali vicissitudinis lege in alterum prompte exorbitans. hinc λιμός καὶ λοιμός, i. pestis & fames: hinc tempestates infandae, aliaque meteora insolita consequuntur. . . Paulatim denique vis ea permanat in hominum sensus & corpora, atque ex lege consortii . . . vertuntur species animorum, & pectora motus concipiunt. Quare cum ex multorum animis una Resp. constet, certum est in illius amplissimam molem vires contagii non nisi longo post tempore pervagari. . .".

[269] Ibidem, 59.

[270] Ibidem, 53: "Non unum ex illo sidere calamitatis genus ingruere suspiciamur,

foreshadowed a long-awaited era of renewal and large-scale mutation: expectantly, he added that "The seeds have been sown, and it seems I can see the day when all the parts of the earth, far and wide, will burn, after the sparks of our fire will have spread everywhere."[271]

Despite the ominous connotations which such a worldwide conflagration obviously carried, Gemma showed himself remarkably optimistic about its probable outcome. Endorsing Polybius' theory about the natural way in which different forms of government were wont to succeed one another[272], our author stated that after having passed through the stages of monarchy, tyranny, dangerous democracy, anarchy or chirocracy, and aristocracy, the state would finally arrive at the best thinkable constitution: namely, the just reign of a single ruler, by whom all power was wielded. However, this monarch was not to rule despotically: his reign would be tempered by attuning his power to the sentiments of the other parts of society, so that "nature would intermingle tones from the middle register with the high and low ones" in order to arrive at perfect harmony.[273] Once again, Gemma compared this ideal constitution with the harmonious structure of the human body, where the head and the stomach did not idly feed themselves while the feet and the arms toiled, but where instead each organ helped and sustained the functions of the other bodily parts. Also, he once again added that the creator of all things had provided a most beautiful example for the organization of the state in every single part of the universe.[274]

Thus, in Gemma's view there could be discerned a kind of natural progression, with different forms of government succeeding each other until in the end the ideal of moderate monarchy was reached.

---

non brevi temporis cursu, non terrae angustis limitibus comprehensum: sed per europae maximam partem, mutationes prorsum infandas, varias, diuturnas, omni miraculo atque expectatione maiores."

[271] Ibidem, 54 "Iacta sunt semina, videre videor illum diem, quo velut scintillis ex nostro incendio dispersis undique ... vicinae terrarum partes & semotae longius flagrent."

[272] See Polybius, *Histories* VI. 3-10.

[273] Gemma, *De prodigiosa specie*, 58: "Sunt enim, teste Polybio, naturaliter conversiones eiusmodi a Monarchia per tyrannidem in Democratiam periculosam, & hinc per Chirocratiam vel Anarchiam potius in Aristocratiam optimam sc. Reipub. partem, donec tandem in unius iustum imperium referatur: cuius videlicet una vis atque potestas vigeat. Non pro arbitrio, sed ad aliarum partium sensus & conspirationem mutuam temperata, uti tonis acutis & gravibus natura medios interiecit, elementis asymbolis alterum utrique consentiens...".

[274] Ibidem.

Although he did not explicitly say so, it seems likely that our author regarded contemporary society as going through the stages of "dangerous democracy" and "anarchy or chirocracy"; after all, in 1577, the royal government of the Netherlands had collapsed, the States had taken matters into their own hands, and in many towns in the southern Netherlands, radical Calvinist regimes had seized power.

Therefore, next to heavenly signs and astrological rules, the 'natural', Polybian theory of succession of governments, too, indicated an impending change. The present disease pervading society was shortly to reach a crisis, out of which, after a general catharsis, Gemma believed a better, more harmonious form of government would come to reign. God intended shortly to scourge his people, and separate the just from the wicked; therefore, he had sent the comet as a sign of his wrath, but at the same time, as a token of hope, for its appearance near the winter solstice, the point at which days once again started to lengthen and light returned, could surely be read as an indication of imminent improvement.[275]

This did not, however, mean that this betterment would come about automatically; it was still necessary for people to do penitence, and to make an effort to restore society. Thus, the population should obey their rulers, while the rulers should reign with paternal benevolence; above all, the middle classes of society should attempt to reconcile those advocating extreme and radical positions.[276] This can undoubtedly be read as an allusion to the moderate stance in matters of religion so ardently championed by the Antwerp humanists. The panegyric to Charity, with which Gemma concluded his tract, is very reminiscent of the central role accorded to this virtue by such members of the Plantin circle as Benito Arias Montano.[277] Our author

---

[275] Ibidem, 61: "Fortassis enim hic nobis cometa in signum erigitur ἀλεξίκακου Θεοῦ... Visus est enim praesens Cometa gradum dirigere statim ab ipso hibernae conversionis initio, ex quo dies augentur noctibus imminutis; animarum ascensus, vitalisque caloris incrementa a Philosophis perhibentur...."

[276] Ibidem, 61-62: "Futurum id facile duco, si populus irritatior iusto parere suis principibus discat, principes paterna benignitate subditis imperare, medii status extrema componere...".

[277] See Hamilton, *Family of Love*, 79. It is true that charity was also one of the central tenets of the Family of Love, but, as Hamilton points out in connection with Arias Montano's emphasis on charity as "the most important cohesive force of society", the use of the word 'charitas' was by no means confined to Familist writings, and Arias Montano, for one, did not employ it in the Familist sense of "an undogmatic attitude, independent of any particular creed".

strongly believed that, in order to restore the peace everyone so ardently longed for, it sufficed simply to remove all the obstacles standing in her way, chief among which was the disappearance of charity from society. After all, without the bond of charity, all good things dissolved, in the same way the universe would dissolve if the sun were removed; and, Gemma was sad to say, in contemporary society, the reign of charity had been usurped by self-love, treason, suspicion, the treacherous breaking of treaties (another allusion to Don Juan and the Perpetual Edict) arrogance, egoism, and jealousy.[278]

Gemma could discern two causes for the demise of charity. First of all, he pointed at the new religion, which had brought about a dangerous dissension within society, because some of its adherents displayed a singular arrogance and contempt of the authorities, and, more often, because pitiable ignoramuses, unacquainted with the truth, followed the bad examples set by the former group.[279] In addition, charity had been driven away by an abundance of vices rife in contemporary society, such as fury, jealousy, vainglory, love of gain, vengefulness, gluttony, and lust.[280] To heal these various wounds of society, Gemma urged his fellow countrymen to strive for a solid, just, and secure peace, the amputation of all disorderly elements in society, and the return of charity, care for the common good, and submission to the authority of the holy Church instead of to private opinions.[281]

We can conclude that Gemma's chief motive in devoting a tract to the comet of 1577 was political commitment. Like most members of the circle of humanists which he frequented, he was greatly horrified by the recent developments in the Low Countries: the treason, suspicion, hypocrisy and civil strife had made a deep impression on our

---

278 Gemma, *De prodigiosa specie*, 64-65.

279 Ibidem, 65: "Hanc autem charitatis extinctionem duae praestant malorum omnium primae atque evidentissimae causae: Altera quidem in principe parte sedem obtinens opinio prava, novae religionis inventrix, animorum divortia in Remp. serens: saepe vel ex fastu singulari contemptuque superiorum, saepe (quae & venia magis quam odio digna est) ignorantia veri exemplorumque pravitate profecta."

280 Ibidem, 65.

281 Ibidem, 65-66: "Quod si ergo instituenda curatio est, atque iam nulla salus bello, sed pacem poscimus omnes; Pacem solidam, rectam atque securam pro fine proponimus: amputentur vitia prima ceu continentes causae: in apistiae locum amnistia procedat: paulatim vero & praecedentes ordine usque ad ultimas auferantur, usque dum constet mutuus amor, & cura publici boni privato commodo vel affectui, communis auctoritas sacrosanctae Ecclesiae, privatis opinionibus atque discursibus praeferatur."

author, and his tract was meant above all as an exhortation to his fellow countrymen to mend their ways. In the same conciliatory, *politique* fashion as the 'middle groups' exemplified by William of Orange, Gemma advocated an end to discord and a striving for peace, to be realized by uniting in obedience to a single ruler, who was, however, to govern in accordance with the consent of the different groups within the population.

The comet, then, served only as one of the components (albeit by no means an unimportant one) of a much more wide-ranging system. Gemma's expectations of an impending cosmic cataclysm, and his yearning for a restoration of peace and unanimity in his afflicted fatherland, lay at the root of his treatise. In trying to demonstrate to his readers that a great political upheaval was imminent, he made use of several different divinatory techniques. Chief among these was a theory of cosmic correspondences of the Stoico-Platonic variety, combined with the well-known medical analogy according to which changes in the macrocosm denoted changes in the microcosm. Gemma extended the latter theory to include also the 'mesocosm', or the level of human society and political development: changes in the body politic, too, were preceded by unusual cosmic occurrences, and they, too, could be cured in the same way as human bodies. The comet of 1577, in Gemma's view, fitted in nicely with such a teratological outlook: it was a supernaturally created sign sent by God as a warning to mankind, as had been the nova of 1572 and the chasms of 1575 which had preceded it.

However, although clearly the teratological aspect of Gemma's approach was the most important, he did not quite abandon the astrological way of reasoning which had become such a significant component of the Aristotelian world view. Since, to him, most comets were sublunary phenomena of the kind described by Aristotle, it seems likely that, like the Stagyrite, he also believed these burning exhalations to be the harbingers of atmospherical conditions of heat and winds on earth. Gemma's remark, that the lowest variety of comets carried significance only for "one region, or a small portion of the earth", seems to endorse this view.[282] Moreover, he professed his belief in the astrological theory according to which a change of triplicities denoted a great alteration in sublunary circumstances, and he subscribed to the reasoning that, since the comet had passed stars

[282] Ibidem, 41; see supra, p. 181.

of the nature of Mars and Mercury, it announced Martial and Mercurial effects. However, his stance vis-à-vis astrology was, on the whole, cautious: he thought the prediction of specific particularities inadmissible, and he was distrustful of the methods used by the astrologers because of the widespread disagreement about them.

The pamphlet devoted to the comet of 1577 by Johannes Heurnius breathed much the same atmosphere as that of his colleague from the southern Netherlands. It, too, had been inspired above all by the tumultuous situation the Netherlands were going through during the months of the hairy star's visibility. In the summer of 1577, a few months before the comet's first appearance, Utrecht, where Heurnius lived and worked as a town physician, had joined the revolt: the civic militia had forced the town government to defect to the side of the rebels during the period of growing polarization that followed the attempted coup of Don Juan. Heurnius, clearly favourably disposed towards the new regime, commenced his pamphlet with a dedication to the "Honourable and Venerable Lords and Citizens of the town of Utrecht". In it, like Gemma, Heurnius pointed out the importance of observing the heavenly bodies for those striving to govern the state correctly: after all, the noble art of astrology had of old been practised by none but the most powerful kings of the earth, and in Egypt, only those skilled at the "harmonious statecraft of the celestial bodies" were admitted to the royal administration.[283]

Heurnius made a distinction between two kinds of men: those who paid attention to God's goodness and wisdom, and who therefore realized that it was essential to observe the movements and substance of the heavens, which God had instituted as an example for man to follow, and on the other hand those who despised the noble astrological art, and who occupied themselves solely with earthly things, not enjoying any spiritual delights. Emphatically, Heurnius reminded his fellow citizens that the former attitude was much to be preferred: after all, God called into being comets and new stars for the sake of mankind; these creatures certainly carried a divine meaning, for God never created anything in vain. Therefore, it was wrong to ignore the appearance of new citizens of heaven; rather, man should follow the

[283] Johannes Heurnius, *De Historie, Natuere, ende Beduidenisse der erschrickelicke Comeet, die geopenbaert is int Jaer ons Heren 1577* (Cologne s.d.): "Ja in Egypten was hy onbequaem geacht tot de Conincklicke administratie, die niet ervaren en was in de eendrachtelijcke Policie der hemelscher Lichamen. . .".

example of the elephants in Ethiopia, who, according to Pliny, regularly met at a certain river to greet the moon. Because, however, many of his contemporaries neglected the study of the heavens, Heurnius had decided to write his tract, as a warning to all men to be aware of the imminent dangers foretold by the comet, and to persuade them to try to mitigate these cometary consequences. Especially those in powerful positions should heed the warning of the comet, our author added, so as to be able to protect their subjects from its harmful consequences.[284]

Further on in his pamphlet, Heurnius provided his idea, that the appearance of a comet should be read as a divine warning to mankind, with a theoretical underpinning quite reminiscent of the one proposed in Gemma's treatise. We have already seen how Heurnius believed the comet to be of an intermediary nature, functioning as a link between the heavens and the earth and formed out of a substance partaking both of the corruptibility characteristic of the sublunary region and of the unchangeability of the celestial realm.[285] Thus, the comet constituted an important link in the chain of correspondences between heaven and earth. With reference to Homer's image of Zeus having embraced heaven and earth with a chain, which acted as a metaphor for a certain power with which all living substances had been impregnated by God, Heurnius asserted that God, whenever an alteration in the sublunary realm was at hand, first of all displayed the signs of such a change at the beginning of the 'chain', that is, in the motions of the celestial bodies. After all, our author added, no notable change could take place on earth without a similar alteration first having occurred in the heavens, in the same way that illnesses in the human body never changed either for the better or for the worse, without being preceded by certain critical signs.[286]

The whole universe, to Heurnius, was permeated with signs: thus,

---

[284] Ibidem: "... als een Waerschouwinge, daer mede ick vermane alle Menschen, ende sonderlinge de Heeren des Volx, dat se met voorsichticheit voorkomen het quaet dat wy toekomende sorgen, om so te minder gequetst te worden, ende om haer selven ende die haer bevolen zijn te beschermen, voor het toekomende Quaet, ...".

[285] Supra, p. 196-198.

[286] Heurnius, *Historie:* "... want natuerlick en konnen wy gheen merckelicke veranderinge opter Erden hebben, so wy oock sien dat geen sware sieckten in ons Lichaem ... ter doot oft tot Leven hen mogen veranderen, sonder voorgaende Teickenen."

fishermen knew by certain signs where they should throw out their nets, and other signs told metallurgists where to dig. If even such coarse substances as earth and water were capable of engendering signs, then, how many more significations should not the heavenly bodies carry?[287] Therefore, our author thought it completely unjustified that most people paid no attention to the majesty of the Lord, as revealed in the celestial signs which exemplified what changes were shortly to occur on earth. This was the more reprehensible, because human beings were endowed with a "natural, rational element", a kind of quintessence enclosed within their bodies, which was closely attuned to the quintessence of the heavens, following it like a magnet followed the pole.[288] Thus, humans were well equipped to discover the divine signs in nature: "With the hands of our reason, we can touch God", Heurnius assured his readers: "Therefore, let us use them."[289]

Chief among the divine signs in nature were comets, which were constituted out of the "astral element" and therefore partook of both celestial and terrestrial qualities. Therefore, the hairy stars certainly carried some kind of meaning; however, Heurnius hastened to add that this meaning was of a signifying, not of a causal variety. He emphasized that comets should be regarded as natural signs, by means of which nature, at God's behest, signalled down to men on earth the causes of her imminent alterations; not, however, as causes of those changes.[290] Thus, in Heurnius' view there existed a close collaboration between God and nature: God, whenever he felt the need to warn mankind of impending upheavals, employed nature as his messenger, setting in motion a chain of correspondences beginning in the heavens and somewhat later reaching the sublunary

---

[287] Ibidem: "Konnen sulcke grove substantien als Erde en Water, sulcke teickenen baren, hoe veel te meer dan sullen de teickenen hem wtgeven in de hemelsche doorschynelicke lichamen?".

[288] Ibidem: "Want alle dinck (naest Godt) hangt syn leven an dese quintum Esse, die in hem is, ende onse quinta Essentia volcht dem quintam Essentiam des Hemels, als een Magneet den Pool."

[289] Ibidem: "Want hy [Julius Caesar] wist wel, dat den alscheppenden Godt door syne wonderwercken gesien wil zyn: ja so nae openbaert hy ons daer in, dat wy hem met de handen onses verstants tasten, wat van synder aert is, Laet ons dan die gebruicken."

[290] Ibidem: "Niet dat ick wil seggen dat Cometen oft dergelicke Teickenen, de thoekomende Veranderingen voortbrengen ofte maken: neen geensins, dan het syn natuerlicke Teickenen, dat de Natuer door Gods gebott haer oorsake der veranderinge tot ons neder seint, want die nu ryp syn."

realm. Moreover, the changes predicted by natural signs were not themselves of a purely 'natural' kind, in the sense that they need not always be necessitated by factors internal to nature herself and independent of man. On the contrary, there was a moral dimension to them: thus, comets were only engendered by nature whenever God was displeased with human conduct, "as if nature herself spoke to us and said: 'Behold, the Lord shows you a comet, which is never seen without subsequent prodigies, to signify, as it were by a natural signalling, acting as a mark, whom His justice decides to punish'."[291]

Like Gemma, then, Heurnius believed in a close correspondence between the different parts of the universe, with changes in one part announcing changes elsewhere. To both authors, a comet was a natural sign, expressly called into being at God's command, in order to forewarn mankind of such imminent alterations. However, again like Gemma, Heurnius also thought it possible for human reason to discover to a certain extent what exactly the comet was meant to presignify. Perhaps because of the fact that he wrote a pamphlet intended for the public at large, instead of a Latin treatise, Heurnius' predictions were much more specific than those of his Louvain colleague. In order to render them more credible, he stated that he had deduced all the consequences he believed the comet to have in store, from the rules drawn up by "the old philosophers", who had arrived at these rules by way of long experience and observation.[292] This remark served to legitimate the use our author proposed to make of the traditional astrological rules of Ptolemy and the Arabs with regard to comets and their consequences.

Sure enough, the second half of Heurnius' pamphlet contained a completely conventional, astrological treatment of the comet's meaning. Thus, the author endorsed Ptolemy's statement, that comets were of the nature of Mars and Mercury, and therefore denoted war, revolt, imprisonment, destruction, princes' anger, horrible mortality,

---

[291] Ibidem: ". . .willen wy het goede, so moeten wy goet zyn, het welcke hoe veer dattet van ons zy, bewyst ons Godt door die natuer, die ons een Monster van gesterte nu voor set, recht of die tot ons spraecke, de Natuer segge ick selven: Siet de Heere stelt u een Comeet voor, die noyt sonder navolgende wonder gesien is, om te beteickenen door natuerlicke wincken, als door een merckteicken, wie zynder rechtferdicheyt belieft te straffen."

[292] Ibidem: "Ende op dat niemand meine day wy dat wt affectie schryvende, wt onsen duym suygen, Willen wy dit wtleggen na die beschryvinge der olden Philosophen. . . Soe is door lange ervarentheit geobserviert, dat een Comeet die hem openbaert in den Oosten, beduit een geboorte des verstants, die in alle Polycie ende Religie, Veranderinge in brenget. . .".

injustice, fire, murder, and changes in government. Also, according to the *Centiloquium,* a comet moving from west to east, like the one under consideration, threatened a foreign invasion. Unlike Gemma, Heurnius did not hesitate to employ the rules drawn up by the Arab astrologer Haly, according to whom the workings of a comet affected the regions and towns subject to those celestial signs through which that comet moved; in the case of the hairy star of 1577, this led to the conclusion that Ghent, Malines, Mons, Jülich, Cleves, Brandenburg, Saxony, Hesse, Thuringia, Hamburg, Bremen, Westphalia, the areas situated near the river Meuse, Sweden, and Denmark would suffer its evil consequences, since all these regions were subject to Capricorn and Aquarius. Optimistically, our author added that, since the comet first appeared in the southwest, and its tail pointed towards the south, Spain would probably also be affected, and that, since the star appeared in the ninth house of the heavens, a reformation was at hand, as well as an "Inquisition inside the Inquisition".[293]

A clear allusion to the offensive Don Juan and Parma were carrying out in the southern Netherlands at the time Heurnius' pamphlet was written, can be read in the remark that "since Saturn and Mercury are the lords of the ascendant, our enemy will come from the southeast."[294] In addition, Heurnius predicted "an unspeakable practice" which would be used to end unanimity among relatives[295]; this may be a reference to the hypocrisy and dissimulation rife in the Netherlands around the year 1577, to which Gemma, too, so eloquently alluded. Furthermore, our author foresaw great sadness in the clerical estate, and the death of a great prelate, as well as misery and fear, bloodshed and war, a great invasion from the east, enormous changes in political, religious, and moral matters, a change of governor, and the death of the representative of a great lord. If the latter remark was intended as a reference to the demise of Don Juan, this prediction was to come true: in fact, the treacherous governor died from the plague on 1 October 1578.

Obviously, these forecasts were all rather predictable; no sooth-

[293] Ibidem: "Ende want hy int negenste Huys des Hemels appariert, so mochter wel een Reformacy comen, ende Inquisitie in de Inquisitie".

[294] Ibidem: "So oock Saturnum ende Mercurium die Heeren zyn des Ascendens, kompt onse Vyant wt den Suydtoosten."

[295] Ibidem: "ende om dat hy is int vyfde huys, beteickent ons een onwtspreckelicke pracktyck, die gebruickt wort om Eendracht van Bloetverwanten te scheiden. . .".

sayer could go wrong on the prognostication of war and changes in political and religious matters, given the chaotic contemporary situation in the Netherlands. However, Heurnius did not confine himself to the prophesying of such local and temporal matters. Like Gemma, he personally knew a number of Familists, and he may even have belonged to the sect himself.[296] It appears that, at the very least, he shared their expectation that the Last Days were at hand: like Gemma, our author ended on the hopeful note that after a period of long and hard struggle, a "golden world" would arise, "changed for the better."[297] This could be deduced from the signifying power of the nova of 1572, which had first been multicoloured, then red, then beautifully golden, then silvery, and finally pale; it could also be calculated from the rules of the art of cabala, according to which the seventh age of the world, the "Year of Rest", was drawing nigh. Therefore, Heurnius urged his readers to mend their ways and to restore brotherly love and charity; after all, Christ himself had prophesied that the end of the world would be announced by unusual signs in the heavens.[298]

c. *The Pamphlets of Portantius, Van Mauden, and Bazelius*

While Gemma, and also Heurnius to a certain extent, attempted to provide their ideas on the comet's significance with quite detailed theoretical underpinnings, their fellow prognosticators Portantius, Van Mauden and Bazelius showed themselves much less interested in elaborate theories of cosmic correspondence and divine agency in nature. Rather, their writings were filled with extensive lists of dire consequences which had followed comets in the past and which would undoubtedly follow the present comet again.

This is not to say that the divine solicitude for man, so emphatically stressed by Gemma and Heurnius, was absent from the accounts of their less learned colleagues. On the contrary, all three

---

[296] Supra, p. 166.

[297] Heurnius, *Historie:* "Dit beteickent ons eerst een groote Turbatie met Confusie, daer in dat twe oft meer partien tegen den anderen worstelen, seer sterck ende lange, daerna een bloetvergietende wreetheit: dan salder een gulden Werrelt opstaen in goeder Veranderinghe."

[298] Ibidem: "Want hy [Christus] selve seit, als gy sulcke Teickens siet, so denckt dattet na by is, ende an de Fygeboom siet men dat de Somer naeckt. Hier mede wil een yder sich selven vermanen, ende sien in hem selven, hoe veel dat in hem van de broederlicke Liefde schynt. . .".

pamphleteers went to great lengths to convince their audiences of the fact that, first and foremost, comets should be regarded as divine signs, messengers sent by God to announce to mankind his anger and to urge them to mend their ways. Although, as we have seen in chapter V, Bazelius, Van Mauden and Portantius all endorsed the Aristotelian theory, according to which comets were burning sublunary vapours, none of them attempted to account for the hairy stars' predictive value by means of the medieval explanation that held them to be parts of a physical chain of causation desiccating the air and heating the atmosphere.

Instead, despite their earthly nature, comets were endowed, by all three authors, with the more elevated status of divine signs. This way of accounting for the consequences they brought in their wake was the most pronounced in the pamphlet by Johannes Portantius. At the very beginning of his tract, he declared that there existed three ways in which God was wont to admonish man on earth: through the Holy Law given to the patriarchs of Israel; through the mortal prophets; and through his signs in the heavens, such as, for example, the rainbow, symbol of the covenant between God and man.[299] Although there were certain "Epicurean, Machiavellian and Lucianic minds" who ridiculed those who regarded comets as messengers of God, Portantius said, this was not to be doubted, since experience taught that they had, throughout history, been sent as signs of God's anger, to warn mankind of their sinfulness. Thus, Christ himself had announced that the end of days would be preceded by prodigious signs, as well as by disasters such as pestilence, earthquakes, and inundations, and the disappearance of charity.

Moreover, the observations made throughout the ages by learned and skilful astrologers testified to the fact that all changes in earthly and monarchical power, both the rises and the falls of empires, had invariably been preceded by presages in the heavens. Portantius

[299] Johannes Portantius, *Beschrijvinghe der nieuwer Cometen, Met aenwijsinge ende vermaninge wat zy bedieden ende voortbrenghen sal* (Antwerp s.d.): "Ende is te verstaene hoe dat hy den menschen door dryen middelen principalijck vermaent, om syn heilighe geboden te onderhouden, om daer door tot die eeuwige salicheyt te comen, namelijck door die Heylighe Wet, den Patriarchen van Israel ghegeven: ten tweeden door die sterffelijcke Propheten, ende ten derden ende lesten door zijn teeckenen des hemels, als Genesis 9. geschreven staet, als dat hij zijn teeckenen des verbonts door zijnen Reghenboghe tusschen syn heylicheyt ende ons vertoonen sal, om dat wy dies ghedachtig wesen souden, ende onse salicheyt overdenckende, onse herten ende handen ten hemel opheffen souden. . .".

added a number of examples to this effect, drawn from both the remotest past (such as the fall of the Persian empire and the destruction of Jerusalem) and from more recent history: thus, after the comet of 1556, great alterations had occurred, especially in religious matters, and the effects of the nova of 1572 could still be felt everywhere.

The present comet too, then, should be regarded as the harbinger of great plagues, our author prophesied. These plagues would be very cruel and horrible, and would be of three kinds. The first would consist of seditions and discord in the cities of the East, South, and West; the second, of the ruin of many great potentates and noble families; and the third, and principal punishments would include great changes in religion, through rare and prodigious treason, in Turkey, Italy, Spain, and France. Moreover, three plagues would arise through the agency of three kinds of people: the Saturnine dissimulators, the Mercurial, ambitious magistrates and officers, and the bloodthirsty tyrants who would be elected by the two former groups. Because of these ominous connotations of the comet, Portantius felt it his duty to warn Flanders and Hainault, and in particular the towns of Ghent and Malines, that stood in danger of being corrupted by "strange minds".[300] By reason of all these dangers, our author concluded, we ought all to pray to God to avert his punishment and to show mercy to the poor sinners on earth.

Van Mauden, too, emphatically regarded comets as harbingers of future calamities, proclaiming God's anger to sinful mankind in order to admonish them to change their ways.[301] The Lord, being merciful, in his paternal solicitude always made sure that his punishments were preceded by signs, so that people would get a chance of doing penitence and thus averting God's wrath. As for the miseries the present cometary apparition had in store for sinful man, van Mauden continued, it portended great revolts and bloody wars, the deaths of many noblemen both through warfare and through ambushes and treacherous ruses; also, many villainous and deceitful

[300] Ibidem: "Daeromme Vlaenderen ende Henegouwe wilt wijsselijck op u hoede zijn. Ghent ende Mechelen wilt neerstich voor u sien, want vremde geesten sullen u soecken te verderven."

[301] David van Mauden, *De geschiedenisse van de nieuwe Comete des Jaers 1577 seer neerstelick g'observeert van datse begonst te schynen tot den eynde toe* (Antwerp 1577): "Soo ist dat wy nu ooc sien een voorbodeken van toecomende plaghen, die ons vercondicht die groote gramschap Gods ende hoe dat syn roede uitgestrekt is over ons, om ons grauwelijcken te straffen.".

councils would be held, seeming friends would turn out to be enemies and traitors, especially after the war had come to an end (doubtless this was another of our 1577 authors' allusions to Don Juan's breaking of the Perpetual Edict), and there would be many earthquakes, strong winds, tempests, snow, ice, cold spells, inundations, mortality among fish and horses, and poverty. In addition, bad fevers and plagues would break out, followed by long illnesses, consumption, quartan fevers, and dropsy. Significantly, our author concluded this comprehensive list of traditional cometary disasters with the Latin remark "Et hec ex caussis Physicis" ("And these are the consequences brought about by physical causes"). Apparently, therefore, van Mauden endorsed the Aristotelian, physical way of explaining at least some cometary concomitants.

However, he immediately added that "Comets, by their own nature, signify many things in an occult way, which cannot be discovered by means of reason, nor understood before we see them work and effectuate their meanings."[302] Thus, next to their traditional, Aristotelian consequences, which remained confined to the physical realm, comets also worked more hidden effects, whose causes could not be elucidated and about which it was unclear whether they were natural or supernatural. Nonetheless, despite his denial of human capability of discovering in advance these effects, van Mauden did venture to further specify them by means of the constellations which the hairy star had passed, and of the rules of the authoritative Arab astrologer, Haly.

Thus, because the comet's course had passed "monstrous and irrational creatures of the heavens", and because in addition, during the past year three monstrous fishes had been washed ashore, our author concluded that the comet probably signified "a monstrous and strange assault"; however, in the end things would turn out well.[303] Next, van Mauden proceeded to quote extensively from Haly's astrological rules as to what effects were proper to comets moving through the signs of Capricorn, Aquarius, and Pisces (as did the present

[302] Ibidem: "Maer aengesien de Cometen uit haer eygene natuer verborgentlijc veel dingen beteekenen, die niet en connen met eenige redenen achterhaelt werden noch begrepen, voor datmen haer siet wercken, en effectueren haer beteekenissen."

[303] Ibidem: "Soo ist wel te presumeren, dat dees Comete een monstreus en vreemden aenslach beteekent: want sij meest is geloopen bij monstreuse en onredelijcke gedierten des Hemels, ende hebben hier dit jaer oock drye mostreuse Visschen de Zee aan landt sien brengen, de welcke met dese Sterre oft Comete wat vreemts sullen voortbrengen: maer het eynde sal goet syn."

comet), as to the countries and cities that would be struck by the hairy star's consequences, as to the individual people that would be affected, and as to the duration of the comet's effects. His answer to the latter question was largely physical in nature: a comet's effects might end either because they ran out of power ("such as happens to all natural things"); because of the appearance of a second comet whose nature was contrary to the first one's; or in the case of the occurrence of an eclipse or conjunction in a place contrary to the comet's position, or of a contrary nature.[304]

Van Mauden concluded his tract with a consideration of the comet's operations with regard to religion, showing himself remarkably optimistic on the issue. Thus, in his opinion the hairy star had no evil in store for the Christian religion, but only good. Conversely, both the Jewish and the Moslem religions would be very badly affected by the comet's workings: due to a number of astrological causes, great alterations would ensue in Judaism, while the Islamic Turkish religion might even be quite rooted out before the comet would cease to operate. Christianity, on the other hand, would thrive: alterations in that religion were only indicated by comets of a Mercurial or Jovian nature, that ran their courses against the order of the zodiac, or in Libra, or near the celestial equator, and none of these circumstances applied to the comet of 1577. Therefore, despite the fact that the immediate consequences of the hairy star would be bad and consist of poverty, war, treason, deception, tempests, and earthquakes, in the end it signified much good, happiness, and joy: all errors in matters of religion would be removed, religious conflicts would cease, and peace would be restored.[305] Clearly, van Mauden here alluded to the contemporary situation in the Netherlands, his message being that

[304] Ibidem: "De cometen einden hen werkingen op drierley manieren. Ten eersten worden hen werkingen geeyndt door henselven, te weten als sy tot den eynde van hen cracht gewracht hebben, gelijc in alle natuerlijcke dingen gebeurt: en so soude dese Comete doch veel jaren wercken. Ten tweeden eynt haer werkinge, wanneer daer compt een ander comete, die haer contrarie is van naturen: oft die door haer cracht contrarie wercken, de werkinge van de eerste Comete te niete doet. Ten derde so wort haer werkinge geeynt, alser eene Eclipsis oft coniunctie geschiet die in een contrarie plaetse is van de Comete, oft van contrary natuer."

[305] Ibidem: "So dat haer eynde sal sijn behoudenisse, blischap en vreucht overmits dat de twee quaede wetten sullen te niet gedaen worden, en alle dwalingen wech genomen sullen worden en geen ander religie, noch veranderingen noch twist van religie sal worden gevonden: maer peys en vrede salt wesen onder de menschen, dewelc ooc de cromheit des steerts schijnt te beteekenen. en ooc dat dese Comete geeynt is *in domo Iovis*, en *ad tropicum Cancri*, daer Jupiter is in sijn verheffinge."

despite the recent tribulations and the failure of the pacification, hope should not be abandoned because in the end peace and unity would prevail.

Nicolaus Bazelius was by no means as optimistic as his Antwerp colleague. His entire pamphlet was saturated with the direst predictions, based above all on the complex rules of Ptolemaic-Arabic astrology. The lunar eclipse which had preceded the appearance of the comet announced strife and discord among friends and relatives, the demise of brotherly love, the apocalyptic scourges of famine, sword, and pestilence, as well as secret conspiracies and plots. Two great and widely renowned potentates would pursue one another with such vengefulness, that one of them, although the best and the most powerful, would have to pay for it by his blood or by exile from the country.[306]

The nobility and the authorities were especially threatened by the unfavourable consequences of the eclipse. Bazelius warned all magistrates to keep vigil day and night, and to make sure not to put on a "masked face"; rather, they should show on their faces and in their actions what lay hidden in their hearts.[307] Thus, like Gemma, Bazelius was much preoccupied with the chaotic circumstances in the Netherlands, with the civil war that had sown discord among former friends and relatives, and with the untrustworthiness of those supposed to promote the common good. Our author foresaw embassies and conferences of great persons being held at the cost of the common people but with little effect; very important topics would be discussed, but nothing would be finally decided. This was probably once again one of the many 1577 allusions to the failed Pacification and the broken Edict. Furthermore, Bazelius added, legates and messengers would be attacked and the letters they carried would be in-

306 Nicolaus Bazelius, *Een nieuwe prognosticatie vanden wonderlijcken ende ellendighen Jare ons Heeren 1578. Midsgaders de beschryvinghe vande Comete des voorleden Jaers* (Antwerp 1578): "...twee groote Potentaten seer verre befaemt, en sullen de taerte niet connen gedeylen, maer sullen elckanderen met sulcker wraeckghiericheyt soo veruolghen, dat den besten ghesinde ende den machtichsten gehouden, sal moeten met zijn laetste bloet, soo wel deur listighe circumventie, als door fortse van wapenen, het selue becoopen, ofte met Schande wt den Lande loopen."

307 Ibidem: "Alle ouerheyt ende Magistraten, moeten dach ende nacht waecken, ende met alder neersticheyt toesien, dat sy gheen ghemasschert aensichte en draghen, maer metten ghelaete ende ghewercken hun ondersaeten betooghen zulcx als int herte besloten is, ende ten sy dat zy in als de eere Godts, ende het ghemeyne weluaeren meer soecken ende voorderen, dan hun singulier profijt, het zalder al qualijck gaen...".

tercepted, so that many artful conspiracies and deceits would be made public[308]; in all probability this should be read as an allusion to the letters of Don Juan to king Philip II, which had been intercepted by Henry of Navarre and sent to William of Orange, and in which the governor had pleaded with his half-brother to resume warfare in the Netherlands.[309]

In addition, Bazelius foresaw great changes in religious and political matters. Love and affection between neighbours, brothers and associates would disappear completely, and the "subjects of Jupiter" would be persecuted far and wide. According to traditional astrology, Jupiter ruled over magistrates, clergy, and jurists; therefore, Bazelius' repeated remarks about the subjects of Jupiter being threatened with banishment and death can be read as prophecies of great upheavals in government. Moreover, it is possible that our author also associated Jupiter with William of Orange: he stated that the Eagle (the constellation of Aquila, thus named after the eagle that carried off the beautiful youth Ganymede for Jupiter) should be careful not to lose his pretty plumage and fall down naked from the sky, a laughing-stock to all the world, which was bound to happen unless he returned the youth Ganymede to Jupiter's palace, from which he had recently abducted him. In this cryptic passage our author may refer

---

308 Ibidem: "Boden, legaten, ende messagiers, sullen wercks ghenoech hebben, maer luttel profyts, want zy int reysen veel aenstoots hebben sullen, ende heurlieder briefuen achterhaelt, waer deure veel listich bedroch sal int openbaer comen tot heurlieder grooten achterdeel."

309 See Emanuel van Meteren, *Belgische ofte Nederlantsche Historie, van onsen tijden. Inhoudende hoe de Nederlanden aenden anderen ghehecht, ende aen Spaengien ghecomen zijn: Met de oorsaken der Inlantsche beroerten, ende oorloghen der selver, Oock de veranderinghe van Princen, van regimenten ende van Religie, mette scheuringhen, verbontenissen ende vredehandelinghen: Mede vervattende eenighe harer ghebueren ende andere Landen handelinghen. Meest onder de regeeringhe van Philippus de II. Coninc van Spaengien, tot synen doot, ende den wtgaenden Iare 1598* (Delft 1599) Book VII, f. 109: "Dese voorschreven saken passeerden daer int openbaer, maer secretelijc met Brieuen ende Boden, heeft by den Coninck totter oorloghe seer verweckt ende ghehist, met sijnen secretaris Escovedo, als naemaels bleeck door haer Brieuen in Gascongien afgheworpen, ende ghesonden aenden Prince van Orangien, vanden Coninck van Navarre." These letters were later deciphered by Marnix van St Aldegonde, read out loud to the States General, and published in print. See also Bor, *Oorsprongk*, 827: "Middelretijd handelde Don Jan mette voorsz Colonellen van de Hoogduitsen, op een geheel ander maniere als de Staten meenden, want hy mette selve eenen aenslag maekte, om hem te voegen tegen de Staten, en practiseerde om hem absoluit meester vanden Lande te maken, en hem te versekeren van de principaelste sterkten vanden Lande, gelijck namaels gebleken heeft by de geintercipieerde brieven, die hy aen de voorsz Hoogduitse Colonellen schreef, en oock mede van de brieven die hy aen den Coninck gesonden hadde, dewelke in Vrankrijk afgeworpen zijn."

to Philip William, William of Orange's son, who had been kidnapped by Philip II, and whose release was the subject of several negotiations during the 1570s.[310]

All the above predictions were based on the astrological rules of the arabicized Ptolemy, with the author associating certain supposedly malignant constellations with certain specific malignant effects (thus, Mercury, Jupiter and Mars being located in the third house of heaven, the house of brotherhood and unity, with Mercury being "burned" by the rays of the sun and "poisoned" by the tail of the Dragon announced discord among friends and relatives). However, some of Bazelius' prognostications were based on a somewhat more physical way of reasoning: thus, Aries and Mars, both of a fiery nature, had ruled over one of the eclipses treated at the beginning of our author's pamphlet; therefore, they would produce hot and fiery results, such as great drought and heat, infertility, scarcity, bad harvests, tempests, earthquakes, and 'hot' illnesses caused by hot, pestilential winds.

The comet, however, was regarded by Bazelius above all as a herald of God's wrath. Although most of the constellations which had preceded the hairy star had been equally ominous, this had not been noticed by the common man, who did not know how to read the signs of the sky. Therefore, God in his mercy had sent a comet, a sign that could not be disregarded by anyone.[311] Its purpose was to

---

310 Bazelius, *Prognosticatie:* "Dese Comete met haer byloopen infortunerende den Arendt ende den Dolphin, en wil die anders niet dan ongeluc ionnen. Zo dat den Arendt wel toe staet te siene ende wachten dat hy geen verlies van syn schoon plumaigen en doe, ende soo naect ende ongepluymt wt den hemel en valle, ende zoo van elcken bespot, want hy sulcx verwachten moet, ten zy dat hy wederom restituere, ende naer Jupiters zale voere dat schoon iongelinc Ganimeden, den welcken hi onlancx looselijc wt Iovis paleis wechgeuoert heeft."

311 Ibidem: "Bouen alle voorgaende waerschouwingen die God den Heere almachtich door de veruaerlijcke Eclipsen van Sonne ende Mane, ende ander constellatien ofte versamingen vande Planeten onder malcanderen, ende met de vaststaende sterren int firmament, de welcke aldermeest vande geleerde ende principalijcke Astrologyns, waer genomen syn, ende van de gemeente veracht, om dat de sulcke dicwils gebeurende, niet en ouerdincken wat datter naer volcht. Soo heeft Godt door syn ongrondeloose bermhertichz ende onsprekelijcke liefde die hij tot dat versteent menschelijc geslachte dragende is, eenen anderen bode ofte voorloper willen wtseynden int firmament, als eenen Herault van zynder straffe ende ontvliedelijcke gramschap, het welcke is dese afgryselijcke ende verschrickelijcke *Comete oft sterre metten steerte,* die haer int hoochste vander locht onder het firmament vertoocht heeft sedert den xiiii. Nouembris, tot den dach van hedent als meer dan xxxiiii. dagen geduert hebbende, de welcke hy zoo lange heeft willen laten schynen, op dat een

incite people to mend their ways and to do penitence. Bazelius corroborated this interpretation of the comet's status with a long list of past comets, which had all been followed by calamitous effects: the invasion of the Turks and the sack of Rome in 1527, the Peasants' Revolt in 1530 [sic], the siege of Vienna by the Turks in 1533, the rise of the evil sect of the Anabaptists and the birth of the Anglican Church in 1533, the deaths of Charles V, his wife, and both his sisters, the religious wars in France: all these evils had been announced by comets; not to speak of the many inundations, earthquakes, droughts and fires, and pestilences which had also been preceded by God's hairy messengers.

Thus, experience taught that comets were always sent with the purpose of notifying men of divine anger. Whenever God could no longer endure the horrible dissimulation, falsehood, hypocrisy and faithlessness of the human race, such as was, in 1577, to be found among all ranks of the population, he poured out scourges over the earth. This was certainly to happen again in the aftermath of the comet of 1577: after all, people were more ambidexterous than ever before, and the "double-faced race of Janus with the two foreheads has been multiplied to such an extent that it is impossible to know with whom one can find sincere loyalty, favour, or friendship".[312] To Bazelius, cometary effects were first and foremost to be associated with the moral order of mankind's sinfulness and God's just punishment; the question of whether they were brought about by natural causes he apparently deemed irrelevant, since he made no attempt to answer it. In the case of the nova of 1572, he even expressly stated that those of its effects that were still to come were unknown even to the angels, and could not be discovered by the human intellect. All that could, in an unspecified "astronomical" way, be gathered about it, was that the change it would bring about would be everlasting, and would come sooner than many people thought. "Therefore, let everyone pray and keep vigil night and day, for it is greatly to be

yegelijc mensche geleert ende ongeleert, groot ende cleyne, schamel ende rycke, die met bedructer herten aenschouwen souden, ende niemant ignorant wesende van syn rechtueerdiche ende groote gramschap...".

[312] Ibidem: "Dat dobbel aensichtich geslachte vanden tweevoorhoofdigen Janus is soo vermenichfuldicht datmen niet en weet waer ofte op wyen oprechte trouwe, ionste, of vriendtschap vinden. Maer want soo van God den Heere ouerlanc voorseydt is, sulcks moet voor het eynde des weerelts gheschieden, souden alle de Prophetien vervult wesen."

feared that the most horrific day of the Lord is very imminent", Bazelius menacingly added.[313]

Although Bazelius believed that the specific effects of comets and other celestial signs could not be accurately discovered by human beings, he did include quite an extensive astrological treatment of the present comet's effects, based on the rules of Ptolemy. Thus, he believed that, because the phenomenon had been of a Martial, Mercurial, and Saturnine nature, it would work the effects proper to those planets. Because its duration had been long (a bit under two months) its consequences would also last for a long time. Other particularities could be deduced from the direction of the tail, the direction of the comet's motion, and the fact that it had stood on the occidental side of the sun. The regions that would be afflicted could be deduced from the zodiacal signs through which the comet had passed; like Portantius, Bazelius specifically warned the city of Ghent to be vigilant in order to prevent being led astray by the "novelty of many things", or being ruined by secret conspiracies and treason.[314]

However, after an exhaustive enumeration of all the consequences that were to be associated with the present comet according to the rules of astrology, Bazelius ended his pamphlet on a rather more optimistic note. He urged his fellow countrymen not to give up hope too quickly; after all, although he had written down all that was to be expected following the teachings of all the ancient astrologers, these

313 Ibidem: "Wat datter geschiet is, sint dat die alderwonderlijcxste claer blinckende sterre, die int iaer 1572 int eerste van Nouember haer vertoochde, de welcke wel xv. Maenden bleef blinckende, altyts vast staende in haer eerste plaetse, eer zy met allen verginc, midts dat wy tot twee diuersche tyden genoech daer af ghe-schreuen hebben, ende ooc midts dat dan de beclagelijcke memorie daer af zeer versch is, en willen wy nu geen verhael doen, wat dat die in toecomende tyden noch door dat teecken gedisponeert ende gedelibereert heeft, te seynden ouer ons ellendige menschen, heeft syne alderhoochste Mt. dat soo in het diepste van zynder verborgelijcxste Thresorie zo besloten dat de Ingelen zelue niet eer weten sullen, dan als dat geschieden sal, hoe wel wy wel Astronomelijck mogen colligieren dat seer gestadich ia eeuwich syn sal, de principaelste veranderinge die sy by bringhen sal, ende dat onuersiens ende al eer, eer menich mensche dat sal connen dincken. Dus elck bidde ende waecke nacht ende dach, want grootelijkcx te vreesen is, dat den alder verschrickelijcksten dach des Heeren seer by nakende is."

314 Ibidem: "Ende ghy Hooftstadt Ghent, zyt met alder besorchsaemheyt nacht ende dach op uwe hoede, ende wakende. Ende siet wel toe dat ghy door de nieuwicheyt van vele saken misleyt niet en wort vander quaetwillicher boose strycken oueruallen, ende dat ghy niet met heymelijcke conspiratien, ende bedecte verraedtschepen (als ghy minst sulcx dunct) ter neder ghesmeten ende gheruineert wort. En wilt niet ouertreden die oude palen ende limiten die uwe voorvaeders u gestelt hebben."

evils were not to happen necessarily. God might still be pacified by ardent prayers, true weeping and penitence over our wickedness, and the adoption of a better way of life, and might even bring about a sudden and unexpected change for the better. After all, "we should most of all put our trust in God the omnipotent Lord, whenever there is no hope of human aid, for he is closest to the unconsoled, and most ready to comfort them."[315]

We can conclude that, despite several significant differences, due above all to the fact that Gemma and Heurnius belonged to the intellectual elite, while the other authors did not, and that Gemma wrote in Latin for a learned audience whereas the others addressed the common man in the vernacular, the Dutch writings on the comet of 1577 breathe much the same atmosphere. All the authors were utterly dismayed at the pitiful state their country found itself in, making many allusions to the warfare and misery rife in their fatherland. Especially the coup of Don Juan, the failure of the Pacification, and the hypocrisy, dissimulation, and discord among former friends and relatives, that are the inevitable concomitants of civil strife, seem to have made a profound impression and were referred to in many of the authors' prophecies. Furthermore, the writings of Gemma, Heurnius, and Bazelius were all marked by an ardent expectation of the imminent Day of Judgement.

All our authors were agreed that comets were to be regarded as divine signs, rather than physical causes, of future calamities. Only van Mauden vaguely alluded to the physical consequences comets could bring about, yet he, too, placed much more emphasis on their status as heralds sent by solicitous divinity to admonish the sinful human race. Despite the fact that every one of our authors, after having expounded this teratological stance, went on to submit the comet to a perfectly conventional astrological treatment, attaching predictive value to the direction of the tail, the colour of the head, the phenomenon's course through the constellations, etcetera, remarkably few attempts were made to connect the rules employed with any

[315] Ibidem: "Het welcke hier voortijts den seer geleerden Philo tot de Christene gesproken heeft, seggende, datmen dan aldermeest goet betrouwen hebben moet op Godt den Heere almachtich, alsser geen hope en is op eenich menschelijc bystandt, want hy dan den ongetroosten menschen alder naest is, ende aldergereetst om dien te consoleren."

form of physical causation. The three pamphleteers, Portantius, van Mauden, and Bazelius, did not show any concern at all for the theoretical underpinnings of the rules they employed; and Heurnius and Gemma, who did make an effort to elucidate the intellectual basis of their predictions, did so in highly Stoico-Platonic terms, stressing the role of divine agency in nature, the unity of the universe, and the sympathetic correspondences between every single part of the great cosmic animal.

4. *The Netherlands, 1618*

a. *Current Affairs: The Parting of the Ways*

When, in late November 1618, an exceptionally bright comet was seen above the Netherlands, its observers were confronted with totally different social and political circumstances from the ones faced by their 1577 colleagues. The forty-one years that separated the two hairy celestial appearances had seen some of the most momentous changes the Low Countries were ever to experience. We have already seen how, in the course of the comet-year 1577, it became increasingly clear that the Pacification of Ghent had failed. The attempted coup of Don Juan brought in its wake a radicalization of the revolt, with Calvinist radicals seizing power, not just in the northern towns that had not yet defected to William of Orange, such as Amsterdam and Utrecht, but also in the Flemish cities of Ghent, Bruges, Ypres, and Arras. The latter development was contrary to the agreement laid down in the text of the Pacification, which held that the religious status quo of 1576 was to be maintained, so that public practice of Protestantism was not to be allowed outside the provinces of Holland and Zealand.

Consequently, fear of the spread of radical Calvinism in the south, especially in the light of the excesses perpetrated by the regime of the Committee of Eighteen in Ghent, induced the States of the Walloon provinces of Hainault and Artois to dissociate themselves from the General Union that had united all the Netherlands provinces at the time of the signing of the Pacification of Ghent. On 6 January, 1579, they concluded a union of their own at Arras, stating their wish to maintain the Roman Catholic faith, as well as to remain loyal to their overlord, the king of Spain. A few weeks later, the rebel provinces, too, reached an agreement among themselves: on 23 January, Holland, Zealand, Gelre, Overijssel, Friesland, and Groningen signed the Union of Utrecht, which, unlike its Arras counterpart, was not intended as a defection from the General Union, but rather as a closer union within the confines of the General Union. They agreed that each province would independently take care of religious matters; however, in matters of war and peace, the signatories were to act as one single province.

Thus, the unity of the Pacification began to crumble. Moreover, Don Juan's successor as Philip II's governor-general of the Netherlands, the Duke of Parma, ably conducted a successful military campaign aimed at forcing the rebel provinces to their knees. Instead of withdrawing all Spanish troops, as agreed with the 'Malcontents' of the Union of Arras, Parma even reinforced them, and went on to conquer substantial parts of Brabant and Flanders. Between 1579 and 1585, the talented general managed to submit, with the aid of his brilliant tactical acumen, all the most important Flemish and Brabant towns: Ypres, Bruges and Ghent were forced to their knees in 1584, Brussels, Malines, and finally Antwerp in 1585.

Although, from 1585 onwards, Parma's offensive stagnated (he was never to accomplish his aim of crossing the great rivers) his campaign brought about an irreversible division between the northern Netherlands, soon to become the Dutch Republic, and the southern Netherlands that were to remain under Spanish rule. Due to largely accidental factors, connected with the fortunes of war, the towns of Bruges, Ghent, and Antwerp, despite having signed the Union of Utrecht, were to be subjected to centuries of Habsburg rule, and to a vigorous programme of re-catholization, while their more northerly allies in, for example, Holland, Zealand, or Overijssel were able to maintain their membership of the Union of Utrecht and to become fully-fledged members of the United Provinces, soon to abjure the sovereignty of Philip II and thus to gain effective independence. Over the years 1577-1588, the failure of the Pacification was to turn into an irreparable rift.

From 1588 onwards, North and South each went their own separate ways.[316] After the death of Parma, Philip II appointed the

---

[316] See Schöffer, "Opstand", 146, who believes the defeat of the Invincible Armada, in 1588, marked the definitive rupture between North and South, and the coming into being of two different states with their own political, economic, social, and cultural characteristics.

brother of the German emperor, the Archduke Albert, as governor-general of the southern Netherlands. Realizing that his son, the future Philip III, was too weak to govern both the Spanish empire and the Netherlands, the king decided to give his daughter Isabella to Albert in marriage, and to present her with the Netherlands for her dowry. After the king's death, in 1598, Albert and Isabella, invariably known as 'The Archdukes', took up residence in Brussels as sovereign rulers. Although, obviously, their regime remained dependent on Madrid to a large extent (thus, the Spanish garrisons in the southern Netherlands remained, and had to swear oaths of allegiance to Philip III, while, if the marriage of the archdukes were to remain childless, their territories would revert to the king of Spain)[317], Albert and Isabella managed to leave their mark on the Spanish Netherlands in a truly decisive way.

During the first decade of the archdukes' reign, the war between Spain and the northern Netherlands continued. Although one of Albert's first actions after his arrival in Brussels was to resume peace-talks with the northern Netherlands, dismayed about the high toll his territories had to pay for the continuous warfare, at the time the Dutch were not yet prepared to abandon their war effort. During the 1590s, Prince Maurice of Orange had ably conducted a series of sieges of towns in the south and east of the Netherlands territory, capturing Breda, Zutphen, Deventer, Nijmegen and Groningen. Although, after about 1598, the Prince's campaign stagnated, and he failed to secure any further notable successes, the northern Netherlands were not yet interested in peace. The advocate of the States of Holland, Johan van Oldenbarnevelt, proposed a campaign in the coastal areas of Flanders, to reduce piracy against Dutch shipping from the ports of Nieuwpoort and Dunkirk. In 1600, Maurice and the troops of the archdukes clashed on the beach at Nieuwpoort; although the battle was won by the Dutch, such heavy losses were inflicted on their forces that they had to retreat to Zealand.[318]

Although the war went on for seven more years after the battle of Nieuwpoort, the Dutch grew less and less enthusiastic about it. In 1603, the brilliant general Ambrogio Spinola arrived in the Spanish Netherlands, financing the siege of Ostend the archdukes were conducting, and forcing the Dutch garrison there to surrender. When,

[317] Israel, *Dutch Republic,* 254.
[318] Parker, *Dutch Revolt,* 234.

three years later, he in addition conducted a reasonably successful campaign in the east of the Netherlands, recapturing a number of towns gained by Maurice in the late 1590s, and making the States General realize that their chain of wooden fortresses along the IJssel was not as impregnable as they had imagined, eagerness to make peace further increased.[319] Both sides were finding the cost of war more and more onerous to bear, and in 1607, a ceasefire was negotiated, followed by a Twelve Years Truce, signed at Antwerp on 9 April 1609.

As far as the parting of the ways between the northern and the southern Netherlands was concerned, the period of the Truce was crucial. Geoffrey Parker even believes that "[the] new 'south Netherlands identity', religious and political, was born under the archdukes and above all during the Truce".[320] Jonathan Israel agrees, that "The Twelve Years Truce proved a crucially formative stage in the recovery and reconstruction of the south Netherlands."[321] After almost forty years of continuing warfare, the inhabitants of Flanders and Brabant were finally able to rebuild their devastated cities, revitalize agriculture and economy, and breathe new life into culture and learning. The university of Louvain, that had suffered the terrible consequences of having garrisons of underpaid, hungry soldiers of different nationalities billeted in the town between 1578 and 1585, was now being restored to prosperity under the auspices of the archdukes[322]; the population increased; agricultural activity expanded; and many of the former industries, above all the Flemish linen industry, revived.[323]

The archdukes managed to achieve what Philip II had never been able to accomplish: to establish a firmly centralized form of government, while entertaining good relationships with both the local nobility and the civic elites. Although they never convened the southern

---

319 Ibidem, 237.

320 Ibidem, 258.

321 Israel, *Dutch Republic,* 410.

322 See Léon van der Essen, "Les tribulations de l'université de Louvain pendant le dernier quart du XVIe siècle", *Bulletin de l'Institut historique belge de Rome* 2 (1922) 61-86; and J. Grisar, "Die Universität Löwen zur Zeit der Gesandschaft des P. Franciscus Toletus (1580) nach bisher unbenützten Quellen des Vatikan. Archivs" in: *Miscellanea in honorem Alberti De Meyer* (Universiteit te Leuven, Publicaties op het gebied der geschiedenis en der philologie, 3e reeks, 23e deel, Leuven/Brussel 1946) 941-968.

323 On the recovery of the southern Netherlands during the reign of the archdukes, see Israel, *Dutch Republic,* 410-420.

States General after 1600, they did consult the various provincial States and allowed them more leverage over the fiscal process.[324] Nonetheless, Albert and Isabella kept matters firmly under control: they issued laws binding in all provinces, and effectively curtailed the power of the nobility, who saw themselves reduced to the status of courtiers without any real influence.[325] The archdukes carried out a number of measures to stimulate economic recovery; thus, they reorganized state finances, and introduced protectionistic tariffs. The improvement in living standards brought about by these and similar measures led to a state of euphoria in the south during the years of the Truce.[326]

Culturally, too, the Spanish Netherlands went through a vigorous revival during the archdukes' reign. Albert and Isabella were ardent supporters of Counter-Reformation culture, pursuing a policy of intensive re-education and confessionalization of the common people, and sponsoring the production of Catholic works of art and architecture. An extensive network of Jesuit colleges soon sprang up throughout the archdukes' territory, whose members firmy applied themselves to the religious education of children from all walks of life, setting up a system of Sunday schooling, producing torrents of devotional literature, and fostering an ideological climate in which the revolt of the northern provinces was regarded as a catastrophe, caused by heresy and corrupt nobles, and in which it was believed that, eventually, the rebel provinces would have to be reabsorbed into the political and ecclesiastical framework administered from Brussels.[327] This was also the period during which Louvain University regained much of her former prestige, especially after the rigorous 'visit' or inspection of the institution by the archdukes, conducted between 1607 and 1617, that brought about a progressive growth both of educational standards and of student numbers.[328]

All in all, then, the Twelve Years Truce was a period of prosperity and revitalization for the inhabitants of the Spanish Netherlands: the reign of Albert and Isabella, though curtailing popular participation in the government process, accomplished an economic and cultural

---

[324] Israel, *Dutch Republic*, 410.

[325] P. Janssens, "Het bestand in de Zuidelijke Nederlanden 1609-1621" in *Nieuwe Algemene Geschiedenis der Nederlanden* VI, 315-324: 317-318.

[326] Ibidem, 318.

[327] Israel, *Dutch Republic*, 418.

[328] Léon van der Essen, *L'université de Louvain (1425-1940)* (Brussels 1945) 62-63.

well-being most of their subjects had never before experienced. Seen in this light, the return of Justus Lipsius to Louvain (from whence, as we have seen above, he had fled in 1578, when the great period of tribulations at the southern university town had started) becomes very understandable. Constantly in search of peace and quiet in order to be able to study and to write, the great humanist was horrified at the never-ending succession of political and religious conflicts with which he saw himself faced at Leiden.[329]

Ever since, in 1581, the signatories of the Union of Utrecht had abjured king Philip II as their sovereign ruler, they had become entangled in a long series of squabbles over the governmental structure the northern Netherlands would have to adopt, now that they had done away with their traditional overlord. William of Orange had, against the will of Holland and Zealand, offered sovereignty to the Duke of Anjou, brother of the king of France, who was a Roman Catholic with *politique* sympathies. In so doing, he had hoped to regain the support of those moderates alienated by Calvinist radicalism, and to reassure Catholics in the south.[330] However, this policy had turned into a miserable failure when Anjou, vexed by his lack of influence brought about by the distrust with which the radical Calvinist faction in the States General had from the outset treated him, had attempted a coup (the so-called 'French Fury' at Antwerp, January 1583). However, just a few years later, the rule of the governor-general appointed by the States General to succeed Anjou, the Earl of Leicester, who was a very orthodox Calvinist, ended in a similar fiasco.

Leicester was backed by the radical Dutch Calvinists, who were continually trying to strengthen Calvinist orthodoxy within the public church, and to increase the influence of that church in state and society.[331] From the moment of his arrival in the Netherlands, however, Leicester was distrusted by the more moderate groups, who were opposed to allowing the church too much influence on state affairs. When Leicester convened a national synod at which a strictly Calvinist church order was drawn up and the claims of the provincial States to supervise the annual meetings of the provincial synods were

---

[329] See Mout, "In het schip", 58-59.
[330] Israel, *Dutch Republic*, 208-209.
[331] Ibidem, 221.

rejected, as were their claims to civil control over the appointment of preachers[332], confrontation between moderates and orthodox Calvinists grew. Several months of political crisis ensued, with the Dutch population deeply divided as to their sentiments towards Leicester. In October 1586, the governor-general purged the Utrecht city council, with the backing of the civic militia, much to the fury of the Holland regents. Affairs came to a head during the summer of 1587, when Leicester attempted an abortive coup, after which he "gave up in disgust and returned to England."[333]

However, he left behind the serious internal tensions that had been growing in the northern Netherlands during his governorship. The Leicester episode was symptomatic of the far-reaching divisions within Dutch society as to, first and foremost, the relationship between church and state. Although, in the short run, Leicester's departure came down to a victory for the Holland regents, led by Oldenbarnevelt, over the more 'popular', orthodoxly Calvinist party that had backed the English governor[334], this did not heal the rift in society, and tensions were to resurface as soon as the signing of the Twelve Years Truce temporarily diverted attention from the war effort.

Lipsius, then, did not feel at home in the northern Netherlands. When he had left Louvain, this had been for the largely negative reason that his peace was being disturbed by the martial conditions there. In the north, he had hoped that it would prove possible to establish a form of government under a single ruler. However, after the failure of Anjou, the death of William of Orange, and finally the fiasco around Leicester, he was bitterly disappointed.[335] What he saw was the emergence of a republican state, made up of seven sovereign provinces, without any powerful central government, ruled by oligarchies of nobles and regents, and rent by internal strife over the relationship between church and state. No wonder, then, that he chose to leave the chaotic, quarrelsome, and, in his opinion, inefficiently governed North to return to his *alma mater,* Louvain, which at the time (1592) found itself on the eve of a period of vigorous recovery

[332] Ibidem, 227.
[333] Ibidem, 230.
[334] Schöffer, "Opstand", 150.
[335] Mout, "In het schip", 62.

under the auspices of a benign princely government capable of assuring the stability and peace necessary for intellectual activity.[336]

While, in the South, such favourable conditions were prevailing during the period of the Truce, in the North the tensions so passionately resented by Lipsius, that had existed from the outset, and that had already come to the surface during the governorship of Leicester, produced a crisis that brought the country to the edge of civil war. The regime led by the Advocate of the States of Holland, Johan van Oldenbarnevelt, religiously tolerant and politically republican and oligarchic, saw itself faced with two important challenges: the antagonism of the stadholder, Maurice, "a potentially quasi-monarchical figure"[337], and the hostility of many Reformed preachers who objected to the government's moderate stance in religious matters.[338]

During the Truce, society in the northern Netherlands grew more and more divided. The factions that had already manifested themselves during the Leicester episode now associated themselves with the protagonists of a squabble between two theologians at the university of Leiden, Arminius and Gomarus, the former of whom held that, despite the Fall, man had retained his responsibility for his actions and could make a free choice between salvation and damnation, while the latter adhered to a strictly interpreted doctrine of predestination. Although the quarrel over predestination was what sparked off the series of discords that were to come to a head in 1618, soon the other issues over which dissension had been simmering for several decades also came to be associated with the two parties of Arminians and Gomarists (usually referred to, from 1611 onwards, as 'Remonstrants' and 'Counter-Remonstrants' respectively). Thus, Arminianism affiliated itself with tolerance in religious matters, the right of civil authority to oversee the church, and Oldenbarnevelt and the States of Holland, while Gomarism upheld the idea that the authorities were not to exercise any power in religious matters.

Rapidly, the controversy, which had first flared up in Holland and Utrecht, spread throughout the Republic. At all levels of society, people took sides; although it has been maintained that, on the whole, the regents chose the Arminian point of view while the common man was Gomarist, this was not the case: in Friesland, Gro-

---

[336] Israel, *Dutch Republic*, 411.
[337] Ibidem, 421.
[338] Ibidem.

ningen, Amsterdam, and Zealand the regents were Counter-Remonstrant, while the lower strata of society often chose to follow the opinion adopted by their own preacher, be it Remonstrant or Counter-Remonstrant.[339] By 1617, Dutch society had become deeply divided: the Counter-Remonstrants issued an 'Act of Separation' forbidding any form of communication with the Remonstrants, and even foreign powers saw themselves involved in the conflict, with England affiliating itself with the Counter-Remonstrants and France with the Remonstrants.[340]

When, in the course of 1617, Prince Maurice publicly sided with the Counter-Remonstrants, the crisis reached its summit. Now, the challenge posed to the power of the Holland regents by the widespread Counter-Remonstrantism was reinforced by the Prince of Orange's open defiance of Oldenbarnevelt's regime. There was a note of rebellion in the air, fostered by a sharp increase in the number of political pamphlets and tracts dealing with the Arminian-Gomarist controversies.[341] In the summer of 1617, after a number of disturbances in several towns during the early months of the year, Oldenbarnevelt felt so threatened that, as a last resort, he empowered Holland towns to raise their own special troops, the *waardgelders,* to maintain order, and asserted that sovereignty in the United Provinces lay entirely with the individual provinces.[342] These decisions, laid down in the so-called 'Sharp Resolution', infuriated Maurice, who regarded them as a threat to his position as captain-general and an affront to his person. In late 1617, he took pains to accelerate the process of defection from the ruling Arminian block in Holland, and during the early months of 1618, he took more concrete action, purging the town councils in several Gelderland and Overijssel towns, and secured the backing of the five provinces of Friesland, Zealand, Gelderland, Overijssel and Groningen against Holland and Utrecht in the States General. When, in July 1618, Maurice's troops were able to disarm the *waardgelders* at Utrecht without resistance, Oldenbarnevelt and his allies abandoned their struggle. Maurice

[339] I. Schöffer, "De Republiek der Verenigde Nederlanden 1609-1702" in I. Schöffer, H. van der Wee & J.A. Bornewasser eds., *De Lage Landen van 1500 tot 1780* (Amsterdam/Brussels 1978) 167-267: 233.

[340] Ibidem, 234.

[341] Israel, *Dutch Republic,* 438-439.

[342] Ibidem, 441.

toured the Holland towns to bring to power Counter-Remonstrant councils, and on 29 August, he had Oldenbarnevelt arrested.

Thus, the conflict ended in the defeat of the Remonstrants. When, in late November 1618, the comet appeared, a National Synod was under way in Dordrecht, the vast majority of whose participants were Counter-Remonstrants. In May 1619, they condemned the Remonstrants as heretics and disturbers of state and Church. Those Arminian preachers prepared to recant would have to subscribe to a formula of submission pledging adherence to the Netherlands Confession and the Heidelberg Catechism, and acceptance of the Acts of the National Synod. Most Remonstrants refused to sign; they were banished from the Republic and settled at Antwerp, where they drew up rules and procedures for a Remonstrant church in exile.[343] Finally, after the Synod had ended, Oldenbarnevelt was tried; charged with high treason, he was beheaded on 13 May 1619.

The authors who committed to paper their thoughts on the comet that remained visible over the Netherlands from November 1618 to January 1619, then, belonged to a different generation from their 1577 colleagues. In that year, although Don Juan's coup and the radicalization of the Revolt that had followed it had seriously diminished hopes of a speedy resolution of the Netherlandish troubles, pacification and unification of all seventeen provinces under their sovereign lord had still seemed a possibility ardently to be pursued. In 1618, by contrast, North and South had taken fundamentally different courses. The Spanish Netherlands, governed by the archdukes, were enjoying peace and prosperity and were well on the way towards rebuilding their devastated economic and cultural life, while the Republic had just experienced a spell of near civil war and a far-reaching division of society. During the comet's visibility, the National Synod at Dordrecht was still being held, and neither the Remonstrants nor Oldenbarnevelt had yet been condemned. However, what had become abundantly clear was that the Stadholder, Maurice, and his chosen party, the Counter-Remonstrants, were now firmly in charge.

In both the North and the South, the years 1618-1619 were marked by a good deal of reflection about what was to happen after the expiry of the Truce in 1621. The archdukes' position on the

---

[343] Ibidem, 462-463.

subject was unequivocal: fearing the consequences renewed warfare would bring to their recovering territories, as well as the enormous financial costs it would entail, they advocated the continuation of the Truce as it stood. In the Republic, sentiments were divided: staunch Counter-Remonstrants favoured a resumption of the war, for religious reasons, as did many of the Holland and Zealand merchants, for economic ones. The land provinces, conversely, were much less eager to recommence hostilities, justifiably fearing that they would be the ones who would have to suffer most of the consequences of a resumption of warfare. And, finally, Madrid was not prepared to renew the Truce on the old terms, demanding that the Dutch evacuate the Indies east and west, that they lift the Scheldt restrictions, and that they tolerate public practice of Catholicism.[344] These differences of opinion were only to be resolved in 1621, when a cunning plan devised by Maurice was to thwart any attempts to renew the Truce. Thus, in the comet-year 1618, as in 1577, the Netherlanders saw themselves faced with a good deal of uncertainty as to their immediate future: the resumption of war was imminent, but not yet inevitable.

b. *The Culmination of the Teratological Tradition: Puteanus*

Erycius Puteanus was an expert at ingratiating himself with powerful protectors. His biographer, Th. Simar, has drawn up a list of some of the humanist's works dedicated, with more or less success, to great persons such as Philip William of Orange, the Duke of Aerschot, and the governors of Philip IV, the marquess of Aytona and don Francisco de Mello.[345] With great zeal, he applied himself to the tracing of the genealogies of some of his envisaged patrons down to extremely ancient ancestors. Moreover, he was very proficient at the art of epitaph-writing, composing eloquent commemoratory poems for such great personages as John of Luxemburg, king of Bohemia, and M.A. Bongratius, the archbishop of Patras. Not only by means of the written word did Puteanus pursue his policies of patron-seeking: he also showed himself very judicious in the choice of godparents for his many children; thus, his daughter Isabella Eugenia had the

[344] Ibidem, 473.

[345] Th. Simar, *Étude sur Erycius Puteanus (1574-1646) considéré spécialement dans l'histoire de la philologie belge et dans son enseignement à l'université de Louvain* (Louvain/Paris/Brussels 1909) 29-33.

archduchess Isabella for a godmother, his daughter Philippina Dorothea was the godchild of Philip William of Orange and the duchess Dorothea of Croy, and his son Urban was even entrusted to the godfatherly care of Pope Urban VIII.[346]

However, sponsorship was not the only thing on Puteanus' mind. Pupil and successor of Lipsius, our author had inherited from his great predecessor an ardent desire for peace and quiet. Although he was a loyal supporter of the archdukes' regime, and a dedicated opponent of Protestantism (thus, he produced a work entitled *Aerarium catholicum,* in which he presented a system he believed would assure inexhaustible financial means for the Church to combat heresy)[347], Puteanus was also a convinced believer in the preferability of peace to any form of warfare. At the beginning of 1609, just before the signing of the Twelve Years Truce, he had written a work called *De induciis belgicis* ("On the Netherlandish Truce"), which he published in 1617 along with a letter by Lipsius on the subject of war and peace. In it, he urged the archduke to conclude a truce with the rebel provinces in order to put an end to the devastation of the country. After the resumption of hostilities, in 1633, our author produced a sequel to this early work, *Belli et pacis statera* ("The Scales of War and Peace"), in which he again spoke out in favour of peace: convinced that it would prove impossible to reconquer the North, he advocated the conclusion of a lasting peace with the Dutch Republic, as well as the renunciation of all Spanish claims to sovereignty over it.[348]

The comet of 1618 constituted an excellent occasion for Puteanus, both to vent his opinions on the destructiveness of warfare and the advantages of peace, and to practise his dexterity at patron-seeking. As we have seen in our previous chapter, his work on the hairy star was dedicated to the archdukes, to whom the author promised a sidereal token of peace and prosperity, instead of a harbinger of war, disaster, and princely deaths. His entire first book had been devoted to demonstrating the comet's superlunary nature; after all, if it were established that the hairy star was located above the moon instead of being composed of Aristotle's filthy sublunary vapours, the entire edifice of cometary consequences thought up by Aristotelian-Ptolemaic comet lore would prove untenable, and its nature as a

---

[346] Ibidem, 34-35.
[347] Ibidem, 28.
[348] Ibidem, 25.

celestial messenger of happiness would be rendered more convincing. Thus, to Puteanus, as to Gemma and Heurnius forty years earlier, the demonstration of the comet's heavenly nature was subservient to the establishment of its significance; the Louvain humanist from the outset made it abundantly clear that the comet entailed an assurance of peace and prosperity, instead of dire disasters.[349]

The second book of Puteanus' tract, which dealt with cometary significance, was, like his first, cosmological book, divided into nine 'paradoxes', which can roughly be subdivided into three categories. After an initial paradox in which the author once again reminded his readers of the happy nature of the comet's predictive value, Puteanus presented his readers with four chapters, each of which was devoted to one of the disasters associated with cometary appearances by the Aristotelian, physical tradition. Thus, in his second paradox our author dealt with the wars allegedly following in the hairy stars' wake; in his third, with pestilences said to be caused or signified by comets; in his fourth, with famine and other disasters; and in his fifth, with the supposedly imminent deaths of kings and falls of empires. In his next chapter, Puteanus took on the rules and prognosticating niceties of Ptolemaic astrology, none of which could find favour in his eyes. Finally, in the last three paradoxes, our author's tract reached its culmination: on the basis of the fact that certain ancient authors had believed comets to be favourable rather than ominous signs, he suggested that the present comet, too, was a hopeful sign foretelling happiness for the king, the archdukes, and the pope, and that only the impious needed to fear its consequences.

In his first paradox, then, Puteanus chided a number of ancient authors for striking terror into the common people by associating the appearance of a comet with all kinds of calamities. At one point, even dreaming about comets was considered a dire omen, our author slightingly pointed out; "What could be more stupid than that?", he wondered, since how could a crop of hair or a beard, that did not even exist in reality, cause any damage?[350] Puteanus believed that the

---

[349] See supra, p. 303.

[350] Erycius Puteanus, *De Cometa anni 1618 novo Mundi spectaculo Libri Duo Paradoxologia, Liber secundus* (Cologne 1619) 69-70: "Viri olim prudentia & doctrina Clari, errorem suum professi, dira omnia & infausta a Cometis derivarunt. An philosophos dicam? sapere se putarunt, si terrerent. . . . Quid stultius? etiam somniari Cometen, pro malorum prodigium habitum, quasi satis in somnio videre, quicquid in caelo timendum esset; noceretque divinitus sive Comae sive Barbae phasma, etiam cum non esset."

fact that so many venerable ancient authors had professed such nonsensical opinions on comets, was due to the circumstance that the poets among them, such as Homer and Virgil, made use of their poetic licence, while the philosophers and historians, such as Seneca, Cicero, and Lucan, had merely described the cometary views of the common man, not their own opinions.[351] Therefore, their utterances on comets ought not to be invested with the venerable authority of the men who had proposed them.

Neither should the biblical passages dealing with celestial omens be associated with cometary appearances, our author confidently continued. After all, the signs referred to by the evangelists were supposed to be new and unusual, and contrary to the laws of Nature, such as the eclipsing of the sun at an unnatural time, the extinction of the moon, and the falling of the stars from heaven. The prophet Isaiah, too, agreed that to announce the Day of the Lord, "the stars of heaven and the constellations thereof shall not give their light: the sun shall be darkened in his going forth, and the moon shall not cause her light to shine."[352] Thus, a comet, being part of the ordinary course of nature and consisting of light, not darkness, to Puteanus could never be regarded as a harbinger of the Day of Wrath.[353] This reading was reinforced by the oft-quoted saying of the prophet Jeremiah, "Learn not the way of the heathen, and be not dismayed at the signs of heaven; for the heathen are dismayed at them."[354]

Much the more sensible attitude, to Puteanus, was exemplified by the emperor Augustus, who had declared sacred the comet that had appeared after the assassination of Julius Caesar. The bad report which comets had had to suffer throughout history due to people's tendency to superstition, was quite unfounded, our author once

---

[351] Ibidem, 72: "Poetas furor excusat, & mysterium carminis, quod aut fictis circumduci vaticiniis debet, aut Carmen esse desinit. Cicero, Seneca, & siqui alii, non suo sed vulgi ore, ac sensu loquuntur."

[352] Isaiah 13:10.

[353] Puteanus, *De Cometa,* 72-73: "Aliud video telum, e sacrarum Litterarum armamentario depromptum. Lucae XXI *Et erunt signa in Sole, & Luna, & Stellis.* Telum, non telum. Etenim Cometas hoc loci designari, adhuc non video. Erunt signa sed nova sed horrenda, & a Naturae lege atque consuetudine aliena, *Arescentibus hominibus prae timore, & expectatione, quae supervenient universo orbi. Nam virtutes caelorum movebuntur.* Quae igitur signa? Aliis Evangelistis expressa: Matth. XXIV Marc. XIII *Sol obscurabitur, & Luna non dabit lumen suum, & stellae cadent de caelo.* Ecce non accendentur nova sidera, sed quae lucent in caliginem non naturalem ibunt. Accedat Isaiae suffragium, cap. XIII *Stellae caeli, & splendor earum non expandet lumen suum: obtenebratus est Sol in ortu suo, & Luna non splendebit lumine suo.*"

[354] Jeremiah 10:2.

again repeated; after all, light and fire were rightly associated with good things, not with evil. Joyous events were always celebrated with fire, flames, and torches; fire was sacred to the ancient Persians, and in Rome, too, the Vestal virgins were charged with the task of guarding the eternal flame of the Empire. If, therefore, on earth fire was invariably connected with happiness, why should the same substance be noxious in the heavens? And why should comets be singled out as especially injurious? After all, the entire admirable edifice of nature could be regarded as a divine token, in which God was revealed to man in various guises: both his menaces and his favour could be detected in the natural world.[355]

Once again, Puteanus stated that lights such as comets were to be praised, since evil omens were invariably connected with darkness, such as solar and lunar eclipses. One of the most common and revealing signs of nature was the daily alternation of day and night, which could be regarded as a token of the transitoriness of human life, teaching us that, like the day, human beings too were destined by fate to be extinguished. Furthermore, the succession of day and night could be seen as a simile instituted by nature to show mankind that human affairs, too, tended to alternate, and that in life sad and happy things invariably succeeded each other. From these observations, Puteanus believed, it could be deduced that human minds were continually being admonished, by the miracles of nature, not to be over-confident, nor to allow themselves to be disheartened by fear.[356]

Thus, like Gemma and Heurnius, Puteanus believed the whole of

---

[355] Puteanus, *De Cometa*, 75-76: "Ne accusemus lumen: rebus laetis assignatum est. Iterum, ne accusemus lumen: praeeunte velut natura, gaudium omne igni, flamma, face, celebratur. Persarum olim Religio nonne sacrum ignem habuit? Apud Romanos, quae Vestalium cura, nisi ut perennem Imperii flammam custodirent? Scilicet, profanum in Caelo noxiumque censebimus, quod in terra curiosa pietas consecravit? Cur vero praecipue Cometae molesti simus? quaecumque videmus & admiramur, summi Numinis monita sunt: in omnibus Deus ostenditur, nunc minax, nunc propitius, ut naturam experimur."

[356] Ibidem, 76-77: "Ut istuc redeam, lumen laudo: a tenebris potius infausta, infesta omina sunt: tegi Solem, languere Lunam, abscondi astra, denique, ut familiarissimum ostentum adiiciam, terrarum umbra diem quotidie frangi. Satis me hercule, satis ab una nocte horroris est: quae quoties terrae incumbit toties mortis legem ac necessitatem ob oculos ponit. Ne mireris igitur fato agi homines & extingui; dies extinguitur. Reliquae fortunae vices eandem imaginem, atque similitudinem subeunt: quia enim lux caliginem excipit, & caligo lucem; variari haec humana discimus, in omni vitae statu laeta tristibus, tristia laetis succedere. Imo discimus, continuis naturae miraculis humanos moneri animos, ne vel nimia se dilatent fiducia, vel terrore comprimantur."

nature to be replete with divine meaning. Through the 'miracles' of nature, which to Puteanus consisted of the orderly workings of the natural laws, such as the endless alternation of day and night, the divinity meant to convey messages to humankind. These divine signs could be read and understood by the human intellect; thus, light was a good sign while darkness announced calamities. Puteanus' scepticism with regard to cometary prognostication, then, should not be regarded as unbelief in the existence of a solicitous godhead caring for man and sending him signs; on the contrary, to the Louvain humanist, the entire universe in all its workings was one giant means of communication between God and men. However, in his view, the traditional methods of explaining cometary significance (Aristotelian physicalism and Ptolemaic astrology) were completely unfounded. He spent his next five paradoxes corroborating this rejection of conventional cometology.

The first of the traditional cometary concomitants Puteanus chose to invalidate was the prediction of war, death, and arson. Why, he asked, should a comet foretell death? The reason usually given was its bloody red colour, or the ruddiness of its tail. But the present comet was not red at all; it rather displayed a pale, yellowish or brown complexion, shining with a feeble light. If, in the beginning, it had been a little bloody, this sanguine colour had disappeared as soon as the light of the sun had become stronger.[357] Another explanation for the comet's deadly connotations given by the astrologers was the possibility that some cruel, malicious star infused it with its noxious influences. However, this option was equally determinedly rejected by our author, who stated that such a theory was not to be conceded to the astrologers.[358]

Puteanus also dismissed the idea that comets announced fires because their rays gave the impression of being intermingled with smoke; after all, the astrologers reasoned, "Where there is smoke, there is fire." Obviously, such a line of reasoning made no sense to our author, who had devoted his entire first book to demonstrating that comets were no burning vapours. "In a comet, there is nothing

[357] Ibidem, 78: "Cur caedem? an quia sanguinis colore sidus horret, aut ferali rubet crine? at hoc nostrum pallidum magis, aut luteum aut subfuscum est, & pigro velut lumine. Si quid tamen sanguinis sub initium fuit, mox elanguit, postquam e longinquo magis lumen Solis venit."

[358] Ibidem: "Repeto: cur caedem? an quia crudele aliquod sidus vires infudit? sed tantum adhuc Astrologis non damus."

but light", Puteanus confidently stated. After all, material that was on fire constantly changed its colour, burning now more brightly, now less. Moreover, the fires that did occur in the lower regions of the heavens, such as chasms, were never followed by conflagrations on earth.[359]

Indeed, the whole question of whether fiery celestial appearances announced wars and their calamitous consequences was irrelevant; after all, wars broke out constantly anyway, whether or not comets appeared. Never had people postponed hostilities until a bloody omen had shone forth; clearly, shedding blood came naturally to humankind, and human beings preferred dying for no good reason to ceasing their fighting. They had even promoted the shedding of other people's blood to the status of a virtue, and there existed peoples that would rather live without a fatherland than without an enemy, rather without a Lar than without Mars.[360] Throughout history, wars had been provoked out of ambition, avarice, or a naturally violent way of life: the Greeks were involved in constant warfare, and the Romans, not content with their own city, threw the entire earth into disarray, subjecting Europe, Asia and Africa to their sway. Nor did the fury of war remain confined to the Mediterranean area; in the more northerly regions, too, incessant conflicts had raged from time immemorial.[361]

This reflection on the ubiquity of warfare throughout history in-

---

359 Ibidem, 78-79: "Ergo cur Flammam? an quia multo fumo radii videntur commisceri? scilicet *Flamma fumo proxima est.* Sed tolle fumum, tolle flammam: in Cometa nonnisi lumen est. Nequid hic peculiare putes: ubique incendii colorem materia variat, nunc puro liquidoque igni, nunc impedito, nec sincero. Et ab igni incendium, si fatale voles, removebo. Subiectam caelo regionem videmus vario & modo, & loco ardere: sed nullam in terra exustionem sequi."

360 Ibidem, 79-80: "Bella, & bellorum aerumnas canant, aut non canant ignes; bella ingruunt, sine quibus genus humanum numquam fuit, numquam erit. Tolle Cometam: in se tamen mutuo mortales saevient, & mitissimum quod possident nomen violabunt. Hactenus enim ut sanguinem funderent, cruentum caeli signum non expectarunt. Collidi bello, tam propemodum humanum est, quam Martiale: quo furore passim ruunt populi & quasi ad feritatem nati ob caussas non caussas perire malunt quam non pugnare. Sic & crudelitas in gloriam olim versa: hausisse sanguinem, decus fuit: quasi per stragem, & caedem virtus inclarescet. Quae virtus? dira foeda, & ab omni humanitate aliena. Si licet dicere, maluerunt aliquae gentes sine patria quam sine hoste vivere: sine Lare, quam sine Marte."

361 Ibidem, 80: "Aliquae [gentes] tamquam pace & felicitate ignavae, bellum semper, iure, sive iniuria, provocarunt. Tantum partim ab ambitione, partim ab avaritia malum fuit, aut ab assueta violentiae natura. . . . Graeci igitur quando unquam quieti? continuis collisi bellis, non terram modo suam, sed & mare crudelitate impleverunt. Ut infelicissimi essent, tantum Cometae deerant. Romani vero non Urbe,

spired (or served as an introduction) for Puteanus to a digression on the contemporary situation in the Netherlands. Observing that, whenever historical wars had temporarily ceased, this had always been in order to prepare an even greater bloodshed, our author stated that, although recently, under the benevolent reign of the most serene Princes, the archdukes, the Netherlanders had been able to enjoy the blessings of peace, it was not yet clear whether this good fortune was shortly to be despised once again.[362] In any case, even happiness could be dangerous: once people grew accustomed to the ceasefire, they were wont to attribute public calamities to the enemy, instead of to a star, and once again to endure the burdens of war as briskly as they had earlier been happy to be at peace.[363]

Puteanus' home town of Venlo, situated in the province of Limburg on the banks of the river Meuse, had recently experienced just how hazardous even peace could be. The beautiful and loyal town had lately been inundated by the river, its eternal enemy; therefore, it now stood in great need of having its walls and dikes repaired. Yet, "the powers of the town could not keep up with its strength of mind", and the construction of a bulwark, on which work had already commenced, had had to be interrupted. Therefore, Puteanus appealed to the archdukes to come to Venlo's aid, shrewdly adding that, if they did so, not just one town, but the entire province would be strengthened, and the town's inhabitants, who had often before had to ward off enemy attacks because of its strategic position, would be all the more likely to fight and even to die for their fatherland and their Princes.[364]

Thus, Puteanus deftly made use of the occasion of bringing to the

---

non Italia contenti, Orbi terrarum graves fuere. . . . Verumenimvero Septentrionem nostrum magis miror, & nationes sub frigidis astris continua rabie incendi, quippe Annalium nostrorum memoria ubique functa est."

362 Ibidem, 82: "Si qua etiam ex intervallo quies, si quod otium, ad maiorem quasi cladem fuit. Hic igitur caelo fidem ardenti habeo: *ipsa tulit bellum, natura per ignes,* pacem quidem his aliquot annis aut pacis bona intelligere, sub optimis Serenissimisque Principibus, Belgae coeperunt: sed an rursus fastidire?"

363 Ibidem: "Quid dicam? interdum & felicitas ipsa molesta est. Quod si etiam posita semel placebunt arma, calamitates publicas non sideri, sed hosti imputabimus, belli onera tam strenue laturi, quam libenter quievimus."

364 Ibidem, 83-84: "Ut sic loquar, mei me imprimis Cives, mea me, in qua natus & educatus sum VENLONA monet. O pulchrum situ, pulchrius fide & constantia, oppidum! sed non satis ornatum nisi & tutum sit. Hei! deficit, & patet qua oppugnari olim solet, ab amne nuper, tanquam novo, perenni tamen hoste deformatum. Instaurari damna posse pericula arceri: egregio coeptum molimine munimentum ostendit.

Archdukes' attention Venlo's cause. However, not all his thoughts were bent on his home town, or even on his fatherland; in a Stoic, cosmopolitan vein, our author went on to pay a fair amount of attention to wars going on in other parts of the world. Thus, he stated that the peace in the Netherlands had been accompanied by hostilities elsewhere: as soon as the Truce had been concluded, conflicts had broken out in Jülich and Cleves (whose problems in Puteanus' opinion had not yet been satisfactorily resolved), between Sweden and Denmark, in France, in Savoy, and most recently in Bohemia. In many parts of the world, then, armies were clashing and towns were being raided; why, then, should the comet be feared as a harbinger of war?

After all, Puteanus pointed out, war was already raging far and wide.[365] It was a mistake very commonly made by the human race, to pay attention to evil and calamities only when these were preceded by signs; when, conversely, no signs appeared, they were wont to bear everything in silence. It was certain, Puteanus declared, that wars did not occur because of the apparition of comets, but that wars that would have broken out anyway were invariably attributed to the hairy stars.[366] If one were to subject ancient history to a detailed examination, one would undoubtedly find that many more wars had occurred than comets had shone forth. The same thing could be said about the present times; thus, although no less than twenty comets had appeared in the course of the sixteenth century, there had been many more wars. In fact, the whole century had seen nothing but wars, so that, if the astrologers' reasonings were correct, each year should have been marked by the appearance of a hairy star.[367]

Why, Puteanus wondered, was it precisely the comet that had been

---

Coeptum utinam feliciter, & ex voto mox absolvendum! Quia animos urbis vires non sequuntur, *Pendent opera interrupta, minaeque/ Ingentes murorum.* Vos succurrite Serenissimi Principes, & iuvate in una urbe? Provincia tota munietur. Sic adiuti & propugnaculi muro cincti Cives, quem saepius pectoribus repulerunt hostem, deinceps non timebunt: sic quoque, si bellum ingruat, discrimen urgeat, stare, pugnare, imo mori pro patria, & Principibus, suave erit."

365 Ibidem, 85: "... In aliis Orbis partibus concurrerunt maximae saepe acies exercitus deleti sunt, urbes expugnatae: & nos nunc tamen homines bellum timemus, quasi nullum fuisset: Cometem timemus, quasi pacis hostem."

366 Ibidem: "Haec mortalium consuetudo est: notant mala, quando signa praecedunt: his destituti, pati malunt, quam conqueri. Non certe quia Cometae fulgent, sequuntur bella, sed quia sequuntur bella, Cometis assignantur."

367 Ibidem, 86-87: "... Expletum quidem nuper seculum decimum quinctum viceni circiter Cometae perterruerunt: sed profecto difficiliori adhuc calculo clades

burdened with the ill repute of being a harbinger of war? After all, the other celestial bodies also sometimes shone with a reddish light, while lightning and thunder, too, were more likely to be associated with warfare. Indeed, winds, storms and clouds presented a more martial aspect than comets, continually moving and clashing.[368] Undoubtedly, the widespread fear of comets had been caused by people's tendency to superstition: people, Puteanus believed, loved to fear, and accordingly invented things to be afraid of, so that, in the end, all natural phenomena were associated with the calamities of war.[369]

Another disaster invariably connected in people's minds with the appearance of a comet, was the outbreak of pestilential disease. However, this to Puteanus was just as unfounded as the comet's alleged indication of wars. Our author proposed to systematically investigate the claims of the astrologers: they might uphold that comets were located in the heavens or that they stood in the air, and again that they were signs of pestilence, or that they caused the disease. First of all, celestial comets being the causes of terrestrial disease was totally unthinkable to Puteanus. After all, the stars in his opinion did not emit any contagious influences or powers; they were pure, simple, and uncontaminated. The celestial bodies, possessing a happier nature than the earthly elements, could never be charged with causing illness. Much more likely, the sublunary element of air was responsible for causing pestilences, since whatever was prone to grow corrupted, was capable of infecting with illness.[370] Nor could heavenly comets be said to be signs of coming diseases. As in the case of war, so too in the case of pestilence it had often occurred that plagues had broken out without any comet being seen. When, in sacred history, God had wished to announce an imminent pestilence

---

computentur. Desunt igitur? imo vero, ut liberius omni caelum culpa liberem, supersunt. Quicquid illud fuit temporis, nil nisi bellum fuit, adeoque tot merito Cometis notandum, quot anni numerantur."

[368] Ibidem, 87-88.

[369] Ibidem, 88.

[370] Ibidem, 90: "Vel igitur in caelo, vel in Aere Cometa est: iterum, vel ut signum, vel ut Causa. De Caelo prius: a quo Caussam audacter removeo, ne divinum corpus humano infici malo fateamur. Nullam profecto sidera tabem spirant, nullumque virus effundunt: pura sunt simplicia, & ab omni prorsus labe aliena. Hoc satis est, sidera sunt imo hoc satis, in Caelo sunt, & felicioris naturae genio dotata. Elementa sua mortales accusent, imo timeant morbis obnoxia; in elementis aerem, mali huius fontem. Timeant, inquam: nam quia putrescere & corrumpi solet, inficit."

to king David, he had done so by means of a prophet, not a comet.[371]

Even if it be conceded to the Aristotelians, that comets were located in the sublunary air, Puteanus still believed that they could not be upheld to be either signs or causes of imminent disease. Everyone, he stated, would agree that sublunary entities were highly unlikely to be divine signs; after all, a sign should be regarded as the voice of God, who spoke out to reveal his power and to admonish mankind. Such a highly dignified status could only be allotted to celestial entities, situated in a region much better and holier than the common air. Moreover, a heavenly location assured ubiquitous visibility and a worldwide audience to the sign in question.[372]

Nor could a sublunary, Aristotelian comet be the cause of pestilential diseases. Commonly, it was held that pestilence was caused by intemperance of the air, brought about by dry and poisonous vapours, expired by the earth, especially in marshy and moist regions. These vapours were supposedly attracted by the force of the sun, and in due course condensed into a ball that grew continually larger, until in the end, moulded and heated, it caught fire.[373] Assuming this to be an accurate description of the generation of a comet, Puteanus proposed to investigate whether the theory of comets as authors of pestilences could be maintained. Since, according to Aristotle, comets were invariably followed by forceful winds, which were also brought about by dry terrestrial exhalations, even before the enflaming of a comet the atmosphere should, on this view, be filled

---

371 Ibidem, 91: "Ab aere quoque, non caelo Apollinis tela sunt: hoc innoxium est, a Caussa tanti mali, nec minus a Signo remotum. Ut bellum non denunciatur Cometa: ita nec Contagio. Saepius enim pestilentia saeviit, nec moniti hoc Signo mortales fuerunt. Deinde cessent signa: moneri Vaticiniis poterunt. Sic Deus olim, ut Sacrae testantur Litterae, Regi Davidi tergeminam calamitatem denunciaturus, non Cometam, sed Prophetam misit. Immisit Pestem, sed tanquam a caelo non profluentem." See II Samuel 12:15.

372 Ibidem, 91-92: "In Aere enim etsi concedam Cometam esse; Signum esse, aut Caussam nego. Et de Signo quidem controversia non erit. Nam quia a Caelo exultat, infra Caelum non consistet, melius Aere atque sanctius *Erunt Signa in Sole, & Luna, & stellis*. Si licet dicere: Signum Dei vox est: Deus, ut homines moneat & potentiam ostendat, Caelo, non Elemento loquitur: & qui in Caelo, ubique sane praesens est: ubique tanquam coram auditur."

373 Ibidem, 94: "Pestis, aiunt, ab aeris intemperie est: haec a sicco & venenoso halitu, quem in locis praesertim palustribus & uliginosis humor ad faecem siccatus exspirat. Atqui ex halitu etiam Cometae, Solis radiis in altum educto. Cogi autem paulatim hanc materiam, velut effumantem, & in molem continuo densari incremento, donec subacta & velut cocta incendatur."

with noxious vapours. The subsequent burning of a comet ought, if anything, to purify the air and consume the pestilential vapours present in it. Thus, the air could be corrupted without any comet appearing, and, conversely, a comet could appear without the air being infected.[374]

Moreover, Seneca, much admired by Puteanus, taught in his *Natural Questions* that pestilences were caused by death-carrying elements hidden in the depths of the earth, which occasionally rose up to infect the atmosphere with unfamiliar diseases.[375] Thus, in the Roman Stoic's view, pestilence had nothing to do with comets. Finally, the association made by Manilius between heavenly torches such as comets and earthly funeral pyres[376], was unconvincing to Puteanus. After all, not everyone burned their dead; most people buried them, and no analogy whatsoever existed between graves and comets.[377]

Apart from wars and pestilences, comets were also commonly charged with causing infertile soils, scarcity, and famine. It was said that, whenever a hairy star shone forth, this indicated that all the juices and powers of the earth were extracted from her body and consumed, so that all the ploughing and sowing of the peasants was to no avail.[378] However, this reasoning made no sense to Puteanus, who stated that infertility of the soil occurred in regions hit by war or pestilence, where the fields were abandoned or devastated. These scourges were caused by human beings, not by comets; our author pessimistically added that people often preferred to claim their food from their enemies instead of from the earth, and to spend their lives shedding blood rather than working. Comets, as the alleged causes of

---

[374] Ibidem, 94-95: "Quid dicam? post haec omnia, vel supersunt halitus, vel assumuntur & absumuntur omnes. Superesse, mirum sit, si magnitudinem Cometae, & molem expendam: rursus, absumi mirum, si reliquam aeris terraeque naturam considerem. Breviter: unde ventorum vis, quae, ut Aristotelicis placet, sequi Cometas solet & ab halitibus proficisci? Supersunt igitur, cum absumi necessum sit: & sparsi aut relicti indunt aeri vitium, priusquam terrere Cometa incipiat. Imo consumuntur cum superesse dixeris: & quia flammam subeunt, expurgantur, adeoque prius nocere desinunt, quam esse. Ad summam: corrumpi aer potest, Cometa vero non nasci: quid nisi ut credas, nasci Cometam, aerem vero non corrumpi?"

[375] See Seneca, *Natural Questions*, VI.27.2.

[376] Manilius, *Astronomica*, I.893-895.

[377] Puteanus, *De Cometa*, 96-97.

[378] Ibidem, 97-98: "Hanc [tellurem] succo suo, & quodammodo vita spoliari volunt, quoties voraci scilicet portento aer incenditur: volunt quicquid magnae matri inest spiritus, quicquid vigoris, in circumfusum elementum extrahi, imo consumi. Hinc casso labore & inverti ab agricolis solum, & semen spargi, & annum impendi."

wars, might even have favourable effects on fertility, Puteanus ironically claimed: if, indeed, they robbed the earth of her juices, they presented it with human blood in return.[379]

Famine and food shortage, like war and pestilence, would occur to exactly the same extent if no comet ever appeared, Puteanus argued. It was only natural that the earth was not equally fertile at all times; therefore, mankind should not expect it to render precisely the same abundance every year. The only thing men could and should do, was to till the land, and make it yield its harvests, with continuous effort; in a word, to avoid that, through their dread of celestial signs or through apprehensive idleness, they were to deserve precisely those evils which they feared prematurely.[380] Human beings should not concern themselves with the laws according to which nature was governed; these matters they should leave to God, whose prescience ensured the harmonious balance between the elements and maintained the order of the cosmos. Thus, in the same way that the microcosm of the human body at times needed bloodletting, so also the macrocosm of the universe needed purging through the exhalation of all noxious vapours and substances. Therefore, the appearance of a comet consisting of sublunary vapours (still on the assumption that such entities existed) was to be regarded as a good sign rather than a bad.[381]

After this resolute invalidation of the Aristotelian theory, that sub-

---

379 Ibidem, 98-99: "Profecto adest Sterilitas, ubi & Bellum, & Pestis saeviunt; ubi aut deseruntur agri, aut vastantur. Hominum haec flagitia sunt, non Cometae: quippe ab hoste malumus, quam a terra victum exigere, sanguine, quam labore vitam tolerare. Alibi quoque ruri propitius Cometa fuit, quia populis (ut id iam concedam) nocuit: subtraxit succum, ut cruorem redderet: & collisit fato acies, ut felicius fruges alerentur."

380 Ibidem, 99-100: "Etenim nascatur, alatur, duret Cometes: homines tamen annonam invenient... Colendi interim, & continuo ad fertilitatem sudore cogendi sunt, ne aut signi metu, aut timido otio mereamur mala, de quibus praematurae querelae sunt. Sane & ipsas vices suas terra habet, impare foecunditati, nisi ex intervallo sterilescat. . . . Desinamus igitur pari lege annos omnes urgere: cessat interdum copia, ut vires sumat: & deficit, ut pleniori exuberet cornu, & spargatur."

381 Ibidem, 100-101: "Abit metus, spes succedat: Felicem iam audebo annum promittere & copiosam annonam: exhalat terra & exspirat quicquid noxium, & velut adustum longioris temporis mora collegit. Vegetiorem quidni putem quam purgari cerno: quemadmodum & corpora humana e gravedine & torpore videntur exsurgere, quibus vel artis subsidio per sectam venam sanguis abundans educitur, vel naturae impulsu per narium, aliosque corporis meatus effluit. Quod in Microcosmo fieri solet, etiam in ipso Cosmo: providit Deus, ut modo suo & ordine rerum Universitatis duraret: ne quid in commercio elementorum & pugna, aut superesset, aut deficeret."

lunary comets caused crop failure and famine, Puteanus stated that in fact, he need not have bothered to provide it: after all, as he had demonstrated in his first book, comets were not burning terrestrial vapours, but celestial bodies, to whom sublunary infertility could not possibly be attributed. For similar reasons, the other calamities usually associated with the hairy stars, such as inundations, fires, shipwrecks and tempests, were equally unfounded: they were caused by coincidence, not by any star.[382] The famous case of the two ancient cities of Buris and Helice, whose flooding by the sea was supposedly announced by a comet, had nothing to do with any celestial appearance, Puteanus added: they were inundated, not *because*, but *after* a hairy star had shone forth.[383] The same could be said about the recent disaster at a Swiss town which had been buried under the mountains after an avalanche. And neither could Puteanus' own recent illness, to which he had fallen victim after the appearance of the comet of 1618, and which his Jesuit friend Bauhusius had in an elegant poem attributed to the comet's noxious influence, have been caused by the hairy star: our author confidently stated that, in recovering, he had conquered, not the comet, but only the disease.[384]

One of the tract's most important chapters, from the point of view of Puteanus' patron-seeking, was that in which he invalidated the traditional idea that comets were the harbingers of princely deaths and the fall of empires. Reverently, our author stated that royal power was attributed to the rulers by God himself, and that therefore whoever violated this power, by means of weapons or by means of words spoken or written, or, like the irreverent soothsayers, by means of the constellations, committed a crime against God.[385] Many were the occasions on which the astrologers had branded the innocent comets with the sobriquet of 'overthrower of empires' or 'changer of governments': Puteanus briefly called to his readers' attention the comets that were associated with the deaths of the emperors Claudius, Constantine, Valentinian I, Charlemagne, Louis II, Henry IV

---

[382] Ibidem, 101-102.

[383] Ibidem, 104-105: "Mirentur iam Scriptores Burin, atque Helicen, Achaiae urbes, a mari haustas: non quia fulserat Cometes, sed postquam fulserat, perierunt."

[384] Ibidem, 106-108: "...Vici tamen luctantem mecum, non Cometam, sed morbum; nisi & hunc Cometen dicas."

[385] Ibidem, 110: "A Deo Sceptra sunt: in Deum nefarius, quisquis sceptra violat; alius (fierine potest?) ferro, alius, ut mitius saeviat, lingua, aut calamo: Divinus autem seu Prognostes caelo."

and Albert I, as well as the princes Giangaleazzo of Milan, Charles the Bold, and Philip the Fair.[386]

To Puteanus, none of these ominous associations made any sense: to him, comets carried no dangerous significance for princes. Once again, he had recourse to the argument that many princely deaths had occurred without the appearance of any comet, and that it was unjustified to turn coincidence into signification.[387] To our author, the story of Pope Urban IV, about whom it was said that he fell ill on the first day a comet was seen and then died three months later on the last day of its visibility, could not be adduced as proof of cometary significance: the pope, in Puteanus' view, had fallen ill because he was afraid of the comet, and had died from this fear.[388] As in the case of wars, Puteanus believed, a careful investigation of history would show that most comets had shone unaccompanied by any funeral, and that most royal funerals had taken place without a comet. The procedure employed by the cometary soothsayers, who from a few examples deduced a necessary connection, was fallacious.[389] Indeed, some of the greatest emperors had died without the admonition of a comet: the deaths of Augustus, Tiberius, Caligula, Nero, and Alexander the Great were not connected with any comet at all, while the hairy star that was associated with Julius Caesar had shone forth after his death, not before it.

The mentioning of the Roman emperors inspired Puteanus to include a digression on the emperor Nero, who, on the occasion of a cometary appearance, had grown so afraid that he had purged the nobility, relying on the Ptolemaic aphorism that a comet in the centre of a king's horoscope portended the death, either of that king himself or of one of his nobles. Nothing could be further from the truth than such futile astrological rules, Puteanus stated. The only reason Nero was murdered was his cruelty: *that* was what needed to be expiated, not the appearance of the comet, as the soothsayers thought. Thus, "kings who were Neros" did have cause to fear a

386 Ibidem, 110-111.

387 Ibidem, 113: "Quid dicam? interierunt alii, tot inquam Caesares, Reges, Principes, quot iam nominati non sunt: nullo tamen sidere moniti: casum cur in signum commutamus?"

388 Ibidem, 114.

389 Ibidem: "Hinc sane fallax observatio est, quae a fortuito paucorum exemplorum consensu necessitatem ducit: fallax, quae pari lege omnia premit. . . . Si quis cum terra caelum componat cum sideribus principes: sine funere Cometas plerosque inveniet: pleraque, sine Cometis funera."

cometary appearance: their injustice and tyranny would certainly be avenged, so that in fact they should be in continual dread of the future.[390]

The fall of many tyrannical regimes had, like the deaths of many kings, failed to be announced by comets; this was not surprising, since for a ruler to be criminal, impious, and tyrannical, to Puteanus was "prodigy enough".[391] This remark seems to suggest that our author had a negative conception of prodigies, as things or circumstances contrary to the orderly, harmonious, arrangement of the universe as laid down in the laws of nature. The same idea can be detected in Puteanus' remarks on prodigies in his first book, where, as we have seen, he defined them as those things that did not at first glance accord with the limited sense of order human beings were capable of perceiving. If, however, men were able to understand how these extraordinary phenomena fitted in with the universal order and harmony, Puteanus believed the need for a concept of 'prodigy' would disappear.[392] Such a point of view is reminiscent of the belief advanced by Lipsius in *De Constantia*, where he wrote that even bad rulers could serve a divine purpose, albeit one not perceived by man.[393]

Puteanus' next 'paradox' was devoted to an extensive invalidation of the manifold rules of Arabo-Ptolemaic astrology. According to the astrologers, cometary prognostication should be based on the form, position, and aspect of the comet in question, so that, for example, a comet in the form of a flute entailed consequences for musicians, a comet moving through the constellation of the Serpent portended poisonings, and a comet forming a triangle or a square with two or three of the fixed stars possessed predictive value with regard to intellectual activity. Puteanus from the outset made it clear that he thought such rules to be nothing but "ingenious nonsense", based on invalid analogies.[394] Nonetheless, he went on to quote extensively the rules for cometary prediction, based on the comets' forms and the

---

390 Ibidem, 120: "Nero igitur quam timidus, tam saevus fuit: dignus quidem tristissimum excipere fatum, quia expiare ausus est: interfectus etiam non aliam ob causam quam quia crudelis fuit. hoc enim erat expiare. Cavete Principes, qui Nerones estis, vobis iniquum infestumque sidus est: timete, vobis minatur."

391 Ibidem, 119 (=121): "... sine prodigio experti sunt [tyranni] quid extremum in rebus humanis esset: prolapsi, priusquam monerentur. An mirum? scelesti, impii, uno verbo tyranni fuerant: hoc satis prodigii erat."

392 See supra, p. 313.

393 See supra, § VI.3.b.

394 Puteanus, *De Cometa*, 120 (=122)-123: "... O ingeniosae nugae!".

signs in which they were born, as expounded by Pliny and by Servius in the tenth book of his commentary on the Aeneid.

However, immediately after these lengthy quotations, our author added that "such were the dreams and figments of the interpreters of the heavens, who believed that everything they attributed to the stars could be applied also to the comet, as if it were a star."[395] Puteanus made no secret of his disapproval of such a method: to him, the astrologers' procedure made a mockery of the eminently beautiful machinery of the universe, subjecting it to imposture, and reducing the renowned comets to the status of enemies of the human race. The fact that soothsayers filled the region of happiness, the heavens, with sad predictions, and forced humankind to fear everything they themselves had branded as terrifying, was symptomatic of a far-reaching alienation of man from the heavens, or of the heavens from man.[396] Puteanus strongly believed in the desirability of liberating nature's innocence from all forms of cleverly fabricated superstition. Admittedly, he said, God did admonish and terrify his people by means of miracles, and thus revealed himself to be God; yet, such miracles were always unusual, surprising, and contrary to the ordinary workings of nature. Such would be the standing still of the sun, the extinction of the stars, or the falling apart of the beautiful structure of the world.[397]

The appearance of a comet, conversely, did not run counter to the orderly course of nature; although its precise identity was as yet unknown, it was a part of nature all the same, and needed to be investigated as such to liberate man from unfounded fear. Puteanus did believe comets to possess a certain influence; however, this influence was as yet occult, since it had never been adequately examined. Because they did not possess any reliable information on the subject, the astrologers took recourse to false, invented ideas not based on

[395] Ibidem, 129: "Ecce quicquid de Astris somniant aut fingunt caeli interpretes, ad Cometen referunt, tanquam ad Astrum."

[396] Ibidem: "Sic & pulcherrima mundi machina imposturis obnoxia fit, & tot illustres ignes in exitium generis humani computantur. Dixerim profecto, aut alienari ab hominibus caelum, aut a caelo homines, dum felicitatis regio tristissimis impletur fatis, & mortale genus quicquid ad terrorem deductum est, timere cogitur."

[397] Ibidem, 130: "Monet fateor, & terret prodigiis Deus: & se Deum ostendit: sed haec tum prodigia sunt, mira, inusitata, a naturae modo aliena. Ut sic dicam, stare Solem, & caelum quiescere: extingui sidera, & nova accendi: deprimi cardinem; & tantae machinae compagem solvi. Sed quid? Haec omnia servent ordinem suum, & nihilominus Auctorem omnium agnoscam Deum. Nam quia Deus est, haec sic fiunt omnia: Sol fertur, caelum movetur, lucent sidera, constat axis, mundus durat."

any thorough investigation. Even the powers of the planets and stars, that were visible daily, had not yet been adequately traced, Puteanus pointed out; how, then, could the astrologers be so bold as to entrust the fates of entire populations and countries to the influences of strange, rare, fleeting 'fires'?[398] The most intolerable and unfounded excess of contemporary cometary prognostication, to Puteanus, was the fact that many soothsayers accorded the comets extensive powers for a long time after their disappearance. Comets might well possess powers, our author admitted; yet, to uphold that these continued after the comet itself had gone was nonsense.[399]

After this decisive abnegation of both the Aristotelian and the Ptolemaic methods of accounting for cometary consequences, Puteanus in his last three paradoxes reached the actual raison d'être of his tract. Having convinced the reader of the untenability of burdening comets with all kinds of horrible effects, he now went on to prove that it was much more likely that the hairy stars boded well for humankind. First of all, the well-read humanist set out to demonstrate that, although many examples could be found of ancient authors describing evil cometary effects (about whom Puteanus had already stated that they did so only out of poetic licence or because they described the sentiments of the common man), there had also been a number of writers from antiquity who had attributed good fortune to cometary appearances.

Such had been the comet that had announced the beginning of the reign of king Mithridates; and such was the famous comet declared sacred by Augustus because it had represented the ascent into heaven of Julius Caesar's soul.[400] Moreover, Puteanus declared, hairs invariably had a favourable connotation on earth; why, then, should we consider them noxious in the heavens? Human beings often grew

[398] Ibidem, 130-131: "Sic & Cometa naturam suam habet, & innoxius est: examinandus, ut terrore liberemur. Naturam, inquam, suam habet: imo & ἀποτέλεσμα suum habet: sed profecto occultum, aut parum adhuc exploratum. Homines igitur quia verum ignorant, affingunt falsum; & fida observatione destituti, corruptam errore opinionem amplectuntur. Nimis profecto arduum est, vim astrorum definire. Grandiori nobis quotidie flamma ipsi Planetae lucent, & tenebras tamen solertiae contemplantium offundunt: errant, cum nequaquam errent, & in errorem praesagia omnia ducunt. Adhuc nescimus, an, quomodo, quando, sive propitii, sive infesti sint, licet quotidiani ac familiares: & peregrinos istis, raris, fugacibus ignibus, non ignibus, omnium & hominum & regionum fata fatui assignamus."

[399] Ibidem, 131-132: "His fidem in futuro quaerere, & Cometis dare vim extinctis, maioris dementiae est. . . .".

[400] Ibidem, 133-134.

their hair as a sign of virtue, and meadows showered with flowers and grass were also, in Latin, sometimes referred to as 'hairy'.[401] Furthermore, Servius, in his commentary on book X of the Aeneid, had deduced, from the fact that according to Pliny comets were associated with the five planets, that their significance could be sometimes good and sometimes bad; thus, a comet proceeding from one of the 'good' planets, Venus and Jupiter, possessed favourable influence.[402] And Avienus, too, had observed that a comet with hair all around its head was benign.[403]

Thus, the authors of antiquity did admit that comets might well possess auspicious characteristics. More specifically, Puteanus stated, a comet was good whenever it proceeded from a conjunction of good things.[404] From this rule, it could be deduced that the comet of 1618 had indeed been favourably disposed. Puteanus at length described the positions of the various planets at the time the comet had first been seen, in a terminology highly charged with allegorical connotations (thus, he spoke of Mars "speeding towards the Lion, and submitting the monster as if in a new Herculean Labour"[405]). The comet, at the time of its first appearance, had found itself near Mercury, and had resembled a harbinger of dawn and a new Phosphorus; however, it had soon departed from the zodiacal circle where the planets stood, and moved northward through a number of constellations which, to Puteanus, were charged with the highest significance.[406]

First of all, the hairy star had passed the constellation of Libra,

---

401 Ibidem, 134-153 (=135): "Ipsae vero in terra Comae laetae sunt: quidni & in caelo? in homine, arbore, prato, herba, quidni in Sidere? De homine palam est: quin & κόμαν, sive comam nutrire, ἀρετῆς σημεῖον, *Virtutis Signum* antiquitus fuisse, Dio Chrysostomus indicat Oratione XXXV. Arboris & prati hoc passim apud Poetas elogium est. Peculiariter etiam κομήτης λειμών, quamadmodum ἀστήρ κομήτης, dicitur *pratum comans*, id est, Laeto, & comante gramine vestitum. Sic igitur & Stella Comans sive Crinita, quae laetos radios spargit. Quod ad herbam, Tithymallus Characias, sive masculus, Κομήτης appellatus, cuius in cacuminibus coma iuncea nascitur, & velut ridet."

402 Ibidem, 153 (=135).

403 Ibidem, 136.

404 Ibidem: "Non mala tantum, sed & bona denotari, satis, si concedat Antiquitas: unde & bonum quandoque Cometam esse, ex eodem dogmate statuas. Bonum scilicet, si ex ambiguo, alterutri bonorum iuncto."

405 Ibidem, 138: "Pari hic paene remotus spatio Mars Gradivus, impetum facere in Leonem visus, novoque labore pone Herculis monstrum urgere."

406 Ibidem, 138-139: "In extremo tandem orbe lucentem barbae iubam novum Sidus explicuit. Aurorae nuncium putes, alterumque quasi Phosphorum: sub cuius radiis Mercurius, agmen velut ex insidiis sequens, latitabat. Furtiva haec vestigia

which was happy and favourable, because being situated on the equinoctial colure, it divided the heavens into two, and marking the beginning of autumn, it separated the year into two equal halves. Because the comet had chosen this lucky sign as its point of departure, it should be considered very propitious. Moreover, the Scales carried the connotation of Justice, and had of old been associated with Rome; Puteanus quoted Manilius to the effect that "Italy belongs to the Balance, her rightful sign; beneath it Rome and her sovereignty of the world were founded, Rome, which controls the issue of events, exalting and depressing nations placed in the scales."[407] Therefore, to Puteanus, the comet's passage through Libra signified good fortune in matters of justice and religion; in particular, it boded happiness for the Pope, who was to establish unanimity in religious matters. He therefore urged his readers to submit their souls to papal authority, pointing out that sometimes, it was necessary to rule sternly, in religion as well as in secular politics.[408]

Next, the comet moved close to the constellation of Hercules. This group of stars Puteanus associated with a combination both of strength and of rational thought and prudence, and with a prudent and judicious management of worldly affairs.[409] Possibly, this should be read as an allusion to the myth of Prometheus being freed from his bonds by Hercules, which enjoyed widespread popularity in Ren-

erant; mox tamen detecta, procurrente in Boreae latus Cometa, & per signa, aut iuxta signa, quae propitia auguror."

[407] See Manilius, *Astronomica*, IV.773-775.

[408] Puteanus, *De cometa*, 140-141: "Mittet felici nunc omine Roma sanctissimae maiestatis radios: & Ecclesiae Romanae Princeps Pontifex maximus, infracto & universali Dogmatum consensu felicitatem Religionis extendet. Submittite animos, quotquot estis in Christianum nomen adoptati: sanctius pietate, quam imperio Orbis regitur. Ne quis tamen imperium desideret, Divinum est, & caelesti nutu gubernatur. Quod ad Discrimina attinet, a Poeta nominata; haec aequitas, non potentia facit: benigna bonis, ne praemium desideres; severa malis, ne supplicium averseris. Regi & confirmari sic quoque Religio debet, aut Religio non erit: Deum exprimere, qui & amantis & irascentis vultum sumit."

[409] Ibidem, 141-142: "Ad Cometam redeo: is a signifero deflectens, mox exadversum Herculi stetit, viresque caeli illuminavit. Praesenti face quidni prudentiam denotem, & summae sic quidem potestati iunctam? Quisquis ille Hercules, ille Heros, qui subiectum caelo mundum regit videt: quisquis ille (& nonne noster?) non impetu sed ratione regitur. Item, quaecumque haec prudentia, quae terrarum Orbem sustinet, oculata est, & ab astris lumen derivat. Sic tanta rerum moles secura est; sic tutae vires, quibus illa sustinetur. Videre, hoc est firmum esse: vis prudentiae expers, caeca est. Huc & poetarum figmenta spectant, qui mihi videntur sceptrum consilio instructum, Argi imagine: destitutum, Cyclopis effinxisse. Frustra affluentis fortunae copia, nisi provide & caute administretur: inani fiducia res saepe maximae dissipatae sunt."

aissance rhetoric and to which Lipsius referred in his *De Constantia*, when he called for the chains of Prometheus to be broken by a new Hercules, so that the bonds imposed on the mind by anxiety could be shaken off.[410]

In any case, Puteanus went on to identify Hercules with both the Habsburg emperor[411] and the king of Spain, so that he could attribute to the comet good fortune for both the German empire and the kingdom of Spain. About the emperor, our author, once again displaying his pacifistic attitude, wrote that he would need to conquer the 'monsters of the age' (undoubtedly he meant the recently erupted Thirty Years War), not through fear and cruelty, but through majesty and clemency.[412] Puteanus honoured King Philip III with lavish praise, calling him the Best and the Greatest, the Mightiest and the Mildest ruler on earth, who was king and commander, lord and father to nearly the whole world, 'Catholic' both with regard to his religion and with regard to his power; our author hastened to add that, indeed, the Catholic religion rightly deserved such 'catholic' power.[413]

The comet further pursued its course through Bootes; indeed, this was the only constellation which it not only passed, but actually crossed. In classical mythology, Bootes was regarded as the guardian of the Great Bear ('Arctophylax'). Puteanus made good use of this myth by identifying Bootes with the archduke Albert (who was ac-

---

[410] See Oestreich, *Neostoicism*, 18.

[411] Since our author had finished his tract on the comet by 4 March 1619, when he wrote it he probably had in mind the emperor Matthias, who, however, died in the course of March 1619, so that Puteanus' readers would think of Ferdinand II, Matthias' successor, as the German emperor.

[412] Puteanus, *De Cometa*, 142: "Mundi Hercules, quis nisi Caesar Augustus AUSTRIACUS, Imperii concordiam & felicitatem, summo auspicio urgens? vere Caesar, & in rerum motu constans, in tempestate fortis: vere Augustus, & Numinis ope aeternitatem nominis stabiliens. Mala & monstra saeculi fugient, non pelle tamen & clava, id est, horrore & saevitia; sed sceptro & diademate, id est maiestate & clementia superanda."

[413] Ibidem, 142-143: "Mundi Hercules, quis nisi novi Orbis Monarcha PHILIPPUS Rex Catholicus? Suo prope unius exemplo ostendit, quid sit in terris Maximum & simul Optimum esse, Potentissimum & Lenissimum. Certat cum humanitate quaedam, ut sic appellem, divinitas: faciliusque terminatam cum Orbe terrarum fortunam regit quam singula alii regna, provincias, urbes, familias. Etenim ubique quasi Rex est, ubique Dominus, ubique Praefectus, ubique Pater. Putes, non Religione tantum, sed potentia nomen titulumque CATHOLICI exprimi. Ne miremur: Religio tantam potentiam meruit: potentia ad summum fastigium cura & consilio provecta, populorum salute confirmata est."

companied by his 'Astraea', or Virgo, Isabella) who, in our author's interpretation, was chasing the fugitive Bear and challenging it for its own good, in order to make it happy.[414] Although Puteanus did not explicitly identify the Bear with the people of the northern Netherlands, he did seem to imply such an interpretation when, immediately following the passage about Bootes/Albert and his Bear, he added: "Why do you hesitate, peoples worthy of a better fate? Return from whence you have deflected. No tranquillity can be found, save under the reign of one king; and the greater and better this king, the more just is his government."[415]

Here, Puteanus finally reached the culmination of his tract: his interpretation of the comet's meaning was, that it had in store goodness and happiness for those governing the southern Netherlands, and that it urged those still resisting them to recognize that by means of this heavenly sign, God admonished all Netherlanders to submit to the reign of the Archdukes and thus to bring the war to an end: "Either everything is vain, or I would say that by means of this celestial fire is designated by whom the Netherlanders should finally be governed, after the turning point in this most miserable war."[416]

Thus, Puteanus did not seem to be very consistent with his own principles. After having rigorously invalidated the rules of astrology, which accorded significance to comets on the basis of their forms, their courses through the heavens, and their positions with regard to planets and stars, he went on to attribute to the present comet a highly dignified meaning on the basis of precisely such considerations, happily making use of the constellations through which the comet had passed and the allegorical associations these carried. He himself realized that some of his readers might believe him to have 'usurped' the predictions of the astrologers to suit his own purposes;

---

414 Ibidem, 145: "Fugientem [ursam sequitur], sed suam: sequitur, ut servet: invitat, ut felicem reddat."

415 Ibidem: ". . . Quid haeretis meliore digni sorte Populi? redite unde descivistis. Non alia tranquillitas est, quam ab uno regi: qui quanto maior meliorque, tanto aequius imperium est."

416 Ibidem, 145-146: "Siqua Prodigio fides, novum quod fulget Sidus fausta omnia pollicetur. A caelo ducendum est omen: illustratur. Deus ipse promittit Principem, quem ostendit. Educite a terra, & erigite oculos: ostendit AUSTRIACUM, ostendit ALBERTUM ostendit vestrum. Date honorem Maiestati, in aetherea regione consecrata est: recipite. Per astra designatur: pacem & felicitatem admittite, utramque Cometa prominente fulgore annunciat. Aut vana omnia sunt aut Caelesti igne designari dixerim, a quo tandem post miserrimi belli conversionem Belgae debeant regi."

however, to him there was an essential difference between his own prognostications and the humbug of the soothsayers.

In Puteanus' opinion, the significance he himself accorded to the comet was pious, while astrological predictions were superstitious, the difference lying in the fact that piety always prophesied good fortune, whereas superstition feared evil and doom.[417] The soothsayers dreaded that whatever they predicted (pestilence, war, famine, and the like) could not be averted. This, to Puteanus, was an impious stance, undoubtedly because according to Roman Catholic religion it was a mortal sin to cease believing in the possibility of divine mercy and salvation. Thus, by giving a religious twist to the matter, our author dexterously managed to discredit the dire predictions, based on Aristotelian physics and Ptolemaic astrology, usually attributed to comets by the astrologers, while yet allowing for a certain auspicious, and to him more elevated, degree of significance proper to the hairy stars.

At the very end of his tract, finally, our author once again articulated his Senecan idea, that comets were only called portents because they were such rare occurrences, and that they were feared because their essences were not known, never having been adequately investigated. Much more dreadful, to Puteanus, were those 'prodigies' which found themselves, not in the heavens (which, as mentioned before, our author believed would be completely devoid of prodigies if only they were investigated more closely) but in the human mind; such mental prodigies, such as avarice, injustice, and impiety, were what caused God, the Law, and Nature to inflame with anger. For whenever God was neglected and despised, for example by those who behaved themselves impiously with regard to the king, the prince, or the magistrate, his fury was aroused, and he would order Nature to act as his avenger.[418] Therefore, our author concluded his tract with a poem in which he pointed out that comets were indeed to be feared, yet these 'comets' were not heavenly bodies such as the one that had shone forth in 1618, but rather the crimes perpetrated by human beings.

We can conclude that, in Puteanus' tract, the teratological tradition revived by Gemma and Heurnius reached its culmination. Like

---

[417] Ibidem, 146-147: "Amovi praesagia, ut tamen usurparem: pietate vicisse superstitionem videor. Pietas bona omnia & fausta ominatur: Superstitio, mala & infausta semper timet."

[418] Ibidem, 148-149.

the two 1577 authors, Puteanus believed in a divinity whose powers pervaded the whole universe and unified it as a well-ordered, harmonious whole, and who was concerned about the well-being of mankind and for this reason at times sent signs to urge them to better their lives in order to avert punishment. However, whereas Gemma and Heurnius still combined this teratological outlook with the use of Ptolemaic rules of astrological prediction, and to a certain extent with the physical consequences attributed to comets by the Aristotelian tradition (thus, Gemma still believed that most comets were burning sublunary vapours, and Heurnius connected the cometary effects with the natures of Mars and Mercury), Puteanus for the most part did away with these traditional explanations. To him, comets were no supernaturally created, divine signs; rather, they were to be regarded as parts of the orderly workings of the natural law that kept the universe together and that made it into a well-balanced whole. Indeed, in his view, the entire cosmos should be seen as one great miracle, in every part of which God's majesty manifested itself. The entities commonly referred to as 'prodigies', such as comets, to Puteanus constituted nothing more or less than parts of the natural order; the only thing that distinguished them from 'non-prodigious' natural effects was the fact that they were as yet inaccessible to human understanding. Thus, Puteanus' view was very reminiscent of that of Saint Augustine, according to whom the whole of nature was one great miracle.[419]

However, despite his efforts to invalidate Ptolemaic astrology and Aristotelian, physical cometary consequences, Puteanus never quite made the final step of stripping comets of all significance whatsoever. In fact, as he himself pointed out, he 'usurped' the self-same astrological rules which he had so vigorously criticized earlier. Comets were ordinary parts of nature, and rules of analogy which stated that certain constellations carried significance with regard to the earthly objects they resembled were nonsense; nonetheless, Puteanus in his final chapters subjected the comet's course through the heavens to an interpretation that would not have been out of place in a perfectly traditional astrological treatise, the only difference being that Puteanus' predictions were favourable while those of the soothsayers were usually gloomy.

Obviously, it is almost impossible to ascertain to what extent

[419] See supra, § VI.2.a.ii.

Puteanus actually believed in his own obsequious auguries. However, bearing in mind his reputation as a deft patron-seeker, it cannot be ruled out that much of the good fortune he associated with the comet was inspired by simple profit seeking. Indeed, apart from the three final chapters, his arguments against cometary prognostication very closely resembled those that were to be proposed, some sixty years later, by the confirmed sceptic of cometary significance, Pierre Bayle. The insistence that wars and the misfortunes engendered by them, were due to nothing but human belligerence; the idea that, if one event followed another, this need not necessarily mean that the first event caused the second; the criticism of the astrological procedure of deducing necessary rules from a limited number of examples; the invalidation of the idea that cometary effects could last after the comet itself had disappeared; the idea (explicit in Bayle, and implicit in Puteanus' statement that it was nonsensical to believe comets announced wars, since in other parts of the world wars were already raging) that there was no reason why a comet should carry significance only for a limited part of the world; the idea that *real* divine miracles were occurrences blatantly contrary to the laws of nature, which the appearances of comets were not; and the acknowledgement of the possibility that comets might equally well announce good fortune instead of disasters; all these criticisms of traditional cometary astrology were put forward both by Bayle and by Puteanus.

Thus, the culmination of the teratological tradition as articulated by Puteanus (at least, in the first three quarters of his tract), according to which the whole of nature was a miracle, since it displayed such miraculous order and harmony, came close to eliminating teratology and divine intervention, precisely because it accorded both an inordinate amount of importance. While Gemma and Heurnius had seen the hand of God at work principally in those occurrences that deviated from the ordinary course of nature, Puteanus, in seeing it at work everywhere, in every natural entity that, in an obvious or in a hidden way, fitted in with the harmonious cosmic pattern, came close to reducing it to marginality. Equating God to nature, to which Puteanus' stance almost amounted, formed a first step on the way to removing God and leaving only nature behind.

c. *The Invalidation of the Astrological Tradition: Fienus and Fromondus*

Puteanus' colleague from the medical faculty, Thomas Fienus, also subjected the comet's possible predictive value to an examination. However, he from the outset made it clear that his was only a limited treatment of the subject; in the terminology of our present chapter, we could say that Fienus proposed to investigate the hairy star's astrological consequences, but not its possible teratological significance. Thus, our author explicitly stated that, despite many ancient testimonies to the contrary, in his opinion comets did not predict anything in a natural way; however, he did not know whether they carried any supernatural predictive value, through divine institution. Fienus, considering himself incompetent, as a professor of medicine, to present any arguments for or against the latter question, announced that he gladly left the elucidation of this supernatural problem to the theologians and the authority of the Holy Fathers.[420] This was not the only occasion on which Fienus chose to dissociate himself from the theological aspects of a natural philosophical question; thus, in a letter of 6 June 1629, Pierre Gassendi praised the Louvain medical professor for treating, in a recent booklet, the philosophical question of the animation of the human foetus in a purely philosophical way, "taking care to preserve the sovereignty of religion."[421]

Fienus well realized that, in the past, cometary consequences had been interpreted in a number of fundamentally different ways. Thus, he quoted Aristotle as a representative of the physical explanation, to the effect that, being dry exhalations, comets announced winds and drought; Ptolemy, as a representative of the astrological tradition, to the effect that comets worked Martial and Mercurial effects; and Pliny, John Damascene, Eusebius, the Sibylline Oracles, Virgil, Lucan, Claudian, and Manilius, as representatives of the teratological tradition that regarded comets as signs of divine wrath and hor-

---

[420] Thomas Fienus, *Dissertatio de Cometa anni 1618* (Antwerp 1619) 59: "His tamen omnibus non obstantibus, opinio mea est, Cometas nihil praesagire naturaliter, seu virtute naturali. An aliquid praesagiant supernaturaliter, & ex instituto Dei, non est meae Professionis disputare; sed id relinquo Theologis disputandum, & auctoritate sanctorum Patrum. . . .".

[421] See F. Sassen, "De reis van Pierre Gassendi in de Nederlanden (1628-1629)", *Mededelingen der Koninklijke Nederlandse Akademie van Wetenschappen, Afdeling Letterkunde, Nieuwe Reeks* 23-10 (Amsterdam 1960) 263-307: 280, who quotes Gassendi's letter to Fienus to the effect that "Tu mihi, vel eo nomine commendandus maxime videris, quod philosophicam materiam pertractes philosophice; id tamen curans interea, ut sua religioni majestas servetur."

rible disasters.[422] He also added that Holy Writ seemed to indicate that such celestial signs always portended God's anger and the imminent punishment of the human race, or even the Day of Judgement; however, his use of the word 'seems' suggests that our author was not very confident about the correctness of such an understanding of the biblical text.[423]

Of these three different methods of interpreting the meaning of comets, Fienus proposed to investigate only the first two: the physical way of Aristotle and the astrological way of Ptolemy, leaving the third, teratological way to the theologians. In order to prove that comets signified nothing in a natural way (for he from the outset made it clear that, from a philosophical point of view, the hairy stars' predictive value was nonexistent) our author proposed six 'proofs', the first, and by far the lengthiest, of which was directed at the Aristotelian way of accounting for cometary effects in the atmosphere, as proposed by the Stagyrite himself and elaborated by his medieval adherents.

To clarify the procedure by which he meant to invalidate the Aristotelian theory to his readers, Fienus first of all reminded them that, for a phenomenon to be regarded as a physical sign of another phenomenon, it had to be either its cause, its co-effect, or its effect.[424] Therefore, for comets to indicate future ills, they would have to be either causes, co-effects, or effects of these ills. The latter option Fienus at once brushed aside: after all, an effect indicated its cause, which necessarily occurred either at the same time, or before, the effect, while, conversely, comets were supposed to presage *future* calamities.[425] Neither could comets be regarded as co-effects of their disastrous concomitants, since comets and their consequences did not

---

422 Fienus, *Dissertatio*, 57-58.

423 Ibidem, 59: "Quin & sacra scriptura variis locis videtur indicare, eiusmodi caeli signa semper Dei iram, & futuram humani generis castigationem indicare; ut 11 Ioelis, & Matth. 24 ubi habetur diem iudicii iudicatum iri eiusmodi, in caelo, ac stellis, portentis praegressis."

424 Ibidem, 59-60: "Si Cometae eiusmodi futura mala indicarent, vel indicarent ea ut causae eorum, vel ut coeffectus; nam ut effectus non possunt indicare. . . . Praeter has signi rationes fere nullae sunt aliae, quia nihil fere est signum alterius Physice, nisi vel ut effectus eius, vel ut coeffectus, vel causa."

425 Ibidem: ". . .nam ut effectus non possunt indicare. Quia quod indicat in ratione effectus, indicat suam causam; & indicat praesens vel praeteritum. Cometa non indicat mala ut suam causam; quia mala subsequentia non sunt causa Cometarum; sed potius Cometae deberent esse ipsorum malorum causae. Et Cometae non indicant praesens, vel praeteritum, sed futurum."

proceed from the same cause. The cackling of chickens, Fienus pointed out, was a co-effect of these chickens' laying an egg, because both circumstances proceeded from one and the same cause (namely, the chicken), and occurred almost simultaneously, because chickens invariably cackled before laying eggs.[426] Comets, conversely, did not proceed from the same cause as their supposed concomitants[427].

In order to prove this, Fienus invalidated the Aristotelian idea, that an abundance of hot and dry exhalations in the air was the material cause of the generation of comets, and at the same time the efficient cause of atmospheric disturbances; from Fienus' point of view, this reasoning was not valid, since comets did not have any dry exhalations as their cause: they were celestial bodies.[428] For the same reason, to Fienus it was unthinkable that comets, being located in the heavens, should be the causes of future terrestrial ills; after all, the entire Aristotelian reasoning about cometary consequences was based on the pernicious effects of dry, fat, hot, sulphurous exhalations burning in the atmosphere.[429]

Indeed, Fienus provocatively added, even if comets did consist of the vapours on fire which the Aristotelians made of them, they would still not indicate any of the sublunary disturbances or calamities that were so often ascribed to them.[430] Before putting forward his reasons

---

426 Ibidem, 60; the example of the cackling and the egg Fienus borrowed from Julius Caesar Scaliger, *Exotericarum Exercitationum Liber XV De Subtilitate, ad Hieronymum Cardanum* (Frankfurt 1612) 296: "Verum sicut aliquis affectus est alterius signum propter cohaerentiam: veluti signum, quod Gallina peperit, cantus. Nam Gallina utriusque & partus, & cantus autor est. Neutrum vero ab altero pendet necessario."

427 Fienus, *Dissertatio,* 60-61: "At Cometae & futuri tristes eventus non sunt effectus, ab una & eadem causa procedentes; ergo non possunt se mutuo indicare; vel unum non potest indicare alterum in ratione coeffectus."

428 Ibidem, 61-62: "Porro quod Cometae, & futura mala etiam non sint coeffectus, probo: Quia non possunt in alia ratione Physice esse coeffectus, quam quod Cometae sint effectus exhalationum in genere causae materialis: & futura mala sint effectus earum in genere causae efficientis; adeoque in diverso genere causae. At futura mala non sunt effectus talium exhalationum ex quibus fit Cometa: quia Cometa non fit ex talibus exhalationibus, sed est corpus caeleste, ut ostensum est: ergo futura mala & Cometae non sunt coeffectus, seu effectus ab eadem causa procedentes."

429 Ibidem, 61: "Quod Cometa non sit causa futurorum eventuum, probo; Quia, quicumque asserunt Cometam esse causam futurorum eventuum, nititur eo fundamento, quod Cometa sit exhalatio fumida, pinguis, calida, sicca, sulphurea in aere accensa. ... At haec doctrina est falsa, ut supra est probatum; ergo falsa quoque est omnis virtus praesagiendi, quae Cometis ex eo capite adscribitur. . .".

430 Ibidem, 64: "Accedit, quod licet verum esset, quod Cometa esset talis exhalatio accensa; tamen nulla ratione posset praesagire pleraque eorum, quae ei adscribuntur."

for believing so, our author offered his readers one of the most complete renditions ever published of the way in which the Peripatetics were wont to burden comets with all kinds of seemingly disparate disasters. The Aristotelians, Fienus stated, reasoned as follows: the exhalations out of which comets are formed, are dry; therefore, they desiccate the air through which they pass. Afterwards, they are kindled, and their huge, burning bodies heat and dry the air, and inflict on it a hot and dry intemperateness. This means that no rains will fall; therefore, the fields used for agriculture will wither, and the air will not be purified (for rainfall is wont to purify the air). Agricultural yields will diminish, and scarcity will ensue. Therefore, people will be forced to feed on less nutritious foods than usual, so that they will fall ill. The same result will ensue from the dry condition of the air, which will give rise to burning fevers and hot diseases, as well as pestilences caused by the smoky, sulphurous qualities spread through the air by the exhalations of which the comet is constituted.

Also, from the fact that not all dry exhalations will have been consumed by the body of the comet at once, those remaining behind in the air will cause winds and tempests; these will violently toss the seas, giving rise to floods. Another portion of the hot and dry exhalations, meanwhile, will have remained behind in the bowels of the earth where they originated; trying to escape, these will bring about earthquakes. And finally, because human bodies are altered according to the qualities of the air, the hot and dry humours, such as choler, inside human beings will increase, rendering them irascible and rebellious. Uprisings and revolts will ensue; princes, being especially delicate, will declare war on each other; in these wars, many of them will die, and this will result in changes of government and religious alterations.[431]

This whole reasoning made no sense at all to Fienus, who attacked it on a number of grounds. First of all, in an argument reminiscent of that put forward two and a half centuries earlier by Henry of Hesse[432], our author stated that for a cometary exhalation to be kindled, it would necessarily have to rise to a very great height to reach the sphere of fire and be carried along by the motion of the *primum mobile*. How, then, could vapours burning at such an immense distance from the earth affect the lower air? Indeed, if, at the time of

431 Ibidem, 62-64.
432 See supra, p. 83-84.

the generation of a comet, all noxious vapours had ascended into the highest reaches of the atmosphere, it should rather be believed that the lower air breathed by human beings was purified, so that no consequences unfavourable to human health would ensue. This reasoning was reinforced by the consideration that the appearance of a comet indicated that all bad exhalations were being consumed by fire, which possessed the quality of exterminating all noxious influences.[433]

Neither was it true that the air was heated to an extraordinary extent by the appearance of a comet. First of all, it was implausible that fire burning so far away from the earth's surface, just below the sphere of the moon, should heat the lower reaches of the atmosphere to any discernible extent. Moreover, the untenability of the Aristotelian idea could also be proved experimentally. If it were true that cometary appearances heated the air, this should be perceptible to the human senses; but "Who has ever sensed exceptional heat or drought in the air during the visibility of a comet?" Fienus rhetorically asked. Rather, the contrary was true, since it was beyond doubt that most comets had appeared during spells of very cold weather.[434] Another obstacle for cometary heat to reach the lower atmosphere was the presence of the middle layer of the air, which was exceedingly cold and should therefore be believed to impede any influences of the comet to reach the earth's surface.[435]

Moreover, the idea that whenever the air was very dry, it did not

---

433 Fienus, *Dissertatio*, 64-65: "Etenim talis exhalatio, necessario accendi deberet, loco altissimo, multis millibus milliariorum a nobis distante, uti supra ostensum est, ex eo quod Cometae motu primi mobilis rapiantur. At exhalationes tam remoto loco accensae, quomodo possent aerem hunc infernum afficere? Si exhalationes pravae omnes adscenderint in suprema aeris loca; ergo aer infima sui parte est depuratus; ergo nullae malae exhalationes hic amplius remanserunt, quae corpora nostra male afficere possint. Et si exhalationes omnes ibi supra sint accensae, ergo sunt exustae; ergo non amplius sunt, nec ullam malignam qualitatem reservarunt. Ignis enim omnem venenatam qualitatem delevit, & extinxit; ut patet in peste, in qua ignes exstruuntur, ut venenata aeris qualitas exstinguatur."

434 Ibidem, 65: "Primum videtur quod non possit [aer] esse calefactus; quia cum ignis ille fuerit tam multis millibus milliariorum a nobis remotus, non potest sic inferiorem aerem multum excalefecisse, vel exsiccasse. . . . Et si aer excalesceret a Cometa, & exsiccaretur, magis calidus esset, cum adhuc ignis, & Cometa durant, quam cum est exstinctus; sed quis in aere a Cometa sensit calorem, aut siccitatem? Non est dubium quin Cometae frequenter apparuerint, quando aer fuit maxime frigidus."

435 Ibidem, 66: "Accedit, quod inter Cometam, & infimam hanc aeris regionem, intercedat media aeris regio frigidissima, quae omnem Cometae in eam activitatem videtur necessario debere impedire."

rain, was equally fallacious. After all, during summer the atmosphere was desiccated, yet only a short while later, in autumn, most rains fell.[436] Also, why should some of the cometary exhalations remain behind in the air to cause winds? Why should they not all ascend? And if they did remain behind, why should they give rise to winds anyway? And if winds did ensue, why should these cause floods? Most tempests raged without any inundations occurring. By a similar reasoning, Fienus thought it implausible that some exhalations should remain behind in the bowels of the earth to generate earthquakes. After all, every year many vapours arose from the depths of the earth, without any earthquakes ensuing. The entire line of thought, to Fienus, was "All too fortuitous and atypical."[437]

Our author's medical expertise was greatly vexed by the Aristotelian assertion, that, because princes' bodies were more delicate than other people's, they were more easily affected by the diseases to which the infected air gave rise. Fienus called this idea "ridiculous". For why should this be the case? Was it not much more reasonable to believe that elderly people, women, and children were more susceptible to being hit by illnesses than princes?[438] Nor was it true that increases in the amount of choler inside human bodies were caused by the condition of the air. They were rather brought about by people's own individual temperaments, their diets, etcetera.[439] Moreover, a small increase in a person's choler did not render him irascible; anger was provoked much more readily by a naturally choleric complexion, by a bad education, by imprudence, or by unjust treatment suffered by the person in question.[440] And wars were caused,

---

436 Ibidem: "Non enim sequitur, aer est exsiccatus ergo nulli fient vapores, & consequenter nullae pluviae. Etenim aestate aer est exsiccatus, & tamen nihilominus paulo post in Autumno fiunt pluviae maximae."

437 Ibidem, 66-67: "... Fortuitum hoc nimis est, & rarum."

438 Ibidem, 67: "Dicunt, quia Principes habent corpora teneriora aliis, ideo prae aliis, ab infecto exhalationibus aere magis afficiuntur, & a morbis, qui tum grassantur, interiuntur. Ridiculum est. Quare enim Principum corpora sunt teneriora corporibus aliorum, aut magis pravis humoribus praedita? Si Principes sint obnoxii Cometis propter corporis sui teneritudinem; ergo multo magis senes, mulieres, pueri; ergo illorum mors a Cometis esset praedicenda potius, quam Regum, aut Principum."

439 Ibidem: "Aer non multum facit corpora biliosa, sed proprium temperamentum, victus ratio, &c."

440 Ibidem, 67-68: " Bilis aliquod adventitium incrementum non multum facit ad hominum iram. Ab aliis causis iracundia procedit; a biliosa complexione naturali, non adventitia, prava educatione, imprudentia, iniuriisque perceptis."

not by the fury of princes, but by lust for power, vengefulness, and similar passions. In a word, Fienus, like Puteanus, believed that the outbreak of wars was due to the free will of princes and of their counsellors, not to any natural causes; and whatever depended on human free will was hardly connected with the influences of the heavens, comets, or the air. Whatever small, fortuitous influences the heavens did exert on such matters, were so uncertain and feeble that it was quite impossible to predict them, Fienus added, implicitly chiding the astrologers.[441]

After this decisive invalidation of the Aristotelian theory of physical cometary causation, Fienus went on to equally resolutely refute the rules of prognostication from comets usually employed by the astrologers. Declaring himself to be in agreement with Julius Caesar Scaliger, who had believed comets to signify nothing, either as signs or as causes, and who in his commentary on Cardano's *De Subtilitate* had stated that "To believe that a comet announces the death of a king is a preposterously weak argument; and even less is it to be supposed that comets overthrow reigns"[442], Fienus presented his readers with a number of quotations of what he considered to be "some more, stupid predictions."[443] Thus, he quite extensively rendered the rules put forward in Ptolemy's *Centiloquium*[444], the maxims presented by Cardano in his *De Rerum Varietate* for prognosticating future events on the basis of such cometary characteristics as colour, speed, and position among the constellations, and the idea put forward by Pliny, that the hairy stars' significance should be deduced from their forms, as well as from the characteristics of the constellations through which they passed.[445] Readers who were interested to learn even more of such "stupid cometary presages", drawn from the "foolish, vain, and superstitious art of divinatory astrology", our author referred to the works of Haly and Albumasar, and the *Aphorisms*

[441] Ibidem, 68: "Nec ira Principum facit bella, sed cupiditas dominandi, amor vindictae, &c. Bella pendent a Principum, & Consiliariorum, aliorumque Principum suadentium libera voluntate. Quae autem a libera hominis pendent voluntate, non sunt obnoxia actionibus neque caelorum, neque Cometarum, neque aeris, nisi valde fortuito, & valde remote. Quorum fortuitorum, & remotorum eventuum nulla potest fieri praedictio, nisi valde temeraria, fortuita, stulta atque incerta."

[442] Ibidem, 68-69: "Iulius Caesar Scaliger ... asserit Cometas nihil praesagire, neque ut causas, neque ut signa: *Etenim,* inquit, *existimare a Cometa Regem interfici, ridiculae dementiae est; tanto minus everti provinciam.*"

[443] Ibidem, 69: "Audiamus tamen & nonnulla alia stulta praesagia."

[444] See supra, p. 58.

[445] See Pliny, *Naturalis Historia* II.23.92.

of Cardano. Indeed, Fienus thought the rules expounded in all these works so utterly ridiculous, that he stated he did not wish to spend any of his time refuting them.[446]

Nonetheless, the rest of his tract was filled with a series of five arguments, which, although they were not directed against any of the astrological rules of cometary divination in particular, still served as weighty rebuttals of the art of astrology in general and of the foundations on which it was based. First of all, he argued that the comet's tail could not work any notable effects. After all, as had been demonstrated earlier, it consisted of nothing but sunlight. So, if comets possessed any powers, these should necessarily reside in their heads. However, cometary heads were very small and insignificant starlets, so that it was not plausible to uphold that they worked any notable effects in the sublunary realm. For, if even one of the fixed stars was not, on its own, capable of effectuating any terrestrial changes, and if, indeed, all the heavenly bodies together exerted but little influence (except for the sun and the moon, who among them produced almost all of the alterations perceived on earth): what, then, could be accomplished by an extremely tiny, fleeting starlet, that was transient and volatile, and lasted for only a short while?[447]

Moreover, even if it were conceded to the astrologers that, apart from the sun and moon, the other celestial bodies, too, sent earthward their own specific influences, this would still not render astrology in general, or cometary prognostication in particular, any more credible to Fienus. After all, the fallaciousness and uncertainty of the predictions made by the art of judicial astrology demonstrated that not even the powers of the ordinary stars, that had been perceptible to man ever since the beginning of the world, and whose influences mankind had been able to observe for almost four thousand years already, were yet adequately known; how, then, should we believe

---

446 Fienus, *Dissertatio*, 70-71: "... Quid stultius his praesagiis excogitari potest? quid stultius me foret, si tempus terere vellem tam stultis, & vanis pronunciatis refellendis, desumptis, ex stulta, vana, & superstitiosa divinatrice Astrologia? ...".

447 Ibidem, 71: "Cauda Cometae nihil agit; quia solum est splendor luminis, ut dictum est. Ergo si est aliqua vis agendi in Cometa, erit in capite eius: at Cometae caput est valde parvum; tam parvum autem sidus, nullam potest magnam in haec inferiora exercere activitatem. Nam si una alia stella sola ex iis quae in caelo sunt, nihil potest notabile efficere in hoc mundo inferiore: immo si omnes stellae caeli simul sumptae parum efficiant, exceptis Sole, & luna, a quibus fere omnis, quae in haec inferiora exercetur actio, procedere percipitur; quid operabitur tam parvum sidus, transiens, nullo loco consistens, & tam exiguo tempore durans?"

the astrologers capable of ascertaining the influences of comets, that appeared so rarely, passed by so quickly, and vanished so soon?[448]

In his next argument, Fienus further elaborated on the theme of the unknowability of the celestial influences, pointing out that, because the stars were so many and came into such close contact with each other, they necessarily had to weaken, alter or disperse one another's powers. This, to him, was the reason that heavenly influences could not be adequately grasped; and, a fortiori, the possible forces exerted by comets were even more difficult to chart, since these bodies conjoined with different stars each day, by which undoubtedly their powers were altered and modified and rendered irrecoverable.[449] The argument the astrologers might propose, that the stellar influences could be discovered by experience, was to no avail: Fienus pointed out that, although it was true that many sad events had throughout history followed cometary appearances, this did not mean that they were actually caused by the hairy stars; more likely, they would have happened in precisely the same manner if no comet had appeared.[450]

Like Puteanus, Fienus pointed out that a careful investigation of history would show that, in the inconstant sublunary realm, both sad and happy events continually occurred, whether or not comets shone forth. Many calamities had erupted during cometless years, and if the annals of history were to be closely scrutinized, the outcome would undoubtedly be that after cometary appearances as many happy events had ensued as disasters had struck. Although at some places, misery might indeed have followed the arrival of a comet, at the same time other regions might well have enjoyed happiness and

[448] Ibidem, 71-72: "Ut ita esset, quod Cometa haberet aliquam in haec inferiora activitatem; qualis ea foret, impossibile esset scire. Nam si scire non possumus, quid aliae caeli stellae operentur, ut ex falsitate, & incertitudine Astrologiae iudiciariae est manifestum; quae tamen semper, & a condito orbe steterunt, & quarum virtus iam quatuor mille paene annis potuit observari; quomodo poterimus cognoscere Cometarum vires, qui raro apparent, statim praetervolant, & cito pereunt?"

[449] Ibidem, 72: "Vires siderum modificantur diversimode propter concursum activitatis tot aliorum siderum, quae suas mutuo vires frangunt, mutant vel intendunt; & haec est causa, cur eorum virtus cognosci nequeat: eadem ratione nec Cometarum poterit: multo minus, ex eo quod Cometae, moveantur, & singulis diebus cum novis sideribus coniungantur, a quibus eorum virtus variatur, modificatur, determinatur, incognoscibilis evadit."

[450] Ibidem: "Dices, experientia cognoscitur; sed illa cognitio nimis est incerta. Nam etsi multa mala & tristes eventus in Mundo post Cometas evenerint; non ideo evenerunt virtute Cometarum: sed vel etiamsi nullus Cometa apparuisset, eodem modo provenissent."

prosperity.[451] The reason that only the sad events were associated with the hairy stars, was that people always paid more attention to disasters than to good fortune, and consequently the former were better remembered. Again like Puteanus, Fienus included several examples of nobles who had died without the accompaniment of a comet, and added that, conversely, many comets had appeared without any notable consequences.[452]

Our author's last two arguments were directed at two specific techniques practised by the astrologers, namely, the determination of the regions that would be affected by the comet's bad influences, and of the moment in time at which these influences were to manifest themselves. As for the regions afflicted, Fienus wondered in what way these could ever be discovered. It could hardly be upheld that comets carried significance for those regions from whence they could be seen, he argued; after all, the comet of 1618, for one, had in 24 hours orbited the whole earth, and had therefore been visible from almost every region. Since the power of the stars was universal, and not directed at one particular country, so, too, was the comet's.[453] Or should we rather say that the hairy star menaced only those regions over whose zeniths it had passed, because there, it had been able to send downwards its rays in a straight line? But, in that case, the same problem presented itself: the comet of 1618 had passed over the zeniths of half of the world, and certainly over those of the regions on this side of the equator, where almost all the principal countries of the earth were situated.[454]

In a final attempt to maintain the theory of cometary significance for particular regions, it might be upheld that the hairy stars carried

---

[451] Ibidem, 72-73: "... Quam multos vidimus tristes eventus hisce annis, quibus tamen nullus Cometa illuxit?... Sicut autem multa tristia eveniunt, ita & laeta; & ideo si quis ex historiis conferret omnia, quae post apparitiones Cometarum in mundo contigerunt, absque dubio non minora laeta, quam tristia reperiret. . . .".

[452] Ibidem, 73-74.

[453] Ibidem, 74: "Si Cometa tantum mala praesagit; cui regioni praesagit? an regionibus a quibus videtur? Atqui 24 horarum spacio hic noster totum mundum circummeavit; & a toto fere mundo visus fuit; an ergo toti mundo minatur? Virtus siderum est universalis, nec ad ullam peculiarem determinatur regionem; ergo erit & Cometarum."

[454] Ibidem, 74-75: "An vero illis regionibus tantum minatur, supra quarum vertice fertur? nam in illas rectis radiis virtutem suam magis demittit. Sed hic Cometa delatus fuit super verticibus mediae partis mundi; & super verticibus omnium mundi regionum, quae sunt cis Aequinoctialem. (Nam ab Aequinoctiali ad minimum eius ortum incepisse non est dubitandum:) proindeque super verticibus omnium fere praecipuarum mundi partium."

predictive value for those countries over whose heads it had hung a long while, or again, that its meaning was intended for those nations over whom it had shone most brighly or had displayed the longest tail.[455] However, these arguments were also deflated by Fienus with reference to the course of the 1618 comet he had himself observed: it had moved very fast over those regions where it had displayed the brightest light and the longest tail, while it had been small and weak over those countries over which it had tarried longest, so that the two circumstances cancelled each other out.[456] Finally, the direction in which the tail pointed, which was often adduced by the astrologers as an indication of the nations to be affected, could not serve as the basis for any prediction: since it was always directed away from the sun, and since the sun travelled round the earth every 24 hours, the tail pointed in every single direction each day.[457]

Nor was it possible to ascertain at what time the effects of the comet would begin to make themselves felt. Many astrologers extended the comet's significance over a time span of many years, so that it was not surprising that, during such a long period, sad events occurred. Also, how could a comet work any effects after it had vanished? Like Puteanus, Fienus pointed out that, after the hairy star's disappearance, its influences ought surely to be destroyed and extinguished by the powers of the other stars, as well as by the blowing of the winds and the variety of the seasons.[458]

Fienus' correspondent on the subject of the comet, Libertus Fromondus, declared himself in total agreement with his older colleague as far as cometary consequences were concerned: in the first sentence

---

[455] Ibidem, 75-76: "Quibus ergo minatur? an iis super quorum capite diutius haesit? an iis super quorum capite, maius lumen, maioremque caudam habuit?"

[456] Ibidem, 76: "Sed nec illi timere debent, quia quibus magno lumine magnaque cauda incubuit, eos cito pertransivit; super quibus diu haesit, eos exiguitate luminis sui, corporisque languore, & debilitate non multum afficere potuit."

[457] Ibidem: "An vero minatur iis, ad quos cauda eius tanquam virga dirigitur, ut Cardanus loco citato, & Plinius lib. II cap. 25 asseruerunt. Illud dici non potest: quia cauda dirigitur ad omnes mundi partes. Cauda enim eius semper est aversa Soli. In qua parte est Sol, in oppositam dirigitur cauda eius. Prout Sol vertitur, ita & cauda eius. Unde cum Sol quotidie totum circummeet globum, necessum est, ut quoque cauda cum ipso circumducatur, & sic 24 horarum spacio, ad omnia caeli puncta semel dirigatur."

[458] Ibidem, 77: ". . . Nam quomodo aget, si periit? videtur quod eius efficacia maxime deberet elucere interim quam durat. Nam postquam periit, evanuit, vel excessit, effectus eius ab aliorum siderum continua, ac perseverante activitate, a ventorum interim orientium diversitate, ab anni temporum alterna varietate, necessario destrui, & exstingui debuisse videtur."

of his very short final chapter, entitled "The predictions of comets", he stated that "You have extensively spoken of comets' predictive value, most illustrious man, so that I would hardly know what to add to it."[459] Thus, apparently, Fromondus was as firmly opposed to the traditional astrological treatment of comets as Fienus.

In one of his earlier chapters, when dealing with the comet's distance from the earth and its size, Fromondus had already made it clear that, to him, the arguments of the astrologers, based on the place in which comets were generated, were "of great fallaciousness and often mistaken."[460] Thus, he could not detect any sound reason for attributing the birth of a comet to any planetary conjunction, as the astrologers habitually did. In the case of the comet of 1618, the astrologers might wish to uphold that its generation was connected with the conjunction of Venus and Jupiter that had taken place on 15 January 1619. However, this to Fromondus would have made sense only if that conjunction had immediately preceded the comet's first appearance, or if it had been instrumental in restoring the hairy star to vigour when it was already languishing. Yet, neither of these circumstances had been the case. The astrologers might try to save their theory by arguing that, since both Jupiter and Venus were benign planets, the cruelty characteristic of the comet might have displeased them, and that therefore their conjunction might have served to destroy the hairy star rather than to revitalize it. However, Fromondus considered these reasonings "arbitrary", and added that anyone would easily be able to invent other, similarly unfounded arguments.[461]

Thus, he was not at all favourably disposed toward the astrological way of reasoning. Since Fienus had already discredited the tradi-

[459] Libertus Fromondus, *Dissertatio de Cometa anni 1618* (Antwerp 1619) 137: "De Cometarum praesagiis largiter a te dictum Clarissime Vir, ut nesciam quid fere possim adnectere."

[460] Ibidem, 129: "Itaque de Cometarum ortu, parum adhuc explorati hactenus vidimus, & Astrologorum praedictiones in magna vanitate & saepe fallaces."

[461] Ibidem, 130: "Pari fortasse ratione, Cometam nostrum coniunctioni Veneris & Iovis 15 Ianuarii anni huius 1619 quis referat acceptum. & colorem res haberet, si immediate praevenisset Cometae ortum, aut si refecisset senescentem iam plane & defectum. Nihil autem horum fuit. Venus suum Iovem convenit, nec tenuiter spirantes adhuc Cometae reliquias recreavit. video effugium, in benignitate blandissimorum Planetarum, quibus invisa diritas Cometarum. Itaque coniunctio haec perniciem potius Cometae nostro festinavit, & impulit diutius fortassis alias inclinaturum. Sed haec arbitraria sunt, & quanta plurima quisque pro ingenio non difficulter confingat."

tional theories of cometary consequences by means of serious arguments, Fromondus in his final chapter chose to invalidate the futile prophecies of the soothsayers by means of a number of examples drawn from recent experience, followed by a short satire. First of all, he stated that the comets that had appeared in recent years had given rise to many wonderful predictions by the Germans as well as by others; and most of these had been proved false.[462] The nova of 1572 had been burdened with being a sign of the end of the world, a harbinger of the coming of Antichrist who was shortly to arrive from Moscovia, and a forerunner of a general alteration and conversion in the Christian Republic (the latter prophecy, Fromondus was convinced, having been invented in order to boost Luther's cause). These developments, according to the astrologers, would be seen only by those who would be alive in the year 1632; Fromondus, however, resolutely dismissed all these vaticinations, as well as other, even more ridiculous ones, as nonsense.[463]

Our author then went on to subject the present comet's significance to a closer examination, which he conducted in a tone of sarcasm throughout. Sardonically, he announced that, in his opinion, the phenomenon indicated more martial conditions than had the comet of 1607, that had brought in its wake the Twelve Years Truce.[464] This 'prediction' sounds rather ironic, if we take into consideration the fact that in two years' time, the Truce was to expire, so that it was almost inevitable that the comet of 1618 would bring in its wake more war than that of 1607, when the ceasefire had been concluded. After this parody of a vaticination, Fromondus subjected the comet's course through the heavens to his sarcasm. Thus, he ironically remarked that the comet had moved through the constellations of the Great Bear and the Dragon, whose influence was malignant, "and, to terrify you even more, when on 29 November the comet

---

462 Ibidem, 137: "Mira Germani, aliique de antecedentium annorum Cometis vaticinati sunt, plerumque falso."

463 Ibidem, 137-138: "De stella illa anni 1572 quam timenda, quam paradoxa! alii, mundum ad metam certissimo venisse, & Antichristum iam iam e Moscovia adventantem, cui stella erat verticalis. alii generalem saltem Reipublicae Christianae conversionem (& scio id favori Lutheri datum.) quam tamen soli visuri essent, qui annum Christi 1632 alii Chymeras alias, & nugas mage maiores confinxerunt."

464 Ibidem, 138: "Sed de Cometa hoc nostro quid dicimus? Video, ut res iam inclinant, maiorem belli subiturum infamiam, quam alterum qui anno 1607 civilium armorum inducias nobis tulit."

was first seen from Louvain, I observed Mars in the 19th degree of Virgo, hurrying and charging straight towards the comet."[465]

Similar irony can be detected in Fromondus' remark, that "Saturn shone in the fourth degree of Gemini, bloody and unusually red. But I was very relieved to find that the good old man was retrograde, and fled from the comet with all the strength which his age allowed him."[466] And it is also impossible to take seriously the question, asked by Fromondus a few sentences later, of "What should we make of the fact that Saturn was timid and receded from the comet? Should we deduce from this that the imminent calamities will, as usual, especially afflict the peasants, beggars, and all miserable people?"[467] Obviously, these were precisely the groups that were always most afflicted by any public misfortune, not because their ruling planet was Saturn, but because they were the most vulnerable classes in society.

Thus, Fromondus ridiculed the methods and prophecies of the astrologers; to him, as to Fienus, comets carried no natural predictive value that could be detected by means of an elaborate system of arbitrary rules such as Ptolemaic astrology. However, unlike Fienus, who as we have seen explicitly stated that he would not speak out on the possible status of comets as divine signs, Fromondus did venture a small excursion into the territory of theology (for which, to be sure, he was better qualified than the medical man Fienus). In his chapter on the comet's distance from the earth and its size, our author, when dealing with the maximum duration of comets, stated that none had ever lasted for more than 180 days. For, he added, the phenomenon that, according to Josephus, was seen over Jerusalem before its destruction and lasted a whole year, was certainly not a comet, but rather a greater prodigy. Because it seemed to threaten only one very particular spot on earth, and because it was immobile, it had been situated in the lower reaches of the atmosphere; otherwise, it would have been set in motion by the movement of the *primum mobile*. For

465 Ibidem: "... Et ut magis palleas, 29 Novembris cum primum Lovanii Cometa visus, Martem in 19 gradu Virginis notavi, directum ad praecipitem, & plane festinantem Cometae occurrere."

466 Ibidem, 138-139: "Saturnus etiam in 4 gradu Gemin. sanguinolentus & insolenter rubeus radiabat. Hoc tamen me recreavit, videre optimum senem retrogradum, & totis quantum aetas ferebat, viribus a Cometa refugientem."

467 Ibidem, 139: "Quid ergo iam Saturnus a Cometa timidus & regrediens? An Agricolis, ut solet, mendicis, ac miseris omnibus praecipue incumbet ista lues?"

such a low, sublunary fire to burn with such a constant flame and with invariably the same form (a sword, which was a figure not usually seen in real comets) the flame necessarily had to be fed by an angel, Fromondus believed. The apparition's constant, swordlike form indicated that the phenomenon had been instituted especially by God, in order to warn the city of its imminent end, and to admonish its inhabitants to do penitence.[468]

Thus, although in Fromondus' view the astrological consequences traditionally associated with comets were unfounded, at times natural phenomena might well be called into being by the solicitous godhead to warn his people of imminent doom. Years later, in his extensive *Meteorology* (first published in 1627), our author counted comets, too, among such phenomena.[469] On the whole, in the *Meteorology* Fromondus showed himself somewhat more traditional in his discussion of cometary significance than in his tract on the comet of 1618. Thus, he admitted that, though at times there had been comets that had boded happiness, on the whole their consequences had been evil, although he did add that there always occurred both good and bad things simultaneously, and that what was good for one party was bad for its opponents.[470]

Having said this, however, Fromondus stated that, nonetheless, comets usually presaged sadness and mourning, not for any astrological reasons, but because they were used by God as signs to urge mankind to do penitence. At times, our author believed, the most benign godhead employed natural effects to admonish the human race. Thus, although he agreed with Fienus's reasoning in his tract on the comet of 1618, that comets were not in any way physical signs or causes of earthly calamities, Fromondus did believe them to be divinely instituted signs. Although they were natural entities, like the

---

[468] Ibidem, 129: "Nam flamma illa quam Iosephus supra urbem Hierosoly. ante excidium annum totum perpendisse narrat, non Cometa indubie, sed prodigium maius fuit. Ut enim certo terrae loco imminere videretur, deprimi valde debuit, immobilis etiam erat: unde in supremo aethere non stetit, secutura alioqui raptum primi mobilis. Igitur citra nubes in materia terrestri & exhalatione arsit, quam identidem deficientem subministrato novo pabulo recreabat Angelus. Formam gladii etiam habuit, quod non solent Cometae; & ut appareat a Deo destinato figuratam, ut de exitio admoneret, & converteret ad poenitudinem."

[469] On Fromondus' *Meteorology*, see Christoph Meinel, "Les *Météores* de Froidmont et les *Météores* de Descartes" in Anne-Catherine Bernès ed., *Libert Froidmont et les résistances aux révolutions scientifiques. Actes du Colloque Château d'Oupeye, 26 et 27 septembre 1987* (Haccourt 1988) 105-129.

[470] Libertus Fromondus, *Meteorologicorum Libri Sex* (2nd edition, Louvain 1646) 193.

rainbow, they possessed a supernatural significance.[471] Divine providence had, from the very beginning of the world, foreseen when, in the course of history, it would prove necessary to warn man of his imminent anger. Thus, he had disposed the natural causes giving rise to comets in such a way, that they would come into existence exactly at the right moments in time.[472] Contrary to what the astrologers believed, then, no specific predictions were possible; the peoples and regions to be affected by the calamities the comet announced, lay hidden in the depths of the divine wisdom that had predisposed all movements and occurrences of the celestial machinery.[473]

We can conclude that the stance of Fienus and Fromondus (on the assumption that the former, in leaving the discussion of comets as divine signs to the theologians, with regard to this problem submitted himself to the judgments of his younger colleague) stood midway between that of Gemma and Heurnius, on the one hand, and that of Puteanus, on the other. Like their humanist colleague at Louvain, they resolutely rejected the astrological way of predicting cometary consequences employing rules based on the hairy stars' colour, form, tail, course through the heavens, etcetera. Although all three authors endorsed the idea that the stars, and the comet, possessed a certain physical influence with which they might in some way or another affect the sublunary realm, they were all agreed that, as yet, this

---

471 Ibidem, 193-194: "Sed non in istum tamen, puto, finem, sed ad luctum & tristia mittitur. Solet enim benignissimum Numen, ubi mensura flagitiorum nostrorum ad plenitudinem venit, & vindictam hic potius temporariam quaṃ aeternam parat, naturalium istorum effectuum occasione uti, & terrore illo ad poenitentiam & sui respectum convertere. Itaque cum Clar. nostro Thomâ Fieno, in Cometa anni 1618 nego, Cometam, maxime caelestem, calamitatum physicam caussam (quomodo enim exhalatio illa, sive Sole tantum perfusa, sive ignis etiam est, ex illo intervallo tantum efficere valeat?) Non nego tamen, signum ex institutione divinâ esse; quod omnium temporum historiae affirmant. Igitur Cometa res quidem naturalis est, ut Iris; utraque tamen supernaturalem a Deo significationem habet."

472 Ibidem, 196: "Materiam eorum ex corporibus elementaribus, & fortasse etiam caelestibus, per solem & caussas alias naturales educi. Sed caussas istas naturales ita a divina sapientia & providentia sic ab initio Mundi dispositas esse, ut tum ex ipsis naturaliter Cometae nascantur, quando Deus imminentium malorum admonere vult mortales."

473 Ibidem: "Nam etsi caussa ista & effectus sint naturales, Deus tamen qui caelestis horologii rotas & motus regit ac temperat, ita olim in antecessum, calamitatum tempora providere & disponere valuit, ut eas solis aut Lunae comitarentur eclipses. Quorumnam tamen hominum distincte & in particulari designentur clades, in abdito divinae sapientiae est, nec certo hoc attingere Astrologorum conjectura potest."

influence had not yet been adequately understood, and that, therefore, no reliable predictions could be based on it. While Puteanus optimistically stated, that some day human beings might be capable of grasping the laws of nature to such an extent that the need for a concept of 'prodigy' would disappear, Fienus and Fromondus were much more cautious about the capabilities of human reason, emphasizing that the stellar powers were so manifold and so complicated that they would probably never be adequately charted. In any case, predictions applying only to one particular region on earth were unfounded: Fienus stressed that the power of the stars was universal, acting on the whole world indiscriminately.

What at least Fromondus, and possibly also Fienus, had in common with the two 1577 authors, was the belief in comets as entities with the specific task of warning mankind of God's impending anger. Whereas Puteanus had reduced such specific divine actions to the fringes of the theatre of nature, seeing the hand of God at work principally in everything that worked in accordance with orderly laws, Fromondus, like Gemma and Heurnius, explicitly accorded superlunary comets a semi-divine status: although their causes were natural, God had endowed them with a specific, supernatural meaning, and their final cause was to admonish mankind. Thus, they were produced with a special purpose at particular moments in time, at which sinful man needed especial warning, and which had been foreseen by God at the Creation. In this sense, comets were extraordinary entities, deviating from the normal course of nature and replete with teratological significance; yet, these deviations had themselves been built into the divine plan in advance.

d. *The Reformation of the Astrological Tradition: Snellius*

Willebrord Snellius did not share his southern colleagues' scepticism with regard to the capability of the human intellect to grasp the various stellar influences. In our previous chapter, we have already seen how the Leiden mathematician was convinced that, if only nature were diligently and constantly observed, it would prove possible to recover much of the wisdom the ancients had possessed and that had been lost in the course of history. We have even seen him refer to the several comets and novae that had appeared during the last quarter of the sixteenth century and the first of the seventeenth, as tokens sent by God to encourage people not to despair, but instead to apply themselves with renewed energy to the investigation of truth and the uncovering of nature's secrets.[474]

Not only the material constitution of comets, and the way in which they were generated, could, in Snellius' view, be discovered through careful investigation of the phenomena. Their significance and the things they portended, too, could be subjected to diligent human scrutiny, and could be accurately charted. Our author from the very beginning of his chapter on cometary significance made it abundantly clear that, to him, it was certain that comets carried a definite predictive value; after all, "Vast bodies are not generated by God in the immense aether, to move among the stars, in vain, or without any effect."[475] In order to discover this meaning, in Snellius' opinion it was necessary to study their matter and nature, as well as to take recourse to what was taught about them by the experience of past cometary appearances, and in combining these two procedures it was possible to arrive at the true significance comets possessed.[476]

In the earlier chapters of his tract Snellius had employed the first of these two methods, carefully studying the present comet's matter and nature; now, in his final chapter, he expounded and criticized the experiences recorded by his predecessors as to the effects the hairy stars had in the past exercised on the terrestrial realm. The Egyptians, Snellius thought, despite the fact that they had been very interested in celestial phenomena, had not given cometary appearances any consideration; while the Chaldaeans had failed to come up with any consistent cometary theory, as was apparent from the fact that the two Greeks who claimed to have studied in Babylonia con-

---

[474] Willebrordus Snellius, *Descriptio Cometae qui anno 1618 mense Novembri primum effulsit* (Leiden 1619) 48: "Non advocabo huc spississimas ignorantiae proximorum seculorum tenebras, quo argumento docemur Summum DEUM nostram socordiam adeo gravi vindicta ultum tanto infortunio mactavisse. Verum ne adeo acerbus casus, tanta calamitas huic deinceps fundo ingruat & a nobis absit, in spem haud vanam tres novae stellae intra tres & triginta annos conspectae omnino me erigunt: modo ipsi nobis posterisque nostris ne desimus. Haud dubie enim sublimia ingenia splendidissimo illorum siderum iubare ad veritatis investigationem & naturae arcanae eruendum excitabuntur. Neque enim unquam temere nullove effectu tantae moles in summo aethere exoriuntur, & a Deo Opt. Max. tanquam signa ad bene sperandum in ipso caelo omnium oculis exposita eriguntur." See supra, p. 351.

[475] Ibidem, 57: "A Deo vastas moles non frustra, aut nullo effectu per immensum aethera cieri, & sidera subterlabi...".

[476] Ibidem, 57-58: "Quid autem hac in parte statuendum sit, ab ipsorum materia & natura, aut ab experientia, aut ab utrisque denique repetendum censeo."

tradicted each other with regard to the Chaldaeans' ideas on the subject of the hairy stars.[477]

Neither were the rules expounded by Ptolemy in his *Tetrabiblos* and his *Centiloquium* very reliable, Snellius thought. After all, the Alexandrian had counted comets among sublunary phenomena, placing them in the same genus as haloes, parelia and similar phenomena.[478] Not surprisingly, this classification could not find favour in Snellius' eyes, since, to him, comets were real, superlunary phenomena consisting of celestial, etherial material similar in kind to that of the sun, while phenomena such as haloes and parelia were mere optical effects.[479] Although he took the trouble of quite meticulously expounding the consequences attributed to comets by Ptolemy, quoting such rules as "When comets appear in the morning, their effects will manifest themselves soon, and when they appear in the evening, late", as well as all the cometary aphorisms of the *Centiloquium,* Snellius categorically stated that he would never agree with those Ptolemaic dicta, which he considered to be "unetiological", not based on any sound cause.[480]

For, although Snellius agreed with Ptolemy that such fiery sights in the sublunary air as "beams, trumpets, and jars"[481] naturally worked Martial and Mercurial effects (either because their appearance denoted a heightened activity of these two planets, or because they themselves were assimilated to the natures of Mars and Mercury), such as wars, heat, winds, revolts, murder, fire, pestilence and the like, he was emphatically opposed to putting comets in the same category as such atmospheric phenomena, as was done by Ptolemy and his adherents.[482] In comparing comets to beams and trumpets, Snellius stated, Ptolemy was quite mistaken.[483]

Thus, neither the scholarship of the Egyptians and Babylonians,

---

477 Ibidem, 58.

478 Ibidem: "Ptolomaeus tamen in tetrabiblo ita de cometarum effectibus iudicat, ut ipsos a virgarum, halonum, pareliorum natura non admodum seiungat, tanquam in aere nostro accenderentur."

479 See supra, p. 347.

480 Snellius, *Descriptio Cometae,* 60: "Caeterum quod cometas significatione stellis transvolantibus trabibus, & id genus aliis propemodum conjungat, nescio quam commode. Ego hic multa desidero, neque enim unquam illis ἀναιτιολογήτοις acquiescam."

481 These were all well-known forms of phenomena seen in the clouds or in the atmosphere, and were regarded either as visions caused by some optical effect or as actual burning meteors.

482 Snellius, *Descriptio Cometae,* 58: "Observandae crinitae (inquit) [Ptolemaeus] seu

otherwise so excellent at explaining the effects of the heavens, nor the elaborate rules devised by Ptolemy, offered the student of cometary significance anything to go on. Snellius therefore once again put forward his opinion, that the principles of the astrological art ought to be more accurately determined. For, if, as was generally maintained, astrology was indeed an art, it should, like all the other arts, be based on experience, investigation, sense perception and induction. If we wish to avoid having to swallow old wives' tales, and the gibberish of madmen, as if they were certain and undisputed, our author announced, we have to base all the arts, including astrology, on the sure footing of these four principles.[484] In astrology as in astronomy, then, Snellius resolutely advocated the Ramist methodology.

To Snellius, there existed a fundamental difference between comets and the different kinds of phantasm (the sort of phenomenon we have seen Gemma refer to as 'chasms') that had appeared in the skies over the Netherlands during recent decades. When a child, our author had heard the adults discuss sights seen in the air several years before the outbreak of the war (which, incidentally, Snellius referred to as "this long and difficult Netherlandish war, which was more vehement as far as the fanaticism of the different parties was concerned, more difficult with regard to the difficulty of the issue, and more lucky with regard to its outcome, than any other war"[485]). These apparitions had represented besieged and raided cities, artil-

in deliquiis, sive alio quolibet tempore fulserint in considerationibus universalibus, quales sunt quae dicitur trabes, tubae, dolia ac huiusmodi. Etenim harum naturae effectio Martiae ac Mercurialis proprietatis, bellorum, aestus, turbulentiae, & eorum quae ista sequi consuevere."; and 60-61: "Non existimo utique, quid sibi velit illa circumlocutione, ista esse Martiae ac Mercurialis naturae me non intelligere, cum ipsius [Ptolemaei] beneficio & interpretatione mihi haec liceat scire, qui Marti aestum & siccitatem, Mercurio autem & siccandi & humectandi vim ex aequo ... attribuat. ... Quae omnia, ut ita in genere vera sint, habent nihilominus aliquid particulare cometae."

[483] Ibidem, 61: "Dum autem istos [cometas] tubis, doliis & id genus phaenomenis comparat, profecto de coeno haurit."

[484] Ibidem: "Vellem equidem huius artis principia paulo accuratius constituta. Nam si ars sit, habebit utique natales & initia ad instar artium reliquarum ἐμπειρίαν, ἱστορίαν, αἴσθησιν, ἐπαγωγήν, experientiam, observationem, sensum & inductionem. in omnibus theorematis veris καὶ ἐντέχνοις istis locus primus ab ipsa origine est attributus. Eadem hic ego principia desidero, & quaero. neque enim aniles fabulas, aut commenta delirantium pro ἀναμφισβητήτοις sibi obtrudi quisquam patietur cui sinciput erit sanum."

[485] Ibidem: "...diuturnum & difficile hoc bellum Belgicum, quo vix aliud partium studiis vehementius, rei magnitudine difficilius, eventu felicius gestum est...".

lery, trumpets, explosions, military camps and the like; in short, our author concluded, they had prefigured exactly the kind of events that were shortly to ensue in reality.[486] Later, in 1588, similar visions had appeared prefiguring the defeat of the Spanish Armada, showing naval battles in the skies above Amsterdam, which had been watched by thousands of people.[487]

However, such atmospheric sights should definitely not be classed in the same category as comets. After all, they, indeed, belonged to the genus of Ptolemaic 'beams, trumpets, and jars', and were to be considered "great signs and tokens in the heavens"; yet, it should not be believed that they actually produced any effects themselves, since they were situated in the inferior region of the lower air. Rather, they should be interpreted as messengers of God's wrath.[488] Comets, conversely, being constituted out of the divine, forceful aether, possessed the same influences as their congeners, the other celestial bodies. Since Snellius considered them to be solar exhalations, formed out of the sun's own matter, he attributed to the hairy stars precisely the same powers as to that most noble and powerful body, stating that they "imitated the nature of the sun".[489] Thus, the comet of 1618 had, in the course of its existence, worked the same effects the sun would have produced: for example, when close to the star Arcturus,

---

486 Ibidem: "Memini me puerum ab aetate provectioribus audivisse, . . . tunc illas aerias Φαντασίας praelusisse secuturis eventibus. Visas nocturno tempore easdem rerum facies, & auditas quoque, quales paulo post obsessis & captis urbibus eveniebant. Tormenta, tubas, sclopetorum displosiones, castrorum metationes, atque alia id genus quam plurima."

487 Ibidem, 62: ". . . memini me admodum puerum . . . audivisse, visas Amstelodami in ipso aere paulo ante Solis occasum navium conflictus maritimos, impressiones hostiles, hos victos illos victores abire, tanquam in puerili naumachi."

488 Ibidem, 62-63 "Illa sunt scilicet τέρατα καὶ σημεῖα ἐν οὐρανῷ μεγάλα. sunt utique istiusmodi signa paulo ἐνεργέστερα evidentiora. Hoc apud Ptolomaeum credo, sunt tubae, dolia, atque id genus alia. Verumenimvero quaenam illis cum cometarum natalibus & exortu communia sint, me omnino ignorare ingenue profiteor. . . . Quamobrem etsi istiusmodi τέρατα ostenta creberrima & horrenda simul cum cometarum fulsionibus superioribus annis visa fuit, illa tamen per se minus virium habent, & potius significare, quam efficere quidquam sunt existimanda. utpote quare in ima aeris regione & nobis propemodum contigua versentur. Nam ita duntaxat pro irae divinae internunciis earum conspectum interpretor."

489 Ibidem, 63: "Atqui cometa cum sit aethereus & materia supernaturali & aetherea prognatus, is effectus habet utique sibi cognatos, pro suae materiae qualitate & natalibus. Namque si solis sit quasi ἀπόρροια eius plane naturam imitabitur, quod in nostro isto manifeste apparuit, qui prout cuique affixarum applicuit, ita Solis naturam & insitam qualitatem expressit."

it had caused a terrible tempest to occur on earth, which the sun, too, was wont to do when in conjunction with Arcturus.[490]

Because of the comet's status as a heavenly body with certain definable influences, Snellius once again put forward his strong belief that, if only each comet were subjected to careful observations, their effects could be discovered. As yet, not enough was known about their qualities; thus, our author refused to lend credence to the common dictum, that the direction of cometary tails indicated what regions would be affected by its consequences, until the cause of the phenomenon had been firmly established. After all, the only thing as yet known about the tails, was the rather vague fact that they were directed by some occult heavenly or solar power.[491] Since comets were brought about by natural causes, they also possessed a certain natural efficiency, which they revealed while they were visible, and which, even as they grew fainter, they sent earthward, although this was not always directly apparent to human perception because the cometary influences could be affected by the forces of the other celestial bodies. These influences would be felt most in places and objects that had already been afflicted by them once before; in regions not predisposed to receive them, they worked no notable effects.[492]

Thus, evidently, Snellius thought of comets as objects capable of actually, physically affecting the earth with their powers; and he believed the way in which these powers acted on the sublunary world to obey certain laws which could, after long experience and diligent observation, be grasped by the human intellect. This makes him the most unadulterated adherent of the physical tradition of cometary causation of all the Dutch authors treated so far. However, the tera-

490 Ibidem: "...cum aliquot diebus post cum horrido Arcturi sidere cooriretur tempestatem horridam & glacialem nobis excitavit, haud aliter quam Sol ipse quando cum illo exoritur...".

491 Ibidem: "Ideo ista posthac accuratius nobis notanda & omnibus diligentius observanda existimo. Porro autem de cauda qoud adjicit in quamcumque tandem plagam illa inclinet, isti malum praesagiri. id verum nec ne sit dubium esse possit, postquam eius causa (qua probabilior saltem a nullo hactenus prolata videtur) inventa est. quippe eam occulta coeli aut Solis vi ita dirigi."

492 Ibidem, 64: "Ita inquam cometarum causae etiamsi naturales sint, negandum non est naturali efficientia praeditos dum conflagrant eandem exserere, & postquam in tenuem auram evanuerunt pro impetu suo ferri, & illinc vim suam etiam cum nobis non appareat, contagione siderum infectam in terras mittere. Maxime autem in loca & res iam ante affectas: nam ubi praevia illa ad recipiendum διάθεσις deest, aut nullam aut imbecillam admodum vim habere, nisi per contagionem malum serpat omnino verissimum est."

tological point of view was not quite absent from his outlook. We have already seen how he regarded the chasms seen in the skies over several Dutch towns as remarkable, divine signs, appearing to warn man of God's anger and prefiguring future events. In addition to these remarks, just before his description of the physical way in which comets affected the earth, he called his readers' attention to two other cases of certain natural phenomena that in his opinion announced future terrestrial events.

The first of these was very comparable to the way the appearance of a comet predicted impending earthly changes. Everyone knew, Snellius stated, that haloes and other such meteors arose from natural and well-understood causes; and it was also generally known that the appearance of a halo portended winds and clear skies. Thus, like comets, sublunary meteors, too, arose from natural causes and in addition possessed a certain predictive value.[493] Although he did not explicitly say so, Snellius by introducing this analogy may have meant to suggest that, since the consequences of haloes and winds were already known from long experience, it was not at all unthinkable that cometary consequences, too, might one day be understood. However, our author's second example was of rather a different order. He declared that very strong gales and tempests, though it was well known that they were caused by nothing but agitated movement of the air, could nonetheless be regarded as signs sent by the irate godhead who wished to put to the test the faith of his people.[494] Clearly, then, to Snellius, phenomena brought about by natural causes might have purely natural effects, but could, in addition, also serve as divine signs. However, as we have seen, in his treatment of the way comets could be said to carry a certain significance, the emphasis was placed much more strongly on the former explanation.

Nonetheless, despite his firm belief in the viability of the astrological enterprise of charting the stellar influences, Snellius was convinced that much in the practice of astrology had still to change before reliable predictions could be made. He pointed out that, up till the present day, the astrological art had been very badly ne-

---

[493] Ibidem: "Ut ventos, halones, aliaque id genus meteora naturalibus & notis causis exoriri nemo est fere qui nesciat, nihilominus ab halonibus constanti augurio ventos & serenitatem praesagiri notissimum est."

[494] Ibidem: "Ventos quoque vehementissimos & turbines navigantibus, satis arboribus noxios irati Numinis aut suorum fidem periclitantis signa esse nemo, opinor, negabit, etsi sola aeris agitatione & motu eosdem cieri notissimum sit."

glected, and that consequently most of the secrets of nature still remained hidden. For this reason, our author did not possess any rules or experiences reliable enough to venture to put forward any predictions of his own on the occasion of the comet of 1618. Still, echoing Seneca, he expressed his confidence that later generations would some day uncover the occult mysteries of nature.[495]

For the time being, however, the rules on which astrologers had to rely were untrustworthy. Why, for example, should we lend credence to the Ptolemaic dictum that a comet that enflamed eleven signs distant from the sun threatened kings and princes in whose horoscope it stood with all kinds of evils, even with death? After all, it had never been observed that any newborn baby in whose horoscope the comet found itself, had died, while babies were most susceptible, both to death and to the radiation of the heavenly bodies.[496] Moreover, not just the rules themselves, but also the predictions based on them were unreliable. Snellius felt the need to point out that events that had taken place before the appearance of a comet (such as, in 1618, the outbreak of the Thirty Years War) could under no circumstances be ascribed to that subsequent comet; apparently, our author had in mind certain prognostications in which just such an unjustifiable procedure had been employed. After all, he added, "It is not difficult to discern those things which are already on our doorsteps and so to speak are thrown before our feet."[497]

Despite his objections to the astrological art as practised by his contemporaries, however, Snellius admitted that though he was not

---

[495] Ibidem, 65: "Veruntamen cum ista quam negligentissime sint habita, & haec adhuc magnam partem in naturae secretioribus adytis lateant recondita, hic nobis subsistendum est: reliqua enim dies & succedens tempus, coelo etiam posteris in haereditatem relicto, in apertum aliquando extrahet."

[496] Ibidem: "Monet Ptolomaeus cometam qui undecim signis a Sole exarserit Regibus & Principibus & Regionibus in quorum centro extiterit, nescio quae mala nominatim interminari, etiam mortes. Sed cum non sit observatum quenquam infantem recens natum in cuius cardine cometa extiterit, exstinctum, qui tamen huic iniuriae oportunissimi sunt, & siderum radiationibus, quam maxime obnoxii, nescio quam constanter in illos, qui iam firmata aetate sint haec ita decretorie tanquam praetorio edicto, a solo cometa, nulla malignorum radiorum applicatione conspirante decerni possint."

[497] Ibidem, 66: "Movebunt forsitan nonnullos etiam nunc imae rerum μεταμορφώσεις in Hispania, Bohemia, alibi. Atqui, o physici Doctores, nemo a casu afficitur ante casum. Illa cometae fulsionem antecesserant, eventus secuturi nobis sunt exspectandi. Istos itaque omnibus paulo accuratius observandos iudico. Neque enim difficile est cernere ea, quae prae foribus, & ante pedes sunt posita."

prepared to accept indiscriminately all prognostications, neither was he inclined to brand them all as nonsense.[498] Sure enough, he himself could not resist the temptation of concluding his tract with a prophecy of his own. While observing the characteristics of the comet of 1618, he had been struck by the strong resemblance of that phenomenon to the comet that had been described by the eighth-century Arab scholar Haly ben-Rhodoan (ʿAlī b. Riḍwān), both with regard to its form and with regard to its movement. Since Haly's comet had been followed by wars, murders, changes of governments, depopulations of regions, drought, scarcity, and pestilence, Snellius now feared similar consequences would accompany the comet of 1618.[499] This fear was reinforced by the fact that, immediately after its birth, the malignant planet Saturn had infected the hairy star with its evil rays; therefore, our author gloomily predicted, the present comet would work very strong Saturnine effects.[500] This cheerless vaticination inspired our author to conclude his tract with a supplicatory prayer to God, to avert all cometary evils from mankind and from the Church.[501]

Thus, of all the authors publishing in Latin on the subject of the comet of 1618, Snellius showed himself by far the most favourably disposed towards the traditional astrological art. Although he agreed with his Louvain colleagues, that the conventional rules of Ptolemaic astrology were unreliable, this did not induce him to reject them altogether. On the contrary: rather than, like Fienus, Fromondus, and Puteanus, abandoning the search for the way in which the celestial bodies physically influenced the earth, and professing scepticism with regard to the human capability of ever discovering any real knowledge on this subject, the Leiden mathematician strongly believed in both the desirability and the possiblity of reforming the art of prediction from the stars. If only the effects worked by comets (which Snellius clearly regarded as actual, physical influences actively

---

[498] Ibidem: "Nolim omnia nimis avide arripi, neque idcirco tamen velim omnia pro nihilo reputari."

[499] Ibidem, 67: "... In nostro cometa natalis soli affinitas, motus vicinitas, & tota similitudo denique me omnino movent."

[500] Ibidem: "Et quod maximi apud me est, quia Saturnus infestis & adversis radiis eius primos ortus exceperit, eumque malignis suis qualitatibus infecerit; ideo Saturninos plane ab huius cometae fulsione effectus, & quidem validos subvereor."

[501] Ibidem: "Quae mala ut DEUS Opt. Ter Max. ab Ecclesia sua, & nostris cervicibus clementissime avertat, supplex eum veneror: quia solus id potest. AMEN."

affecting the sublunary region and capable of being charted) were observed more accurately, and if only the art of cometary prognostication were to be founded on induction on the basis of a large number of observations, our author was convinced future generations would be able to predict cometary concomitants as easily as, in his own time, clear weather could be predicted from the appearance of haloes.

e. *The Continuation of the Astrological Tradition: Mulerius*

In his Dutch-language broadside on the comet of 1618, the Groningen professor of medicine, Nicolaus Mulerius, showed himself much less critical of the conventional practice of astrology than his Leiden colleague Snellius. His pamphlet did not show a single trace of disapproval of the rules employed, and predictions made, by astrologers in the Ptolemaic tradition. Nor did it contain any acknowledgment of the current debate on the viability of the art of prognosticating, in which Snellius had taken such a resolute stance. The reason for these omissions may well have to do with the kind of audience Mulerius was addressing: after all, writing in Dutch, he must have intended his tract for the common man, who, it can be assumed, was not much interested in theoretical niceties.

Nonetheless, Mulerius started his chapter on cometary significance with a short elucidation of the theoretical foundations on which he believed prognostication from comets could be based. Aristotle, our author maintained, had rightly pointed out that every natural phenomenon was called into existence with a purpose: God and Nature created nothing in vain.[502] Thus, rain served to fertilize the soil, snow, to protect the grass and the young crops from the cold, and thunder to purify the air; and similarly, every single member of the human body had its own special use. In a perfectly Aristotelian vein, Mulerius continued that nothing could be generated out of the terrestrial elements, if the powers and influences of the heavens did not act on the sublunary realm; from this circumstance could be deduced as a consequence that whenever something unusual appeared in the

[502] Nicolaus Mulerius, *Hemelsche Trompet Morghenwecker, ofte Comeet met een Langebaert Erschenen anno 1618. in Novembri ende Decembri* (Groningen 1618) chapter V: "*Aristoteles* heeft wel te recht geseyt, dat Godt ende die Natuyr niet met allen voorbrengen te vergheefs. Alle dinck heeft sijn waerom."

heavens, this would bring in its wake extraordinary occurrences below the moon as well.[503]

The greater the size of such a novel celestial body, Mulerius declared, the more powerful would be its influences.[504] This remark indicates that, like Snellius, our author conceived of celestial and cometary powers as actual physical entities whose potential effects could, at least to a certain degree, be discovered. Comets, being very large objects (some of them, Mulerius thought, were even bigger than one hundreth of the terrestrial globe) certainly possessed their own specific powers and significance. It did not matter whether they were eternal bodies or newly created entities, or whether they were etherial or elementary, our author maintained: it was beyond doubt that they caused sublunary changes.[505] After all, the moon, too, by her waxing and waning brought about very notable terrestrial effects, causing alterations in the sea, the air, the trees and even in human bodies. Therefore, Mulerius could see no objection whatsoever to the reasoning that comets, too, brought about alterations in the elements as well as in human beings. He explicitly endorsed the Aristotelian theory, so emphatically rejected by Fienus, that the hairy stars by their influences modified conditions in the atmosphere, so that storms and pestilential winds would arise, and people would fall ill and grow distempered and prone to revolts and rebellions.[506]

Such a reasoning, our author maintained, did not in any way infringe on divine providence; after all, God was wont to act through

---

[503] Ibidem: "Ende dewijl het kennelijck is dat de Elementen, aerde, water, lucht gestuyrt worden van den Hemel, also dat wt die Elementen niet voortgebrocht en wordt sonder toe doen vande Hemelsche influentien ende crachten: So volcht oock daer uyt dat so wanneer yet nieuws ende onghewoon inden Hemel erschijnt, dat hier beneden oock wat bysonders anden dach comen sal."

[504] Ibidem: "Ende hoe groter schijnsel hoe meerder werckinghe."

[505] Ibidem: "Een Comeet als boven gheseyt is, can soo groot wesen als die honderste deel vanden aerdtschen cloot, sommighe min, sommighe oock groter. Het sy dan dat die Cometen natuyrlijcke lichamen sijn van anbegin des Werelts gheschapen, oft datse nieuws gheworden sijn uyt Hemelsche ofte uyt Elementsche stoffe, dat doet al even veel, sy hebben even wel haer cracht ende beteeckinge."

[506] Ibidem: "Waeromme sullen wy dan niet bekennen dat die Cometen so groote lichamen zijnde, een alteratie mede maecken inde lucht, int water, ende oock op der aerden? Waer door onse lichamen eerst inde natuyrlijcke gheesten ende dan int'ghebloet, een ander qualiteyt ofte temperament becomen? Ende dat hier deur die lucht onstelt can worden, het sy met stormwinden ofte Pestilentialen ende veninighen qualiteyten, ende dat daer door onse lichamen worden ghedisponeert tot cranckten, ende die gemoederen opweckt tot onrust ende onvreden?"

intermediate causes.[507] Thus, like Fromondus, Mulerius saw comets as natural entities that were employed by God in order to work the effects he desired. The difference between these two stances was, that in Fromondus' view comets were merely signs of divine wrath; if in addition they also possessed certain operative powers of their own, these in any case remained unknowable to the human mind. Mulerius, conversely, thought of cometary influence as perfectly explicable by means of common sense arguments based on Aristotelian physics.

Furthermore, the fact that all comets brought in their wake consequences of the war-famine-pestilence variety, could also be proven by an appeal to past experience. All the peoples of the earth, Mulerius maintained, were agreed that after the appearance of a comet, great calamities and punishments ensued, such as earthquakes, floods, infertility, war, pestilence, deaths of kings and princes, and falls of empires.[508] For the benefit of his unschooled audience, he followed this remark by a list of historical disasters that had been accompanied by comets, such as the flooding of Buris and Helice, the destruction of Corinth and Carthage, the fall of Jerusalem, and more recently the Sack of Rome, the Anabaptist occupation of Münster, and the All Saints Day flood of 1570 in Holland and Frisia. The comet of 1577 Mulerius, not surprisingly, associated above all with the Dutch Revolt and its bloody consequences: thus, he sadly recounted how Flanders had been "completely destroyed by foreign nations, such as Spaniards, Italians, Albanians, Walloons, French, English, Scots etcetera", so that many people had had to flee the country. Most of the refugees, Mulerius continued, had moved to Holland, "where they found a good home"[509]; this appreciative remark was undoubtedly inspired by our author's own experience as a Flemish immigrant in the northern Netherlands.

The comet of 1607 had portended the conclusion of the ceasefire

[507] Ibidem: "Dit en strijdt niet teghen Gods voorsienicheyt ende regieringhe. Want Godt werckt ordinaris door middelen."

[508] Ibidem: "Ende wat die Cometen aengaet, t'gemeen ghevoel van alle volckeren is geweest, dat na de Cometen grote plaghe ende gemeenen straffen volgen, als aertbevingen, hoog wateren, onvruchtbaer jaren, crygh ende oorloghen, pestilentien, sterfte vande Coninghen ende grote Potentaten, destructien van steden, ende omsettinghen van Coninckrijcken ende gewalt."

[509] Ibidem: "Flander worde alheel verwoest door vreemde Natien, als Spagnaerden, Italianen, Albanoisen, Waelen, Fransen, Enghelschen, Schotten etc. alsoo dat' meeste volck gaf t'landt ten besten, ende vertrock in vreemde landen, doch meest in Hollandt, daer sy een goede herberghe ghevonden hebben."

which had preceded the Twelve Years Truce, Mulerius went on. Mars, having been exiled from the Netherlands, had taken refuge in Italy, France, and Sweden; however, our author ominously announced, "since he cannot get a firm grip anywhere, he may want to return to his former post; yet, in a far different guise than before."[510] This may be an allusion to the disputes between the Remonstrants and the Counter-Remonstrants, which had in recent months brought the northern Netherlands to the brink of civil war; after all, our author was a staunch Counter-Remonstrant and a good friend of Gomarus. When, in December 1618, Mulerius wrote his tract, the situation was still rather unsettled; although, with hindsight, we now know that Maurice's coup had lifted the danger of civil war, this probably did not at once become clear to Mulerius and his contemporaries, especially since at the time, neither the trial of Oldenbarnevelt nor the Synod of Dordrecht had yet been concluded.

As far as consequences of the comet of 1618 itself were concerned, Mulerius was not very specific. He clearly believed in its predictive value, and indicated several possible methods of discovering its message. Thus, he declared that the fact that the comet had moved along more or less the same course among the constellations as the hairy star of 1607, was certainly very significant.[511] This may imply that he was optimistic about the present comet's meaning; after all, the 1607 one had marked the conclusion of the ceasefire with the southern Netherlands. Yet, Mulerius himself did not explicitly draw this conclusion. Another way of making sense of the 1618 appearance, our author went on, was to apply to it the rules laid down by Ptolemy in his *Centiloquium*. After all, Philip Melanchthon, a "learned and God-fearing theologian", had recommended Ptolemy's astrology as more trustworthy than that of other astrologers, "who write a great deal of humbug."[512] According to Mulerius, the Ptolemaic aphorisms about

---

510 Ibidem: "Nae t'jaer 1607. t'Nederlandt creeg een opschorsinge van wapenen na dat de Oorloch geduyrt hadde 40. jaren langh. Mars hier uyt ghebannen synde voer na Italien, na Vranckrijck, na Sweden: Maer de wijl hy nergens stant can grypen, sal misschien sijn oude plaetse willen versoecken. Doch alheel op een ander wyse als te voren."

511 Ibidem: "Merckt nu wel an hoe dat elf jaer daer na is onse Comeet uyt het selve gat oft daer ontrent uytgheborsten, ende is langhs deselve Cours gevaren. N. Noordwest aen, tot an des voorsz. Beers sterte, al waer men se noch sach den 17. ende 18. Decemb. seer cleyn, als een schip dat bycans uyt het gesichte gevaren is. Dit heeft mede sijn besonder insien ende beduydinge."

512 Ibidem: "*Philippus Melanthon* [sic] een geleerde ende Godtvruchtich *Theologant*

comets, which he quoted in his own Dutch translation from the Greek, offered a "key to open the secret chamber, and to see in a great mirror what news this comet brings for Germany, France, Spain, Italy, etcetera."[513]

However, despite his self-professed confidence in the capability of Ptolemaic astrology to elucidate the comet's significance, our author did not go on to apply the rules he advocated, and to make any specific predictions. Instead, he included a moralizing passage about the characteristics of good government, declaring that those countries whose governors possessed wisdom, piety, and authority, invariably did well, since these three virtues were the pillars on which the common good rested. If all three pillars were weak, the government would crumble, and great lords would be overthrown, at least, in countries that were peopled, not by slaves, but by free men who had of their own free will subjected themselves to the command of a just regime. The governing of such a people required wisdom; if only they were ruled wisely, the regime would need no foreign aid against its own subjects.[514] This can undoubtedly be read as an allusion to the bad government of Philip II and his abjuration, which was often justified by the idea that the king had failed to serve the common good, for which purpose he had been appointed.[515] However, the remark about good rulers needing no foreign aid may well be directed also at Johan van Oldenbarnevelt, who had during the years

raet datmen meer houden sal van *Ptolemaei Astrologia,* als van d'anderen die veel beuselinghen schryven."

513 Ibidem: "Hier hebt ghy eenen sleutel om t'Secreet Camer te openen, ende sien daer in een grote spieghel wat dese Comeet voor nieuwe tydinge brengt voor Duytslandt, Vranckrijc, Spagnien, Italien, ect. [sic]".

514 Ibidem: "T'gaet den Landen wel daer die Regierders begaeft sijn met drie deuchden, Wijsheyt, Vromicheyt, ende Autoriteyt. Dit sijn de drie vaste pylaren daer t'Gemeen welvaer op rust. Wanneer een pylaer ontbreeckt, so lopet risko van om te vallen, ten sy dat het geholpen wort. Maer hoe willet toe gaen daer alle drie swack zijn? Even also als een steenen huys ghebout sonder anckers. So haest al daer swaerte ende ladinghe op comt, t'huys stort neer onder die voeten. Dit wil gemeenlijck een oorsake wesen, waer door grote Heeren van hun stoel afglyden. Om over Slaven te heerschen heeftmen geen grote wijsheyt van doen. Want slaven moeten met hun Heer te vrede wesen, hoedanich hy oock zij. Maer vrye Luyden die hun selven uyt vrye wille onderworpen hebben onder t'ghebiet van een billig Overicheyt, die moeten met wijsheyt gheregiert worden. Ende wanneer sulcx gheschiet, soo heeft de Overicheyt gheen uytheemsche steunsel van doen teghen sijn ondersaten."

515 See Martin van Gelderen, "Introduction" in Martin van Gelderen ed., *The Dutch Revolt* (Cambridge Texts in the History of Political Thought, Cambridge 1993) esp. pp. xxiv-xxv.

leading up to 1618 been pursuing a pro-French policy, to which the Counter-Remonstrants, to which party Mulerius belonged, were strongly opposed.

After this moralistic digression, which in fact had nothing to do with the comet, our author made a few very vague predictions, declaring that it was well possible that the hairy star's operations would be reinforced by those of the two novae that had shone several years earlier. It could not easily be ascertained where the comet's consequences would be felt; although it had passed over Persia, Turkey, Hungary, Italy, Bohemia, Poland, Germany, the Netherlands, and Denmark, it was not to be believed that all these regions would be affected. What was, however, definitely to be feared, was that, in addition to being a trumpet to warn people, the phenomenon was also to be regarded as a death bell.[516] Nonetheless, the only relatively concrete prediction of doom Mulerius dared to include in his pamphlet was directed at the Holy See, which, he believed, might well come to suffer the consequences of the fact that its motto was, that "It is good to make peace and to conclude treatises, but woe to him that keeps them."[517] On this anti-popish note, our author concluded his tract.

f. *Prudent Predictions: The Pamphlets*

The other Dutch pamphlets devoted to the comet of 1618 can be treated as a single category, because the ideas put forward by all five authors (Cats, Velsius, Saxus Fontanus, Nicolai, and an anonymous author) were very similar indeed. Like the pamphleteers of 1577, they all saw the comet first and foremost as a divine messenger notifying mankind of God's anger and imminent punishment; despite the fact that most of them (the exception being Cats, as we shall see shortly) adhered to the Aristotelian theory of cometary generation from hot, burning exhalations, they did not make any attempts to explain their consequences in physical terms such as those proposed by Mulerius. The main difference between the pamphlets of 1577

[516] Mulerius, *Hemelsche Trompet*: "Wat voorder onse Comeet angaet t'is te vresen, dat niet alleen die Trompet an boort sal, maer oock de Dootclocken."

[517] Ibidem: "Sullen die elck daer wat van hebben, dat waer te veel. Ic mene neen. S. Pieters stoel mochte wel een spaec ofte stal verliesen. Dan cant noch een drietal blyven. Vraegt ghy waerom? Om dat om die stoel voor een advijs staet: *Bonum est pacisci & facere foedera, sed vae illi qui servat.* T'is goet verdragen ende verbontenissen te maken, dan t'is een slechte bloet diese houdt."

and those of 1618 was, that the latter put much greater emphasis on the limitations of the astrological art, piously declaring that only God knew the precise nature of the calamities to come.

The ideas put forward by all five 1618 pamphleteers are best summed up by a description of the broadside composed on the occasion of the comet by Jacob Cats, the well-known Dutch poet and politician, who provided his readers with a most systematic and orderly exposition of the theoretical foundations of cometary prediction, which his colleagues treated in a much more chaotic and piecemeal fashion. Cats's pamphlet was composed in December, 1618, while the National Synod was under way in Dordrecht, as a "New Year's gift addressed at all right-minded wits". The title page promised Cats's audience an "instruction on how to correctly and piously explain all heavenly signs, and strange stars"[518]; thus the author from the outset made it clear that, in his opinion, comets were to be regarded as signs, on the same footing as other unusual heavenly occurrences.

In the poem with which he commenced his tract, entitled "Thoughts, occasioned by the present tailed star, on the meeting of the Synod currently assembled"[519], Cats further emphasized the comet's divine and miraculous status, by associating it with two famous historical precedents, the star of Bethlehem and the nova of 1572. In the first stanza, the poet dealt with the star that had announced Christ's birth, calling it "A star, a wondrous star, whose movement ran counter to that of all the other celestial bodies, and confused and silenced all those many people who thought they understood the motions of the heavens."[520] Another miraculous star, to which Cats devoted his second stanza, had appeared in 1572, and had shone upon the earth with "golden, unusual rays", in whose wake had followed the horrible Saint Bartholomew Massacre.[521] However, the

---

518 Jacob Cats, *Aenmerckinghe op de tegenwoordige Steert-Sterre, ende den loop deser tijden so hier als in ander Landen, Met aenwijsinge vande rechte wetenschap om alle teykenen des Hemels, ende vreemde Gesterten wel ende loffelijck uyt te leggen, alles tot een Nieuwe Iaer gifte alle rechtsinnige Verstanden toe-geeygent* (Middelburg 1619).

519 Ibidem: "Bedenckinghe By ghelegentheyt van de jegenwoordighe STEERT-STERRE, Op de vergaderinghe van de huydendaechsche SYNODE".

520 Ibidem: "Een Sterr', een wonder Sterr', die tegens ander Sterren, Quam stieren haren loop, en ging den geest verwerren, En sluyten toe den mont van al den grooten hoop, Die meynden te verstaen des hemels hoogen loop."

521 Ibidem: "Een ander nieuwe Sterr' heeft haer vergulde stralen, Met ongewoonen glans, op 't Aertrijck laten dalen, Als door het Fransche lant om Godes suyver woort, Wiert over al gepleecht dien grouwelijcken Moort."

events announced by this nova, like those denoted by the Star of Bethlehem, had in the end been good, Cats stated; after all, through God's miraculous works, the martyrs' blood had turned into the seed of the Church. The nova had been sent to urge people to adopt the right faith, and God had so to speak been born once again in the humble consciences of those who believed in him.[522]

Finally, in the third stanza, Cats came to treat the present comet, another unusual star revealed by God in the skies. Since both stars dealt with in the previous stanzas had, each in its own way, announced the coming of Christ, Cats implored heaven that the comet of 1618, too, might become the harbinger of the descent of the Son of God among men, this time in the guise of a return of peaceful conditions on earth and the driving away of the strife and discord afoot in the contemporary Dutch Republic.[523] By thus suggesting a close analogy between the star of Bethlehem, the nova of 1572, and the 1618 comet, Cats presented his readers with two implicit ideas: firstly, that, like the two previous apparitions, the present comet, too, was a supernatural miracle that ran counter to the ordinary laws of nature, and that had been rendered visible by God for a specific purpose; and secondly, that the phenomenon's message was to urge people to restore peace and unity, and unanimously to admit the true faith in their hearts. In connection with the poem's title, in which as we have seen Cats mentioned the Synod of Dordrecht, these remarks can be read as a call for a speedy resolution of the affairs being discussed there.

The rest of Cats's pamphlet consisted of a detailed elaboration of

---

[522] Ibidem: "Even-wel nochtans, door Godes wonder wercken, Der Martelaren bloet was als het zaet der Kercken, En Godes heylich volck, int midden van de doot, Wiert tot het recht geloof noch meer als oyt genoot., Wat bracht dees Sterre goets? God quam sijn uyt-vercoren, Begroeten op een nieu, en wert als nieu geboren, Niet in der ossen stal, maer in een sacht gemoet, Verweckt door sijne doot, gereynicht door sijn bloet."

[523] Ibidem: "God laet nu wederom haer thoonen in de wolcken, Een ongewoone Sterr', dies is het hert der volcken, Verslagen en verbaest; dus staet een yder stil, En wacht met vrees en angst wat dit beduyden wil., Coom Godes ware Soon, coom ware Son, coom weder, Coom naedert uwe Kerck, en dael nu immers neder, Met dijnen weerde geest, en Goddelijcke kracht, Ter plaetse daer u werck met vlijt nu wert betracht., Geeff dat dit nieuwe Jaer, en dese nieuwe Sterre, Vernieuw' ons ouden mensch; drijff, lieve God, drijff verre, Drijff uyt ons Neder-lant dien vreemden Tuymel-geest, Die over al het hert der vromen maeckt bevreest., Coom ware Vrede-vorst, en stel doch in de stede, van desen langhen twist dijn Goddelijcken vrede, Geeff dat der mensen hert verhart gelijck een steen, Door dijnen sachten geest mach weder werden een."

the former of these two ideas; the notion, that is, that comets were supernatural entities, on a par with other miraculous, unusual natural phenomena, that carried a meaning that was at once very specific and very general. Although no particularities could be predicted from them, because only God knew the exact nature of the punishments which he intended to send earthward, yet it was clear that comets signified only one thing: an urgent appeal to mankind to mend their lives. This general idea Cats conveyed in two different ways.

First of all, he composed a poetic "Dialogue about the present Comet", in which initially two and later three fictional characters discussed the plausibility of cometary divination. The character opening the conversation, aptly named 'Wel-hem'[524], wanted to know what exactly was the meaning of the cometary apparition he had seen "today between six and seven o'clock", and about which everyone spoke all day long.[525] This question met with an irritated response from Wel-hem's interlocutor, 'Gheen-aert'[526], who emphatically denied that the signs placed in the heavens by the astrologers carried any meaning: "I know nothing about rams or bulls whirling about in the heavens, because I prefer the ox that stands in its stable. I do not know any fishes, other than those that are served on dishes, and the only crabs and lobsters I have ever heard of are those we find in the water."[527]

Wel-hem, however, was not so easily to be put off. He declared that the astrologers had told him, that all comets were invariably followed by evil consequences, such as scarcity, famine, deaths of princes, changes of government, and the like. Well then, he continued, are we not currently being confronted with all kind of momentous changes? Thus, a new class was rising throughout the country (a clear allusion to Maurice's purges and the ascendancy to power of

---

524 An untranslatable pun on the Dutch name 'Willem' (William), with the first syllable 'Wel' being a Dutch word denoting agreement with a given statement.

525 Cats, *Aenmerckinghe:* "Heden tusschen ses en seven, Sach ick 't nieu Ghesterte sweven, Daer nu yder, vrouch en laet, Door de gantsche stadt van praet., Soo 't u niet en sou mis-haghen, 'k Woud' u hier wel wat van vragen, 'k Wou wel hooren vast besluyt, Wat dit teycken doch beduyt."

526 A name equally untranslatable as 'Wel-hem', with the first syllable 'Gheen' meaning 'None'.

527 Cats, *Aenmerckinghe:* "Ic en heb my noyt vermeten, Yet van s'hemels loop te weten, Of van teyckens te verstaen, Die de Sterren wijsen aen., Ic en weet van Ram, noch Stieren, Die ontrent den hemel swieren, Want ick houde meest van al, Vanden os die staet op stal., Ick en ken geen ander Visschen, Als die goet zijn voor de disschen, Ic en weet van Crabb' of Creeft, Als die ons het water geeft."

the Counter-Remonstrants), and abroad, too, alterations were being carried through: in Spain, the Duke of Lerma had been unhorsed, war had broken out in Bohemia, and the emperor Matthias might well shortly come to suffer the same fate as Philip II. After all, Wel-hem moralistically continued, "to break ancient liberties, to preach the word of God with the aid of a sword, to exert pressure on conscience by means of violence and human blood, sternly to speak of nothing but coercion, selfishly to accept foreign counsel, to break treaties, to ignore the complaints of the population, all these things dispossess Princes of their properties."[528] Clearly, in the Dutch Republic of 1618, the bloody conflicts that had raged before the Truce had not yet been forgotten.

Gheen-aert, however, could not be persuaded, pointing out that people constantly looking upwards to study the stars stood in danger of not minding where they were walking and thus of falling into holes or puddles. More alarmingly still, those who wished to "measure the heavens, and to predict many secrets" were to be regarded as "great idiots worthy of being mocked", because in continually scrutinizing the skies they tended not to notice the evil with which they were confronted on earth.[529] Wel-hem strongly objected to such a point of view, declaring that human beings ought not, like swine, to root about in the earth and concern themselves only with earthly matters, since God had created them to contemplate his wondrous works.[530]

The deadlock created by Wel-hem's and Gheen-aert's irreconcilable points of view was broken by the arrival of a third interlocutor, named Reyn-hart ('Pure-heart', a pun on the Dutch name Reinaert)

---

[528] Ibidem: "Oude vryheyt te verbreken, Mettet sweert Gods woort te preken, Door gewelt en menschen bloet, Dwang te stellen aan 't gemoed, Al met strengheyt in te dringen, Niet te spreken als van dwingen, Aen te nemen vreemden Raedt, Omghevoert door eygen baet, Hant-gevesten te verachten, Niet te passen op de clachten, Die de goeGhemeynte doet, Set de Princen uyt haer goet."

[529] Ibidem: "Lestmael hoord' ick eens vertellen, Datter een van dees gesellen, Die de Sterren, Son en Maen, Willen op haer duym verstaen, Datter een ging in de Sterren, Kijcken, gapen, sien en werren, En ter wijl hy besich was, Viel hy in een diepe plas., Seg my, zijn 't niet groote sotten, En wel weerdich om bespotten, Die soo verre, die soo hooch, Loopen met haer ydel oogh?, Die den hemel willen meten, En voor-seggen veel secreten, End' en mercken niet het quaet, Datter voor haer deure staet?".

[530] Ibidem: "Wel hoe dus? zijn dan de menssen, Anders niet dan drec en penssen?, Sal ons hooft, gelijck een swijn, Staech na d'eerd' gebogen zijn?, Neen, wy zijn van God geschapen, Niet om eeckels op te rapen, Off te sitten by den heert, Off te wroeten in der eert., Maer veel eer om aen te mercken, Godes wonderlijcke wercken, Daer in dient den tijt besteet, Alsmen niet te diep en treet."

who expounded what was, in Cats's eyes, the correct stance vis-à-vis astrology. Reyn-hart declared that it was impossible for human beings to correctly interpret the signs of the heavens, whose secrets lay hidden in the depths. Only God knew their precise meaning, and he was displeased at any human attempt to fathom his occult decisions. All we need to know, the wise Reyn-hart assured his audience, was that the appearance of a comet signifies impending divine punishment, which may still be averted if we mend our ways.[531]

To further clarify this central idea, so eloquently advanced in the poetical dialogue, Cats's pamphlet also comprised a short tract called "More detailed remarks about the comets and other celestial signs."[532] He once again declared, that there existed three kinds of people: the Wel-hems, who believed that every single human action was caused by the stars, and could be predicted by observing them: the Gheen-aerts, who were only interested in material, earthly things, and paid no attention whatsoever to the heavens; and the Reyn-harts, who chose the golden mean and advocated the study of the celestial bodies, up to a certain limit.

The first point of view, Cats stated, was to be rejected by all pious men as running counter to divine providence, as having been prohibited by Holy Scripture[533], and as harmful for a civil way of life (by which he probably meant that, if all actions were to be ascribed to the stars, no morality would be left and thieves and murderers would be able to excuse themselves with an appeal to the bad constellation

---

531 Ibidem: "Wie sal seggen int bysonder, Wat dat ons bediet dit wonder?, Wie derff seggen over-luyt, Wat dit teycken doch beduyt, ... Wie can doch door menschen vonden, 's Hemels diepten recht doorgronden, Zijn dees saken niet te hooch, Voor des menschen duyster oogh?, 's Hemels teyckens te bedijden, Dunckt u dat oock niet te strijden, Tegens 't woort en hooch gebodt, Van den alleen wijsen God?, God en is niet wel te vreden, Dat den dommen mensch wil treden, In sijn diep-verborgen raet, Die ons verr' te boven gaet. . . . 't Is genouch voor ons te weten, Dat de steerten der Cometen, Voor gewis ons beelden aff, Teykens van des Heeren straff., ... *Dan wilt ghy een regel weten, Hoemen teykens en Cometen, Wt moet legghen, en verstaen, Hoort ick sal 't u wijsen aen:, Wilt sulck een bedietsel geven, Dat verbetering van leven, Mach ontstaen in elck gemoet, Soo is u uytlegging' goet."*

532 Ibidem: "Naerder op-merckinge over de Cometen ende andere teykens des hemels."

533 In Jeremiah 10:2: "Thus saith the Lord, learn not the way of the heathen, and be not dismayed at the signs of heaven: for the heathen are dismayed at them" and in Deuteronomy 18:9: "When thou art come into the land which the Lord thy God giveth thee, thou shalt not learn to do after the abominations of those nations" (with the following verses containing prohibitions to employ divination, or to consult astrologers, wizards, witches, necromancers etc.).

under which they were born). Moreover, most soothsayers were interested in nothing but material gain. Yet Gheen-aert's stance, rejecting all contemplation of the heavens, was equally inadmissible, because the glory of the Creator had been revealed above all in the celestial region, which constituted the highest, and consequently noblest part of the cosmos. Therefore, the attitude to be embraced was Reyn-hart's golden mean.

The rest of Cats's tract was devoted to defending the praiseworthiness of such an attitude. First of all, he declared, the study of the ordinary course of the heavens was a commendable and useful activity, since it allowed men to distinguish the seasons, predict the tides, navigate, and anticipate the weather conditions. And as for the unusual celestial appearances, these should be regarded as indubitable signs notifying men of God's intention to make something unusual happen on earth as well. Such had been the standing still of the sun in the days of Joshua, the receding of the sun during king Hezekia's reign, the star of Bethlehem, and the extraordinary solar eclipse during Christ's crucifixion.[534] Moreover, Holy Scripture also taught that the Second Coming of Christ was to be preceded by signs in the heavens. And in addition, a host of ancient poets and historians could be cited to the effect that comets were invariably followed by sad events; therefore, our author interpreted the hairy stars, with their fiery, bloody rays, as 'God's rod'.[535]

Thus, Cats associated the appearance of comets with other celestial phenomena, whose miraculous and supernatural character was undisputed. Sure enough, he did mention the fact that some people thought comets to be brought into existence by natural causes, and could not therefore be said to possess any significance.[536] These people usually appealed to Jeremiah 10:2, where God admonished his people not to fear the signs of heaven as the heathen feared them.

---

[534] Joshua 1:13, Isaiah 38:8, St Matthew 2:2 and St Matthew 27:45 respectively.

[535] Cats, *Aenmerckinghe:* "Soo en ist niet vreemt dat van alle oude hercomen Sterren met Steerten, en dierghelijcke teykenen by de treffelickste verstanden van alle eeuwen zijn gheacht gheweest als voorbeelden van eenighe sonderlinghe veranderinghe onder den menschen, meestendeel hellende tot eenighe droevige uyt-coomste, en midts dien met hare vierighe en bloedige straelen als Godes Roede inde lucht verthoonende. . .".

[536] Ibidem: "Men vint onder-en-tusschen luyden die van ghevoelen zijn, dat de Cometen, die sy lieden segghen uyt natuyrlicke oorsaken voort te comen, niet met allen en souden bedieden, ende datmen daer op geen acht ter werelt behoort te nemen. . .".

However, Cats argued, this verse did not at all imply that heavenly signs possessed no significance; it only meant to convey the idea that they ought not to be feared *in the way the heathen feared them,* i.e., by losing all faith in the merciful promises of God, and by ascribing all subsequent calamities to physical necessity. Rather, Cats declared, we, as Christians, have to keep faith in God, who is capable of rendering inefficacious the signs of the soothsayers, since he has dominion over the heavens.[537] Thus, comets should not be deprived of their status as divine signs, nor should the calamities they brought in their wake be ascribed to mere physical causality.

Moreover, Cats added, it had not been conclusively proved that the comets themselves were due to natural causes, and even if they were, they might still in addition possess a divine significance. After all, the rainbow, too, was brought about by natural causes yet had a supernatural meaning. And again, the idea that if comets possessed any power, this should manifest itself immediately, because it could not be assumed to continue being effective after the comet's disappearance, was also based on the faulty assumption that comets themselves worked the disasters they announced.[538] To Cats, comets were not due to natural causes themselves, nor did they possess any effective physical power: they were neither more nor less than harbingers

---

537 Ibidem: "Ten anderen, en ghebiedt den Propheet niet datmen de teykenen niet achten en soude, maer datmen sich daer voor *niet vreesen* en soude, te weten *als hun de Heydenen vreesen,* 't welck niet anders te seggen en is, dan datmen sich soo niet ontsetten en moet int aensien van dusdanighe teyckenen, datmen sich mistrouwen soude aen de genadige beloften van den goeden God, toe-schrijvende alle toecomende swaricheden tot de cracht ende de werckinghe der teykenen, als de beweghende oorsaken van de selve, maer veel eer datmen vast vertrouwen moet dat de Heere, als oock heersende over de ghesterten, de teykenen der Waerseggheres te niete can maken."

538 Ibidem: "Aen-gaende de bewijs-reden die van de natuyrlicke oorsake der Cometen wert ghenomen, de selve en gaet mede niet vast, want schoon men het voorbrenghen der Cometen int gheheel de natuyrlicke oorsaken al wilde toe-schrijven ('t welck alsnoch niet claer ghenouch bewesen is) soo en soude daerom terstont niet volghen, dat de selve niet yet wes en soude beduyden, ofte aff-beelden. Den Reghenboghe wert buyten twijffel door natuyrelicke oorsaken voort-ghebracht, even-wel nochtans isset een teyken des verbonts tusschen God ende alle levende gedierte, Genes. 9.12. Andere meynen dat indien de Cometen eenighe cracht hadden, dat de werckinghe van de selve dadelick, ende ter wijl sy staen ende ghesien werden, haer souden moeten openbaren: Want, segghense, sy en connen niet verstaen werden te wercken, alsse al vergaen ende te niete zijn. Dan die luyden bemercken de Cometen, als oorsaken van de swaricheden die na de selve volghen, 't welck ongheriemt is: alsoo de selve alleenlick moeten ghenomen werden als voor-boden van Godes toecomende straffe, ende niet als off den geessel selfs uyt haren steert, als van haer oorsaecke, voort-quame. . .".

of God's impending punishment. For this reason, no precise predictions could be made from cometary appearances, since they did not obey any laws susceptible to human reason; only God himself, or those expressly inspired by him, knew exactly what calamities each hairy star portended.[539]

The central ideas proposed by Cats were repeated, in less systematic fashion, by his four fellow pamphleteers, Velsius, Nicolai, the anonymous, and Saxus Fontanus; with one difference: whereas Cats defended the view that comets were supernatural beings miraculously created at God's command, the others did not explictly address the question of whether the tailed stars were natural or supernatural entities. From Velsius' remarks, that the comet of 1618 "pointed upwards its terrible, fiery, martial tail, or rays, because of copious feeding material, which happens because of the movements of some of the celestial bodies, which attract many fiery vapours upwards and collectively set them on fire", and that the comet would remain visible for a long or a short time, "according to its feeding material, which may be either little or much"[540], we can deduce that this author, at least, adhered to the orthodox, Aristotelian theory of cometary generation, and that, therefore, to him they were natural entities. The anonymous author acknowledged the fact that some astronomers thought the comet to be located higher up than the moon, but was not himself prepared to go any further than to place the hairy star at roughly the same height as the moon, thereby suggesting that, like Aristotle, he believed it to be situated in the sphere of fire, just below the moon's orb.[541] Saxus Fontanus suggested that he, too, regarded comets as natural entities, when he stated that

---

[539] Ibidem: "Wat de Comete, boven 't gene wy hier int gemeen geseyt hebben, beduyden mach, houden wy alleen aen Gode, ende den genen die desenthalven eenighe bysondere openbaringhe van Gode mach hebben ontfangen, bekent te zijn."

[540] Johannes Velsius, *Prognosticatie van de nieuwe ende seer schrickelijcke Comeet ofte harige Ster Anno 1618. verschenen* (Franeker 1618): "... so sal ons dese gelijc andere Cometen niet minder quaet te wege brengen, dewijl hy zijn schrickelijcke vuyrige *martiaelse* steert ofte stralen is opwerts stekende door groote aendrievende, ende voedende matery, twelcke door eeniger Hemelscher lichamen beweginghe gheschiet, waer door veel vuyrige dampen ghetrocken ende versamelt aenghesteken worden..."; "... na dat zijn aenvoedende matery weynich ofte overvloedich zijnde al brandende verteert wort...".

[541] Anonymous, *Corte beschrijvinghe ende Affbeeldinghe des seer grooten ende schrickelijcken Comeets, de welcke den 27. November in 't Iaer 1618. ons verschenen is* (2nd ed., Leiden 1619): "Daerom indien dese Comeet inde selve Circkel staet daer de Maen in loopt, (alhoewel sommighe Astronomi getuygen de selve hoger te staen)...".

comets were "*not always* the causes of such disasters"[542]. Only Balduinus Nicolai seemed to share Cats's view of supernatural comets, declaring that "Comets do not appear or show themselves by their own nature, but at God's command"[543]; this remark, however, does leave open the possibility that the generation of comets was, indeed, natural, but that only their becoming visible was due to divine agency.

Unlike Cats, then, his four fellow pamphleteers were not very interested in the question of whether the comet was a natural entity or a divine prodigy; however, probably they still adhered to the Aristotelian theory of cometary generation. Yet, their outlooks were for the most part very similar to Cats's. Thus, they all strongly emphasized the status of comets, be they natural or supernatural, as harbingers of God's wrath, made to appear at divine command to warn people of impending disasters and to admonish them to mend their ways while there was yet time. Thus, like Cats, the anonymous author stated that those people who laughed at the significance attributed to comets, because they were convinced that the hairy stars were due to natural causes, were to be regarded as impious, since, although it was true that comets in themselves did not produce any disasters, they certainly did possess the status of divine messengers portending evil consequences.[544] Nicolai, too, stressed the hairy stars' divine function, comparing their tails to rods appearing to warn mankind, and declaring that the calculation of the lengths of these tails in order to

---

542 Saxus Fontanus, *Bediedenisse vande nieuwe Comete ofte Sterre metten Steerte, die int laetste vande maendt November, sich eerst aenden Hemel geopenbaert heeft* (The Hague 1619): "Ende hoe wel sulcke verschijninghe der Cometen niet even altijdts de oorsaken en zijn sulcx ongheluck...".

543 Balduinus Nicolai, *Verklaringhe Om yegelijcken te adverteeren van de dreyghementen Godts, die wy aenmercken in 't aenschouwen vanden Comeet, gesien in 't Jaer na de geboorte onses Heeren ende Salich-makers Jesu Christi 1618 (na de konst der Astronomie van oude Doctooren daer op gheschreven, te weten, nae dat de vertooninghe geschiet, in 't Oosten ofte Westen, oock nae de teeckenen des Hemels ofte Zodiac, daer de Comeet in gesien wort, als mede nae dat de steert lanck ofte kort is* (Amsterdam s.d.): "...daerom verstaen wy, dat de Comeeten niet uyt eyghen natuere en komen ofte haer vertonen, maer uyt Godts bevel."

544 Anonymous, *Corte beschrijvinghe:* "Daer worden oock wel seer veel godtloose menschen ghevonden dwelck alsulcke teeckenen ende voorboden des Godtlicken toorens seer kleyn achten, offt oock wel gheheel bespotten als dwelck haere natuerlicke oorsaecke hebben, bekennen wy dat de Cometen in haer selven gene ongheluck wercken ende veroorsaecken, maer Godt de Heere die sulck ongheluck over ons laet comen laet alsulcke teeckenen als voorboden der toekomende straffe voorschiinen, op dat de menschen door tydighe verbetering haeres levens de selve mogen ontvlieden...".

be able to make predictions on the basis of the outcome of this calculation (as the astrologers were wont to do) was of no use, because such physical characteristics did not make any difference. The only thing man needed to know was the divine message comets delivered: God is angry and people need to repent.[545] And Fontanus, too, saw the rod of God in the comets' tails, and pointed out that God never sent any punishments without previous warnings.[546]

Thus, all four authors agreed with Cats that there was only one admissible way of interpreting cometary appearances. In general, they meant impending divine wrath; however, the particularities of the imminent calamities, of what kind these would be, and which regions would be affected, could not be foreseen by human minds. The pamphleteers were very cautious in order to avoid the charge of impiety, stressing that they well realized that only God knew the details of his providential plan, and that men should not attempt to uncover them. Thus, both Fontanus and Velsius warned their readers that they could not provide them with any certain predictions, since "our intellect always stumbles with regard to divine and supernatural matters, and is full of doubt".[547] Nicolai even concluded

---

545 Nicolai, *Verklaringhe:* "Wy verstaen dat haer lange steert ons beteeckent een Roede, die Godt vertoont om ons te straffen ende castijden, so wy dan ons niet beteren, so sal de straffinghe wel te langher mogen duren."; ". . . dese calculatie [of the comet's length] loopt al wat hooch, maer daer en is so veel niet aen gelegen, al en verstaet de gemeene man dese calculatie niet. Daer is meer aen gelegen te belijden dat God ons daer mede te kennen geeft, dat syn Godlijcke Majesteyt vertoornt is op 't Menschelijcke geslachte, ende datmen oprechte boete doet."

546 Saxus Fontanus, *Bediedenisse:* "Dat de straffen ende plaghen comen om der sonden wille, is kenlicker dan dat 'tselve eenighe demonstratie ofte bewijsinghe van doen heeft, ende dat Godt noyt eenighe plaghen en heeft gesonden, sonder den menschen daer af eerst langhe te vooren gewaerschout te hebben, soude ick met vele schriften connen doen blijcken, dan also ick meen 'tselve by niemanden gheloochent ofte ontkent en werdt, soo ist onnoodich daer af veel te verhalen. Maer dewijle wy nu wederom een voorbode sien van toecomende plaghen, die ons verkondicht de groote gramschap Godes, ende hoe dat sijn roede uyt ghestreckt is over ons om ons te straffen, so wy ons niet en beteren noch en bekeeren; soo heb ick tot waerschouwinghe van een yeghelicken, de bediedinghe vande Comete, die ons nu nieulicx is verschenen int laetste van November deses Jaers 1618. hier willen stellen, ten eynde wy tot Godt keeren, ende door oprecht berou ende beteringhe onses levens, mogen ontcomen die straffen ende plaghen die ons de Heere door dese Comete is voorsegghende."

547 Velsius, *Prognosticatie:* "Niet dat ick my derf aennemen ofte vermeten een seecker ordel ofte uytlegginghe daer van te doen (dewijle onse verstandt in de Godlijcke ende boven natuyrlijcke dinghen altijdt struckelt, ende vol onseeckerheyt is). . ."; Fontanus, *Bediedenisse:* "Het schijnt vermetenheyt te wesen van sulcke voorteeckenen yet wat seeckers te willen difinieren, de wijle onse verstandt inde Godtlicke ende boven natuerlicke dinghen altijdt struykelt, ende vol onsekerheyt is. . .".

his pamphlet with a separate "Warning not to misuse the art of astronomy", in which he told his readers that astrology was in itself a laudable and commendable enterprise, which had been practised by such venerable predecessors as king David and the three Magi. However, the art could easily be abused: thus, thieves or murderers might try to exculpate themselves by an appeal to the malignant star under which they had been born. This, Nicolai stated, was a fallacious argument, since the planetary influences, though certainly powerful, could nonetheless never compel human beings to act, because all planets obeyed to God's command.[548] For this reason, Nicolai himself, in order to avoid being branded a "false prophet, or a person called soothsayer", stressed that all he did in his pamphlet was to expound the rules of the ancient astronomers.

All authors included lists of past cometary appearances, with their consequences, in order to demonstrate that history taught that no comet had ever shone forth without calamities following in their wake. Not surprisingly, the recent sad events in the Netherlands figured prominently in these lists, being associated with the comet of 1577 and the nova of 1572. Especially Fontanus paid much attention to near-contemporary history, referring to the chastising of Zutphen and Naarden and the capture of Haarlem by Don Fadrique, the battle of the Mokerheide, the Spanish Fury, the coup d'état of Don Juan, and the battle of Gembloux with Parma's subsequent successes in the southern Netherlands. Nicolai, Velsius, and Fontanus in addition cited a number of rules of cometary divination, drawn from the traditional astrology of Ptolemy, Haly, Albumasar, and even Butsahan of India. However, they made sure to emphasize once again God's power over the stars: he could render null and void all the elaborate theories of the astrologers. Fontanus declared that the particularities to be attributed to the present comet could not be foreseen, since these depended on human behaviour in the near fu-

548 Nicolai, *Verklaringhe:* "De Astronomische konst is goet ende profijtelijck, wanneer men deselve niet en misbruyckt... Het is kennelijc ende openbaer, datter veel zijn, die menen als sy quaet doen, datse dan van eenich Planete daer toe beweecht worden, om het selve quaet te doen: Jae een Dief ofte Moordenaer sal wel derven seggen, dat hy geboren is in een quade Planeet, ende dat hy niet anders doen en can: (dat is valsch) Want Godt en heeft gantsch niet gheschapen tot verderfenisse ofte verdoemenisse vande Menschen. D'influentie vande Planeten en mogen niet dwinghen den Mensche die Godt vreest, om quaet te doen. Hoe-wel de Planeeten niet eenderley operatie en doen, ende datse oock niet eenderley inclinatie en hebben, soo ist nochtans, datse alle Godt gehoorsaem zijn."

ture; and Nicolai stated that comets left whenever God ordered them to, since he was stronger and more powerful than the stars.[549]

Despite their cautious stance, the pamphleteers did include a number of more or less specific predictions for the comet of 1618. Thus, Velsius believed that, because of its martial tail, and because it had been generated by a conjunction of Mars and Saturn, the phenomenon would bring in its wake great cruelty and warfare, deaths of great lords, mutiny and bad harvests, especially in the countries under Scorpio's rule, such as Norway and Sweden, but also some of its effects would be felt in regions ruled by Cancer, Leo, and Aquarius. In addition it announced civil strife, fires, drought, tempests, shipwreck, pestilence, famine, fevers and miscarriages. According to Fontanus, the comet's operations would begin to make themselves felt in March 1619, and would last for several years. They would include the death of one or more kings in the southwest (this would seem to imply Spain), and also great wars and seditions, ambushes, and treacherous councils, strong winds, snow and ice, floods, fevers, and changes in government and religion. All these effects, however, could still be averted by doing penitence.

Although he, too, tried his hand at a few prognostications, stating that the comet presaged great changes in Holland and other provinces within the next three to four years, and would also have an effect on Spain (not a very risky prediction, since of course he knew that in 1621 the Truce was to expire), Nicolai was not so much interested in the events of the near future, but rather chose to examine the circumstances that had induced God to send a comet in the first place. He declared that the present situation in the Netherlands was very sad: many who used to be friends had now come to hate each other, and the country was rife with slander and deceit. The

---

549 Fontanus, *Bediedenisse:* " Ick repetere noch eens dat alle Landen beschenen van dese Comete, veranderinghe sullen lijden, jae die oock treffelick ende groot sullen sijn: ende voeghe daer by, tot goet ofte quaet, na dat wy sijn; willen wy het goede, soo moeten wy goet syn. . ."; Nicolai, *Verklaringhe:* "Hier uyt volcht dan, so wy waerachtich boete doen, met vasten ende bidden, met schreyen ende weenen, gelijc als die van Ninive deden, dat God ooc medeleyden met ons hebben sal. Daer en is geen middel om de gramschap Godts te stillen, dan dat wy ons verootmoedighen voor syn H. Majesteyt met verslagen herten. Dit doende ick en twijffel niet in sijn Barmherticheyt. Wanneer Christus de zee bekeven heeft, de stormen hebben opgehouden. De Cometen ooc also haest dat God haer gebiet te vertrecken, soo vertrecken sy, niet dat 't daer mede genoech is als sy vertrocken syn, wy moeten continueeren tot boete, soo wy ontkoomen willen vande dreygementen die wy voor onsen oogen sien.".

author was also very critical of the many pamphlets issued by Remonstrants and Counter-Remonstrants to vilify each other, branding these as "extraordinary counterfeits and ugly depictions to render people ugly and hated". Nicolai declared himself impartial, saying that whether one were a Remonstrant or a Counter-Remonstrant, in any case it was unchristian to hate each other. The two parties might well believe that the consequences of the comet would hit the opposing group; however, Nicolai denied this to be the case, urging both sides to examine their own consciences.[550]

We can conclude that, in 1618 as in 1577, the Dutch-language pamphlets not written by university professors expressed the view that comets were divine signs, rather than physical causes, of impending calamities. Like their predecessors forty years earlier, the 1618 pamphleteers unanimously expounded the teratological stance, that hairy stars were messengers called forth by God, who was dismayed by human misbehaviour and at the same time solicitous for the well-being of mankind, and whose anger could be averted by prayer and penitence. Although they did quote and apply the rules of traditional Ptolemaic-Arabic astrology, the predictions they based on these were very vague and general, and not at all physically underpinned. All the authors took great pains to avoid being accused of impiety, stressing the tentative character of their prognostications several times. Not the physical power exercised by the hairy stars, but the omnipotence and vigilance of God, who could go against the influences of the stars, was the principal thought on the minds of the 1618 pamphleteers.

---

[550] Nicolai, *Verklaringhe:* "Ick ben bedroeft dat de liefde also verkont is, ende de ongerechticheyt in menigherhande maniere alle dage vermeerdert, ende d'overhant neemt. Het is te beweenen datter veel zijn die goede vrienden geweest hebben, ende malcanderen haten. Het is nu alreede een tijt geleden, datmen in verscheyden steden ende plaetsen niet en hoort dan lasteringen, verachtingen ende calomnien. Men siet ooc extraordinaire Conterfeytsels ende leelicke afbeeldingen, om de menschen leelick ende hatelijck te maken: Ick spreke als onpartijdich, ick spreke aen beyde zijden, ofte Remonstranten ofte Contra-Remonstranten genaemt zijn, ick segge dat 't niet wel noch Christelijck gedaen is malkanderen te haten."; ibidem: "De Contra-Remonstranten sullen oock mogen meenen dat de voorsz Comeet soude beduyden een straffinghe voor de Remonstranten, ofte ten minsten voor de gene die gevangen zijn van wegen 't Lants saken. Desgelijcx sullen de Remonstranten oock mogen dencken, dat dese dreyghementen beteeckenen straffen der Contra-Remonstranten, ende veel anderen moghen dencken, gelyck sommighe wel derven seggen, dat de beduydenisse soude alleenlic voor andere verre landen zijn. Het soude veel beter ende nootsaeckelijcker zijn, dat een yegelijc zijn eygen conscientie examineerde."

5. *Conclusion*

What, then, were the Dutch authors' views on the nature of cometary significance, and on the raisons d'être of the hairy stars? First of all, it can with a fair degree of certainty be pointed out that the authors themselves regarded the possible meaning of the unusual phenomenon as the most important reason for devoting their attention and their time to describing it. The comet's final cause, so to speak, itself constituted the final cause of the cometary tracts. This certainly holds true for all the pamphlets, in which hardly any attention was devoted to the phenomenon's physical characteristics, the way in which comets were generated, or the place where they were located, with the pamphleteers focusing almost exclusively on the comets' significance for man on earth. However, each of the learned authors, too, addressed the question of the tailed stars' meaning; to Gemma, Heurnius, and Puteanus, the formulation of an answer to this question even constituted the sole motivation for studying the phenomenon's physical characteristics, while Fienus, Fromondus, and Snellius also devoted separate chapters of their tracts to the comet's predictive value. In 1577 as in 1618, it was unthinkable for a learned author to write a tract on any natural phenomenon, let alone one traditionally so highly charged with significance as a comet, without committing to paper his thoughts on its possible final cause.

The preoccupation with significance was undoubtedly enhanced by the tumultuous political situation in the contemporary Netherlands. The general tendency throughout the sixteenth century, and throughout Europe, had been to devote substantial attention to unusual natural occurrences, such an attitude being fostered by the religious upheavals of the era of Reformation and Counter-Reformation, as well as by the revival of ancient philosophies such as Neoplatonism and Stoicism. By the final quarter of the century, religious and political tensions had reached unprecedented peaks in the Netherlands, with all seventeen provinces being ravaged by warfare, civil strife, religious conflict, dissimulation and deceit, treacherous governors, tyrannical rulers, broken treaties, failed attempts at reconciliation, on top of more conventional early modern inconveniences such as plague, scarcity, and famine. No wonder, then, that some of the members of the humanist circle around Christopher Plantin at Antwerp, with whom both Gemma and Heurnius maintained friendly relationships, expected the impending end of the world, and strove to reconcile the manifold religious and political differences by advocating (at least, among themselves) the uncovering of the divine plan, as exemplified in nature, as a means of reaching truth. Also, given the martial and oppressive conditions of the age, it need not surprise us that Justus Lipsius saw an analogy between the contemporary Netherlands and imperial Rome, and resorted to the teachings of the great Roman Stoic Seneca to provide him with moral guidance.

Under these political and intellectual conditions, it would have been remarkable indeed if we had been unable to detect any traces of Stoicism in the Dutch cometary tracts of 1577 and 1618. After all, Gemma and Heurnius, the two most important 1577 authors, certainly knew Lipsius, and belonged to the same circle of humanists with *politique* convictions; and our prominent 1618 author Puteanus was Lipsius's best pupil as well as his successor at Louvain. Sure enough, these are the three authors in whose views on cometary significance the greatest amount of influence of Stoic divinatory ideas can be found. Yet the other authors, too, were affected by the revival of non-Aristotelian ideas: Fienus and Fromondus both refused to accept the traditional, Aristotelian way of accounting for cometary consequences, and in Fromondus' 1627 *Meteorology*, his ideas on the significance of at least superlunary comets bore a marked resemblance to the Stoic ideas of Puteanus, so that it seems plausible that, had he addressed the question of cometary significance in other than strictly natural terms in his tract on the comet of 1618, he would there, too, have advocated a stance similar to the one propounded by Puteanus in *his* work on the comet.

Thus, the teratological stance, put forward by such authors as Seneca, Cicero, and Saint Augustine, can be said to have exercised a fair amount of influence on the Dutch authors, both in 1577 and in 1618. Central to teratology were two main tenets: the idea that the cosmos was one single living entity, with each and every part corresponding to all the other parts by means of an unspecified and unquantifiable world soul, or pneuma, and that therefore changes in one part of the cosmic organism were invariably accompanied by changes elsewhere; and the belief that this great cosmic whole was governed by a benevolent, omnipotent God, whose providence assured the harmonious, regular course of nature, and whose solicitude for mankind induced him to provide them with natural signs whenever a cosmic upheaval was imminent, so that they might foretell the future and take their precautions accordingly.

These teratological ideas had been elaborated above all by the ancient Stoics and Neoplatonists, whose works were being rediscovered during the sixteenth century. However, as in the case of cosmology, with which we have dealt in our previous two chapters, in divination, too, the revived non-Aristotelian ideas from Antiquity collided, during the sixteenth and seventeenth centuries, with a firmly established and highly esteemed world system: the Aristotelian world view. With regard to the prediction of the future, as with regard to the construction of the cosmos, the Aristotelian outlook differed from that of its ancient rivals. Aristotle himself had gone no further than to allow for the prediction of certain imminent changes in sublunary nature by means of physical powers exercised by the celestial bodies, most notably the sun and moon, on the terrestrial realm. This purely physical stance had been elaborated by Ptolemy, and later by the Arab astrologers, who had attempted to include the influences of all the heavenly bodies in an elaborate scheme of astrological causation, laid down in a great number of complex rules. When, during the twelfth and thirteenth centuries, the Latin West had rediscovered the Aristotelian corpus, the astrological way of predicting the future had also been enthusiastically adopted by the scholastics; although they needed to allow for a certain measure of human freedom and divine omnipotence, on the whole their explanations of sublunary occurrences was based on Ptolemaic-Arabic astrology and Aristotelian physics.

Thus, in answering the question of how the appearances of the comets of 1577 and 1618 affected the Aristotelian world view in the Netherlands, we need to establish whether the Dutch authors who addressed the problem of cometary significance (which, as we have seen, they all did, more or less elaborately) had abandoned, or were in the process of abandoning, the Aristotelian, physical way of accounting for comets' consequences in favour of its teratological counterpart. The answer to this question cannot possibly be straightforward: although, as mentioned above, it is certainly true that Stoic and Neoplatonic ideas strongly appealed to our authors, because of their humanistic interests as well as because of the turbulent political situation, we need to be very careful in assigning each of the authors to one of the two divinatory strands exclusively. If, in the following, we shall nonetheless attempt to make a tentative classification, the reader has to bear in mind that the most basic premises of both the teratological and the astrological stances were accepted unhesitatingly by all authors.

The premise lying at the root of all teratology, the idea that there existed a godhead who controlled all the powers of the universe and who took an interest in the behaviour of man on earth, was obviously accepted, as a tenet necessarily true, by all Christians. Thus, to all our authors, in theory the divine omnipotence could, at all times, go against the ordinary course of nature, and consequently also against the influences of the celestial bodies. On the other hand, none of our authors denied that the stars did indeed possess such influences; the basic premiss of astrology, that there existed certain physical powers, analogous to those of the sun and moon, which were emitted by all heavenly bodies and affected the sublunary realm, was still universally accepted. Differences of opinion, then, for the most part occurred with regard to the extent to which God chose to exercise his omnipotence, and to deviate from his *potentia ordinata,* on the one hand, and with regard to the extent to which the universally acknowledged powers of the stars could be grasped and charted by the human intellect, on the other.

The ways in which our Dutch authors answered these questions were quite varied. Although, roughly, it can be said that the outlooks of Gemma, Heurnius, and Puteanus, and also Fromondus in his *Meteorology,* were more teratological, while those of Snellius and Mulerius were more astrological, with Fienus and Fromondus on the comet of 1618 constituting a middle group, since they were much opposed to astrology but refused to speak out on the comet's possible teratological value, the differences between the various outlooks were gradual rather than fundamental. To what extent the fact that, in 1618, the two authors from the northern Netherlands, Snellius and Mulerius, were more astrologically inclined while their southern colleagues preferred the teratological stance, is due to any fundamental difference between the intellectual climates of the universities of the Dutch Republic, on the one hand, and Louvain on the other, is very difficult to answer and would require further study. In any case, Snellius' predilection for meticulous observation and quantifiable relations, inspired by his Ramism, naturally fitted in better with the observable physical powers of the astrologers than with the all-pervading world soul of the teratologists, while Mulerius' orthodoxly Aristotelian stance can be attributed to the fact that Groningen university, to which institution he was strongly committed, had, in its *leges academicae,* professed its adherence to the philosophy of the Stagyrite.[551]

Beyond the shadow of a doubt, Mulerius' explanation of cometary consequences was the most conventionally Aristotelian of all. In orthodoxly Peripatetic fashion, he declared that the powers of the celestial bodies exercised a definite influence on the sublunary realm. Therefore, he thought, comets, too, affected the terrestrial region with their powers: like the moon, they were capable of modifying sublunary atmospheric conditions, which modifications in turn caused pestilences and unfavourable weather conditions, as well as certain changes in the human body, with all the consequences this entailed, such as irascibility, seditiousness, war, deaths of princes, etcetera. Thus, Mulerius explained all the traditional cometary concomitants by a purely physical way of reasoning.

Snellius, too, agreed that, necessarily, the hairy stars, being created out of the self-same ether out of which the stars were formed, had to be believed to be capable of physically affecting the earth, just like the other celestial bodies. Indeed, since in his opinion, comets were to be regarded as exhalations of the sun, and therefore were constituted of solar material, they should be thought of as endowed with the same powers the sun, too, possessed. Consequently, they affected the earth in a similar way as the sun. However, these cometary influences had never been accurately charted; for this reason, no certain predictions could as yet be made from their appearances. However, Snellius, like many of his contemporaries, strongly believed in the possibility of improving and reforming the astrological art. If only more accurate observations were to be made, and cometary influences were to be meticulously charted, he thought the precise nature of their powers would be discovered. Thus, Snellius' cometology was overwhelmingly astrological in nature. However, he did acknowledge the existence, and the high significance, of teratological, divine signs in nature: to him, chasms, fiery battles seen in the clouds, and the like, could be regarded as prefigurations of imminent earthly events. However, unlike comets, these portents, though highly meaningful, did not exercise any physical powers, since, physically speaking, they were nothing but optical illusions.

Gemma and Heurnius, both writing in 1577, did attribute terato-

---

551 See supra, p. 322.

logical value to the hairy stars. To them, the superlunary comet of 1577 had been called into being at the express command of God; natural and supernatural causes had combined in its generation. Both authors endorsed the Stoic and Neoplatonic view, that the cosmos was a great living being whose parts corresponded to each other. An unusual celestial occurrence, such as a superlunary comet, could on this view be regarded as a warning of impending earthly changes, directed at mankind by the divine providence ruling the universe and watching over the harmonious balance between its parts. Such a divine sign indicated a state of cosmic imbalance, brought about by the moral depravity of the human race. Whenever God detected such human deviations from the harmonious, natural state in which the cosmos was supposed to find itself, he employed natural causes to create special natural entities to admonish people to return to a moral way of life in keeping with the order of the cosmos. Despite this overwhelmingly teratological stance, Gemma and Heurnius did, up to a certain limit, make use of the rules of astrology to somewhat further specify the changes indicated by the comet. However, Gemma warned that the best way of predicting the future was from the divine signs in nature, not the rules of the astrologers, and he warned against making prognostications of particulars, such as the exact regions that were to be affected.

The aim of Fienus and Fromondus, unlike that of their 1577 predecessors, was not so much to admonish their fellow countrymen on the basis of teratological considerations (obviously, the need for such moral remonstrances had greatly diminished by 1618, when the Truce was in vigour and Louvain was thriving), but rather to invalidate the rival astrological tradition. Although, clearly, the way in which medieval scholastics had explained cometary consequences, by regarding comets as burning vapours in the atmosphere, had been rendered completely untenable to our authors, who both resolutely placed at least the present comet above the moon, Fienus still took pains to convince his readers that, even if they were sublunary, comets would still not cause any terrestrial changes. On the basis of a series of common sense arguments based on the Aristotelian constitution of the atmosphere, Fienus decisively invalidated the entire reasoning of comets as links in a chain of sublunary, physical causation producing drought, wind, tempests, pestilences, famine, war, earthquakes, and all the other traditional cometary concomitants. Next, he directed his scathing criticism at the rules of Ptolemaic astrology, thereby rendering equally untenable the idea that the effects of comets, as celestial bodies, could be predicted. His arguments against astrology were mostly based on the conviction that, although the stars did exercise certain influences, these were undetectable by the human intellect, since they were too manifold and far too complex to grasp.

Fromondus was equally convinced of the untenability of the astrological tradition. However, unlike Fienus, he did venture a very short excursion into the realm of theology, stating that the comet that had hung over Jerusalem for a year on end, had been divinely called into being and had been kept burning by an angelic intelligence. Thus, clearly, to Fromondus at least, the teratological strand was not as unfounded as the astrological. The same idea, now generalized to include all superlunary comets as well, can be found in Fromondus' later *Meteorology*, where he declared comets that were located in the superlunary realm to arise from natural causes, but nonetheless to have been called forth at God's behest. However, unlike Gemma, Fromondus did not believe such prodigious comets to be created on the spot by God at the moment they were needed; rather, he thought they had been built into the cosmic machinery at the time of the Creation, since God by his providence had already foreseen at what moments in history their appearance would prove necessary.

This Fromondian idea comes very close to the stance of Puteanus, which in a sense can be regarded as the culmination of the teratological tradition. While, to Fromondus, superlunary comets, though produced by natural causes which had been built into the laws of the universe at the Creation, were still to be considered extraordinary entities, called forth by God only on very special occasions, his colleague Puteanus regarded them as examples of the regular, ordinary course of nature. To him, God showed himself, his omnipotence, and his solicitude for mankind, not principally in unusual, supernatural or extraordinary occurrences, as Gemma, Heurnius and even Fromondus believed, but rather in the harmonious unity of the cosmos, in which every single part obeyed its own proper laws, and was yet in complete harmony with all the other parts. To Puteanus, the entire universe was replete with divine signs, the most potent of which were not necessarily the most rare. Thus, to him, one of the most significant ways in which God showed mankind how to live in accordance with nature, was the daily alternation of day and night, light and darkness, in which he revealed to mortal man that nothing in nature could last forever, and that the fact that every being, including man, would one day perish, was no more than natural.

Comets, to Puteanus, were perfectly ordinary natural entities that obeyed the laws of nature just like all the other components of the cosmos. There was nothing especially prodigious about them: if only the human intellect were capable of understanding the regularities according to which nature operated, their apparent extraordinarity would vanish. What was most to be wondered at, as the greatest prodigy of all, was the harmonious regularity of the cosmos. In this sense, Puteanus can be regarded as the heir of Seneca, who had declared that the daily rising and setting of the sun was a much greater prodigy than the occasional appearance of a comet, and of Saint Augustine, to whom all the operations of nature, be they rare or common, were miraculous and revealed the power of God. Puteanus would have nothing to do with the astrological, physical explanation of cometary concomitants: events such as wars and rebellions were due to the free will of men, and therefore not to be explained by celestial or cometary influences; wars, pestilences, and deaths of princes occurred anyway, whether a comet appeared or not; natural disasters such as the flooding of Buris and Helice could not be attributed to a comet that happened to have appeared at the same time as the disaster, because if two things succeeded each other, this did not necessarily mean that the former was the cause of the latter; and incidental scarcity and famine were only natural, since they were brought about by a certain universal regularity.

These and similar Puteanian arguments were very similar indeed to those put forward against cometary divination, roughly sixty years later, by Bayle. Moreover, many of Bayle's arguments, such as the idea that cometary powers could not last after the comet itself had disappeared, the idea that it could not be ascertained exactly what the effects of comets might be, assuming they had any, the idea that it was implausible to uphold that comets carried more significance for one region than for another, and the idea that wars and the like were due to human free will, not to any heavenly influences, had also been advanced by Fienus. Clearly, Bayle, in invalidating cometary superstition, drew on a tradition of astrological scepticism already very much alive at the beginning of the seventeenth century, which itself, in its turn, was firmly based on much older precedents.

Bayle's arguments against comets as causes of calamities on earth, and against comets as parts of a physical, causal nexus leading to such calamities, then, were not original and were well-known to, and eagerly employed by, our Dutch authors. However, his idea that comets could not be said to be miraculous, divine signs, either, found a precedent only in the ideas of Puteanus. Both he and Bayle stated that, whenever God *really* wished to admonish mankind, which happened only rarely, he used manifestly counternatural means, such as a solar eclipse at a naturally impossible moment. And both authors agreed, that God's power and magnificence manifested themselves above all in the ordinary, regulated course of nature, to which comets, too, belonged. Thus, by expanding teratology to such an extent that the whole of nature, in all its regular maifestations, could be regarded as a divine sign, they went a long way towards removing divine agency from nature.

There can be little doubt that Puteanus derived his ideas on the harmony of nature from the Stoics. In our previous chapter, we have demonstrated that his cosmological ideas were overwhelmingly Stoic in nature, based for the most part on Seneca. Furthermore, Puteanus was the pupil and successor of Lipsius, whose life's work had been the uncovering of the Stoic philosophy and the application of Stoic ideas to the contemporary situation. And, finally, the whole outlook conveyed by Puteanus' tract bears a clear resemblance to the divinatory ideas of the Stoics, as expounded above in section 2.a. The idea that everything that happened in nature was due to natural causes, which were part of the causal nexus inscribed in the cosmos by divine providence, and that of a divine power present in all parts of the universe and regulating it, attributed to the Stoics by Cicero in his first book *On Divination,* clearly have a great affinity with the Puteanian outlook.

Thus, we may tentatively propose the following answer to the question of how, in the course of the seventeenth century, comets slowly but surely came to lose their special significance (at least, in some quarters): the return to prominence of the divinatory theory of the ancient Stoics, teratological in character and placing great emphasis on the divine agency regulating the cosmos, made a great impact on the sixteenth century, which because of the Reformation and the many religious disputes was already much concerned with God's use of nature as a means of communication with man (as opposed to the Aristotelian way of accounting for stellar influences by an appeal to physical causation) anyway. Some scholars (probably most of them) adopted the teratological outlook in order to explain the occurrence of extraordinary phenomena deviating from the ordinary course of nature: these, to them, were the signs sent by God to admonish man. However, some scholars went a step further, endorsing the idea put forward by Fromondus, that, though an unusual celestial occurrence such as a superlunary comet deviated from the ordinary course of nature, nonetheless these deviations had been built into the divine plan from the outset. Still further went those, like Puteanus and Bayle, who adopted the entire Stoic outlook with its great emphasis on regularities in nature and on extraordinary phenomena as conforming to these regularities, thereby all but eliminating God's agency, which in the outlook of authors such as Gemma and Heurnius had been so prominently present, from nature.

Thus, Newton and Halley were not needed to eliminate the significance attached to cometary appearances. Although, obviously, the computation of cometary orbits did much to reinforce the idea that comets were perfectly ordinary natural entities obeying certain fixed laws of nature, in some quarters this view had been taken much earlier, due to the influence of the newly revived ideas of ancient, non-Aristotelian philosophers. In the same way that, in order to reconcile new astronomical findings with the tenets of their traditional, Aristotelian world views, our authors adapted these world views by means of an appeal to the authorities of ancient thinkers other than Aristotle, so, too, with regard to final causality, Aristotelian, physical explanations were, to a greater or lesser extent, replaced with the teratological views of the ancient Stoics and Neoplatonists. Both processes were gradual; yet, slowly but surely, they engendered an entirely new world view. Thus, in order to be able to look forward, the authors who dealt with the comets of 1577 and 1618 needed first of all to look back.

CHAPTER SEVEN

CONCLUSION: THE AGE OF TWO-FACED JANUS

"The two-faced race of Janus with the double forehead has been multiplied to such an extent that it is impossible to know with whom still to find sincere loyalty, favour, or friendship", complained Nicolaus Bazelius in his pamphlet on the comet of 1577. Obviously, he intended this remark as an allusion to the extremely insecure social and political circumstances with which his country saw itself faced during those difficult years 1577-1578: the hardships of war, the uncertainty as to what was to happen now that the Pacification of Ghent had failed, the clash of different religious and political convictions that divided the population and gave rise to a sense of dissimulation and deceit rife throughout society, and the insecurity brought about by the constantly changing military situation. However, unwittingly, Bazelius, with his observation that his time was one of conflicts of loyalties, during which it was impossible to unequivocally assign people to one or another of the many different positions that could be taken vis-à-vis the problems of the age, appropriately characterized the period on another level as well.

For, indeed, a striking similarity can be detected between the politico-religious conflict, on the one hand, and the intellectual battle being fought simultaneously inside the heads of the protagonists of the present study, on the other. In the same way that the majority of the Netherlandish population desired nothing so much as reconciliation and pacification, without irrevocably breaking with their Spanish overlord, so too, the authors of the cometary tracts strove to maintain their loyalty to the traditional authority of the 'Master of Them That Know' by reconciling the new discoveries made around 1600 to his ideas. And in the same way that many Netherlanders, instead of either unconditionally upholding the old, Roman Catholic faith in all its manifestations, or overnight adopting an orthodox form of Calvinism, chose to a certain extent to modify their religious convictions in accordance with the new ideas advanced by the reformers, without explicitly professing adherence to the Protestant denomination, so too the alterations suffered by our authors' world views were piecemeal and gradual rather than abrupt.

Indeed, it would have been very surprising if any of our learned authors, on being confronted with the comets of 1577 or 1618, had unconditionally upheld the Aristotelian theory of cometary generation, or had advanced an opinion on the constitution of the cosmos that was entirely in conformity with the ideal type of the Aristotelian world view (that is, a geocentric universe divided into two fundamentally dissimilar regions each obeying its own set of physical laws). Even the Dutch authors writing in 1577-78, well before the publication of Tycho's findings on the comet's superlunary nature, were not prepared to categorically endorse such a conventional world view.

In our fourth chapter, we have shown why this need not surprise us. First of all, the Aristotelian world view had never been as rigid and monolithic as has often been assumed. Instead, the cosmology of the Stagyrite never reached western Europe in its original form; from the outset, scholars were faced with many variants contrived by the Byzantine and Arabic commentators, who had, each in his own idiosyncratic way, tried to elucidate whatever cosmological or physical details Aristotle had failed to address, or on which his utterings were equivocal. Furthermore, when the Stagyrite's teachings were received in the Latin West, they had from the outset been confronted by rival doctrines; Christianity in particular, the pillar on which western civilization was founded, could not be reconciled to an unmodified Aristotelian world view. Most importantly, Christian scholars had to allow for the possibility that God, being omnipotent and at any time capable of wielding his absolute power, might go against the ordinary rules of Aristotelian physics and might work unexpected effects that were inexplicable by the human intellect. However rarely such divine interventions might occur, still they needed to be reckoned with, and they imparted to the late medieval world view its dynamic character.

This dynamism revealed itself most pronouncedly in the *quaestiones*-structure of scholasticism. Since God could theoretically go against any of the rules in accordance with which nature normally behaved, such divine actions, which from the point of view of the ordinary course of nature were absurdities, needed nonetheless to be seriously contemplated. Thus, such fourteenth-century authors as Oresme and Buridan discussed the possibility of God creating other worlds besides our own, each possessing its own centre of gravity, so that heavy bodies finding themselves in those worlds would tend to move towards that centre of gravity, rather than towards our earth.

Although, in the end, such ideas were invariably rejected as absurdities, with late medieval scholastics clinging to the conviction that, despite God's omnipotence, he nonetheless acted in the vast majority of cases in accordance with the laws of the Aristotelian universe, still the arguments in favour of the contrary position were explicitly put forward and elaborately substantiated.

During the sixteenth century, a subtle yet momentous change of attitude began to make itself felt. Slowly but surely, some of the unorthodox positions that had always been discarded as absurdities began to be affirmed as true. The most famous example of such a realistically interpreted, non-Aristotelian stance was the idea of Copernicus, that the earth possessed a threefold motion, while the sun stood still in the centre of the universe. Whereas the theoretical possibility of a moving earth had been argued, in the fourteenth century, by Oresme, he had in the end rejected it. In 1543, Copernicus upheld that it was the true state of affairs in the universe. The reactions with which this Copernican idea met, during the second half of the sixteenth century, are typical of the spirit of the age. First of all, in most quarters the medieval attitude of discussing the merits of theories believed to be impossible in reality continued unabated in the guise of the so-called 'Wittenberg Interpretation', with scholars employing the Copernican scheme as a calculational tool without believing it to correspond to reality. Secondly, and in the present context most importantly, efforts were made by a number of thinkers, Tycho Brahe among them, to reconcile Copernican model of the planets to a geocentric universe.

The fortunes of the Copernican theory are exemplary of the intellectual atmosphere in which the Dutch cometary tracts came into being. The scholarly reaction to Copernicanism had two important characteristics in common with the Netherlandish authors' reactions to the appearance of the comets of 1577 and 1618, as expounded in the present study: first, considerable efforts were made to make new ideas conform to the existing world view (as in the attempts at reconciling the Copernican scheme to Aristotelian physics); and secondly, any deviations that *were* made needed to be legitimated by an appeal to ancient authorities (in the same way that the heliocentric theory was often associated not only with Copernicus, but also with such ancient heliocentrists as the Pythagoreans Philolaus and Ecphantus).

Copernicanism was not the only novel (or: revived) theory to make an inroad upon the traditional cosmos during the decades prior to the appearance of the comet of 1577. The ancient philosophy of Stoicism, too, went through a revival, and Stoic cosmology became a popular topic of concern. Although remarkably similiar to Aristotelianism in several of its tenets (like Aristotle, the Stoics believed the earth to be located at the centre of the universe, surrounded by several concentric layers in which the elements of water, air, and fire were located, and in which atmospheric phenomena such as winds and rainclouds were generated; and like Aristotle, they endorsed the rectilinearity of sublunary motions, believing the heavy elements of earth and water to move straight down and the light elements of air and fire to move straight up) there was one very important difference between the two outlooks: to the Stoics, the cosmos as a whole formed a single unity. It was a living being whose entire body was pervaded with a life-giving substance, the pneuma, which consisted of a mixture of fire and air, and which constituted a physical means of contact between the heavens and the earth. Aristotle's impassable boundary between sublunary and superlunary regions, and his concomitant strict separation of celestial and terrestrial physics, each obeying their own set of laws, was strikingly absent from the Stoic account, as was the Aristotelian quintessence and its attendant postulate, that the heavenly regions were incorruptible.

While, by the second half of the sixteenth century, the idea of cosmic unity had become very popular even with perfectly traditional thinkers, like Robert Bellarmine, who adopted it into their world views on the authority of the Church Fathers as well as the ancient Stoics, the traditional, Aristotelian account of cometary generation had become untenable to some. Because it was discovered that the tails of comets were always antisolar, scholars such as Gemma Frisius and Girolamo Cardano returned to the ancient theory, legitimated by the authority of the Presocratics Hippocrates of Chios and Aeschylus, that comets were spherical lenses whose tails consisted of refracted sunlight. Although this did not necessarily mean that they were located beyond the moon, Cardano strongly believed that they were superlunary; to him, this idea did not pose any problem since he had already adapted his world view to the Stoic tenet that ordinary air extended all the way to the heavens, and that there existed no celestial quintessence. Similar views were put forward by Jean Pena.

Thus, by the time the comet of 1577 appeared, the Dutch authors who observed it were already familiar with the idea of comets being located beyond the moon. Rather than regarding it as a reprehensible deviation from the Aristotelian world view, they counted the concept of celestial comets existing next to, or even to the exclusion of, elementary ones, among those piecemeal adaptations of the traditional constitution of the cosmos that were much in vogue throughout the sixteenth century. The Stoic idea of celestial and terrestrial regions forming a single continuum that stretched out from the earth to the fixed stars had already gained a fair deal of respect and adherence before 1577. Because such an outlook did not impose the need for a sublunary location for comets (after all, on this view the heavens were as alterable as the terrestrial region), it now became possible to contemplate the possibility that the hairy stars were celestial objects, like the planets and stars.

Needless to say, not everyone wholeheartedly endorsed the Stoic outlook, which violated the time-honoured ideas of celestial incorruptibility and strict separation of celestial and terrestrial physics. Yet, Stoic notions were clearly 'in the air', and during the 1570s they were to prove their mettle as suitable reinforcements capable of replacing those Aristotelian tenets that were rendered blatantly untenable by the novel celestial appearances of the nova and the comet. This, indeed, is precisely the way in which the two learned authors of 1577, Gemma and Heurnius, employed them. Both men believed the comet of that year to be located beyond the moon; and both men tried to carefully readjust, rather than fully discard, their world views in the light of this belief. For, indeed, it should be dubbed a 'belief' rather than a 'discovery': neither author employed any observations or parallax calculations accurate enough to convincingly demonstrate the hairy star's superlunary position. Gemma arrived at a small parallax value more or less by chance (witness the derogatory way in which Tycho Brahe evaluated his Louvain colleague's 'calculations'), while Heurnius did not calculate any parallax value at all; a much more important reason for placing the comet in the heavens seems to have been the analogy its motion and colour displayed with those of the planet Mercury (Gemma) or the role the comet was supposed to fulfil in transmitting to mankind the messages of the godhead (Heurnius). Clearly, Tycho Brahe's accuracy was not needed to render convincing the idea of superlunary comets to an age well acquainted with Stoic ideas of fundamental similarity between heaven and earth, and cosmic correspondence.

The most important deviation from the traditional world view that Gemma and Heurnius came up with as a result of the comet's appearance, was the postulation of the existence of a special kind of matter, neither completely terrestrial and corruptible nor entirely celestial and incorruptible, out of which comets could be formed. Both authors located this intermediary matter in a special sphere of its own, situated among the Aristotelian spheres in whose existence they still believed: while Gemma, relatively conventionally, proposed to insert it between the spheres of the air and that of the moon, so that it amounted to not much more than a replacement of the Aristotelian sphere of fire, Heurnius situated his 'intermediary' sphere, which he dubbed the 'astral element', above the moon's orb; he probably conceived of it as filling the entire area between the moon and the fixed stars. However, Gemma's celestial region, too, was subject to alterations to a certain extent: he put forward the idea that, very occasionally, the stars might, at God's behest, produce offspring that remained hidden from human view until the most propitious moment had arrived for them to display themselves. Such new celestial bodies were very rare, and when they appeared they were charged with the highest significance; such had been the comet of 1577.

Thus, the learned authors of 1577 went some way towards abandoning the fundamental difference between celestial and terrestrial matter by postulating intermediate zones filled with a kind of matter that partook of both celestial and terrestrial qualities. Also, they allowed for the possibility of a certain measure of change taking place in the heavens. Gemma quoted a number of Biblical texts as well as the Church Fathers in favour of this idea; almost exactly the same quotations had been employed five years earlier by Robert Bellarmine in his defence of Stoic cosmology in his Louvain Lectures. Thus, in 1577, the Aristotelian world views of Gemma and Heurnius were having to contend with certain modifications; especially the strict dichotomy of sublunary and superlunary regions was suffering a severe attack, with the idea of the existence of at least certain specific areas of the cosmos where celestial and terrestrial qualities were allowed to meet and mingle proving particularly appealing to replace it. The Stoic idea of cosmic unity was clearly beginning to make itself felt.

By the time the comet of 1618 appeared, this idea had gained considerable momentum, for very weighty reasons: its relevance had been enormously enhanced by the telescopic observations made by Galileo in 1609-1610. It is a truly remarkable coincidence that these observations should have been made precisely during an era which had, for quite some time already, been rediscovering Stoicism and readjusting its world views in order to accommodate the Stoic tenet of fundamental similarity between terrestrial and celestial regions. The telescopic sights had, as it were, had their way prepared by the gradual unearthing of the Stoics, to whom the heavens had been as changeable as the earth. When, therefore, Galileo published his findings, the idea that, indeed, the Stoics had been right, must have crossed several people's minds. Especially book seven of Seneca's *Natural Questions* must have appeared remarkably well-suited for providing the new discoveries with a venerable, ancient legitimation: in it, the famous Stoic had eloquently defended the idea, not only that comets were superlunary bodies just like the planets, but also that in fact the whole universe was full of entities not visible to man, hidden during most of their existence, and that these, though seemingly erratic, nonetheless obeyed certain divinely preordained laws.

These ideas, the relevance of which was greatly enhanced because of the telescopic discoveries, which had shown that, indeed, many more celestial bodies existed than had previously been believed, exercised considerable influence on the Dutch authors who wrote about the comet of 1618. To them, the Aristotelian dogma of celestial incorruptibility had become quite untenable; only Mulerius still paid lip service to it. All the other learned authors in 1618 believed that the entire region between the moon's orb and the firmament of the fixed stars was fluid, orbless, and corruptible to a certain extent: the zone of intermediary matter postulated in 1577 by Gemma and Heurnius had, forty years later, come to occupy the entire region of the planetary heavens. Primarily responsible for this change in outlook had been the sights seen by Galileo through the telescope: all Dutch authors referred to these as highly significant novelties, and all of them unconditionally believed them to correspond to the real state of affairs in the universe (apart, perhaps, from Mulerius, who, at least verbally, was still reluctant to abandon Aristotle's celestial incorruptibility). Galileo's findings, then, served to render convincing to our authors the idea that the heavens suffered changes like the terrestrial region; yet, the acceptability of such a concept was undoubtedly facilitated by the fact that the same view had already been put forward, during the previous decades, by the adherents of the newly revived Stoic cosmology. Also, it was helpful that the Church Fathers, themselves under the influence of Stoicism, had on several occasions referred to the heavens as corruptible and fluid; no wonder that most of the Dutch authors enthusiastically quoted these Patristic utterances to render their own views more convincing.

By 1618, the discoveries of Galileo had also converted several scholars in the Netherlands to the Tychonic world system: the fluidity of the heavens, the moons of Jupiter, and the phases of Venus had greatly enhanced its credibility. Puteanus and Fromondus explicitly professed their adherence to the geoheliocentric scheme; Snellius, though favourably mentioning it, nonetheless would not explicitly endorse any more than the Capellan system, in which only Venus and Mercury orbited the sun; Fienus and Mulerius did not speak out on the issue, and the most we can gather from their tracts is that they were still geocentrists. None of our authors expressed an allegiance to Copernicanism; only Fromondus devoted a fair amount of—sympathetic—attention to the heliocentric world system, but in the end cautiously sided with the Tychonic arrangement.

The world views of our 1618 authors, then, had changed considerably in comparison with those proposed by Gemma and Heurnius in 1577. By 1618, the solid spheres had been abandoned, the heavenly region was corruptible and fluid, and to most authors, the geoheliocentric world system had proved its credibility. However, although obviously the 1618 deviations from traditional Aristotelianism were more significant than the 1577 ones, it still cannot be maintained that, by 1618, Dutch scholars had abandoned the Aristotelian world view in any significant way. The new ideas that were proposed in the aftermath of the telescopic findings, though by no means insignificant, could relatively easily be reconciled to the traditional world system. After all, they only derogated from one of the fundamental Aristotelian tenets, that of the strict dichotomy between sublunary and superlunary regions. The other components of the Stagyrite's world view (geocentricity, division of the sublunary regions into different layers, each filled with one of the sublunary elements, circularity of heavenly motions, explanation of atmospheric phenomena) could be reconciled to the new discoveries without much difficulty. By adopting the compromise world system of Tycho Brahe, the earth could remain at rest in the centre of the universe, so that the basic tenets of Aristotelian physics could remain intact.

The new elements our authors were forced, by the discovery of cometary superlunarity and the telescopic findings, to adopt into their world views, then, were not earth-shaking. They were all related to just one of the Aristotelian dogmas: that of celestial incorruptibility and strict division between celestial and terrestrial physics. These were precisely those tenets that had already been challenged, throughout the sixteenth century, by those favourably disposed towards the newly·rediscovered cosmology of the Stoics. The fact that the Stoic tradition, as well as the formidable authority of the Church Fathers, sanctioned celestial alterability, undoubtedly made it easier for our authors to reconcile themselves with the recent novelties.

Seneca's idea, that the heavens were full of bodies normally invisible to man, and that nobody knew what lay hidden in the depths of the universe, popped up again and again in the Dutch cometary writings, and served as a means of providing the telescopic findings with ancient precedents, and to show that, though seemingly novel, they were well capable of being embedded in ancient tradition. Though Seneca's own cometary theory, that the hairy stars were eternal bodies, was not particularly popular with the Dutch authors (only Mulerius provisionally adopted it, and Fienus was sympathetically inclined towards it without, however, endorsing it as true), his cometary writings, and especially the idea of cosmic unity and fundamental similarity between the celestial and the terrestrial realms, that they conveyed, exercised a strong appeal. Both Fromondus and Puteanus eagerly made use of the analogy between the terrestrial region *as a whole* and the celestial region *as a whole,* both of which comprised corruptible and incorruptible elements alike, in order to render convincing the idea that new things could be born in the heavens. Fromondus even allowed a certain intrusion into the heavens of laws that, in the Aristotelian view, belonged to the terrestrial region, when he stated that celestial comets might lose and gain heaviness, and accordingly move down and up again in straight lines.

Not only did the world views of our authors, in the narrow sense of their opinions on the material constitution of the universe, suffer a certain measure of readjustment as a result of the new discoveries; their entire mental outlook, too, was affected by the re-evaluation of traditional tenets. Here, too, the Stoic idea of cosmic unity and sympathy played a significant role. These tenets lay at the root of what we have defined as the teratological stance with regard to divination, according to which the cosmos was a single living entity in possession of a soul that permeated the entire body of the universe, so that every part was in close contact with every other part, be it located in the heavens or in the terrestrial atmosphere. This idea was coupled with a firm belief in the existence of a benevolent godhead, who was concerned for the welfare of mankind, and who therefore made use of the laws of cosmic sympathy to convey signs of impending events to the earth-dwellers. This theory of divination, as expounded above all by Seneca, by Manilius, and by Cicero in the Stoic half of his dialogue *On Divination*, had during antiquity exercised a strong appeal in imperial Rome. The Roman historians Tacitus, Dio and Suetonius paid much attention to the signs with which important events in history were accompanied, among which comets played a prominent role. Although this divinatory strand in cometology had during the later Middle Ages been superseded by its ancient rival, the astrological theory of Ptolemy and Aristotle, who had explained the consequences of comets by means of certain meteorological conditions, such as wind or drought, with the revival of ancient Neoplatonism and Stoicism in the sixteenth century the divinatory way of elucidating cometary consequences made a significant comeback.

Although none of the Dutch authors completely did away with the basic tenets of Ptolemaic astrology (thus, they all believed that the heavens affected the sublunary regions with some kind of physical power) the teratological stance had made considerable headway, both in 1577 and in 1618. Mulerius was the only one of the authors to profess his adherence to the Aristotelian way of accounting for cometary consequences by means of a chain of physical causes, maintaining that the appearance of a comet was due to hot and dry atmospherical conditions, which in turn gave rise to all kinds of terrestrial inconveniences such as earthquakes, storms, and drought. The other authors were all more or less emphatically opposed to this naturalistic, medieval reasoning: some of them, such as Fienus and Puteanus, fulminated at length against this, to them, ridiculous line of thought. Others, such as Gemma and Heurnius, did not completely do away with the whole elaborate machinery of Ptolemaic astrology, but in their outlooks this was to a large extent superseded by the idea that, since the cosmos was to be regarded as a great living body, a surer way of predicting the future was to study the divine signs which appeared, not only in the heavens, but throughout the universe, and to inquire what cosmic 'diseases' such signs denoted, in the same way as ordinary physicians studied the signs of the different parts of the human body to diagnose disease. Snellius, though sympathetic in principle to the Ptolemaic idea that it ought to be possible to chart all the physical influences emitted by the stars, and to predict their terrestrial consequences (however much time and diligence this would still require), nonetheless also endorsed the teratological view of nature being pervaded with divine signs sent by God to warn mankind. Among such signs, he counted 'chasms' and other fiery appearances seen in the clouds in the form of battles and the like; not, however, comets, which to Snellius were purely physical entities (small parts of solar matter) with purely physical (solar) effects.

The author who pursued the teratological line of reasoning to its outermost limits was Puteanus: whereas Gemma and Heurnius had believed that the signs sent by God to warn mankind of impending changes were primarily to be sought in those occurrences that deviated from the ordinary course of nature, Puteanus instead advocated the Senecan stance, according to which God showed himself, his omnipotence, and his solicitude for mankind, in the harmonious unity of the cosmos, in which every single part obeyed its own proper laws, and was yet in complete harmony with all the other parts. In this view, the entire universe was replete with divine signs, the most potent of which were not necessarily the most rare. Thus, the appearance of a comet did not necessarily carry more predictive value, just because it occurred more rarely than the daily rising and setting of the sun. To Puteanus, as to Seneca and Saint Augustine before him, what was most to be wondered at, as the greatest prodigy of all, was the harmonious regularity of the cosmos. In this sense the Louvain humanist can be regarded as a worthy precursor of physico-theology.

During the period under consideration, then, we can pinpoint a tendency to explain the repercussions of cometary appearances in terms of divine solicitude for the wellbeing of mankind, instead of referring them to physical causes whose effects were perfectly well explicable by rational deduction. It is tempting to suggest that this interpretation of comets as signs, sent by the godhead to warn of impending alterations on the political and social level, was brought about at least partly as a result of the chaotic political circumstances in contemporary Europe. It is a well-known fact that the upheaval pervading Europe in the wake of the Reformation (the religious wars and the civil strife going on in several European countries) was accompanied by an enormously expanding market for broadsides, which became a very popular means of spreading new ideas. During the sixteenth and early seventeenth centuries, many of these pamphlets dealt with the way in which God was believed to employ unusual natural occurrences, comets among them, as a means of revealing to mankind his intentions.

Although most of the present study has dealt with learned tracts, written by men belonging to the academic elite, rather than with popular broadsides, nonetheless we have been able to detect traces of a similar concern with the intentions of the godhead, as manifested in natural phenomena, in the writings of these learned authors as well. Especially Gemma, Heurnius, and Puteanus concerned themselves at length with the consequences the appearance of a comet might have in the terrestrial realm. It need not surprise us that it was precisely these three authors who explicitly regarded their comets in terms of messengers imparting to the human race the intentions of the deity. After all, all three men were closely connected to currents of thought in which it was customary to closely monitor happenings that took place on every single level of the cosmic whole, to attempt to trace the connections they felt must exist between these happenings, and to relate them all to a single divine plan.

Gemma and Heurnius both belonged to the same circle of humanists with *politique* convictions as Justus Lipsius, who elaborated the idea that nature revealed the majesty of God by harking back to the ancient philosophy of the Stoics, according to which the study of nature and its laws was propaedeutic to attaining the calm disinterestedness of the Stoic sage, a condition much to be desired in an age of constant warfare, clashing opinions, and dissimulation. Tellingly, Lipsius on several occasions compared the current state of affairs in his fatherland to the conditions in imperial Rome, as described by Tacitus; and, as we have seen, imperial Rome was precisely the place where the teratological way of interpreting phenomena such as comets had been very much in vogue. Puteanus in his tract on the comet of 1618 showed himself the manifest heir of his former teacher Lipsius: the main ideas the author tried to convey were the inconstancy of terrestrial events, dependent as they were on nothing but the fickle will of man and on mere coincidences, as opposed to the harmonious regularity of the cosmic laws, in which the hand of God could be detected. Similarly, Gemma and Heurnius one generation earlier had advocated the stance that the dissension and conflict rife in contemporary society had brought on a 'disease' of the cosmos, which needed to be remedied; the comet of 1577 had fulfilled the role of a messenger sent as a warning to mankind.

Thus, both in 1577 and in 1618, our Dutch authors (with the exception of Mulerius) displayed a marked preference for the view of comets as divine signs denoting cosmic imbalance, rather than as phenomena bringing about a well-defined set of physically explicable atmospheric effects. This change in attitude cannot be explained merely by pointing out that all the authors placed the comets above the moon, and that superlunary comets were much less likely to cause sublunary heat and drought than burning terrestrial vapours. After all, earlier in the sixteenth century comets had been superlunary entities to Cardano; yet he came up with an elaborate reasoning to the effect that, nonetheless, they could be regarded as signs of dry and hot weather conditions and all the consequences these entailed according to the Ptolemaic-Aristotelian account. Moreover, both Fienus and Puteanus took great pains to demonstrate that, even if comets were, indeed, to be regarded as fiery terrestrial exhalations, still the effects they were traditionally assumed to work on earth could not be maintained. Therefore, it is much more likely that our authors' teratological outlook was inspired at least in part by the chaotic political and intellectual conditions with which they were confronted; these were the seeds from which the theories of the Family of Love, and the Lipsian revival of Stoicism, sprang. It would be fascinating to see to what extent Stoic cosmology and teratology influenced the writings on the comets of 1577 and 1618 published in other European countries; in other words: was the appeal of Stoicism especially pronounced in the Netherlands because of the Lipsian tradition and the war?

Bazelius came up with a fitting simile in comparing his contemporaries to Janus, the Roman god with the double forehead, who contemplated both times past and times to come, encompassing both the past and the future in his all-embracing glance. For our purposes, the metaphor can be interpreted in two different ways. First, we can take 'the past' to refer to the conventional Aristotelian world view as it had come to dominate the cosmological thought of the later Middle Ages. Although serious inroads had been made on this view already during the earlier half of the sixteenth century, it still predominated the way in which our authors regarded the constitution of the cosmos, and continued to do so throughout the period under consideration. However, 'the future', too, was tangibly present in their outlooks: they were confronted with novelties, in the guise of superlunary comets and previously unheard-of sights seen through the telescope, which they could not pass by in silence. Nonetheless, rather than completely abandoning the old Aristotelian cosmos, they chose to reconcile it, by means of readjustments as minimal as possible, to the new findings.

However, this procedure was much facilitated by the existence of ancient ideas that lent credibility to the many celestial novelties, and to the new way of regarding the cosmos that these novelties necessitated. Especially the philosophy of the Stoics, which had for a number of reasons gained great popularity among the intellectual elite of the Netherlands around the year 1600, proved well-suited to legitimize any deviations from the Aristotelian world view which our authors saw fit to introduce. The outlook of Seneca served as an invaluable tool, both on the cosmological and on the divinatory levels. In this sense, then, one of Janus's faces was directed at the very ancient past of the Stoics and other non-Aristotelian philosophers from antiquity, while the other already looked ahead at the developments which, in the not too distant future, were to lead to the complete overthrow of the Aristotelian world view, the ascendancy of the Newtonian system, and the cessation of dialogue. On this view, the period under consideration was a period that looked back in order to be able to look ahead, and was truly an Age of Two-Faced Janus.

BIBLIOGRAPHY

Printed Primary Sources

Aegidius of Lessines, *De essentia, motu et significatione cometarum* in Lynn Thorndike ed., *Latin Treatises on Comets between 1238 and 1368 AD* (Chicago 1950) 103-184.

Albert the Great, *Liber I Meteororum, Tractatus III,* capita 1-3 (translated by Lynn Thorndike in his *Latin Treatises on Comets between 1238 and 1368 AD* (Chicago 1950)) 18-23.

Anonymous, *Aliud iudicium de cometis* (1338), in Lynn Thorndike ed., *Latin Treatises on Comets between 1238 and 1368 AD* (Chicago 1950) 226-233.

Anonymous (Jacob Cats), *Aenmerckinghe op de tegenwoordige Steert-Sterre, ende den loop deser tijden so hier als in ander Landen, Met aenwijsinge vande rechte wetenschap om alle teykenen des Hemels, ende vreemde Gesterten wel ende loffelijck uyt te leggen, alles tot een Nieuwe Iaer gifte alle rechtsinnige Verstanden toe-geeygent* (Middelburg 1619).

Anonymous (Cornelius Gemma), *Des vremde sterrens nu eerst hem verthoonende, ende vast gheduerighens observatie: Waer door die wonderlijcke crachte der voorsichticheyt Gods, ende die Maiesteyt sijnder glorie, den menschen ooghen haer claerlyck openbaert* (Louvain 1573).

Anonymous, *Corte beschrijvinghe ende Affbeeldinghe des seer grooten ende schrickelijcken Comeets, de wvelcke den 27. November in 't Iaer 1618. ons verschenen is* (2nd ed., Leiden 1619).

Anonymous, *Liber de significatione cometarum* (Eastern Spain, c. 1238) in Lynn Thorndike ed., *Latin Treatises on Comets between 1238 and 1368 AD* (Chicago 1950) 16-61.

Anonymous (Peter of Limoges?), *Iudicium de cometis* (1301), in Lynn Thorndike, *Latin Treatises on Comets between 1238 and 1368 AD* (Chicago 1950) 202-207.

Anonymous, *Van die wonderlijcke nieuwe Sterre openbarende in den hemel, die noyt voor desen tijt ghesien is ghewеest, Wt den Latijne in Duytsch ouerghesedt.* (Antwerp 1572).

Aquinas, Thomas, *Commentarium in Librum Aristotelis Meteorologicorum,* Cap. VI-VII, Lectiones 9, 10 & 11 in *Sancti Thomae Aquinatis Doctoris Angelici Opera Omnia* III (Rome 1886) 348-356.

Aristotle, *Meteorology* (Greek text + English translation by H.D.P. Lee, Loeb Classical Library, London/ Cambridge Mass. 1962).

Aristotle, *On Generation and Corruption* (translated by C.J.F. Williams, in J.L. Ackrill, *A New Aristotle Reader* (Oxford 1987) 143-158).

Aristotle, *On the Heavens* (Greek text + English translation by W.K.C. Guthrie, Loeb Classical Library edition, Cambridge Mass./London 1986).

Augustine, Saint, *City of God* (translated by Henry Bettenson, Penguin Classics edition, London 1984).

Bacon, Francis, *Aphorisms,* in Fulton H. Anderson ed., *Francis Bacon: The New Organon and Related Writings* (Indianapolis 1960).

Bacon, Roger, *Opus majus,* in John Henry Bridges ed., *The "Opus majus" of Roger Bacon* (Oxford 1897).

Bailly, Jean Sylvain, *Histoire de l'astronomie moderne depuis la fondation de l'école d'Alexandrie jusqu'à l'époque de 1782* (Paris 1782).

Bayle, Pierre, *Pensées diverses sur la Comète* (1683, edited with an introduction by A. Prat, re-edited by Pierre Rétat, Paris 1984).

Bazelius, Niclaes, *Een nieuwe prognosticatie vanden wonderlijcken ende ellendighen Jare ons Heeren 1578. Midsgaders de beschryvinghe vande Comete des voorleden Jaers. Ghecalculeert by M.Niclaes Bazelius, Medecijn ende Chirurgijn ordinaris der stede ende Casselrye van Sinte VVinox-Bergen in V.Vest-Vlaenderen* (Antwerp 1578).

Bekker, Baltasar, *Ondersoek van de Betekeninge der Kometen* (Amsterdam 1683).

Bellarmine, Robert, *Louvain Lectures* in U. Baldini & G.V. Coyne eds. & trans., *The Louvain Lectures of Bellarmine and the Autograph Copy of his 1616 Declaration to Galileo* (= Studi Galileiani, vol.1, no.2; Vatican City, 1984).

Bor, Pieter, *Oorsprongk, begin, ende vervolgh der Nederlandsche Oorlogen, beroerten, en burgerlyke oneenigheden, beginnende met d'opdracht der selve Landen, gedaen by keyser KAREL den Vijfden, aen sijnen Soon Konink PHILIPPUS van Spanjen, en eindigende met het einde van 't jaer MDC, Waerachtelijk en onpartijdelijk beschreven* (1595, reprint Amsterdam 1679).

Brahe, Tycho, *De mundi aetherei recentioribus phaenomenis* (Uraniborg 1588) in J.L.E. Dreyer, *Tychonis Brahe Dani Opera Omnia* IV (Copenhagen 1922) 5-377.

Brahe, Tycho, *Von der Cometten Uhrsprung was die alten vnnd neuen Philosophi inn denselben vermaint vnnd dauon zuhalten sei* (1578; first published by J.L.E. Dreyer in *Tychonis Brahe Dani Scripta Astronomica* IV (Copenhagen 1922) 381-396).

Cardanus, Hieronymus, *De Rerum Varietate Libri XVII* (Basle 1557).

Cardanus, Hieronymus, *De Subtilitate Libri XXI nunc demum recogniti atque perfecti* (Basle 1554).

Cassius Dio, *Roman History* (Greek text + English translation by Earnest Cary, Loeb Classical Library edition, Cambridge Mass./London 1914-27).

Cicero, Marcus Tullius, *De divinatione* (Latin text + English translation by William Armistead Falconer, Loeb Classical Library edition, Cambridge Mass./London 1923).

Cicero, Marcus Tullius, *De natura deorum* (Latin text + English translation by H. Rackham, Loeb Classical Library edition, Cambridge Mass./London 1933).

Clairaut, Alexis Claude, *Théorie du mouvement des comètes. Dans laquelle on a égard aux altérations que leurs orbites éprouvent par l'action des Planètes. Avec l'application de cette théorie à la Comète qui a été observée dans les années 1531, 1607, 1682 & 1759* (Paris s.d.).

Comte, Auguste, *Cours de philosophie positive* II: *La philosophie astronomique et la philosophie de la physique* (Paris 1830; reprint Paris 1924).

Copernicus, Nicolaus, *Nicolai Copernici Torinensis Astronomia Instaurata, Libris Sex Comprehensa, qui* de Revolutionibus orbium caelestium *inscribuntur. Nunc demum post 75 ab obitu authoris annum integritati suae restituta, Notisque illustrata, opera & studio D. Nicolai Mulerii Medicinae ac Matheseos Professoris Ordinarij in nova Academia quae est Groningae* (Amsterdam 1617).

Diogenes Laertius, *Lives of eminent philosophers* (Greek text + English translation by R.D. Hicks, Loeb Classical Library edition, Cambridge Mass. 1925).

Diodorus of Sicily, *Library of History* (Greek text + English translation by C.H. Oldfather, Loeb Classical Library, London/Cambridge Mass. 1968).

Du Séjour, Dionis, *Essai sur les Cometes en général, et particulièrement sur celles qui peuvent approcher de l'orbite de la terre* (Paris 1775).

Effigies & vitae Professorum Academiae Groningae & Omlandiae (Groningen 1654).

Emmius, Ubbo, Letter to Sibrandus Lubbertus, 29 September 1608, in H. Brugmans ed., *Briefwechsel des Ubbo Emmius* II (The Hague 1923) 51-52.

Fas, J.A., *Verhandeling over twee Comeeten uit de berekeningen van de Heer Halley* (Leiden 1787).

Fienus, Thomas, *De cometa anni 1618 dissertatio* in *De Cometa anni 1618. dissertationes Thomae Fieni in Academia Lovaniensi Medicinae et Liberti Fromondi Philosophiae Professorum. In quibus tum istius motus, tum aliorum omnium essentia, effectus, & praesagiendi facultas declarantur* (Antwerp 1619) 3-78.

Fienus, Thomas, *Thomae Fieni Disputatio An Caelum moveatur & Terra quiescat, ad generosos et nobiles viros DD. Thobiam Matthaei et Georgium Gays*, in *De Cometa anni 1618. dissertationes Thomae Fieni in Academia Lovaniensi Medicinae et Liberti Fromondi Philosophiae Professorum. In quibus tum istius motus, tum aliorum omnium essentia, effectus, & praesagiendi facultas declarantur* (Antwerp 1619) 141-153.

Fontanus, Saxus, *Bediedenisse vande nievwe Comete ofte Sterre metten Steerte, die int laetste vande maendt November, sich eerst aenden Hemel geopenbaert heeft* (The Hague 1619).

Fromondus, Libertus, *Dissertatio de Cometa anni 1618* in *De Cometa anni 1618. dissertationes Thomae Fieni in Academia Lovaniensi Medicinae et Liberti Fromondi Philosophiae Professorum. In quibus tum istius motus, tum aliorum omnium-essentia, effectus, & praesagiendi facultas declarantur* (Antwerp 1619) 79-140.

Fromondus, Libertus, *Meteorologicorum Libri Sex* (2nd ed., Louvain 1646).

Fromondus, Libertus, *Saturnalitiae Caenae, Variatae Somnio, sive Peregrinatione Caelesti* (Louvain 1616).

Fromondus, Libertus, *Vesta, sive Ant-Aristarchi Vindex, Adversus Iac. Lansbergium Philippi F. Medicum Middelburgensem. In quo Decretum S. Congregationis S.R.E. Cardinalium anno MDCXVI & alterum anno MDCXXXIII adversus Copernicanos terrae motores editum, iterum defenditur* (Antwerp 1634).

Galilei, Galileo, *Dialogue Concerning the Two Chief World Systems* (Florence 1632, translated by Stillman Drake, Berkeley/Los Angeles/ London 1967).

Galilei, Galileo, *Sidereus Nuncius* (1610, translated by Stillman Drake in *Discoveries and Opinions of Galileo* (Garden City 1957)).

Galilei, Galileo, *Sidereus Nuncius* (1610, translated by Stillman Drake in *Telescopes, Tides, & Tactics. A Galilean Dialogue about the "Starry Messenger" and Systems of the World* (Chicago/London 1983)).

Gemma, Cornelius, *De naturae divinis characterismis seu raris & admirandis spectaculis, causis, indiciis, proprietatibus rerum in partibus singulis universi, libri II* (Antwerp 1575).

Gemma, Cornelius, *De prodigiosa specie, naturaque Cometae, qui nobis effulsit altior lunae sedibus, insolita prorsus figura, ac magnitudine, anno 1577. plus septimanis 10. Apodeixis tum Physica tum Mathematica. Adiuncta his explicatio duorum chasmaton anni 1575. nec non ex cometarum plurium phaenomenis epilogistica quaedam assertio de communi illorum natura, generationum causis atque decretis supra quam hactenus a Peripateticis annotatum est* (Louvain 1578).

Geoffrey of Meaux, *De stellis comatis* (1315), in Lynn Thorndike ed., *Latin Treatises on Comets between 1238 and 1368 AD* (Chicago 1950) 208-214.

Geoffrey of Meaux, tract on the comet of 1337, in Lynn Thorndike ed., *Latin Treatises on Comets between 1238 and 1368 AD* (Chicago 1950) 219-225.

Gerard de Silteo, *Summa de astris*, Part I, dist. 23, cap. 4, in Lynn Thorndike ed., *Latin Treatises on Comets between 1238 and 1368 AD* (Chicago 1950) 194-195.

Grant, Edward ed., *A Source Book of Medieval Science* (Cambridge Mass. 1974).

Grosseteste, Robert, *De cometis* in Ludwig Baur ed., *Die Philosophischen Werke des Robert Grosseteste, Bischofs von Lincoln* IX (Münster 1912) 36-41.

Grosseteste, Robert, *De generatione stellarum* in Ludwig Baur ed., *Die philosophischen Werke des Robert Grosseteste, Bischofs von Lincoln* IX (Münster 1912) 32-36.

Guillemin, Amédée, *Les comètes* (Paris 1887).

Halley, Edmond, *Astronomical Tables, with Precepts both in English and Latin for computing the places of the Sun, Moon, Planets, and Comets* (London 1749; English version 1752).

Hancock, Blyth, *The Astronomy of Comets. In Two Parts* (London 1786).

Henry of Hesse, *Questio de cometa* (1368) in Hubert Pruckner ed., *Studien zu den astrologischen Schriften des Heinrich von Langenstein* (Berlin 1933) 89-138.

Heurnius, Johannes, *De Historie, Natuere, ende Beduidenisse der erschrickelicke Comeet, die geopenbaert is int Jaer ons Heren 1577.* (Cologne s.d.).

Hevelius, Johannes, *Cometographia* (Danzig 1668).

Heymensz, Pieter, *Een Prognosticatie, ofte voorsegginghe op de Komeet-Sterre met de steert, welcke haer verthoont ende gheopenbaert heeft den xvi. Julij na de nieuwe stijl, in dit teghenwoordige Jaer M. D. XCVI. Ghepractiseert door een Liefhebber van de edele Conste van Astrologie, M. Pieter Heymensz. Doctoor inde Medecijnen tot Delff* (Delft, s.d.).

Isidore of Seville, *Isidori Hispalensis Episcopi Etymologiarum sive originum libri 20* I (Books 1-10) (ed. & comm. by W.M. Lindsay, London/Oxford 1971).

Jacchaeus, Gilbertus, *Institutiones physicae* (2nd ed., Leiden 1624).

John of Damascus, *Exposition of the Orthodox Faith* (Ed. Eligius M. Buytaert, New York/ Louvain/Paderborn 1955).

John of Legnano, *Tractatus de cometa* (1368) in Lynn Thorndike ed., *Latin Treatises on Comets between 1238 and 1368 AD* (Chicago 1950) 234-259.

Josephus, Flavius, *Jewish War* (Greek + English translation by H. St. J. Thackeray, Loeb Classical Library, London/ New York 1928).

Kaiser, F., *De komeet van Encke en hare naderende verschijning* (Leiden 1838).

Kepler, Johannes, *New Astronomy* (Prague 1609; translated by William H. Donahue, Cambridge 1992).

Lalande, Jérôme de, *Réflexions sur les Comètes qui peuvent approcher de la terre* (Paris 1773).

Laplace, P.S. de, *Exposition du système du monde, par LA PLACE, suivie d'un précis de l'histoire de l'astronomie* (6th ed. Brussels 1829).

Lipsius, Justus, letter to Hieronymus Berchemius in *Iusti Lipsi Epistolae*, A. Gerlo/ M.A. Nauwelaerts/H.D.L. Vervliet eds. (Brussels 1978) I, 75. 76 10 13.

Mach, Ernst, *Die Mechanik in ihrer Entwickelung historisch-kritisch dargestellt* (2nd edition, Leipzig 1889).

Manilius, Marcus, *Astronomica* (Latin + English translation by G.P. Goold, Loeb Classical Library, Cambridge Mass./ London 1977).

Mauden, David van, *De geschiedenisse van de nieuwe Comete des Jaers 1577 seer neerstelick g'observeert van datse begonst te schynen tot den eynde toe. Door M. David van Mauden Doctor inder Medicijnen, Lecteur der Chirurgijnen en gesworen Stadt Medecijn deser Stadt van Antwerpen* (Antwerp 1577).

Maupertuis, Pierre-Louis Moreau de, *A letter upon comets* (London 1769).

Meteren, Emanuel van, *Belgische ofte Nederlantsche Historie, van onsen tijden. Inhoudende hoe de Nederlanden aenden anderen ghehecht, ende aen Spaengien ghecomen zijn: Met de oorsaken der Inlantsche beroerten, ende oorloghen der selver, Oock de veranderinghe van Princen, van regimenten ende van Religie, mette scheuringhen, verbontenissen ende vredehandelinghen: Mede vervattende eenighe harer ghebueren ende andere Landen handelinghen. Meest onder de regeeringhe van Philippus de II. Coninc van Spaengien, tot synen doot, ende den wtgaenden Iare 1598* (Delft 1599).

Montucla, J.E., *Histoire des mathématiques* (2nd ed. Paris 1799-1802).

Mulerius, Nicolaus, *D.D. Nicolaum Mulerium, van Brugghe, Doctor inde Medicynen van weghen der stadt Groeningen ende omme-landen. Zijne Practica. Met de beschryvinghe vande Comeet. Als oock mede vande Pasquil ende ghenoechlijck discours, hoe dat wy door Gods genade aens den Nederlandtschen Vrede-handel zijn ghecomen* (Groningen 1608).

Mulerius, Nicolaus, *Hemelsche Trompet Morghenwecker, ofte Comeet Met een Langebaert Erschenen anno 1618. in Novembri ende Decembri* (Groningen 1618).

Nicolai, Balduinus, *Verklaringhe Om yegelijcken te adverteeren van de dreyghementen Godts, die wy aenmercken in 't aenschouwen vanden Comeet, gesien in 't Jaer na de geboorte onses Heeren ende Salich-makers Jesu Christi 1618 (na de konst der Astronomie van oude Doctooren daer op gheschreven, te weten, nae dat de vertooninghe geschiet, in 't Oosten ofte Westen, oock nae de teeckenen des Hemels ofte Zodiac, daer de Comeet in gesien wort, als mede nae dat de steert lanck ofte kort is)* (Amsterdam s.d.).

Orderic Vitalis, *Ecclesiastical History* in Marjorie Chibnall ed., *The Ecclesiastical History of Orderic Vitalis* III (Books V and VI) (Oxford 1972).

Orthius, Theodorus, *Theses physicae de meteoris* (Leiden 1600).

Peter of Limoges, *Iudicium* on the comet of 1299, in Lynn Thorndike, *Latin Treatises on Comets between 1238 and 1368 AD* (Chicago 1950) 199-201.

Petit, Pierre, *Dissertation sur la nature des cometes au Roy. Avec un Discours sur les Prognostiques des Eclipses & autres Matieres curieuses* (Paris 1665).

Pilius, Petrus, *Disputatio Physica de Cometis* (Leiden 1598).

Pingré, Alexandre Guy, *Cométographie ou Traité historique et théorique des Comètes* (Vol. I Paris 1783, vol. II Paris 1784).

Plato, *Timaeus* (translated by Desmond Lee (Penguin Classics edition, *Plato: Timaeus and Critias* (Harmondsworth 1979)).

Pliny the Elder, *Natural History* (Latin + English translation by H. Rackham, Loeb Classical Library, London/ Cambridge Mass. 1938).

Plotinus, *Enneads* (translated by Stephen MacKenna, Penguin Classics edition, London 1991).
Polybius, *The Histories* Books V-VIII (Greek text + English translation by W.R. Paton, Loeb Classical Library edition, Cambridge Mass. 1923).
Portantius, Johannes, *Beschrijvinghe der nieuwer Cometen, Met aenwijsinge ende vermaninge wat zy bedieden ende voortbrenghen sal* (Antwerp s.d.).
Pseudo-Plutarch (Aetius), *De placitis philosophorum libri V* in G.N. Bernardakis ed., *Plutarchi Chaeronensis Moralia* V (Leipzig 1893) 264-372.
Ptolemy, Claudius, *Tetrabiblos* (Greek text + English translation by F.E. Robbins, Loeb Classical Library, London/ Cambridge Mass. 1980).
Puteanus, Erycius, *De Cometa anni 1618 novo Mundi Spectaculo Libri Duo Paradoxologia* (Cologne 1619).
Riccioli, Ioannes Baptista, *Almagestum novum* (Bologna 1651).
Scaliger, Julius Caesar, *Julii Caesaris Scaligeri Exotericarum Exercitationum Liber XV De Subtilitate, ad Hieronymum Cardanum* (Frankfurt 1612).
Seneca, Lucius Annaeus, *Naturales Quaestiones* (Latin text + English translation by Thomas H. Corcoran, Loeb Classical Library, London/Cambridge Mass. 1972).
Snellius, Willebrord, *Descriptio Cometae qui anno 1618 mense Novembri primum effulsit* (Leiden 1619).
Stobaeus, Johannes, *Eclogae physicae et ethicae* in Curtius Wachsmuth & Otto Hense eds., *Anthologii libri duo priores qui inscribi solent Eclogae physicae et ethicae*, Vol. 1 (s.l. 1884).
Struyck, Nicolaas, "Korte beschrijving van alle de Comeeten of Starrtsterren die ik in de geschiedenissen heb kunnen vinden" in: *Inleiding tot de algemene Geographie, benevens eenige sterrekundige en andere Verhandelingen* (Amsterdam 1740).
Suetonius, *Lives of the Caesars* (translated by Robert Graves, Penguin classics, 1957).
Tacitus, *Annals* (translated by Michael Grant, Penguin classics, 1956).
Velsius, Johannes, *Prognosticatie van de nieuwe ende seer schrickelijcke Comeet ofte harige Ster Anno 1618. verschenen* (Franeker 1618).
Venerable Bede, *De natura rerum* in *Bedae Venerabilis Opera* (*Corpus Christianorum Series Latina*) (Turnhout 1975).
Venerable Bede, *Ecclesiastical History of the English People* (Bertram Colgrave and R.A.B. Mynors eds., Oxford 1969).
Virgil, *Georgics* (ed. & comm. by R.A.B. Mynors, Oxford 1990).
Whewell, William, *A history of the inductive sciences, from the earliest to the present time* (3rd ed., London 1857).

Secondary Literature

Allen, Don Cameron, *The star-crossed Renaissance. The quarrel about astrology and its influence in England* (New York 1966).
Archenhold, F.S., *Alte Kometeneinblattdrucke* (Berlin 1911).
Ariew, Roger, "Theory of comets at Paris during the seventeenth century", *Journal of the history of ideas* 53 (3) (1992) 355-369.
Baelde, M., "De Nederlanden van Spaanse erfopvolging tot beeldenstorm, 1506-1566" in I. Schöffer, H. van der Wee & J.A. Bornewasser eds., *De Lage Landen van 1500 tot 1780* (Amsterdam/Brussels 1978) 38-101.
Baldwin, Martha R., "Magnetism and the anti-Copernican polemic", *Journal for the history of astronomy* 16 (1985) 155-174.
Barker, Peter, "Jean Pena (1528-58) and Stoic physics in the sixteenth century" in Ronald H. Epp ed., *Spindel Conference 1984: Recovering the Stoics*, Supplement to *The Southern Journal of Philosophy* 23 (1985) 93-107.
Barker, Peter, "The optical theory of comets from Apian to Kepler", *Physis* 30 (1993) 1-25.

Barker, Peter, "Stoic contributions to early modern science", in Margaret J. Osler ed., *Atoms, 'pneuma', and tranquillity. Epicurean and Stoic themes in European thought* (Cambridge 1991) 135-154.

Barker, Peter and Bernard R. Goldstein, "The role of comets in the Copernican Revolution", *Studies in the history and philosophy of science* 19 (1988) 299-319.

Barker, Peter & Bernard R. Goldstein, "Is seventeenth-century physics indebted to the Stoics?", *Centaurus* 27 (1984) 148-164.

Barton, Tamsyn, *Ancient astrology* (London/New York 1994).

Berkel, K. van, "De geschriften van Rudolf Snellius. Een bijdrage tot de geschiedenis van het ramisme in Nederland", *Tijdschrift voor de geschiedenis van de geneeskunde, natuurwetenschappen, wiskunde en techniek* 6 (1983) (4) 185-194.

Berkel, K. van, "Universiteit en natuurwetenschap in de 17e eeuw, in het bijzonder in de Republiek" in H.A.M. Snelders & K. van Berkel eds., *Natuurwetenschappen van de Renaissance tot Darwin. Thema's uit de wetenschapsgeschiedenis* (The Hague 1981) 107-130.

Berkel, K. van, *In het voetspoor van Stevin. Geschiedenis van de natuurwetenschap in Nederland 1580-1940* (Meppel/Amsterdam 1985).

Berry, Arthur, *A short history of astronomy. From the earliest times through the nineteenth century* (Cambridge 1898).

Biagioli, Mario, *Galileo, courtier. The practice of science in the culture of absolutism* (Chicago/London 1993).

Biographie Nationale de Belgique (28 vols. + suppl., Brussels 1866-1944).

Blake, Ralph M., "Theories of hypothesis among Renaissance astronomers" in Ralph M. Blake, Curt J. Ducasse and Edward H. Madden eds., *Theories of scientific method* (Seattle 1960) 22-49.

Boll, Franz, *Studien über Claudius Ptolemäus* (Leipzig 1894).

Bosmans, Henri, *Documents inédits sur Grégoire de Saint-Vincent* (Brussels 1903).

Bouwsma, William J., "The two faces of humanism: Stoicism and Augustinianism in Renaissance thought" in Heiko A. Oberman & Thomas A. Brady Jr. eds., *Itinerarium Italicum. The profile of the Italian Renaissance in the mirror of its European transformations. Dedicated to Paul Oskar Kristeller on the occasion of his 70th birthday* (Leiden 1975) 3-60.

Brandt, John C. and Robert D. Chapman, *Introduction to comets* (Cambridge 1981).

Brown, Peter Lancaster, *Comets, meteorites and men* (London 1973).

Bruin, C.C. de, "Radicaal spiritualisme te Leiden", *Rondom het woord* 17 (3) (1975) 66-81.

Burke, Peter, "Tacitism, scepticism, and reason of state" in J.H. Burns & Mark Goldie eds., *The Cambridge history of political thought 1450-1700* (Cambridge 1991) 479-498.

Capp, Bernard, *English almanacs, 1500-1800: Astrology and the popular press* (London/Boston 1979).

Céard, Jean, *La nature et les prodiges. L'insolite au XVIe siècle, en France* (Geneva 1977).

Ceyssens, L., "Diepere gronden van het Jansenisme", *Tijdschrift voor theologie* 6 (1966) 395-420.

Ceyssens, L., "Le drame de conscience augustinien des premiers jansénistes" in: *Augustinus magister. Communications du congrès international augustinien* (Paris 1954) Part II, 1069-1076.

Ceyssens, L., "Le Janséniste Libert Froidmont (1587-1653)", *Bulletin de la Société d'Art et d'Histoire du Diocèse de Liège* 43 (1963) 1-46.

Chambers, George F., *The story of the comets simply told for general readers* (Oxford 1909).

Christianson, J.R., "Tycho Brahe's German treatise on the comet of 1577: A study in science and politics", *Isis* 70 (1979) 110-140.

Claeys Bouuaert, F., *L'ancienne université de Louvain. Études et Documents* (*Bibliothèque de la revue de l'histoire ecclésiastique*, fasc. 28, Louvain 1956).

Cohen, H. Floris, *The Scientific Revolution. A historiographical inquiry* (Chicago 1994).
Cohen, I. Bernard, *Revolution in science* (Cambridge Mass. 1985).
Collingwood, R.G., *The idea of history* (Oxford 1946; paperback edition Oxford 1994).
Coopland, G.W., *Nicole Oresme and the astrologers. A study of his Livre de Divinacions* (Liverpool 1952).
Copenhaver, Brian P., "Astrology and magic" in Charles B. Schmitt & Quentin Skinner eds., *The Cambridge history of Renaissance philosophy* (Cambridge 1988) 264-300.
Copenhaver, Brian P., "Natural magic, hermetism, and occultism in early modern science" in Robert S. Westman & David C. Lindberg eds., *Reappraisals of the Scientific Revolution* (Cambridge 1990) 261-301.
Copenhaver, Brian P. & Charles B. Schmitt, *Renaissance philosophy* (*A history of western philosophy*, III, Oxford 1992).
Crowe, Michael J., *Theories of the world from Antiquity to the Copernican Revolution* (New York 1990).
Curry, Patrick, *Prophecy and power. Astrology in early modern England* (Cambridge 1989).
Dales, Richard C., "The origin of the doctrine of the double truth", *Viator* 15 (1984) 169-179.
Daston, Lorraine, "Marvelous facts and miraculous evidence in early modern Europe", *Critical Inquiry* 18 (1) (1991) 93-124.
Daston, Lorraine, "Wunder, Naturgesetze und die wissenschaftliche Revolution des 17. Jahrhunderts" in *Jahrbuch der Akademie der Wissenschaften in Göttingen* (1991) 99-122.
Dear, Peter, "The Church and the new philosophy" in Stephen Pumfrey, Paolo L. Rossi & Maurice Slawinski eds., *Science, culture and popular belief in Renaissance Europe* (Manchester 1991) 119-139.
Dibon, Paul Auguste Georges, *L'enseignement philosophique dans les universités néerlandaises à l'époque pré-cartésienne (1575-1650)* (Leiden 1954).
Dibon, P., "L'influence de Ramus aux universités néerlandaises du 17e siècle", *Actes du XIème Congrès International de Philosophie, Bruxelles, 20-26 Août 1953* XIV (Amsterdam/Louvain 1953) 307-311.
Dicks, D.R., *Early Greek astronomy to Aristotle* (Cornell 1970).
Dijksterhuis, E.J., *De mechanisering van het wereldbeeld* (Amsterdam 1950).
Dijksterhuis, E.J., *Simon Stevin. Science in the Netherlands around 1600* (The Hague 1970).
Donahue, William H., *The dissolution of the celestial spheres 1595-1650* (New York 1981).
Donahue, William H., "The solid planetary spheres in post-Copernican natural philosophy" in Robert S. Westman ed., *The Copernican achievement* (Berkeley 1975) 244-275.
Dorsten, J.A. van, *The Anglo-Dutch Renaissance. Seven essays* (J. van den Berg & Alastair Hamilton eds., Leiden/ New York/ Copenhagen/ Cologne 1988).
Dorsten, J.A. van, "Een humanistische universiteit" in *Leidse universiteit 400. Stichting en eerste bloei 1575-ca. 1650* (Amsterdam 1975) 10-11.
Dorsten, J.A. van, *The radical arts. First decade of an Elizabethan Renaissance* (Leiden 1970).
Dorsten, J.A. van, "*Temporis filia veritas*: wetenschap en religievrede", *Tijdschrift voor geschiedenis* 89 (1976) 413-419.
Drake, Stillman, *Galileo: Pioneer scientist* (Toronto 1990).
Drake, Stillman & C.D. O'Malley eds., *The controversy on the comets of 1618. Galileo Galilei, Horatio Grassi, Mario Guiducci, Johann Kepler* (Philadelphia 1960).
Dreyer, J.L.E., *A history of astronomy from Thales to Kepler* (2nd ed. Dover 1953).
Dudley Sylla, Edith, "The Oxford Calculators" in Norman Kretzmann, Anthony Kenny & Jan Pinborg eds., *The Cambridge history of later medieval philosophy. From the rediscovery of Aristotle to the disintegration of scholasticism 1100-1600* (Cambridge 1982) 540-563.
Duhem, Pierre, *To save the phenomena: An essay on the idea of physical theory from Plato to Galileo* (trans. Chicago 1969; first published in French 1908).

Eamon, William, *Science and the secrets of nature. Books of secrets in medieval and early modern culture* (Princeton 1994).

Einstein, Albert, "Foreword" to *Galileo Galilei: Dialogue concerning the two chief world systems - Ptolemaic & Copernican* (translated by Stillman Drake, Berkeley/Los Angeles/London 1953; second revised edition 1967).

Elaut, L., "Nicolaus Mulerius uit Brugge, de eerste medische hoogleraar te Groningen (1564-1630)", *Scientiarum Historia* 1 (1959) 3-13.

Ernst, Germana, "From the watery trigon to the fiery trigon: Celestial signs, prophecies and history" in Paola Zambelli ed., *Astrologi hallucinati. Stars and the end of the world in Luther's time* (Berlin/ New York 1986) 265-280.

Essen, Léon van der, "Les tribulations de l'Université de Louvain pendant le dernier quart du XVIe siècle", *Bulletin de l'Institut historique belge de Rome* 2 (1922) 61-86.

Essen, Léon van der, *L'Université de Louvain (1425-1940)* (Brussels 1945).

Evers, G.A., "Voorteekenen in de lucht", *Historia* 12 (1947) 249-254.

Federici Vescovini, Graziella, "Peter of Abano and astrology" in Patrick Curry ed., *Astrology, science and society* (Bury St Edmunds 1987) 19-39.

Fontaine Verwey, H. de la, "Het Huis der Liefde en zijn publikaties" in idem, *Uit de wereld van het boek I: Humanisten, dwepers en rebellen in de zestiende eeuw* (Amsterdam 1975) 85-111.

Freyburger, Pierre, "Le problème du fatalisme astral dans la pensée protestante en pays germaniques" in *Divination et controverse religieuse en France au XVIe siècle* (Cahiers V.L. Saulnier, 4, Paris 1987) 35-55.

Funkenstein, Amos, "The dialectical preparation for scientific revolutions. On the role of hypothetical reasoning in the emergence of Copernican astronomy and Galilean mechanics" in Robert S. Westman ed., *The Copernican achievement* (Berkeley 1975) 165-203.

Funkenstein, Amos, *Theology and the scientific imagination from the Middle Ages to the seventeenth century* (Princeton 1986).

Garin, Eugenio, *Astrology in the Renaissance. The zodiac of life* (Bury St Edmunds 1984).

Gascoigne, John, "A reappraisal of the role of the universities in the Scientific Revolution" in Robert Westman & David C. Lindberg eds., *Reappraisals of the Scientific Revolution* (Cambridge 1990) 207-260.

Gaukroger, Stephen, "Introduction: the idea of Antiquity" in Stephen Gaukroger ed., *The uses of Antiquity: The Scientific Revolution and the classical tradition* (Dordrecht/ Boston/London 1991) ix-xvi.

Gelderen, Martin van, "Introduction" in Martin van Gelderen ed., *The Dutch Revolt* (Cambridge Texts in the History of Political Thought, Cambridge 1993) ix-xxxiv.

Gillespie, C.S. ed., *Dictionary of Scientific Biography* (18 vols., New York 1970-1981).

Gingerich, Owen, "From Copernicus to Kepler: Heliocentrism as model and as reality", *Proceedings of the American Philosophical Society* (1973) 513-522.

Gingerich, Owen, "Johannes Kepler" in René Taton & Curtis Wilson eds., *Planetary astronomy from the Renaissance to the rise of astrophysics. Part A: Tycho Brahe to Newton* (Cambridge 1989) 54-78.

Gingerich, Owen and Robert S. Westman, *The Wittich connection: Conflict and priority in late sixteenth-century cosmology* (*Transactions of the American Philosophical Society*, lxxviii, Part 7; Philadelphia 1988).

Grant, Edward, "Aristotelianism and the longevity of the medieval world view", *History of science* 16 (1978) 93-106.

Grant, Edward, "The condemnation of 1277, God's absolute power, and physical thought in the late Middle Ages", *Viator* 10 (1979) 211-244.

Grant, Edward, "Cosmology" in D.C. Lindberg ed., *Science in the Middle Ages* (Chicago 1978) 265-302.

Grant, Edward, "The effect of the condemnation of 1277" in Norman Kretzmann, Anthony Kenny & Jan Pinborg eds., *The Cambridge history of later medieval philosophy.*

From the rediscovery of Aristotle to the disintegration of scholasticism 1100-1600 (Cambridge 1982) 537-539.

Grant, Edward, *Planets, stars, and orbs. The medieval cosmos 1200-1687* (Cambridge 1994).

Grant, Edward, "The reaction of the universities and theological authorities to Aristotelian science and natural philosophy" in Edward Grant ed., *A source book of medieval science* (Cambridge Mass. 1974) 42-52.

Grant, Edward, "The translation of Greek and Arabic science into Latin" in Edward Grant ed., *A source book of medieval science* (Cambridge Mass. 1974) 35-41.

Grant, Edward, "Ways to interpret the terms 'Aristotelian' and 'Aristotelianism' in medieval and Renaissance natural philosophy", *History of science* 25 (1987) 335-358.

Grant, Robert M., *Miracle and natural law in Graeco-Roman and early Christian thought* (Amsterdam 1952).

Grisar, J., "Die Universität Löwen zur Zeit der Gesandschaft des P. Franciscus Toletus (1580) nach bisher unbenützten Quellen des Vatikan. Archivs" in *Miscellanea Historica in honorem Alberti De Meyer* (*Universiteit te Leuven, Publicaties op het gebied der geschiedenis en der philologie*, 3e reeks, 23e deel, Louvain/Brussels 1946) 941-968.

Hahm, David E., *The origins of Stoic cosmology* (Ohio 1977).

Hall, A.R., "William Wotton and the history of science", *Archives internationales de l'histoire des sciences* 28 (1949) 1047-1062.

Hall, A. Rupert, *From Galileo to Newton 1630-1720* (London 1963, Dover edition New York 1981).

Hall, Marie Boas, *The scientific Renaissance 1450-1630* (London 1962, Dover edition New York 1994).

Hamilton, A., *The Family of Love* (Cambridge 1981).

Hammerstein, Helga Robinson, "The battle of the booklets: Prognostic tradition and proclamation of the word in early sixteenth-century Germany" in Paola Zambelli ed., *Astrologi hallucinati. Stars and the end of the world in Luther's time* (Berlin/New York 1986) 129-151.

Hansen, Bert, *Nicole Oresme and the marvels of nature: The "De causis mirabilium"* (Toronto 1985).

Hansen, Bert, "Science and magic" in David C. Lindberg ed., *Science in the Middle Ages* (Chicago/ London 1978) 483-506.

Harline, Craig E., *Pamphlets, printing and political culture in the early Dutch Republic* (Dordrecht 1987).

Hazard, Paul, *La crise de la conscience européenne 1680-1715* (Paris 1935).

Hellman, C. Doris, "The gradual abandonment of the Aristotelian universe: a preliminary note on some sidelights" in *Mélanges Alexandre Koyré* (Paris 1964) 283-293.

Hellman, C. Doris, "Kepler and comets" in A. Beer & P. Beer eds., *Kepler: Four hundred years. Vistas in astronomy* 18 (1975) 789-796.

Hellman, C. Doris, "The rôle of measurement in the downfall of a system: some examples from 16th-century comet and nova observations" in A. Beer ed., *New aspects in the history and philosophy of astronomy* (Oxford 1967) 43-52.

Hellman, Clarisse Doris, *The comet of 1577: Its place in the history of astronomy* (New York 1944).

Hess, Wilhelm, *Himmels- und Naturerscheinungen in Einblattdrucken des XV.-XVIII. Jahrhunderts* (Leipzig 1911).

Hooykaas, R., *Humanisme, science et réforme. Pierre de la Ramée (1515-1572)* (Leiden 1958).

Hooykaas, R., "The reception of Copernicanism in England and the Netherlands" in *The Anglo-Dutch contribution to the civilization of early modern society. An Anglo-Netherlands symposium* (Oxford 1976) 33-44.

Huisman, G.C., "Boeken en brieven van Nicolaus Mulerius" in *Bibliotheek, wetenschap en cultuur. Opstellen aangeboden aan mr W. R. H. Koops bij zijn afscheid als bibliothecaris der Rijksuniversiteit te Groningen* (Groningen 1990) 283-296.

Israel, Jonathan I., *The Dutch Republic. Its rise, greatness, and fall 1477-1806* (Oxford 1995).

Janssens, P., "Het bestand in de Zuidelijke Nederlanden 1609-1621" in *Nieuwe Algemene Geschiedenis der Nederlanden* VI, 315-324.

Jardine, Nicholas, "How to appropriate a world system" (essay review of Owen Gingerich & Robert S. Westman, *The Wittich connection*), *Journal for the history of astronomy* 21 (1990) 353-357.

Jardine, Nicholas, *The birth of history and philosophy of science. Kepler's "A Defence of Tycho against Ursus" with essays on its provenance and significance* (Cambridge 1984).

Jardine, Nicholas, "Demonstration, dialectic, and rhetoric in Galileo's *Dialogue*" in D.R. Kelley & R.H. Popkin eds., *The shapes of knowledge from the Renaissance to the Enlightenment* (Dordrecht 1991) 101-121.

Jardine, Nicholas, "Scepticism in Renaissance astronomy: A preliminary study" in Richard H. Popkin & Charles B. Schmitt eds., *Scepticism from the Renaissance to the Enlightenment* (Wiesbaden 1987) 83-102.

Jardine, Nicholas, "The significance of the Copernican orbs", *Journal for the history of astronomy* 13 (1982) 168-193.

Jarrell, Richard A., "The contemporaries of Tycho Brahe" in René Taton & Curtis Wilson eds., *Planetary astronomy from the Renaissance to the rise of astrophysics. Part A: Tycho Brahe to Newton* (Cambridge 1989) 22-32.

Jervis, Jane L., *Cometary theory in fifteenth-century Europe* (Dordrecht/Boston/Lancaster 1985).

Jonckbloet, W.J.A. ed., *Gedenkboek der Hoogeschool te Groningen, ter gelegenheid van haar vijfde halve eeuwfeest* (Groningen 1864).

Jones Schofield, Christine, *Tychonic and semi-Tychonic world systems* (New York 1981).

Jorink, Eric, "Hemelse tekenen. Nederlandse opvattingen over de komeet van 1618", *Gewina* 17 (1994) 68-81.

Kassler, Jamie C., "The paradox of power: Hobbes and Stoic naturalism" in Stephen Gaukroger ed., *The uses of Antiquity: The Scientific Revolution and the classical tradition* (Dordrecht/Boston/London 1991) 53-78.

Kennedy, E.S., "Comets in Islamic astronomy and astrology", *Journal of near eastern studies* 16 (1957) 44-51.

Kluyskens, J., "Justus Lipsius' levenskeuze: het irenisme", *Bijdragen en mededelingen betreffende de geschiedenis der Nederlanden* 88 (1) (1973) 19-37.

Kocher, Paul H., *Science and religion in Elizabethan England* (San Marino, California 1953).

Köhler, Hans-Joachim, "The *Flugschriften* and their importance in religious debate: a quantitative approach" in Paola Zambelli ed., *Astrologi hallucinati. Stars and the end of the world in Luther's time* (Berlin/New York 1986) 153-175.

Korolec, J.B., "Free will and free choice" in Norman Kretzmann, Anthony Kenny & Jan Pinborg eds., *The Cambridge history of later medieval philosophy. From the rediscovery of Aristotle to the disintegration of scholasticism 1100-1600* (Cambridge 1982) 629-641.

Kragh, Helge, *An introduction to the historiography of science* (Cambridge 1987).

Kuhn, Thomas S., *The Copernican Revolution. Planetary astronomy in the development of western thought* (Cambridge, Mass./London 1957).

Leff, Gordon, *Medieval thought: Saint Augustine to Ockham* (London 1959).

Lemay, Richard, "The true place of astrology in medieval science and philosophy: towards a definition" in Patrick Curry ed., *Astrology, science, and society. Historical essays* (Bury St Edmunds 1987) 57-73.

Lewis, C.S., *The discarded image. An introduction to medieval and Renaissance literature* (Cambridge 1964).

Lindberg, David C., *The beginnings of western science: The European scientific tradition in philosophical, religious, and institutional context, 600 B.C. to A.D. 1450* (Chicago 1992).

Lindberg, David C., "Conceptions of the Scientific Revolution from Bacon to

Butterfield: a preliminary sketch" in David C. Lindberg and Robert S. Westman eds., *Reappraisals of the Scientific Revolution* (Cambridge 1990) 1-26.

Maier, Anneliese, "The achievements of late scholastic natural philosophy" in Steven D. Sargent ed., *On the threshold of exact science. Selected writings of Anneliese Maier on late medieval natural philosophy* (Philadelphia 1982) 143-170.

Meinel, Christoph, "Les *Météores* de Froidmont et les *Météores* de Descartes" in Anne-Catherine Bernès ed., *Libert Froidmont et les résistances aux révolutions scientifiques. Actes du Colloque Château d'Oupeye, 26 et 27 septembre 1987* (Haccourt 1988) 105-129.

Moesgaard, Kristian Peder, "Copernican influence on Tycho Brahe" in Jerzy Dobrzycki ed., *The reception of Copernicus' heliocentric theory. Proceedings of a symposium organized by the Nicolas Copernicus Committee of the international union of the history of science* (Dordrecht/ Boston 1972) 31-55.

Molhuysen, P.C. & P.J. Blok eds., *Nieuw Nederlandsch biografisch woordenboek* (Leiden 1911-37).

Monchamp, Georges, *Galilée et la Belgique. Essai historique sur les vicissitudes du système de Copernic en Belgique* (Saint-Trond/Brussels/Paris 1892).

Moore, Patrick, *The comets. Visitors from space* (Cambridge 1973).

Moran, Bruce T., *The alchemical world of the German court. Occult philosophy and chemical medicine in the circle of Moritz of Hessen (1572-1632)* (Stuttgart 1991).

Moreau, J.J., *Honderd veertien Nederlandse brieven van Erycius Puteanus aan de astronoom Michael Florent Van Langren* (Antwerp/Amsterdam 1957).

Mout, M.E.H.N., "Humanists and the horrors of war: Neostoicism and the Revolt of the Netherlands" in H. de Schepper & P.J.A.N. Rietbergen eds., *España y Holanda. Ponencias de los coloquios Hispano-Holandeses de historiadores 1984-1988* (Madrid/ Nijmegen 1993) 109-118.

Mout, M.E.H.N., "In het schip: Justus Lipsius en de Nederlandse Opstand tot 1591" in S. Groenveld, M.E.H.N. Mout & I. Schöffer eds., *Bestuurders en geleerden. Opstellen over onderwerpen uit de Nederlandse geschiedenis van de zestiende, zeventiende en achttiende eeuw, aangeboden aan Prof. Dr. J.J. Woltjer bij zijn afscheid als hoogleraar van de Rijksuniversiteit te Leiden* (Amsterdam 1985) 55-64.

Mout, M.E.H.N., "Prognostica tijdens de Nederlandse Opstand" in C. Augustijn et al. eds., *Kerkhistorische opstellen aangeboden aan Prof. dr. J. van den Berg* (Kampen 1987) 9-19.

Mout, N., "The Family of Love (Huis der Liefde) and the Dutch Revolt" in A. Duke & C.A. Tamse eds., *Britain and the Netherlands* VII: *Church and state since the Reformation* (The Hague 1981) 76-93.

Müller-Jahncke, Wolf-Dieter, "Der Paracelsische Weg zu Astrologie und Magie" in Udo Benzenhöfer ed., *Paracelsus* (Darmstadt 1993) 98-136.

Nalis, H.J. & J.L. Salman, "'Wie sal ick dan wes goedes konnen prognosticeren?' Deventer almanakken en prognosticaties in roerige tijden (1555-1610)", *Deventer Jaarboek* (1994) 6-39.

Noordegraaf, Leo & Gerrit Valk, *De Gave Gods. De pest in Holland vanaf de late middeleeuwen* (Amsterdam 1988).

North, John D., "Celestial influence-the major premiss of astrology" in Paola Zambelli ed., *Astrologi hallucinati. Stars and the end of the world in Luther's time* (Berlin/ New York 1986) 45-100.

North, John, "Medieval concepts of celestial influence: A survey" in Patrick Curry ed., *Astrology, science and society. Historical essays* (Bury St Edmunds 1987) 5-16.

Oestreich, Gerhard, *Neostoicism and the early modern state* (Cambridge 1982).

Ong, W.J., *Ramus. Method and the decay of dialogue* (Cambridge, Mass./London 1958).

Ortroy, Fernand Van, *Bio-Bibliographie de Gemma Frisius, fondateur de l'école belge de géographie, de son fils Corneille, et de ses neveux les Arsénius* (Brussels 1920; reprint Amsterdam 1964).

Park, Katharine and Lorraine J. Daston, "Unnatural conceptions: the study of mon-

sters in sixteenth- and seventeenth-century France and England", *Past and Present* 92 (1981) 20-54.
Parker, Geoffrey, *The Dutch Revolt* (revised ed., London 1985).
Parker, Geoffrey, "Mutiny and discontent in the Spanish army of Flanders, 1572-1607" in idem, *Spain and the Netherlands 1559-1659. Ten studies* (revised ed., Glasgow 1990) 106-121.
Pauly, A.F. von, *Paulys Real-Encyclopädie der klassischen Altertumswissenschaft* (Stuttgart 1894-1940).
Pennington, D.H., *Europe in the seventeenth century* (1970; revised ed., London 1989).
Petry, M.J., "Burgersdijk's physics" in E.P. Bos & H.A. Krop eds., *Franco Burgersdijk (1590-1635). Neo-Aristotelianism in Leiden* (Amsterdam/Atlanta 1993) 83-118.
Poppi, Antonino, "Fate, fortune, providence and human freedom" in Charles B. Schmitt & Quentin Skinner eds., *The Cambridge history of Renaissance philosophy* (Cambridge 1988) 641-667.
Pruckner, Hubert, "Inhaltsübersicht des Tractatus contra astrologos coniunctionistas" in Hubert Pruckner, *Studien zu den astrologischen Schriften des Heinrich von Langenstein* (Leipzig/ Berlin 1933) 43-72.
Pumfrey, Stephen, "Magnetical philosophy and astronomy, 1600-1650" in René Taton & Curtis Wilson eds., *Planetary astronomy from the Renaissance to the rise of astrophysics. Part A: Tycho Brahe to Newton* (Cambridge 1989) 45-53.
Quetelet, Ad., *Histoire des sciences mathématiques et physiques chez les Belges* (Brussels 1864).
Radding, Charles M., "Superstition to science: Nature, fortune, and the passing of the medieval ordeal", *American historical review* 84 (1979) 945-969.
Rekers, B., *Benito Arias Montano (1527-1598)* (London/Leiden 1972).
Robinson, James Howard, *The great comet of 1680. A study in the history of rationalism* (Northfield, Minn. 1916).
Roets, Achiel, "Libertus Fromondus (1587-1653) en het kerkelijk gezag", *Sacris erudiri* 2 (1949) 335-366.
Rooses, M., *Christophe Plantin, imprimeur anversois* (Brussels 1883).
Rosen, Edward, *Kepler's Conversation with Galileo's Sidereal Messenger* (New York/ London 1965).
Rosen, Edward, *Three imperial mathematicians. Kepler trapped between Tycho Brahe and Ursus* (New York 1986).
Ruestow, Edward G., *Physics at seventeenth and eighteenth century Leiden: Philosophy and the new science in the university* (The Hague 1973).
Russell, Bertrand, *History of western philosophy* (2nd. ed., London 1961).
Russell, John L., "Catholic astronomers and the Copernican system after the condemnation of Galileo", *Annals of science* 46 (1989) 365-386.
Russell, John L., "The Copernican system in Great Britain" in Jerzy Dobrzycki ed., *The reception of Copernicus' heliocentric theory. Proceedings of a symposium organized by the Nicolas Copernicus Committee of the international union of the history of science* (Dordrecht/ Boston 1972) 189-239.
Sambursky, S., *Physics of the Stoics* (Princeton 1959, paperback 1987).
Sandbach, F.H., *The Stoics* (2nd ed., London/Indianapolis 1989).
Sassen, Ferdinand, *Het oudste wijsgeerig onderwijs te Leiden* (Mededeelingen van de Koninklijke Nederlandse Akademie van Wetenschappen, Afd. Letterkunde, Nieuwe Reeks, 4 (Amsterdam 1941)).
Sassen, F., "De reis van Pierre Gassendi in de Nederlanden (1628-1629)", *Mededelingen der Koninklijke Nederlandse Akademie van Wetenschappen, Afdeling Letterkunde, Nieuwe Reeks* 23-10 (Amsterdam 1960) 263-307.
Saunders, Jason Lewis, *Justus Lipsius. The philosophy of Renaissance Stoicism* (New York 1955).
Schaffer, Simon, "Authorized prophets: Comets and astronomers after 1759", *Studies in eighteenth-century culture* 17 (1987) 45-74.

Schaffer, Simon, "Newton's comets and the transformation of astrology" in Patrick Curry ed., *Astrology, science and society. Historical essays* (Woodbridge/Wolfeboro 1987) 219-243.

Schechner Genuth, Sara, "Devils' hells and astronomers' heavens: Religion, method, and popular culture in speculations about life on comets" in Mary Jo Nye, Joan L. Richards & Roger H. Stuewer eds., *The invention of physical science. Intersections of mathematics, theology and natural philosophy since the seventeenth century* (Dordrecht/Boston/London 1992) 3-26.

Schechner Genuth, Sara, *From monstrous signs to natural causes: the assimilation of comet lore into natural philosophy* (Unpubl. diss. Harvard 1988; University Microfilms, 1990).

Schofield, Christine, "The Tychonic and semi-Tychonic world systems" in René Taton & Curtis Wilson eds., *Planetary astronomy from the Renaissance to the rise of astrophysics. Part A: Tycho Brahe to Newton* (Cambridge 1989) 33-44.

Schotel, G.D.J., *Vaderlandsche volksboeken en volkssprookjes van de vroegste tijden tot het einde der 18e eeuw* I (Haarlem 1873).

Schmitt, Charles B., *Aristotle and the Renaissance* (Cambridge Mass./ London 1983).

Schmitt, Charles B., *Cesare Cremonini, un Aristotelico al tempo di Galilei* (Venice 1980) 3-21.

Schmitt, Charles B., "Perennial philosophy: from Agostino Steuco to Leibniz", *Journal of the history of ideas* 27 (1966) 505-532.

Schmitt, Charles B., "Prisca theologia e philosophia perennis: due temi del Rinascimento italiano e la loro fortuna" in *Il pensiero italiano del Rinascimento e il tempo nostro. Atti del V Convegno internazionale del Centro di studi umanistici, Montepulciano, 1968* (Florence 1970) 211-236.

Schöffer, I., "De opstand in de Nederlanden, 1566-1609" in I. Schöffer, H. van der Wee & J.A. Bornewasser eds., *De Lage Landen van 1500 tot 1780* (Amsterdam/ Brussels 1978) 103-165.

Schöffer, I., "De Republiek der Verenigde Nederlanden 1609-1702" in I. Schöffer, H. van der Wee & J.A. Bornewasser eds., *De Lage Landen van 1500 tot 1780* (Amsterdam/Brussels 1978) 167-267.

Shea, William R., "Galileo, Scheiner, and the interpretation of sunspots", *Isis* 61 (1970) 498-519.

Shumaker, Wayne, *The occult sciences in the Renaissance. A study in intellectual patterns* (Berkeley/Los Angeles 1972).

Simar, Th., *Étude sur Erycius Puteanus (1574-1646) considéré spécialement dans l'histoire de la philologie belge et dans son enseignement à l'université de Louvain* (Louvain/Paris/Brussels 1909).

Smet, A. de, 'Copernic et les Pays-Bas', *Janus* 60 (1973) 13-23.

Snelders, H.A.M., "Alkmaarse natuurwetenschappers uit de zestiende en zeventiende eeuw" in *Van Spaans beleg tot Bataafse tijd. Alkmaars stedelijk leven in de zeventiende en achttiende eeuw* (Alkmaarse historische reeks 4, Zutphen 1980) 101-122.

Snelders, H.A.M., "Science in the Low Countries during the sixteenth century. A survey", *Janus* 70 (1983) 213-227.

Swart, K.W., *Willem van Oranje en de Nederlandse Opstand 1572-1584* (edited by R.P. Fagel, M.E.H.N. Mout & H.F.K. van Nierop, The Hague 1994).

Tester, S.J., *A history of western astrology* (Bury St. Edmunds 1987).

Thomas, Keith, *Religion and the decline of magic. Studies in popular belief in sixteenth and seventeenth-century England* (London 1971).

Thomson, S.H., "The text of Grosseteste's De Cometis", *Isis* 19 (1933) 19-25.

Thoren, Victor E., "The comet of 1577 and Tycho Brahe's system of the world", *Archives internationales de l'histoire des sciences* 29 (1979) 53-67.

Thoren, Victor E., *The lord of Uraniborg. A biography of Tycho Brahe* (Cambridge 1990).

Thorndike, Lynn, "Albumasar in Sadan", *Isis* 45 (1954) 22-32.

Thorndike, Lynn, *A history of magic and experimental science* (8 vols., New York 1923-1958).

Thorndike, Lynn ed., *Latin treatises on comets between 1238 and 1368 AD* (Chicago 1950).

Universiteit te Leuven, de, 1425-1985 (Louvain 1986).

Van Helden, Albert, "Galileo, telescopic astronomy, and the Copernican system" in René Taton & Curtis Wilson eds., *Planetary astronomy from the Renaissance to the rise of astrophysics. Part A: Tycho Brahe to Newton* (Cambridge 1989) 81-105.

Van Helden, Albert, *The invention of the telescope* (*Transactions of the American Philosophical Society* 67, 4 (Philadelphia 1977)).

Van Helden, Albert, "The telescope in the seventeenth century", *Isis* 65 (1974) 38-58.

Verbeke, Gerard, "Ethics and logic in Stoicism" in M. Osler ed., *Atoms, pneuma and tranquillity: Epicurean and Stoic themes in European thought* (Cambridge 1991) 11-24.

Vermij, R.H., "Het copernicanisme in de Republiek: een verkenning", *Tijdschrift voor geschiedenis* 106 (1993) 349-367.

Voet, L., *The golden compasses: A history and evaluation of the printing and publishing activities of the Officina Plantiniana at Antwerp* (Amsterdam 1969).

Vollgraff, J.A., "Pierre de la Ramée (1515-1572) et Willebrord Snel van Royen (1580-1626)", *Janus* 18 (1913) 595-625.

Walker, D.P., *The ancient theology* (London 1972).

Ward, Benedicta, *Miracles and the medieval mind. Theory, record and event, 1000-1215* (Philadelphia 1982).

Webster, Charles, *From Paracelsus to Newton. Magic and the making of modern science* (Cambidge 1982).

Westman, Robert, "The comet and the cosmos: Kepler, Mästlin and the Copernican hypothesis" in Jerzy Dobrzycki ed., *The reception of Copernicus' heliocentric theory. Proceedings of a symposium organized by the Nicolas Copernicus Committee of the International Union of the History and Philosophy of Science, Torun, Poland, 1973* (Dordrecht/Boston 1972) 7-30.

Westman, Robert S., "The astronomer's role in the sixteenth century: A preliminary study", *History of science* 18 (1980) 105-147.

Westman, Robert S., "The Melanchthon Circle, Rheticus, and the Wittenberg interpretation of the Copernican theory", *Isis* 66 (1975) 165-193.

Westman, Robert S., "Three responses to the Copernican theory: Johannes Praetorius, Tycho Brahe, and Michael Mästlin" in Robert S. Westman ed., *The Copernican achievement* (Berkeley 1975) 285-345.

Westman, Robert S. and David C. Lindberg, "Introduction" in Robert S. Westman & David C. Lindberg eds., *Reappraisals of the Scientific Revolution* (Cambridge 1990) xvii-xxvii.

Westra, Frans, *Nederlandse ingenieurs en de fortificatiewerken in het eerste tijdperk van de Tachtigjarige Oorlog, 1573-1604* (Groningen 1992).

Whitaker, Ewen A., "Selenography in the seventeenth century" in René Taton & Curtis Wilson eds., *Planetary astronomy from the Renaissance to the rise of astrophysics. Part A: Tycho Brahe to Newton* (Cambridge 1989) 119-143.

White, A.D., "A history of the doctrine of comets", *Papers of the American Historical Association* 2 (New York 1887) 109-147.

Winter, P.J. van, *Hoger beroepsonderwijs avant-la-lettre. Bemoeiingen met de vorming van landmeters en ingenieurs bij de Nederlandse universiteiten van de 17e en 18e eeuw* (Verhandelingen der Koninklijke Nederlandse Akademie van Wetenschappen, Afd. Letterkunde, Nieuwe Reeks, deel 137, Amsterdam 1988).

Wolf, Rudolf, *Geschichte der Astronomie* (Munich 1877).

Woltjer, J.J., "Opstand en onafhankelijkheid" in idem, *Tussen vrijheidsstrijd en burgeroorlog. Over de Nederlandse Opstand, 1555-1580* (Amersfoort 1994) 9-63.

Woltjer, J.J., "De Vrede-makers", *Tijdschrift voor geschiedenis* 89 (1976) 299-321.

Woltjer, J.J. & M.E.H.N. Mout, "Settlements: the Netherlands" in Thomas A. Brady, Heiko A. Oberman & James D. Tracy eds., *Handbook of European history 1400-1600: Late Middle Ages, Renaissance and Reformation. Volume II: Visions, programs and outcomes* (Leiden/ New York/ Cologne 1995) 385-415.

Wright, M.R., *Cosmology in Antiquity* (London/New York 1995).
Wrightsman, Bruce, "Andreas Osiander's contribution to the Copernican achievement" in Robert S. Westman ed., *The Copernican achievement* (Berkeley 1975) 213-243.
Yeomans, Donald K., *Comets. A chronological history of observation, science, myth, and folklore* (New York 1991).
Zambelli, Paola, "Fine del mondo o inizio della propaganda?" in Leo S. Olschki ed., *Scienze, credenze occulte, livelli di cultura* (Florence 1982) 291-368.
Zambelli, Paola, "Many ends for the world. Luca Gaurico instigator of the debate in Italy and Germany" in Paola Zambelli ed., *Astrologi hallucinati: Stars and the end of the world in Luther's time* (Berlin/New York 1986) 239-263.
Zanta, Léontine, *La Renaissance du Stoicisme au XVIe siècle* (Paris 1914).

INDEX

www.ingramcontent.com/pod-product-compliance
Lightning Source LLC
LaVergne TN
LVHW010631300926
849162LV00001B/117